ADVANCED CALCULUS

Second Edition

by David V. Widder

Professor of Mathematics, Emeritus
Harvard University

DOVER PUBLICATIONS, INC., *New York*

Bibliographical Note

This Dover edition, first published in 1989, is an unabridged, corrected republication of the ninth (corrected) printing of the Second Edition (1961; in the "Prentice-Hall Mathematics Series") of the work originally published by Prentice-Hall, Inc., Englewood Cliffs, N.J., 1947.

Library of Congress Cataloging-in-Publication Data

Widder, D. V. (David Vernon), 1898
    Advanced calculus / by David V. Widder.—2nd ed.
        p. em.
    Reprint. Originally published: Englewood Cliffs, N.J. : Prentice-Hall, 1961. (Prentice-Hall mathematics series).
    Includes index.
    ISBN-13: 978-0-486-66103-2
    ISBN-10: 0-486-66103-2
    1. Calculus. I. Title. II. Series: Prentice-Hall mathematics series.
QA303.W48 1989
    515-dc20   89-33578
                                                                          CIP

Manufactured in the United States by Courier Corporation
66103213
www.doverpublications.com

PREFACE TO THE FIRST EDITION

This book is designed for students who have had a course in elementary calculus covering the work of three or four semesters. However, it is arranged in such a way that it may also be used to advantage by students with somewhat less preparation. The reader is expected to have considerable skill in the manipulations of elementary calculus, but it is not assumed that he will be very familiar with the theoretic side of the subject. Consequently, the book emphasizes first the type of manipulative problem the student has been accustomed to and gradually changes to more theoretic problems. In fact, the same sort of crescendo appears within the chapters themselves. In certain cases a fundamental theorem, whose meaning is easily understood, is stated and used at the beginning of a chapter; its proof is deferred to the end of it.

Believing that clarity of exposition depends largely on precision of statement, the author has taken pains to state exactly what is to be proved in every case. Each section consists of definitions, theorems, proofs, examples, and exercises. An effort has been made to make the statement of each theorem so concise that the student can see at a glance the essential hypotheses and conclusions.

Three of the chapters involve the Stieltjes integral and the Laplace transform, topics which do not appear in the traditional course in advanced calculus. The author believes that these subjects have now reached the stage where a knowledge of them must be part of the equipment of every serious student of pure or applied mathematics.

The book may be used as a text in various ways. Certainly, the usual college course of two semesters cannot include so much material. The author's own procedure in his classes has been to present all of any chapter used but to offer different chapters in different years. Another method, which might be particularly useful for the engineering student or for the prospective applied mathematician, would be to use the first two thirds of each chapter. The final third could then be used for reference purposes. It should be observed that the separate chapters are more or less independent. Subject to the fact that the latter half of the book is more difficult than the first, the order of presentation may be greatly varied. For example, Chapter IV might follow Chapter I, or indeed the material from both might be judiciously combined. The instructor would then have to supply some of the elementary material about tangent planes to surfaces. A suggested shorter course could be based on suitable portions of Chapters I, IV, VI, VII, VIII, IX, X, XII.

D.V.W.

PREFACE TO THE SECOND EDITION

In this revision of the text the main features of the first edition are preserved. There follows below a list of the more important changes.

The dot-cross notation for vector operations has been substituted for the dash-roof system. This change has necessitated the use of some distinctive designation for vectors, and an arrow over a letter representing a vector is now used. This convention seems to be in very general use among lecturers. (Bold-faced type, though satisfactory in a text, is not easily transferred to a blackboard.)

Certain theorems have been sharpened, where this could be done without too much sacrifice of simplicity. For example, the class of differentiable functions of several variables has been interpolated between C and C^1.

The treatment of Stieltjes integrals has been altered somewhat with the purpose of making it more useful to the student who is not very familiar with the basic facts about the Riemann integral. In fact, such a student may, if he wishes, correct his deficiency without studying the Stieltjes integral by concentrating on sections 1, 6, and 7 of Chapter 5.

The material on series has been augmented by the inclusion of the method of partial summation, of the Schwarz-Hölder inequalities, and of additional results about power series. There has also been added a brief discussion of general infinite products and an elementary derivation of an infinite product for the gamma function.

Many new exercises have been added, some of which are intended to be of the easier variety needed for developing initial skills. Answers to some of the exercises are in a final section.

The author wishes to acknowledge here his debt to the many persons who have given him suggestions for improvement of the text. At the risk of unintentional omissions there follows a list of their names: R. D. Accola, L. V. Ahlfors, Albert A. Bennett, Garrett Birkhoff, B. H. Bissinger, R. E. Carr, R. P. Boas, D. L. Guy, I. Hirschman, L. H. Loomis, K. O. May, E. L. Post, D. G. Quillen, A. E. Taylor, H. Whitney, J. L. Walsh.

D.V.W.

CONTENTS

9 Infinite Series (Continued)

I

Partial

Differentiation

§1. *Introduction*

We shall be dealing in this chapter with real functions of several real variables, such as $u = f(x, y)$, $u = f(x, y, z)$, etc. In these examples the variables $x, y, z, \ldots$ are called the *independent variables* or *arguments* of the function, u is the *dependent variable* or the *value* of the function. Unless otherwise stated, functions will be assumed *single-valued;* that is, the value is uniquely determined by the arguments. Multiple-valued functions may be studied as combinations of single-valued ones. For example, the equation

$$(1) \qquad\qquad u^2 + x^2 + y^2 = a^2$$

defines two single-valued functions,

$$(2) \qquad\qquad u = +\sqrt{a^2 - x^2 - y^2}$$

$$(3) \qquad\qquad u = -\sqrt{a^2 - x^2 - y^2} \qquad\qquad x^2 + y^2 \leqq a^2$$

A function of two variables clearly represents a surface in the space of the rectangular coordinates x, y, u. In the study of functions of more than two variables, geometrical language is often retained for purposes of analogy, even though geometric intuition then fails.

1.1 PARTIAL DERIVATIVES

A partial derivative of a function of several variables is the ordinary derivative with respect to one of the variables when all the rest are held constant. Various notations are used. The partial derivatives of $u = f(x, y, z)$ are

$$\frac{\partial u}{\partial x} = f_1(x, y, z) = \frac{\partial f}{\partial x} = \frac{\partial}{\partial x} f(x, y, z)$$

$$\frac{\partial u}{\partial y} = f_2(x, y, z)$$

$$\frac{\partial u}{\partial z} = f_3(x, y, z).$$

An important advantage of the subscript notation is that it indicates an operation on the function that is independent of the particular letters employed for the arguments. Thus, if $f(x, y, z) = xz^y$, we have

$$f_2(x, y, z) = xz^y \log z$$
$$f_2(r, s, t) = rt^s \log t$$

It shares this advantage with the familiar $f'(x)$ for the derivative of a function of one variable. The notations for the value of a derivative at a point are illustrated by

$$\frac{\partial u}{\partial y}\bigg|_{x=x_0, y=y_0, z=z_0} = \frac{\partial f}{\partial y}\bigg|_{(x_0, y_0, z_0)} = f_2(x_0, y_0, z_0)$$

For example,

$$f_2(x_0, y_0, z_0) = \frac{d}{dy} f(x_0, y, z_0)\bigg|_{y=y_0}$$

EXAMPLE A. $f(x, y) = x^{xy}$

$$\frac{\partial f}{\partial x} = x^{xy}(y \log x + y), \qquad \frac{\partial f}{\partial y} = x^{xy+1} \log x$$

EXAMPLE B. $f(x, y, z) = x \sin (yz)$
$$f_3(a, 1, \pi) = -a$$

1.2 IMPLICIT FUNCTIONS

The example of §1 serves to illustrate how a function may be defined implicitly. Thus, equation (1) defines the two functions (2) and (3), which are said to be defined *implicitly* by (1) or *explicitly* by (2) and (3). In other cases, a function may be defined implicitly even though it is impossible to give it explicit form. For example, the equation

$$(4) \qquad u + \log u = xy$$

defines one single-valued function u of x and y. Given any real values of the arguments, the equation could be solved by approximation methods for u. Yet u cannot be given in terms of x and y by use of a finite number of the elementary functions.

The partial derivatives of a function defined implicitly may be obtained without using an explicit expression for the function. One has only to differentiate both sides of the defining equation with respect to the independent variable in question, remembering that the dependent variable is really a function of the independent ones. For example, differentiating equation (1) gives

$$2x + 2u \frac{\partial u}{\partial x} = 0, \qquad \frac{\partial u}{\partial x} = -\frac{x}{u}$$

$$2y + 2u \frac{\partial u}{\partial y} = 0, \qquad \frac{\partial u}{\partial y} = -\frac{y}{u}$$

These results can be checked directly by use of equation (2) or of equation (3). From equation (4) we would have

$$\frac{\partial u}{\partial x} = \frac{uy}{u+1}, \qquad \frac{\partial u}{\partial y} = \frac{ux}{u+1}$$

The method applies equally well if several functions are defined by simultaneous equations.

EXAMPLE C. $\begin{cases} v + \log u = xy \\ u + \log v = x - y \end{cases}$

$$\begin{cases} \dfrac{1}{u} \dfrac{\partial u}{\partial x} + \dfrac{\partial v}{\partial x} = y \\ \dfrac{1}{v} \dfrac{\partial v}{\partial x} + \dfrac{\partial u}{\partial x} = 1 \end{cases} \qquad \frac{\partial u}{\partial x} = \frac{\begin{vmatrix} yu & u \\ v & 1 \end{vmatrix}}{\begin{vmatrix} 1 & u \\ v & 1 \end{vmatrix}} = \frac{u(y - v)}{1 - uv}$$

One could also solve for $\dfrac{\partial v}{\partial x}$. To obtain the derivatives with respect to y, one has only to differentiate the defining equations with respect to that variable.

1.3 HIGHER ORDER DERIVATIVES

Partial derivatives of higher order are obtained by successive application of the operation of differentiation defined above. The notations employed will be sufficiently illustrated by the following examples.   If $u = f(x, y, z)$,

$$\frac{\partial^2 u}{\partial x\, \partial y} = \frac{\partial}{\partial x}\left(\frac{\partial u}{\partial y}\right) = f_{21}(x, y, z)$$

$$\frac{\partial^3 u}{\partial z^2\, \partial y} = \frac{\partial}{\partial z}\left(\frac{\partial^2 u}{\partial z\, \partial y}\right) = f_{233}(x, y, z)$$

$$\frac{\partial^4 u}{\partial x\, \partial y\, \partial z^2} = \frac{\partial}{\partial x}\left(\frac{\partial^3 u}{\partial y\, \partial z^2}\right) = f_{3321}(x, y, z)$$

A function of two variables has two derivatives of order one, four of order two, and 2^n of order n.   A function of m independent variables will have m^n derivatives of order n.   Later we shall see that many of the derivatives of a given order will be equal under very general conditions. In fact, the number of distinct derivatives of order n is then the same as the number of terms in a homogeneous polynomial in m variables of degree n:

$$\binom{n + m - 1}{n} = \frac{(n + m - 1)!}{n!(m - 1)!}$$

EXAMPLE D. $u = \log (x^2 + y)$

$$\frac{\partial^3 u}{\partial y^2\, \partial x} = \frac{\partial^3 u}{\partial x\, \partial y^2} = \frac{\partial^3 u}{\partial y\, \partial x\, \partial y} = \frac{4x}{(x^2 + y)^3}$$

EXAMPLE E. $u + \log u = xy$

$$\frac{\partial u}{\partial x} = \frac{uy}{u + 1}$$

$$\frac{\partial^2 u}{\partial y\, \partial x} = \frac{u}{u + 1} + \frac{y}{(u + 1)^2}\frac{\partial u}{\partial y}$$

$$= \frac{u}{u + 1} + \frac{xyu}{(u + 1)^3} = \frac{\partial^2 u}{\partial x\, \partial y}$$

EXERCISES   (1)

1. Find $\dfrac{\partial}{\partial x}\dfrac{\sin xy}{\cos (x + y)}$.

2. Find $\dfrac{\partial^2}{\partial r^2} \log (r^2 + s)$.

3. Find $f_1(x, y)$ and $f_2(1, 2)$ if $f(x, y) = \tan^2 (x^2 - y^2)$.

4. If $z^2 = x^2 - 2xy - 1$, find $\dfrac{\partial z}{\partial x}$ when $x = 1$, $y = -2$, $z = -2$ first by the explicit and then by the implicit method.

5. If

$$u - v + 2w = x + 2z$$

$$2u + v - 2w = 2x - 2z$$

$$u - v + w = z - y$$

find $\dfrac{\partial u}{\partial y}, \dfrac{\partial v}{\partial y}, \dfrac{\partial w}{\partial y}$.

6. If $f(x, y) = x \tan^{-1}(x^2 + y)$, find $f_1(1, 0)$, $f_2(x, y)$.

7. If $f(x, y, z) = x \log y^2 + ye^z$, find $f_1(1, -1, 0)$, $f_2(x, xy, y + z)$.

8. If $u = x^{y^z}$, find $\dfrac{\partial u}{\partial x}, \dfrac{\partial u}{\partial y}, \dfrac{\partial u}{\partial z}$.

9. If $u = x^u + u^y$, find $\dfrac{\partial u}{\partial x}, \dfrac{\partial u}{\partial y}$.

10. If

$$u^2 + x^2 + y^2 = 3$$

$$u - v^3 + 3x = 4$$

find $\dfrac{\partial u}{\partial x}, \dfrac{\partial u}{\partial y}, \dfrac{\partial v}{\partial x}, \dfrac{\partial v}{\partial y}$.

11. If $u = x^y$, show that

$$\frac{\partial^3 u}{\partial x^2 \, \partial y} = \frac{\partial^3 u}{\partial x \, \partial y \, \partial x}.$$

12. Prove the statement in the text about the number of terms in a homogeneous polynomial.

13. In Example C find $\dfrac{\partial^2 u}{\partial x^2}$.

14. If

$$2u + 3v = \sin x$$

$$u + 2v = x \cos y,$$

find $u(x, y)$ explicitly and then $u_1(\pi/2, \pi)$. Find the same derivative by the implicit function method.

15. If

$$u^2 + v^2 = x^2$$

$$2uv = 2xy + y^2$$

find $u_1(x, y)$ when $x = 1$, $y = -2$, $u = 1$, $v = 0$, by the two methods of the previous problem.

§2. *Functions of One Variable*

We recall here certain notions about functions of one variable, which the student is assumed to have met before, perhaps in a less precise form. We shall also introduce certain abbreviating notations that will facilitate the statement of theorems.

2.1 LIMITS AND CONTINUITY

A function $f(x)$ approaches a limit A as x approaches a if, and only if, for each positive number ϵ there is another, δ, such that whenever $0 < |x - a| < \delta$ we have $|f(x) - A| < \epsilon$. That is, when x is near a (within a distance δ from it), $f(x)$ is near A (within a distance ϵ from it). In symbols we write

$$\lim_{x \to a} f(x) = A$$

EXAMPLE A. $\lim_{x \to 1} \sqrt{x} = 1$

For, in this example, we may choose δ equal to the given ϵ. We have

$$|f(x) - A| = |\sqrt{x} - 1| = \frac{|x - 1|}{\sqrt{x} + 1} \qquad 0 < x < 2$$

If $0 < |x - 1| < \delta = \epsilon$, we obtain

$$|\sqrt{x} - 1| < \frac{\epsilon}{\sqrt{x} + 1} < \epsilon$$

EXAMPLE B. $f(x) = \sin (1/x)$ $x \neq 0$

Here $f(x)$ has no limit as x approaches zero. Since $f(x)$ takes on the values -1 and $+1$ infinitely often in every neighborhood of the origin, it is certainly not within a distance less than 1 from any number throughout any neighborhood of the origin.

If, in the definition of limit, the first inequalities are replaced by $0 < x - a < \delta$ $(0 < a - x < \delta)$, we say that $\lim f(x) = A$ as x approaches a from above (below) and write

$$\lim_{x \to a+} f(x) = A, \qquad (\lim_{x \to a-} f(x) = A)$$

EXAMPLE C. $f(x) = \dfrac{1}{1 + e^{1/x}}$ $x \neq 0$

$$\lim_{x \to 0-} f(x) = 1, \qquad \lim_{x \to 0+} f(x) = 0$$

It is now easy to formulate what is meant by a continuous function. Let us first introduce the following symbols:

    $\in$ — "belongs to" or "is a member of."

    $\Rightarrow$ — "implies."

    $\Leftrightarrow$ — "implies and is implied by" or "if, and only if."

    C — "the class of continuous functions."

    $/$ — "not."

Definition 1. $f(x) \in C$ at $x = a \Leftrightarrow \lim_{x \to a} f(x) = f(a)$

This may be read, "$f(x)$ belongs to the class of functions continuous at $x = a$" (or "$f(x)$ is continuous at $x = a$") "if, and only if, the limit of $f(x)$ is $f(a)$ as x approaches a."

In Example A, the function $\sqrt{x}$ is continuous at $x = 1$, since also $\sqrt{1} = 1$. Observe that the last equality in Definition 1 is equivalent to

$$\lim_{x \to a} f(x) = f(\lim_{x \to a} x)$$

For a function to be continuous at $x = a$, it certainly must be defined there. Thus, $f(x) = (\sin x)/x$ is not continuous at $x = 0$ in the first instance, since division by zero is undefined. However, if $f(0)$ is defined as 1, $f(x)$ becomes continuous at $x = 0$. In Example B, $f(x)$ is discontinuous at $x = 0$ on two counts: $f(0)$ is undefined, and the limit involved does not exist. No choice of definition for $f(0)$ could make $f(x)$ continuous at $x = 0$.

If in Definition 1 "$x \to a$" is replaced by "$x \to a+$" ("$x \to a-$"), $f(x)$ is said to be continuous on the right (left) at $x = a$. Thus, in Example C, $f(x)$ is continuous on the right at $x = 0$ if $f(0) = 0$. We say that

$$f(x) \in C, \quad a < x < b, \quad \Leftrightarrow f(x) \in C$$

at each x of the interval $a < x < b$. Further,

$$f(x) \in C, \quad a \leq x \leq b, \quad \Leftrightarrow f(x) \in C, \quad a < x < b$$

and

$$\lim_{x \to a+} f(x) = f(a), \quad \lim_{x \to b-} f(x) = f(b)$$

In Example C, with $f(0) = 0$,

$$f(x) \in C \hspace{6cm} 0 \leq x \leq 1$$

EXAMPLE D. $f(x) = \dfrac{1}{x}$ $x \neq 0$

$$f(x) \notin C \hspace{6cm} -1 < x < 1$$

2.2 DERIVATIVES

We now introduce further classes of functions, those which have derivatives of certain orders.

Definition 2. $f'(a) = \lim\limits_{\Delta x \to 0} \dfrac{f(a + \Delta x) - f(a)}{\Delta x}$

$$f'_+(a) = \lim\limits_{\Delta x \to 0+} \dfrac{f(a + \Delta x) - f(a)}{\Delta x}$$

$$f'_-(a) = \lim\limits_{\Delta x \to 0-} \dfrac{f(a + \Delta x) - f(a)}{\Delta x}$$

These three numbers are called, respectively, the derivative, the derivative on the right, and the derivative on the left of $f(x)$ at $x = a$. For example, if $f(x) = |x|$, then $f'(0)$ does not exist, but $f'_+(0) = 1$ and $f'_-(0) = -1$. Distinguish between $f'_+(a)$ and $f'(a+) = \lim\limits_{x \to a+} f'(x)$.

EXAMPLE E. $f(x) = x^2 \sin(1/x)$ $x \neq 0$

$f(0) = 0$

$f'_+(0) = \lim\limits_{\Delta x \to 0+} \Delta x \sin(1/\Delta x) = 0$

$f'(0+) = \lim\limits_{x \to 0+} [2x \sin(1/x) - \cos(1/x)]$

The latter limit clearly does not exist.

Higher derivatives are defined in the obvious way by successive application of Definition 2.

Definition 3. $f(x) \in C^n \Leftrightarrow f^{(n)}(x) \in C$ $n = 1, 2, \dots$

It is easy to see that when $f'(x)$ exists then $f(x) \in C$. Hence, if $f(x) \in C^n$, we also have

$$f(x) \in C^k \quad \text{for} \quad k = 0, 1, 2, \dots, n - 1 \ (C^0 = C)$$

EXAMPLE F. $\begin{cases} f(x) = 0 & x < 0 \\ f(x) = x & x \geq 0, \end{cases} \quad f(x) \in C, \quad f(x) \notin C^1 \ -1 < x < 1$

$\begin{cases} f(x) = 0 & x < 0 \\ f(x) = x^2 & x \geq 0, \end{cases} \quad f(x) \in C^1, \quad f(x) \notin C^2 \ -1 < x < 1$

These examples show how to construct a function $f(x) \in C^n$ for which $f(x) \notin C^{n+1}$. Note the difference between Example E and the first case under Example F. These two functions fail to belong to C^1 for different reasons. The first has a derivative at every point but this derivative is not continuous at $x = 0$, the second has no derivative at $x = 0$. This suggests that it would be profitable to define a class of functions "between" C and C^1. This is, in fact, the case (see §3.6). In the interests of simplicity, we shall not use a special letter for the class.

2.3 ROLLE'S THEOREM

Theorem 1 (Rolle).   1. $f(x) \in C$ $a \leqq x \leqq b$

2. $f'(x)$ exists $a < x < b$

3. $f(a) = f(b) = 0$

$\Rightarrow$ $f'(\xi) = 0$ for some $\xi, a < \xi < b$

CASE I. $f(x) \equiv 0$.   Then $f'(x) = 0$ for all x.

CASE II. $f(x) \not\equiv 0$.   Then there is a number c, $a < c < b$, where $f(c) \neq 0$. If $f(c) > 0 \, (< 0)$, then $f(x)$ has an absolute maximum* (minimum) at a point $\xi, a < \xi < b$.   Hence,

$$\frac{f(\xi + \Delta x) - f(\xi)}{\Delta x} \leqq 0 \qquad (\geqq 0) \qquad \xi < \xi + \Delta x < b$$

$$\frac{f(\xi + \Delta x) - f(\xi)}{\Delta x} \geqq 0 \qquad (\leqq 0) \qquad a < \xi + \Delta x < \xi$$

Allowing Δx to approach zero, we see by hypothesis 1 that both quotients approach $f'(\xi)$, which must therefore be nonnegative and nonpositive. Hence $f'(\xi) = 0$.   Observe that hypotheses 1 and 2 are satisfied if $f(x) \in C^1$ in $a \leqq x \leqq b$.   The latter weaker hypothesis is often sufficient for applications.

2.4 LAW OF THE MEAN

Theorem 2 (Law of the mean).   1. $f(x) \in C$ $a \leqq x \leqq b$

2. $f'(x)$ exists $a < x < b$

$\Rightarrow$ $f(b) - f(a) = f'(\xi)(b - a)$ for some $\xi, a < \xi < b$

The function

$$\varphi(x) = f(x) - f(a) - \frac{f(b) - f(a)}{b - a}(x - a)$$

satisfies all hypotheses of Theorem 1.   The conclusion $\varphi'(\xi) = 0$ leads at once to the desired result.   One can easily see the origin of the function

---

* This fact is obvious geometrically.   A proof by use of Definition 1 alone will be found following Theorem 7 of Chapter 5.

$\varphi(x)$ by observing that it gives the length of the line segment AB in Figure 1.

If we set $a = c$, $b = c + h$ or if we set $b = c$, $a = c + h$ $(h < 0)$, the law becomes in either case

$$f(c + h) - f(c) = hf'(c + \theta h)$$
$$0 < \theta < 1$$

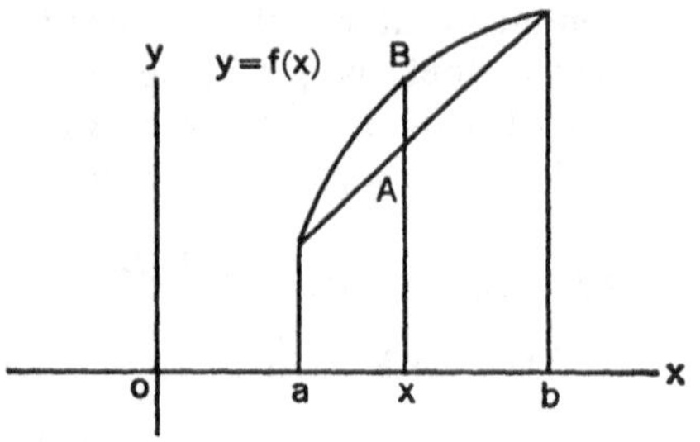

Fig. 1.

EXAMPLE G. $f(x) = x^3$,

$a = 1$, $b = 2$, $\xi = \sqrt{\tfrac{7}{3}}$, $1 < \sqrt{\tfrac{7}{3}} < 2$

$c = 2$, $h = -1$, $\theta = 2 - \sqrt{\tfrac{7}{3}}$ $0 < 2 - \sqrt{\tfrac{7}{3}} < 1$

EXAMPLE H. $f(x) = \sin x$, $a = \pi$ $0 < h < \pi/2$

$$\sin (\pi + h) = \sin (\pi + h) - \sin \pi = h \cos (\pi + \theta h)$$ $0 < \theta < 1$

$$-h < -\sin h < -h \cos h$$

$$1 < \frac{\tan h}{h} < \sec h$$

$$\lim_{h \to 0+} \frac{\tan h}{h} = 1$$

EXERCISES (2)

1. Evaluate the following limits if they exist:

   (a) $\lim\limits_{x \to 0} e^{-1/|x|}$,   (b) $\lim\limits_{x \to 0+} e^{-1/x}$,   (c) $\lim\limits_{x \to 0-} e^{-1/x}$.

2. If $[x]$ means "the largest integer $\leq x$," is the function $[x]$ continuous on the right or continuous on the left at $x = 2$? Does

$$[1] = \lim_{x \to 0} [1 - |x|] ?$$

3. The function $[x]$ is continuous in which of the following intervals:

$$0 < x < 1, \quad 0 \leq x \leq 1, \quad 0 \leq x < 1, \quad 0 < x \leq 1?$$

4. Does the function $f(x) = |x|^3$ belong to C^2?   Compute $f_+'''(0)$ and $f_-'''(0)$.

5. Define $f(x)$ of Example C as 0 at $x = 0$ and compute $f_+'(0)$. You may assume that $te^{-t} \to 0$ as $t \to +\infty$.

6. Find ξ in the law of the mean if

$$f(x) = \tan^{-1} x, \quad a = 1, \quad b = \sqrt{3},$$

and show that it lies in the required interval.

7. Find ξ in Rolle's theorem for $f(x) = x^3(1 - x)^5$, and show that it lies in the required interval.

8. Find θ in the law of the mean for $f(x) = Ax^2 + Bx + C, (A \neq 0)$, and show that $0 < \theta < 1$.

9. If $f(x) = \sqrt{x - x^2}$, show that $f(x)$ satisfies the hypotheses of Rolle's theorem for $a = 0$, $b = 1$. Find ξ. Show that $f(x) \notin C^1$ in $0 \leq x \leq 1$.

10. If $f(x) = x \sin (1/x), x \neq 0, f(0) = 0$, show that $f(x)$ satisfies the hypotheses of Rolle's theorem for $a = 0$, $b = 1/\pi$. How many values of ξ exist? *Hint:* Apply Rolle's theorem on smaller intervals.

11. In the previous exercise does $f(x) \in C^1$ in $0 \leq x \leq 1/\pi$?

12. Formulate exact definitions for the following:

$$\lim_{x \to +\infty} f(x) = A, \quad \lim_{x \to -\infty} f(x) = A, \quad \lim_{x \to a+} f(x) = +\infty.$$

13. Construct $f(x)$ such that $f(x) \in C^n, f(x) \notin C^{n+1}$.

14. Prove by the method employed for Example H that

$$\lim_{h \to 0-} \frac{\tan h}{h} = 1.$$

15. Prove from the definition of limit that a function cannot have two different limits as its independent variable approaches a limit.

16. Prove that the existence of $f'(a)$ implies the continuity of $f(x)$ at $x = a$.

17. If $f(x) = [x]$ as in Exercise 2, show that $f'(2-) = 0$ and that $f'_-(2)$ does not exist.

18. If $f(x) \in C^1$ in $a \leq x \leq b$, show that $f'_+(a) = f'(a+)$. *Hint:* Use the law of the mean.

19. If $[x]$ is defined as in Exercise 2, is $[1 - [x]]$ continuous on the right (left)? Answer the question for each x.

§3. *Functions of Several Variables*

We now proceed with a systematic treatment of partial differentiation. We develop first the method of differentiating composite functions analogous to

$$[f(g(x))]' = f'(g(x))g'(x)$$

$$\frac{du}{dx} = \frac{du}{dy}\frac{dy}{dx}$$

for functions of one variable.

3.1 LIMITS AND CONTINUITY

We begin by defining the limit of a function of two variables.   A function $f(x, y)$ approaches a limit A as x approaches a and y approaches b,

$$\lim_{\substack{x \to a \\ y \to b}} f(x, y) = A,$$

if, and only if, for each positive number ϵ there is another, δ, such that whenever $|x - a| < \delta$, $|y - b| < \delta$, $0 < (x - a)^2 + (y - b)^2$ we have $|f(x, y) - A| < \epsilon$.   That is, when (x, y) is at any point inside a certain square with center at (a, b) and width 2δ (except at the center), $f(x, y)$ differs from A by less than ϵ.

EXAMPLE A. $f(x, y) = x^2 + y^2$

Given ϵ, we may choose $\delta = \sqrt{\epsilon/2}$.   For, the inequalities $|x| < \sqrt{\epsilon/2}$, $|y| < \sqrt{\epsilon/2}$ imply $(x^2 + y^2) < \epsilon$.   Hence,

$$\lim_{\substack{x \to 0 \\ y \to 0}} (x^2 + y^2) = 0$$

EXAMPLE B.   If $f(x, y) = \dfrac{x - y}{x + y}$ $\qquad\qquad\qquad x \neq -y$

$$f(x, y) = 1 \qquad\qquad\qquad\qquad\qquad x = -y$$

then $f(x, y)$ approaches no limit as (x, y) approaches the origin.   For $f(x, y)$ is as large as we like at points near the line $x = -y$.   On the other hand, observe that

$$\lim_{x \to 0} \left[\lim_{y \to 0} f(x, y) \right] = 1, \qquad \lim_{y \to 0} \left[\lim_{x \to 0} f(x, y) \right] = -1$$

Definition 4. $f(x, y) \in C$ at $(a, b) \Leftrightarrow \lim_{\substack{x \to a \\ y \to b}} f(x, y) = f(a, b)$

Any collection of points (x, y) is called a *point set*.   The set of points $|x - a| < \delta$, $|y - b| < \delta$ is known as an *open square* or *two-dimensional interval* or a *δ-neighborhood* of the point (a, b).   A point (a, b) is a *limit point* of a set S if every δ-neighborhood of (a, b) contains points of S different from (a, b).   A set S is *closed* if it contains all its limit points.   A point is an *interior point* of S if it is the centre of a δ-neighborhood composed entirely of points of S.   A set is *open* if it is composed entirely of interior points.   For example, if S is the set of points (x, y) for which $x^2 + y^2 < a^2$, S is open. Limit points of this set not in it are those for which $x^2 + y^2 = a^2$.   The *boundary* of a set is the set of all limit points not interior points.

A *domain* is an open set, any two of whose points can be joined by a broken line having a finite number of segments, all of whose points belong to the set. A *region* is either a domain or a domain plus some or all of its boundary. If it contains all of its boundary, it is a *closed region*.

We say that $f(x, y) \in C$ in a domain D if, and only if, $f(x, y) \in C$ at each point of D. Also $f(x, y) \in C$ at a boundary point (a, b) of a region R where $f(x, y)$ is defined if, and only if,

$$\lim_{\substack{x \to a \\ y \to b}} f(x, y) = f(a, b) \qquad\qquad (x, y) \in R$$

That is, the point (x, y) approaches (a, b) only through points of R. This corresponds to one-sided approach for functions of one variable. Then $f(x, y) \in C$ in R if $f(x, y) \in C$ at each point of R.

3.2 DERIVATIVES

We now define the classes C^n for functions of several variables. We first give limit definitions of the partial derivatives described in §1. We use the letter R to indicate a *region*.

Definition 5. $f_1(a, b) = \dfrac{\partial f}{\partial x}\bigg|_{(a,b)} = \lim_{\Delta x \to 0} \dfrac{f(a + \Delta x, b) - f(a, b)}{\Delta x}$

$\qquad\qquad f_2(a, b) = \dfrac{\partial f}{\partial y}\bigg|_{(a,b)} = \lim_{\Delta y \to 0} \dfrac{f(a, b + \Delta y) - f(a, b)}{\Delta y}$

Definition 6.

$$f(x, y) \in C^n \text{ in } R \Leftrightarrow \frac{\partial^n f}{\partial x^n}, \quad \frac{\partial^n f}{\partial x^{n-1} \partial y}, \quad \dots, \quad \frac{\partial^n f}{\partial y^n} \in C \text{ in } R.$$

It can be shown that if $f(x, y)$ satisfies the condition of this definition, then $f(x, y) \in C^k$ ($k = 0, 1, 2, \dots, n - 1$), just as for functions of a single variable.

3.3 A BASIC MEAN-VALUE THEOREM

We are now able to establish a result of fundamental importance in the theory of partial differentiation. It may be considered analogous to Theorem 2, the law of the mean for functions of a single variable. We shall use the letter D to indicate a *domain*.

Theorem 3.   1. $f(x, y) \in C^1$ *in* D

$\qquad\qquad$ 2. *The circle* $(x - a)^2 + (y - b)^2 \leqq \delta^2$ *lies in* D

$\Rightarrow \quad f(a + \Delta x, b + \Delta y) - f(a, b)$

$\qquad\qquad = f_1(a + \theta_1 \Delta x, b)\Delta x + f_2(a + \Delta x, b + \theta_2 \Delta y)\Delta y$

where $\Delta x^2 + \Delta y^2 < \delta^2$ *and* $0 < \theta_1 < 1, 0 < \theta_2 < v.$

Set

(1) $$\Delta f = f(a + \Delta x, b + \Delta y) - f(a, b)$$

and rewrite it as follows:

$$\Delta f = [f(a + \Delta x, b) - f(a, b)] + [f(a + \Delta x, b + \Delta y) - f(a + \Delta x, b)].$$

Here we have added and subtracted $f(a + \Delta x, b)$ on the right-hand side of equation (1). Now apply Theorem 2 to the function $f(x, b)$ of the single variable x. Its derivative is $f_1(x, b)$. We thus obtain for the first bracket above

Fig. 2.

$$f(a + \Delta x, b) - f(a, b) = f_1(a + \theta_1 \Delta x, b) \Delta x$$

$$0 < \theta_1 < 1$$

Next apply the same theorem to the function $f(a + \Delta x, y)$. We thus obtain

(2) $$\Delta f = f_1(a + \theta_1 \Delta x, b) \Delta x$$
$$+ f_2(a + \Delta x, b + \theta_2 \Delta y) \Delta y$$

$$0 < \theta_2 < 1$$

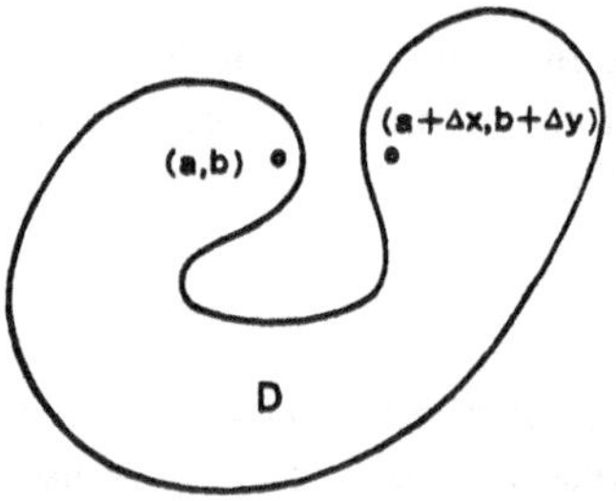

There is no reason to suppose that $\theta_1 = \theta_2$, and in general these two numbers will be different. A more symmetric form of the law of the mean, a form involving a single θ, will appear in §9, equation (4).

Observe the force of hypothesis 2. If we were to replace it by the hypothesis that (a, b) and $(a + \Delta x, b + \Delta y)$ are both points of D, equation (2) might not be true. A glance at Figure 2 will show why.

EXAMPLE C. $f(x, y) = x^2 + y^2 + x^3$

$(a, b) = (1, 2)$

$f(1 + \Delta x, 2 + \Delta y) - f(1, 2)$

$\quad = 5\Delta x + 4\Delta x^2 + 4\Delta y + \Delta y^2 + \Delta x^3$

$\quad = [2(1 + \theta_1 \Delta x) + 3(1 + \theta_1 \Delta x)^2] \Delta x + [4 + 2\theta_2 \Delta y] \Delta y$

We can determine for this particular example the exact values of θ_1 and θ_2, $\Delta x > 0$,

$$\theta_1 = \frac{-4 + \sqrt{16 + 12\Delta x + 3\Delta x^2}}{3\Delta x}, \qquad \theta_2 = \frac{1}{2}.$$

3.4 COMPOSITE FUNCTIONS

We use the result of Theorem 3 to differentiate a function of functions, one case of which is stated in the following theorem.

Theorem 4.   1. $f(x, y), g(r, s), h(r, s) \in C^1$

(3) $\Rightarrow$ $\dfrac{\partial}{\partial r} f(g, h) = f_1(g, h) g_1(r, s) + f_2(g, h) h_1(r, s)$

(4) $\dfrac{\partial}{\partial s} f(g, h) = f_1(g, h) g_2(r, s) + f_2(g, h) h_2(r, s)$

The regions in which the given function $\in C^1$ are not stated, in the interests of simplicity. It is understood, of course, that the region for (r, s) and the one for (x, y) must be such that the functions g, h can be substituted in $f(x, y)$ to form

$$f(g(r, s), h(r, s))$$

From the definition of a partial derivative, we have

$$\frac{\partial f}{\partial r}\bigg|_{(r_0, s_0)} = \lim_{\Delta r \to 0} \frac{\Delta f}{\Delta r}$$

$$\Delta f = f(g(r_0 + \Delta r, s_0), h(r_0 + \Delta r, s_0)) - f(g(r_0, s_0), h(r_0, s_0)).$$

Now apply Theorem 3, setting

$$g(r_0 + \Delta r, s_0) = x_0 + \Delta x, \qquad x_0 = g(r_0, s_0)$$

$$h(r_0 + \Delta r, s_0) = y_0 + \Delta y, \qquad y_0 = h(r_0, s_0)$$

By the continuity of g and h, we see that Δx and Δy tend to zero with Δr. We have

$$\frac{\Delta f}{\Delta r} = f_1(x_0 + \theta_1 \Delta x, y_0) \frac{\Delta x}{\Delta r} + f_2(x_0 + \Delta x, y_0 + \theta_2 \Delta y) \frac{\Delta y}{\Delta r} \qquad 0 < \theta_1, \theta_2 < 1$$

Now let Δr approach zero and make use of Definition 5 and Definition 6 to obtain

$$\frac{\partial f}{\partial r}\bigg|_{(r_0, s_0)} = f_1(x_0, y_0) g_1(r_0, s_0) + f_2(x_0, y_0) h_1(r_0, s_0)$$

Replacing x_0, y_0 by their values and dropping subscripts, we have equation (3). Equation (4) is obtained in a similar way. The results are easily remembered by putting them in the following form, analogous to the second equation of this section:

$$\frac{\partial f}{\partial r} = \frac{\partial f}{\partial x} \frac{\partial x}{\partial r} + \frac{\partial f}{\partial y} \frac{\partial y}{\partial r}$$

$$\frac{\partial f}{\partial s} = \frac{\partial f}{\partial x} \frac{\partial x}{\partial s} + \frac{\partial f}{\partial y} \frac{\partial y}{\partial s}$$

EXAMPLE D. $f(x, y) = xy, \qquad f_1 = y, \quad f_2 = x$

$$\frac{\partial}{\partial r} gh = \frac{\partial}{\partial r} f(g(r, s), h(r, s)) = yg_1 + xh_1 = hg_1 + gh_1$$

Thus the rule for differentiation of a product is the same whether the factors are functions of one or of two variables, a fact which is also evident from the definition of a partial derivative.

3.5 FURTHER CASES

The following cases are proved in a manner analogous to that used for the proof of Theorem 4:

CASE I.

$$u = f(x, y, z), \quad x = g(r, s), \quad y = h(r, s), \quad z = k(r, s)$$

$$\frac{\partial u}{\partial r} = \frac{\partial u}{\partial x}\frac{\partial x}{\partial r} + \frac{\partial u}{\partial y}\frac{\partial y}{\partial r} + \frac{\partial u}{\partial z}\frac{\partial z}{\partial r}$$

CASE II.

$$u = f(x), \quad x = \varphi(r,.s, t)$$

$$\frac{\partial u}{\partial s} = \frac{du}{dx}\frac{\partial x}{\partial s} = f'(\varphi(r, s, t))\varphi_2(r, s, t)$$

CASE III.

$$u = f(x, y, z), \quad x = \varphi(t), \quad y = \psi(t), \quad z = \omega(t)$$

$$\frac{du}{dt} = \frac{\partial u}{\partial x}\frac{dx}{dt} + \frac{\partial u}{\partial y}\frac{dy}{dt} + \frac{\partial u}{\partial z}\frac{dz}{dt} = f_1\varphi' + f_2\psi' + f_3\omega'$$

To prove these and analogous results, one must use Theorem 3 and suitable modifications thereof (Theorem 2 or Exercise 6 of the present section).   Distinguish carefully between total and partial derivatives.

EXAMPLE E. $u = \sin(e^x + y), \quad x = f(t), \quad y = g(t)$

$$\frac{du}{dt} = [\cos(e^x + y)]e^x f'(t) + [\cos(e^x + y)]g'(t)$$

EXAMPLE F.

$$u = f(x, y), \quad x = g(r, s), \quad y = h(r, s), \quad r = \varphi(t), \quad s = \psi(t)$$

$$\frac{du}{dt} = f_1[g_1\varphi' + g_2\psi'] + f_2[h_1\varphi' + h_2\psi']$$

3.6 DIFFERENTIABLE FUNCTIONS

For a more delicate analysis it is useful to introduce a class of functions lying between C and C^1, the class of *differentiable functions*.   The name might suggest that $f(x, y)$ is differentiable if f_1 and f_2 exist.   Actually it is desirable to require somewhat more. We define $f(x, y)$ to be *differentiable* at $(a, b) \Leftrightarrow f_1(a, b)$, $f_2(a, b)$ exist and if

(5) $\quad f(a + \Delta x, b + \Delta y) - f(a, b)$

$$= f_1(a, b)\,\Delta x + f_2(a, b)\,\Delta y + \varphi(\Delta x, \Delta y)\,\Delta x + \psi(\Delta x, \Delta y)\,\Delta y,$$

where $\varphi(\Delta x, \Delta y)$ and $\psi(\Delta x, \Delta y) \to 0$ as $(\Delta x, \Delta y) \to (0, 0)$. For example, if

$$f(x, y) = 2 - y + 2x^2 - x^2 y$$

f is differentiable at every point. In particular at $(0, 0)$, $f = 2$, $f_1 = 0$, $f_2 = -1$, and

$$f(\Delta x, \Delta y) - f(0, 0) = f_2(0, 0)\,\Delta y + (2\Delta x - \Delta x\,\Delta y)\,\Delta x.$$

Hence we may take $\varphi(\Delta x, \Delta y) = 2\,\Delta x - \Delta x\,\Delta y$ and $\psi(\Delta x, \Delta y) = 0$. Or we could choose $\varphi = 2\,\Delta x$, $\psi = -(\Delta x)^2$, and in either case φ and ψ have the desired limit zero.

Let us observe first that if $f(x, y) \in C^1$ in D then $f(x, y)$ is differentiable at every point of D. For, if we set

$$\varphi(\Delta x, \Delta y) = f_1(a + \theta_1\,\Delta x, b) - f_1(a, b)$$
$$\psi(\Delta x, \Delta y) = f_2(a + \Delta x, b + \theta_2\,\Delta y) - f_2(a, b)$$

equation (5) becomes equation (2), and from the assumed continuity of f_1 and f_2 the functions φ and ψ tend to zero as $(\Delta x, \Delta y) \to (0, 0)$.

EXAMPLE G. $f(x, y) = |x|\,(1 + y) \in C$ at $(0, 0)$ but is not differentiable there. For $f_1(0, 0)$ does not exist.

EXAMPLE H. $f(x, y) = x$ when $|y| < |x|$, $f(x, y) = -x$ when $|y| \geqq |x|$. Here $f_1(0, 0) = 1$, $f_2(0, 0) = 0$. Now if $f(x, y)$ were differentiable at $(0, 0)$, equation (5) would become when $\Delta y = \Delta x$

$$f(\Delta x, \Delta x) = -\Delta x$$
$$= \Delta x + \Delta x\,\varphi(\Delta x, \Delta x) + \Delta x\,\psi(\Delta x, \Delta x).$$

But this is a contradiction, as one sees by canceling Δx and letting $\Delta x \to 0$. Hence $f(x, y)$ is not differentiable at $(0, 0)$. It is continuous there.

EXAMPLE I. $f(x, y) = g(\sqrt{x^2 + y^2})$, $\qquad g(x) = x^2 \sin(1/x)$, $\qquad g(0) = 0$.

This function is differentiable at $(0, 0)$ (see Exercise 20). It does not belong to C^1 there. To prove this, it is enough to show that $f_1(x, x)$ has no limit as $x \to 0+$. But

$$f_1(x, x) = \frac{1}{\sqrt{2}}\,g'(x\sqrt{2}) \qquad\qquad x > 0$$

and we saw in Example E, §2, that $g'(0+)$ does not exist.

Let us summarize the results just obtained.

I. $f \in C^1 \Rightarrow f$ is differentiable (Theorem 3).

II. f is differentiable $\Rightarrow f \in C$ (Exercise 17).

III. There exist continuous functions not differentiable (Example G).

IV. There exist nondifferentiable functions having partial derivatives (Example H).

V. There exist differentiable functions not belonging to C^1 (Example I).

EXERCISES (3)

1. If

$$u = x^2 + y^2, \quad x = r \cos \theta, \quad y = r \sin \theta$$

compute $\dfrac{\partial u}{\partial r}$ and $\dfrac{\partial u}{\partial \theta}$ first by Theorem 4 and then by eliminating x and y.

2. Same problem if $u = x^y$.

3. Compute the derivative of u with respect to t if

$$u = xy^z, \quad x = t, \quad y = t^2, \quad z = t^3$$

4. Find the numbers θ_1 and θ_2 of Theorem 3, if

$$f(x, y) = x^2 + 3xy + y^2, \quad a = b \doteq 0, \quad \Delta x = 1, \quad \Delta y = -1$$

5. Same problem if

$$f = e^{xy}, \quad a = b = \Delta x = \Delta y = 1$$

6. Prove a theorem analogous to Theorem 3 for a function of three variables.

7. Prove Case I.

8. Prove Case II.

9. Prove Case III.

10. If

$$u = f(x, y), \quad x = g(r, s), \quad y = h(t)k(r)$$

find $\dfrac{\partial u}{\partial r}, \dfrac{\partial u}{\partial s}, \dfrac{\partial u}{\partial t}$

11. If $u = f(x, y)$, $x = r \cos \theta$, $y = r \sin \theta$, show that

$$\left(\frac{\partial u}{\partial x}\right)^2 + \left(\frac{\partial u}{\partial y}\right)^2 = \left(\frac{\partial u}{\partial r}\right)^2 + \frac{1}{r^2}\left(\frac{\partial u}{\partial \theta}\right)^2$$

Explain the exact meaning of the equation, dissolving the mystery of a function of (x, y) equated to a function of (r, θ).

12. $\dfrac{d}{dx} f\left(\dfrac{g(x)}{h(x)}\right) = ?$

13. $\dfrac{\partial}{\partial y} \log f(y, g(x, y)) = ?$

14. In Example C, compute the limit of θ_1 as $\Delta x \to 0$.

15. If $f(x, y) \in C$ at (a, b), show that $f(x, b)$ is continuous at $x = a$ and $f(a, y)$ is continuous at $y = b$.

16. If $f(x, y) = 1$ when $xy \neq 0$ and $= 0$ when $xy = 0$, show that $f(x, 0) \in C$ for all x and $f(0, y) \in C$ for all y but that $f(x, y) \notin C$ at $(0, 0)$. Show that

$$\lim_{x \to 0} (\lim_{y \to 0} f(x, y)) = \lim_{y \to 0} (\lim_{x \to 0} f(x, y))$$

17. Show that if $f(x, y)$ is differentiable at a point it is continuous there. *Hint:* Let $(\Delta x, \Delta y) \to (0, 0)$ in (5).

18. If

$$f(x, y) = \frac{x^6 - 2y^4}{x^2 + y^2} \qquad x^2 + y^2 \neq 0$$

$$f(0, 0) = 0$$

show that $f(x, y)$ is differentiable at $(0, 0)$.

19. If $f(x, y) = \sqrt{|xy|}$, show that $f(x, y)$ is not differentiable at $(0, 0)$.

20. In Example I show that $f_1(0, 0) = f_2(0, 0) = 0$. Show that $f(x, y)$ is differentiable at $(0, 0)$. *Hint:*

$$\varphi = \Delta x \sin (\Delta x^2 + \Delta y^2)^{-1/2}, \quad \psi = \Delta y \sin (\Delta x^2 + \Delta y^2)^{-1/2}$$

21. Theorem 4 is still true if f is only differentiable and if g and h possess partial derivatives. Prove.

§4. *Homogeneous Functions. Higher Derivatives*

A polynomial in x and y is said to be homogeneous if all its terms are of the same degree. For example,

$$f(x, y) = x^2 - 2xy + 3y^2$$

is homogeneous. It is easy to generalize the property so that functions not polynomials can have it. Observe, in the above example, that

$$f(\lambda x, \lambda y) = \lambda^2 f(x, y)$$

for any positive number λ. We use this characteristic of homogeneous polynomials to make the generalization. The definition is stated for a function of two variables, but it is easily altered to apply to a function of any number of variables.

4.1 DEFINITION OF HOMOGENEOUS FUNCTIONS

Definition 7. *A function $f(x, y)$ is homogeneous of degree n in a region R if, and only if, for (x, y) in R and for every positive value of λ*

(1) $$f(\lambda x, \lambda y) = \lambda^n f(x, y)$$

The number n is positive, negative, or zero and need not be an integer. In the above example $n = 2$ and R is the whole xy-plane. The region R must be such that $(\lambda x, \lambda y)$ is a point of it for all positive λ whenever (x, y) is a point of it. For example, R may be an angular region between two infinite rays emanating from the origin or the whole plane.

EXAMPLE A. $f(x, y) = x^{1/3}y^{-4/3}\tan^{-1}(y/x)$

Here $n = -1$; R is any quadrant without the axes.

EXAMPLE B. $f(x, y) = 3 + \log(y/x)$

This function is homogeneous of degree 0; R is the first or third quadrant without the axes.

EXAMPLE C. $f(x, y) = (\sqrt{x^2 + y^2})^3$

Here $n = 3$; R is the whole plane. Observe that if λ is a negative number, equation (1) is not satisfied for this function. For,

$$f(\lambda x, \lambda y) = |\lambda|^3 f(x, y)$$

EXAMPLE D. $f(x, y) = x^{1/3}y^{-2/3} + x^{2/3}y^{-1/3}$

This function is not homogeneous.

4.2 EULER'S THEOREM

Theorem 5 (Euler).   1. $f(x, y) \in C^1$ (x, y) in R

                 2. $f(x, y)$ is homogeneous of degree n in R.

(2) $\Rightarrow$ $f_1(x, y)x + f_2(x, y)y = nf(x, y)$ (x, y) in R

To prove this, differentiate equation (1) partially with respect to λ,

$$xf_1(\lambda x, \lambda y) + yf_2(\lambda x, \lambda y) = n\lambda^{n-1}f(x, y)$$

Finally, set $\lambda = 1$.

We point out in passing that certain authors* define homogeneity in a different way, demanding that equation (1) should hold for *all* real values of λ. With this definition the function of Example C is not homogeneous. But this definition would have the disadvantage that the converse of Euler's theorem would be false, whereas we shall now prove that the converse is valid under Definition 7.

Theorem 6.   1. $f(x, y) \in C^1$ (x, y) in R

                 2. $xf_1 + yf_2 = nf$ (x, y) in R

    $\Rightarrow$ $f(x, y)$ is homogeneous of degree n (x, y) in R

---

* See, for example, A. Del Chiaro, "Sulle Funzioni Omogenee," *Atti della Reale Accademia dei Lincei*, Series 6, vol. 13 (1931), p. 475.

It is to be understood in this theorem that R is the type of region described under Definition 7. Choose (x_0, y_0) an arbitrary point of R, and form the function

$$\varphi(\lambda) = f(\lambda x_0, \lambda y_0)$$

defined for all positive values of λ. Then by hypothesis 2

$$\varphi'(\lambda) = x_0 f_1(\lambda x_0, \lambda y_0) + y_0 f_2(\lambda x_0, \lambda y_0)$$
$$nf(\lambda x_0, \lambda y_0) = \lambda x_0 f_1(\lambda x_0, \lambda y_0) + \lambda y_0 f_2(\lambda x_0, \lambda y_0)$$
$$\lambda \varphi'(\lambda) = n\varphi(\lambda)$$

Now differentiate $\varphi(\lambda)\lambda^{-n}$ with respect to λ, and obtain

$$[\varphi(\lambda)\lambda^{-n}]' = \varphi'(\lambda)\lambda^{-n} - n\varphi(\lambda)\lambda^{-n-1}$$

The right-hand side of this equation is zero by virtue of the previous equation. Hence,

$$\varphi(\lambda)\lambda^{-n} = C$$

where C is a constant which may be determined by setting $\lambda = 1$,

$$f(x_0, y_0) = C.$$
$$f(\lambda x_0, \lambda y_0) = \lambda^n f(x_0, y_0)$$

Since (x_0, y_0) was an arbitrary point of R, the theorem is proved.

4.3 HIGHER DERIVATIVES

Higher derivatives of composite functions may be computed by the principles already at our disposal. As an example, let us compute the three derivatives of order two for the function $u = f(\varphi(r, s), \psi(r, s))$. We assume that the three functions involved belong to C^2.

$$\frac{\partial u}{\partial r} = f_1 \varphi_1 + f_2 \psi_1, \qquad \frac{\partial u}{\partial s} = f_1 \varphi_2 + f_2 \psi_2$$

Differentiating again, remember that f_1 and f_2 are themselves composite functions:

$$\frac{\partial^2 u}{\partial r^2} = f_1 \varphi_{11} + f_2 \psi_{11} + \varphi_1[f_{11}\varphi_1 + f_{12}\psi_1] + \psi_1[f_{21}\varphi_1 + f_{22}\psi_1]$$

$$\frac{\partial^2 u}{\partial s\, \partial r} = f_1 \varphi_{12} + f_2 \psi_{12} + \varphi_1[f_{11}\varphi_2 + f_{12}\psi_2] + \psi_1[f_{21}\varphi_2 + f_{22}\psi_2]$$

(4)

$$\frac{\partial^2 u}{\partial r\, \partial s} = f_1 \varphi_{21} + f_2 \psi_{21} + \varphi_2[f_{11}\varphi_1 + f_{12}\psi_1] + \psi_2[f_{21}\varphi_1 + f_{22}\psi_1]$$

$$\frac{\partial^2 u}{\partial s^2} = f_1 \varphi_{22} + f_2 \psi_{22} + \varphi_2[f_{11}\varphi_2 + f_{12}\psi_2] + \psi_2[f_{21}\varphi_2 + f_{22}\psi_2]$$

We have omitted the arguments in these functions to save space. In each φ or ψ with any subscript, they are (r, s); in each f they are $(\varphi(r, s), \psi(r, s))$. If we admit that $f_{12} = f_{21}$, $\varphi_{12} = \varphi_{21}$, $\psi_{12} = \psi_{21}$, facts that we shall prove later, we see that $\dfrac{\partial^2 u}{\partial r\, \partial s} = \dfrac{\partial^2 u}{\partial s\, \partial r}$. This will also be evident later without computation.

EXAMPLE E.

$$u = f(x, y) = e^{xy} \qquad x = \varphi(r, s) = r + s \qquad y = \psi(r, s) = r - s$$
$$f_1 = y e^{xy} \qquad\qquad \varphi_1 = 1 \qquad\qquad\qquad \psi_1 = 1$$
$$f_2 = x e^{xy} \qquad\qquad \varphi_2 = 1 \qquad\qquad\qquad \psi_2 = -1$$
$$f_{11} = y^2 e^{xy} \qquad\quad \varphi_{11} = \varphi_{12} = 0 \qquad\quad \psi_{11} = \psi_{12} = 0$$
$$f_{12} = f_{21} = (1 + xy)e^{xy} \quad \varphi_{21} = \varphi_{22} = 0 \qquad\quad \psi_{21} = \psi_{22} = 0$$
$$f_{22} = x^2 e^{xy}$$

From the formulas above, we have, for example,

$$\frac{\partial^2 u}{\partial r\, \partial s} = -4rs e^{r^2 - s^2}$$

This result can be checked directly by eliminating x and y before differentiating.

EXAMPLE F. $u = f(g(t), h(t))$

$$\frac{du}{dt} = f_1 g' + f_2 h'$$

$$\frac{d^2 u}{dt^2} = f_1 g'' + f_2 h'' + g'[f_{11} g' + f_{12} h'] + h'[f_{21} g' + f_{22} h']$$

This result could also be obtained from equation (4) by replacing $\varphi(r, s)$ by $g(t)$, $\psi(r, s)$ by $h(t)$, etc.

EXERCISES (4)

1. Verify Theorem 5 for Examples A and B by computing both sides of Euler's equation directly.

2. Which of the following functions are homogeneous?

   (a) $\sqrt{x} - \sqrt{y}$

   (b) $\log y - \log x$

   (c) $(x^3 + y^3)^{2/3}$

   (d) $\left[\dfrac{x + y}{xy} + x^{2/3} e^{x/y} \right] y^{-5/3}$

   (e) $xf(y/x) + yg(x/y)$.

   Determine R and n for the homogeneous ones.

3. Do Exercise 1 for the homogeneous examples of Exercise 2.

4. Define homogeneity for $f(x, y, z)$, and show that it implies

$$f(x, y, z) = x^n f(1, y/x, z/x) \qquad (x > 0)$$

Illustrate by an example.

5. Prove Euler's theorem by use of the equation of Exercise 4.

6. If

$$u = e^v, \qquad v = \sin(xyz),$$

find $\dfrac{\partial^2 u}{\partial y\,\partial z}$.

7. In Example E find $\dfrac{\partial^2 u}{\partial r^2}$ first by using equations (4) and then by eliminating x and y.

8. Same problem for $\dfrac{\partial^2 u}{\partial s^2}$.

9. If $f(x, y)$ is homogeneous of degree n, show that

$$x^2 f_{11} + xy f_{12} + xy f_{21} + y^2 f_{22} = n(n - 1) f$$

What continuity assumption are you making?

10. Show that when $f(x, y)$ is homogeneous of degree n any derivative of order k is homogeneous of degree $n - k$.

11. Find $f''(t)$, if $f = e^x \sin y$, $x = t^2$, $y = 1 - t^2$, first by the method of the text, then by eliminating x and y before differentiation.

12. $\dfrac{\partial^2}{\partial x\,\partial y} f(x^2 - y, x + y^2) = ?$

13. $\dfrac{\partial^3}{\partial y\,\partial z\,\partial y} f(g(x, y, z)) = ?$

14. If for all positive values of λ

$$u(\lambda x, \lambda y, z/\lambda) = u(x, y, z)$$

prove $xu_1 + yu_2 = zu_3$. Illustrate with $u = e^{zx + v^2 z^2}$.

15. In the previous exercise show that $\dfrac{\partial^n u}{\partial z^n}(x, y, 0)$ is homogeneous of degree n.

Use the illustration of that exercise for $n = 3$.

§5. *Implicit Functions*

In section 1 we sketched briefly the method of obtaining the derivatives of functions defined implicitly. We now discuss the method in more detail. An equation of the form

(1) $$F(x, y, z) = 0$$

cannot necessarily be solved for one of the variables in terms of the other two.   For example, the equation

$$x^2 + y^2 + z^2 + a^2 = 0$$

has no solution if $a \neq 0$.   Even if $a = 0$, the equation does not define z as a function of (x, y) in any domain but only at the point $(0, 0)$.   We shall give later a sufficient condition that there should be a solution.   For the present, we shall discuss the method of finding the derivatives of the implicit function if it is known to exist.   That is, we shall assume that $z = f(x, y)$ exists and satisfies equation (1)

$$F(x, y, f(x, y)) \equiv 0$$

and we shall seek to compute the partial derivatives of $f(x, y)$ in terms of F.

5.1 DIFFERENTIATION OF IMPLICIT FUNCTIONS

Theorem 7.   1. $f(x, y)$, $F(x, y, z) \in C^1$

2. $F(x, y,\ \ f(x, y)) \equiv 0$ $\qquad\qquad\qquad\qquad$ (x, y) *in* D

3. $F_3(x, y,\ \ f(x, y)) \neq 0$ $\qquad\qquad\qquad\qquad$ (x, y) *in* D

$\Rightarrow$
$$f_1(x, y) = -\frac{F_1(x, y, f(x, y))}{F_3(x, y, f(x, y))}$$

$$f_2(x, y) = -\frac{F_2(x, y, f(x, y))}{F_3(x, y, f(x, y))}$$

The proof is immediate.   We have only to differentiate the equation of hypothesis 2.   We obtain

$$F_1 + F_3 f_1 \equiv 0, \qquad F_2 + F_3 f_2 \equiv 0$$

The result is now obtained by dividing the equation by the nonvanishing function F_3.

EXAMPLE A. $F(x, y, z) = x^2 + y^2 + z^2 - 6$
Equation (1) now defines the two explicit functions

$$z = \sqrt{6 - x^2 - y^2}, \qquad z = -\sqrt{6 - x^2 - y^2}$$

Compute $\dfrac{\partial z}{\partial x}$ at $(1, -1, 2)$.   By Theorem 7 we have

$$F_1(x, y, z) = 2x, \qquad F_1(1, -1, 2) = 2,$$

$$F_3(x, y, z) = 2z, \qquad F_3(1, -1, 2) = 4, \qquad \frac{\partial z}{\partial x} = -\frac{2}{4}$$

By the explicit method,

$$\frac{\partial z}{\partial x} = \frac{-x}{\sqrt{6 - x^2 - y^2}}, \qquad \frac{\partial z}{\partial x}\bigg|_{(1,-1,2)} = -\frac{1}{2}$$

5.2 OTHER CASES

The equation

(2) $F(x, y) = 0$

treated in elementary calculus, can now be handled by the present method. If this equation defines y as a function of x, we can compute its derivative in terms of F. For, remembering that y is a function of x, we have

$$F_1 + F_2 \frac{dy}{dx} = 0$$

(3) $\dfrac{dy}{dx} = -\dfrac{F_1}{F_2}$ $F_2 \neq 0$

EXAMPLE B. $u = f(x, u)$.   Find $\dfrac{du}{dx}$.

This is a special case of equation (2) where $F(x, u) = f(x, u) - u$.

$$\frac{du}{dx} = -\frac{f_1(x, u)}{f_2(x, u) - 1} \qquad f_2(x, u) \neq 1$$

EXAMPLE C. $u = f(g(x, u), h(y, u))$.   Find $\dfrac{\partial u}{\partial x}, \dfrac{\partial u}{\partial y}$.

This is a special case of equation (1) where

$$F(x, y, u) = f(g(x, u), h(y, u)) - u$$

$$\frac{\partial u}{\partial x} = -\frac{f_1 g_1}{f_1 g_2 + f_2 h_2 - 1}, \quad \frac{\partial u}{\partial y} = -\frac{f_2 h_1}{f_1 g_2 + f_2 h_2 - 1}$$

$$f_1 g_2 + f_2 h_2 - 1 \neq 0$$

5.3 HIGHER DERIVATIVES

One may compute the higher derivatives of functions defined implicitly. For example, let us compute $\dfrac{d^2 y}{dx^2}$ for equation (2). We have only to differentiate both sides of equation (3), and to remember that the arguments on the right are x and y and that y itself is the function of x defined by equation (1). Then

$$\frac{d^2 y}{dx^2} = -\frac{F_2(F_{11} + F_{12}y') - F_1(F_{21} + F_{22}y')}{F_2^2}$$

But y' is given by equation (3), so that

$$\frac{d^2y}{dx^2} = -\frac{F_{11}F_2^2 - (F_{12} + F_{21})F_1F_2 + F_{22}F_1^2}{F_2^3}$$

In like manner, we could compute the higher derivatives for Examples B and C.

We observed at the beginning of this section that it is possible to give sufficient conditions that a given equation should have a solution. The essential feature of the condition is precisely the nonvanishing of the functions which appear in the denominators when computing the first partial derivative. Thus, for equation (1) it is $F_3 \neq 0$; for equation (2), $F_2 \neq 0$. For Example B the condition is $f_2 \neq 1$, and for Example C it is $f_1g_2 + f_2h_2 - 1 \neq 0$. The student should be careful to insist explicitly on the nonvanishing of every denominator. Observe that it may be possible to solve a given equation for any one of the variables appearing. One can be certain which is intended in a given problem if any derivative is written. Thus, if $\dfrac{\partial x}{\partial y}$ is required in connection with equation (1), we may be sure that x is the dependent variable; y and z, the independent variables. We find

$$\frac{\partial x}{\partial y} = -\frac{F_2}{F_1}, \qquad \frac{\partial x}{\partial z} = -\frac{F_3}{F_1} \qquad\qquad F_1 \neq 0$$

$$\frac{\partial y}{\partial x} = -\frac{F_1}{F_2}, \qquad \frac{\partial y}{\partial z} = -\frac{F_3}{F_2} \qquad\qquad F_2 \neq 0$$

EXERCISES (5)

1. If $u = \log(x + u)$, find $\dfrac{du}{dx}$ and check by Example B.

2. Find $\dfrac{d^2u}{dx^2}$ in the previous exercise.

3. If $\log uy + y \log u = x$, find $\dfrac{\partial y}{\partial u}$ and $\dfrac{\partial y}{\partial x}$.

4. Find $\dfrac{\partial^2 y}{\partial x^2}$ in the previous exercise.

5. If $\sin zy = \cos zx$, compute $\dfrac{\partial z}{\partial x}$ when $z = \pi$, $x = \frac{1}{3}$, $y = \frac{1}{6}$.

6. If $xy + yz - xz = 2$, find $\dfrac{\partial z}{\partial x}, \dfrac{\partial z}{\partial y}$ by the method of the present section and also by first solving for z.

7. Find $\dfrac{d^2u}{dx^2}$ for Example B. Verify your result by the explicit method if $f = x + u^2$.

8. If $x^2 + u^2 = f(x, u) + g(x, y, u)$, find $\dfrac{\partial u}{\partial x}, \dfrac{\partial u}{\partial y}$.

9. If $u = f(x, y, u)$, find $\dfrac{\partial x}{\partial u}, \dfrac{\partial x}{\partial y}$.

10. If $z(z^2 + 3x) + 3y = 0$, prove that $\dfrac{\partial^2 z}{\partial x^2} + \dfrac{\partial^2 z}{\partial y^2} = \dfrac{2z(x-1)}{(z^2 + x)^3}$.

11. If $u = f(x + u, yu)$, find $\dfrac{\partial u}{\partial x}$ and $\dfrac{\partial u}{\partial y}$.

12. In Exercise 11, find $\dfrac{\partial x}{\partial u}, \dfrac{\partial x}{\partial y}$.

13. In Exercise 11, find $\dfrac{\partial y}{\partial u}, \dfrac{\partial y}{\partial x}$.

14. In Exercise 11, set $y = g(x)$ and find $\dfrac{du}{dx}$.

15. In Exercise 11, set $u = g(x, y)$ and find $\dfrac{dy}{dx}$.

§6. *Simultaneous Equations. Jacobians*

The method of the previous section applies equally well to functions defined implicitly by a number of simultaneous equations. Here again we do not discuss the solubility of the system of equations but only the method of finding the derivatives of the solutions, assumed to exist. The student should be familiar with the elements of the theory of determinants. In particular, he will need Cramer's rule for solving simultaneous linear equations and Laplace's method of expanding a determinant by means of minors.

6.1 TWO EQUATIONS IN TWO UNKNOWNS

Theorem 8. 1. $F(u, v, x, y)$, $G(u, v, x, y)$, $f(x, y)$, $g(x, y) \in C^1$

2. $F(f(x, y), g(x, y), x, y) \equiv 0$
   $G(f(x, y), g(x, y), x, y) \equiv 0$

3. $\Delta = \begin{vmatrix} F_1 & F_2 \\ G_1 & G_2 \end{vmatrix} \neq 0$

$\Rightarrow \qquad f_1 = -\dfrac{\begin{vmatrix} F_3 & F_2 \\ G_3 & G_2 \end{vmatrix}}{\Delta}, \qquad g_1 = -\dfrac{\begin{vmatrix} F_1 & F_3 \\ G_1 & G_3 \end{vmatrix}}{\Delta}$

$f_2 = -\dfrac{\begin{vmatrix} F_4 & F_2 \\ G_4 & G_2 \end{vmatrix}}{\Delta}, \qquad g_2 = -\dfrac{\begin{vmatrix} F_1 & F_4 \\ G_1 & G_4 \end{vmatrix}}{\Delta}$

The proof is similar to that of the previous theorem.   Differentiating with respect to x, we obtain

$$F_1 f_1 + F_2 g_1 + F_3 = 0$$

$$G_1 f_1 + G_2 g_1 + G_3 = 0$$

Solving these for f_1 and g_1 by Cramer's rule, we have the first half of our result. To obtain the other half, differentiate with respect to y. Hypothesis 3 is, of course, needed for the application of Cramer's rule.

6.2 JACOBIANS

Determinants like those above, whose elements are partial derivatives, occur so frequently that it is worth while having a notation for them.   This is particularly desirable when the order of the determinants is higher than two.   Let us illustrate the notation by the use of three functions F, G, H of six variables u, v, w, x, y, z, appearing in that order.   The *Jacobian of F, G, H with respect to u, w, z*, for example, is

$$\frac{\partial(F, G, H)}{\partial(u, w, z)} = \begin{vmatrix} F_1 & G_1 & H_1 \\ F_3 & G_3 & H_3 \\ F_6 & G_6 & H_6 \end{vmatrix}$$

As a further example, suppose we add a fourth function K of the same six variables.   Then

$$\frac{\partial(G, F, K, H)}{\partial(w, x, z, u)} = \begin{vmatrix} G_3 & F_3 & K_3 & H_3 \\ G_4 & F_4 & K_4 & H_4 \\ G_6 & F_6 & K_6 & H_6 \\ G_1 & F_1 & K_1 & H_1 \end{vmatrix}$$

It is important to observe how the order of appearance of the functions and variables in the notation makes itself evident in the defining determinant.   We could express the results of Theorem 8 in Jacobian notation:

$$f_1 = \frac{\partial u}{\partial x} = -\frac{\partial(F, G)}{\partial(x, v)} \bigg/ \frac{\partial(F, G)}{\partial(u, v)}$$

Although the notation provides no economy in this simple case, it does give a convenient memory rule for the results.   Except for the sign one could obtain the left side from the right by treating the symbols algebraically and canceling the marks ∂, ( ), F, G, v.   Note that one has the same rule in Theorem 7:

$$\frac{\partial z}{\partial x} = -\frac{\partial F}{\partial x} \bigg/ \frac{\partial F}{\partial z}, \qquad \frac{\partial z}{\partial y} = -\frac{\partial F}{\partial y} \bigg/ \frac{\partial F}{\partial z}$$

6.3 FURTHER CASES

As another example consider the system

$$F(u, v, w, x) = 0$$
$$G(u, v, w, x) = 0$$
$$H(u, v, w, x) = 0$$

Let u, v, w be the dependent variables, x the independent variable. The method gives us $\dfrac{du}{dx}, \dfrac{dv}{dx}, \dfrac{dw}{dx}$, the derivatives being total since there is a single independent variable. We obtain

$$\frac{du}{dx} = -\frac{\partial(F, G, H)}{\partial(x, v, w)} \bigg/ \frac{\partial(F, G, H)}{\partial(u, v, w)}, \qquad \frac{dv}{dx} = -\frac{\partial(F, G, H)}{\partial(u, x, w)} \bigg/ \frac{\partial(F, G, H)}{\partial(u, v, w)}$$

$$\frac{dw}{dx} = -\frac{\partial(F, G, H)}{\partial(u, v, x)} \bigg/ \frac{\partial(F, G, H)}{\partial(u, v, w)}, \qquad \frac{\partial(F, G, H)}{\partial(u, v, w)} \neq 0$$

Note that the same memory rule applies.

If the four functions of §6.2 are set equal to zero, we would have, if we considered u, v, w, x as dependent variables, for example,

$$\frac{\partial x}{\partial z} = -\frac{\partial(F, G, H, K)}{\partial(u, v, w, z)} \bigg/ \frac{\partial(F, G, H, K)}{\partial(u, v, w, x)}, \qquad \frac{\partial(F, G, H, K)}{\partial(u, v, w, x)} \neq 0$$

Observe that the number of dependent variables is equal to the number of simultaneous equations.

6.4 THE INVERSE OF A TRANSFORMATION

A set of equations of the form

$$u = f(x, y, z)$$
$$v = g(x, y, z)$$
$$w = h(x, y, z)$$

is known as a *transformation*. It *transforms* a point with coordinates (x, y, z) into another with coordinates (u, v, w). If these equations can be solved for x, y, z, we have three functions of u, v, w. The three corresponding equations constitute the *inverse* of the original transformation. They would give explicitly the point or points (x, y, z) from which (u, v, w) could have come in the original transformation.

The present method enables us to obtain the derivatives of x, y, z, with respect to u, v, w, without actually knowing the inverse transformation. For, we have only to set

$$F(u, v, w, x, y, z) = u - f(x, y, z)$$

$$G(u, v, w, x, y, z) = v - g(x, y, z)$$

$$H(u, v, w, x, y, z) = w - h(x, y, z)$$

and proceed as before.   For example,

$$\frac{\partial y}{\partial w} = -\frac{\partial(F, G, H)}{\partial(x, w, z)} \bigg/ \frac{\partial(F, G, H)}{\partial(x, y, z)} = \frac{\begin{vmatrix} f_1 & g_1 & h_1 \\ 0 & 0 & 1 \\ f_3 & g_3 & h_3 \end{vmatrix}}{\begin{vmatrix} f_1 & g_1 & h_1 \\ f_2 & g_2 & h_2 \\ f_3 & g_3 & h_3 \end{vmatrix}}$$

$$= -\frac{\partial(f, g)}{\partial(x, z)} \bigg/ \frac{\partial(f, g, h)}{\partial(x, y, z)} \qquad \frac{\partial(f, g, h)}{\partial(x, y, z)} \neq 0$$

EXAMPLE A. $x = 4u + 3v$

$$y = 3u + 2v$$

Find $\dfrac{\partial u}{\partial y}$.   It is more convenient to differentiate the equations directly than to apply the above formulas.

$$0 = 4\frac{\partial u}{\partial y} + 3\frac{\partial v}{\partial y}$$

$$1 = 3\frac{\partial u}{\partial y} + 2\frac{\partial v}{\partial y}, \qquad \frac{\partial u}{\partial y} = 3$$

In this simple case, we may check by obtaining explicitly the inverse transformation

$$u = -2x + 3y$$

$$v = 3x - 4y$$

EXAMPLE B. $F(u, v, g(u, v, x)) = 0$

$$G(u, v, h(u, v, y)) = 0$$

To find $\dfrac{\partial u}{\partial y}$, for example, we must solve

$$F_1 \frac{\partial u}{\partial y} + F_2 \frac{\partial v}{\partial y} + F_3 \left[g_1 \frac{\partial u}{\partial y} + g_2 \frac{\partial v}{\partial y} \right] = 0$$

$$G_1 \frac{\partial u}{\partial y} + G_2 \frac{\partial v}{\partial y} + G_3 \left[h_1 \frac{\partial u}{\partial y} + h_2 \frac{\partial v}{\partial y} + h_3 \right] = 0$$

for $\dfrac{\partial u}{\partial y}$. This will be possible if

$$\begin{vmatrix} F_1 + F_3 g_1 & F_2 + F_3 g_2 \\ G_1 + G_3 h_1 & G_2 + G_3 h_2 \end{vmatrix} \neq 0$$

EXERCISES (6)

1. If

$$F = xu + v - y$$
$$G = u^2 + vy + w$$
$$H = zu - v + vw$$

compute $\dfrac{\partial(F, G, H)}{\partial(u, w, v)}$.

2. In the previous exercise compute

$$- \frac{\partial(G, H, F)}{\partial(x, y, w)} \bigg/ \frac{\partial(G, H, F)}{\partial(x, y, u)}$$

Formulate a problem to which this is the answer.

3. If

$$u + v + w = x$$
$$u^2 + v^2 + w^2 = 2x - 1$$
$$u^3 + v^3 + w^3 = 3$$

find $\dfrac{dv}{dx}$ first directly and then by use of the Jacobian formulas of §6.3.

4. In Example B complete the computation of $\dfrac{\partial u}{\partial y}$.

5. If

$$x = r \cos \theta$$
$$y = r \sin \theta$$

find $\dfrac{\partial r}{\partial x}$ by use of Jacobians and then by using the explicit inverse of the transformation.

6. Find the derivative of u with respect to x if

$$xu + uv = u - x$$
$$v^2 + xv = u + x$$

Is the derivative total or partial?

7. If

$$u^2 + v^2 - xy + y^2 = 1$$

$$ux - vy + uv - v^2 = y$$

find $\dfrac{\partial v}{\partial y}$ first by differentiating the equations and then by use of the formulas of Theorem 8.

8. Show that Theorem 8 is not applicable to the system of equations

$$u^2 + v^2 + x^2 = y^2$$

$$\log (u^2 + v^2) + y^2 = x^2$$

by showing that the Jacobian of the system vanishes identically.   Show directly that the system can have a solution if, and only if (x, y) lies on a certain rectangular hyperbola.   Hence, the system cannot define a pair of functions u, v in any domain.

9. Find $\dfrac{\partial u}{\partial x}, \dfrac{\partial u}{\partial y}$ by use of Jacobians if

$$u = f(u, v, x)$$

$$v = g(u, v, y)$$

10. Find $\dfrac{du}{dx}$ if

$$u = f(v, w, x)$$

$$v = g(w, u, x)$$

$$w = h(u, v, x)$$

11. If $F(x, y, u) = 0$, is $\dfrac{\partial u}{\partial x}$ equal to the reciprocal of $\dfrac{\partial x}{\partial u}$?

12. Same problem if

$$F(x, y, u, v) = 0$$

$$G(x, y, u, v) = 0$$

Take u, v as dependent variables when computing $\dfrac{\partial u}{\partial x}$; take x, y as dependent variables when computing $\dfrac{\partial x}{\partial u}$.

13. In the transformation

$$u^2 + v^2 = 4xy$$

$$uv = 2xy - 2y^2$$

solve explicitly for u and v.   Find all the points (u, v) which correspond to $(x, y) = (-1, -1)$.

14. In the previous problem find $\dfrac{\partial u}{\partial x}$ when $(x, y) = (-1, -1)$, $(u, v) = (-2, 0)$ by the implicit function method.

15. Same problem using the explicit solution obtained in Exercise 13.

§7. Dependent and Independent Variables

In the previous sections, we have been more or less consistent in our notation, using the letters $u, v, w, \ldots$ for dependent variables and the letters x, y, z, t for independent variables. In the statement of a given problem involving several variables, it is not always possible to determine from the notation which variables are intended to be independent and which dependent. One must then state clearly what one is assuming the situation to be, or else one must treat all possible cases. We shall take the latter point of view in the present section. If a partial derivative, such as $\dfrac{\partial y}{\partial x}$, appears in the statement of a problem, we may be sure that one of the dependent variables is y and one of the independent variables is x. We shall illustrate by use of a number of examples.

7.1 FIRST ILLUSTRATION

Find $\dfrac{\partial u}{\partial x}$ if

(1)
$$u = f(x, y)$$
$$y = g(x, z)$$

Since u is dependent and x independent, and since there must be two dependent variables corresponding to the two equations, we can have only two cases.

CASE I. Dependent variables u, z; independent variables x, y. Differentiate the given equations with respect to x.

$$\frac{\partial u}{\partial x} = f_1, \qquad 0 = g_1 + g_2 \frac{\partial z}{\partial x}$$

Hence,
$$\frac{\partial u}{\partial x} = f_1, \qquad \frac{\partial z}{\partial x} = -\frac{g_1}{g_2} \qquad\qquad g_2 \neq 0$$

CASE II. Dependent variables, u, y; independent variables x, z.

Here
$$\frac{\partial u}{\partial x} = f_1 + f_2 \frac{\partial y}{\partial x}, \qquad \frac{\partial y}{\partial x} = g_1$$

Hence,
$$\frac{\partial u}{\partial x} = f_1 + f_2 g_1, \qquad \frac{\partial y}{\partial x} = g_1$$

The following notation is sometimes employed to distinguish between such cases:

CASE I. $\dfrac{\partial u_{x,y}}{\partial x}, \dfrac{\partial z_{x,y}}{\partial x}$.

CASE II. $\dfrac{\partial u_{x,z}}{\partial x}, \dfrac{\partial y_{x,z}}{\partial x}$.

The independent variables are used as subscripts against the dependent ones.

7.2 SECOND ILLUSTRATION

Find $\dfrac{\partial u}{\partial y}$ if equations (1) are given.

CASE I. $\dfrac{\partial u_{y,z}}{\partial y}, \dfrac{\partial x_{y,z}}{\partial y}$.

$$\frac{\partial u}{\partial y} = f_1 \frac{\partial x}{\partial y} + f_2$$

$$1 = g_1 \frac{\partial x}{\partial y}$$

$$\frac{\partial u_{y,z}}{\partial y} = f_2 + \frac{f_1}{g_1}$$

$$\frac{\partial x_{y,z}}{\partial y} = \frac{1}{g_1} \qquad\qquad g_1 \neq 0$$

CASE II. $\dfrac{\partial u_{y,x}}{\partial y}, \dfrac{\partial z_{y,x}}{\partial y}$. In this case, the two equations are independent of each other. The first defines u; the second defines z.

$$\frac{\partial u_{y,x}}{\partial y} = f_2$$

$$\frac{\partial z_{y,x}}{\partial y} = \frac{1}{g_2} \qquad\qquad g_2 \neq 0$$

7.3 THIRD ILLUSTRATION

Find $\dfrac{\partial y}{\partial x}$ if

(2)
$$v = f(x, y, z)$$
$$x = g(y, u, v)$$

CASE I. $\dfrac{\partial y_{x,u,v}}{\partial x}$.   The second equation alone is sufficient.

$$1 = g_1\frac{\partial y}{\partial x}, \qquad \frac{\partial y}{\partial x} = \frac{1}{g_1} \qquad\qquad g_1 \neq 0$$

CASE II. $\dfrac{\partial y_{x,z,v}}{\partial x}$.   The first equation alone is sufficient.

$$f_1 + f_2\frac{\partial y}{\partial x} = 0$$

$$\frac{\partial y}{\partial x} = -\frac{f_1}{f_2} \qquad\qquad f_2 \neq 0$$

CASE III. $\dfrac{\partial y_{x,z,u}}{\partial x}$.   Both equations are necessary.

$$f_2\frac{\partial y}{\partial x} - \frac{\partial v}{\partial x} = -f_1, \qquad g_1\frac{\partial y}{\partial x} + g_3\frac{\partial v}{\partial x} = 1$$

Then

$$\frac{\partial y_{x,z,u}}{\partial x} = \frac{1 - f_1 g_3}{g_1 + f_2 g_3} \qquad\qquad g_1 + f_2 g_3 \neq 0$$

EXERCISES (7)

1. If

$$x^2 + y^2 + z^2 + u^2 = 1$$
$$xy - zu = 2$$

compute $\dfrac{\partial z_{xy}}{\partial x}$ and $\dfrac{\partial z_{xu}}{\partial x}$.

2. In the previous exercise compute $\dfrac{\partial x}{\partial u}$ (all meanings).

3. If

$$F(r, s, t, u) = 0$$
$$G(r, s, t, u) = 0$$

compute $\dfrac{\partial s}{\partial u}$.

4. Same problem for $\dfrac{\partial r}{\partial s}$.

5. Find $\dfrac{\partial u}{\partial x}$ if

$$u = x^2 + y^2$$
$$y = x^z$$

Check by use of the results of §7.1.

6. In both cases of the illustration of §7.1, find the two derivatives with respect to the other independent variable.

7. For equations (1) find $\dfrac{\partial^2 u}{\partial x^2}$.

8. Find $\dfrac{\partial v}{\partial t}$ if

$$f(x, v, t) = 0$$
$$g(t, u, x) = 0$$

9. Find $\dfrac{du}{dx}$ if

$$f(u, v, w) = x^2$$
$$g(u, v, x) = \log w$$
$$h(u, v, w, x) = 0$$

10. For equations (2), enumerate all cases in which both equations are necessary.

11. In Exercise 3 find $\dfrac{\partial^2 s}{\partial t\, \partial u}$.

12. In Exercise 3 find $\dfrac{\partial^2 s}{\partial u^2}$ (both meanings).

13. If

$$F(u, r, s, x, y) = 0$$
$$x = g(r, s)$$
$$y = h(r, s)$$

find $\dfrac{\partial u_{rs}}{\partial r}$.

14. Same problem for $\dfrac{\partial u_{xy}}{\partial x}$.

15. Show that

$$\frac{\partial y}{\partial x} = 1 \bigg/ \frac{\partial x}{\partial y}$$

in all three cases of §7.3.

16. Given three equations in the seven variables t, u, v, w, x, y, z, show that

$$\frac{\partial v_{uwyz}}{\partial y} = 1 \bigg/ \frac{\partial y_{uvwz}}{\partial v}$$

if certain Jacobians are not zero.

§8. *Differentials.   Directional Derivatives*

We shall introduce briefly the idea of the differential of a function of several variables. Just as for functions of one variable, one could build the whole technique of differentiation on the differential. On the other hand, the differential can always be obtained from the derivative, which we have already learned to compute, by recourse to the very definition of the differential. It is this latter point of view which we shall adopt.

8.1 THE DIFFERENTIAL

It will be sufficient to give our definitions for functions of two variables. Let $u = f(x, y)$ be a function of C^1, x and y being independent variables. Form the following function of four variables:

$$\varphi(x, y, r, s) = f_1(x, y)r + f_2(x, y)s$$

If $r = \Delta x$, $s = \Delta y$ are variables whose range is a neighborhood of $r = 0$, $s = 0$, then the differential of u, du, is defined as $\varphi(x, y, \Delta x, \Delta y)$:

$$(1) \qquad du = \varphi(x, y, \Delta x, \Delta y) = f_1(x, y)\,\Delta x + f_2(x, y)\,\Delta y$$

Thus there is associated with each point (x, y) where $f(x, y) \in C^1$, a differential which is itself a linear function of two variables Δx, Δy.

EXAMPLE A. $u = f(x, y) = \dfrac{x}{y}, \quad f_1 = \dfrac{1}{y}, \quad f_2 = -\dfrac{x}{y^2}$

$$\varphi(x, y, r, s) = \frac{r}{y} - \frac{xs}{y^2}$$

$$du = \frac{\Delta x}{y} - \frac{x\,\Delta y}{y^2}$$

EXAMPLE B. $u = f(g(x, y), h(x, y))$

$$(2) \qquad du = (f_1 g_1 + f_2 h_1)\,\Delta x + (f_1 g_2 + f_2 h_2)\,\Delta y$$

It would be a simple matter to deduce the fundamental rules for obtaining the differentials of sums, products, quotients, etc. In fact, such a procedure would produce a slightly simpler technique than the one we have already developed, insofar as it concerns composite functions. We illustrate by Example B above. Here, from the definition of the differential, we have

$$dg = g_1\,\Delta x + g_2\,\Delta y$$
$$dh = h_1\,\Delta x + h_2\,\Delta y$$

Substituting in equation (2), we have

$$du = df = f_1\,dg + f_2\,dh$$

Observe now the close similarity of this result with the definition in equation (1). It is precisely this sort of similarity which could be exploited to effect the simplification referred to above.

8.2 MEANING OF THE DIFFERENTIAL

The student is familiar with the fact that the equation of the tangent plane to the surface $z = f(x, y)$ at the point (x_0, y_0, z_0) of the surface is

$$z - z_0 = f_1(x_0, y_0)(x - x_0) + f_2(x_0, y_0)(y - y_0)$$

By definition, dz at (x_0, y_0) is

$$dz = f_1(x_0, y_0)\,\Delta x + f_2(x_0, y_0)\,\Delta y$$

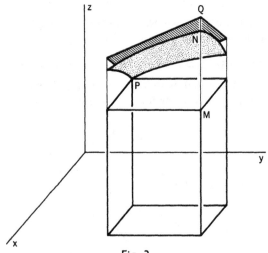

Fig. 3.

so that the point Q, Figure 3, with coordinates $(x_0 + \Delta x, y_0 + \Delta y, z_0 + dz)$ lies on that plane.   If

$$\Delta z = f(x_0 + \Delta x, y_0 + \Delta y) - f(x_0, y_0)$$

then the point N, $(x_0 + \Delta x, y_0 + \Delta y, z_0 + \Delta z)$, lies on the surface $z = f(x, y)$. Hence $MN = \Delta z$, $MQ = dz$.   That is, $|dz|$ is the length of the ordinate $x = x_0 + \Delta x$, $y = y_0 + \Delta y$ cut off between the tangent plane and the plane $z = z_0$.   It is clear from the defining property of a tangent plane that dz will be nearly equal to Δz for small values of Δx and Δy.   Since dz is usually so much more easily computed than Δz, the former is frequently used in place of the latter in approximate computations.

We have hitherto assumed for simplicity that $f(x, y) \in C^1$.   But if we assume only that $f(x, y)$ is differentiable at (a, b), the differential df is equally well defined at (a, b) by equation (1).   Then equation (3) of §3.6 becomes

$$\Delta z = \Delta f = dz + \Delta x \varphi(\Delta x, \Delta y) + \Delta y \psi(\Delta x, \Delta y)$$

and this shows, without any appeal to geometry, in what sense dz is nearly equal to Δz when $(\Delta x, \Delta y)$ is near $(0, 0)$. Thus

$$\frac{|\Delta z - dz|}{|\Delta x| + |\Delta y|} \leq |\varphi(\Delta x, \Delta y)| + |\psi(\Delta x, \Delta y)|$$

$$\lim_{\substack{\Delta x \to 0 \\ \Delta y \to 0}} \frac{\Delta z - dz}{|\Delta x| + |\Delta y|} = 0$$

EXAMPLE C. Find approximately how much $x^2 + y^3$ changes when (x, y) changes from $(1, 1)$ to $(1.1, .9)$.

$$d(x^2 + y^3) = 2x \, \Delta x + 3y^2 \, \Delta y, \qquad \Delta x = .1, \quad \Delta y = -.1$$

$$d(x^2 + y^3)|_{(1,1)} = 2 \, \Delta x + 3 \, \Delta y$$

Approximate change in $(x^2 + y^3)$ is

$$|2(.1) + 3(-.1)| = .1$$

Actual change in $(x^2 + y^3)$ is $.061$.

8.3 DIRECTIONAL DERIVATIVES

We now introduce a natural generalization of partial derivatives. In the definition of $f_1(x_0, y_0)$, the numerator of the difference quotient used involves the values of $f(x, y)$ at two points $(x_0 + \Delta x, y_0)$ and (x_0, y_0). As Δx approaches zero, the first point approaches the latter along the line $y = y_0$. For $f_2(x_0, y_0)$ a point $(x_0, y_0 + \Delta y)$ approaches (x_0, y_0) along the line $x = x_0$. We now replace these two special lines by an arbitrary line through (x_0, y_0).

A direction ξ_α is defined as the direction of any directed line which makes the angle α with the positive x-axis (positive angles measured in the counterclockwise sense as usual). Thus the line segment directed from the point $(0, 0)$ to the point $(-1, -1)$ has the direction $\xi_{5\pi/4}$ or $\xi_{-3\pi/4}$.

Definition 8. *The directional derivative of $f(x, y)$ in the direction ξ_α at* (a, b) *is*

$$\frac{\partial f}{\partial \xi_\alpha}\bigg|_{(a,b)} = \lim_{\Delta s \to 0} \frac{f(a + \Delta s \cos \alpha, b + \Delta s \sin \alpha) - f(a, b)}{\Delta s}$$

EXAMPLE D. $f(x, y) = x^2 - 2y$, $a = 1$, $b = 2$, $\alpha = 3\pi/4$.

$$\frac{\partial f}{\partial \xi_{3\pi/4}}\bigg|_{(1,2)} = \lim_{\Delta s \to 0} \frac{\left(1 - \dfrac{\Delta s}{\sqrt{2}}\right)^2 - 2\left(2 + \dfrac{\Delta s}{\sqrt{2}}\right) + 3}{\Delta s} = -2\sqrt{2}$$

At each point (x, y) a function has infinitely many directional derivatives so that $\dfrac{\partial f}{\partial \xi_\alpha}$ is a function of the three variables x, y, α. In computing a directional derivative of higher order, the variable α must, of course, be held constant. For example, if

$$\frac{\partial f}{\partial \xi_\alpha} = x \cos \alpha + y \sin \alpha$$

then

$$\frac{\partial^2 f}{\partial \xi_\alpha^2} = \frac{\partial}{\partial \xi_\alpha}\left(\frac{\partial f}{\partial \xi_\alpha}\right)$$

$$= \lim_{\Delta s \to 0} \frac{(x + \Delta s \cos \alpha)\cos \alpha + (y + \Delta s \sin \alpha)\sin \alpha - x \cos \alpha - y \sin \alpha}{\Delta s}$$

$$= \cos^2 \alpha + \sin^2 \alpha = 1$$

Observe that

$$\frac{\partial f}{\partial \xi_0} = f_1, \qquad \frac{\partial f}{\partial \xi_{\pi/2}} = f_2, \qquad \frac{\partial f}{\partial \xi_\pi} = -f_1, \qquad \frac{\partial f}{\partial \xi_{3\pi/2}} = -f_2$$

Theorem 9. 1. $f(x, y) \in C^1$

$$\Rightarrow \qquad \frac{\partial f}{\partial \xi_\alpha} = f_1(x, y) \cos \alpha + f_2(x, y) \sin \alpha$$

By Theorem 3 we have

$$\frac{f(a + \Delta s \cos \alpha, b + \Delta s \sin \alpha) - f(a, b)}{\Delta s} = f_1(a + \theta_1 \Delta s \cos \alpha, b) \cos \alpha$$
$$+ f_2(a + \Delta s \cos \alpha, b + \theta_2 \Delta s \sin \alpha) \sin \alpha$$

where $0 < \theta_1 < 1, 0 < \theta_2 < 1$. Now, when Δs approaches zero, we obtain the desired result.

This theorem enables one to compute directional derivatives without reverting to the defining limiting process. In Example D, we have

$$\frac{\partial f}{\partial \xi_\alpha} = 2x \cos \alpha - 2 \sin \alpha$$

for any point (x, y) and any direction α. In particular for $x = 1, y = 2$, $\alpha = 3\pi/4$, the derivative is $-2\sqrt{2}$ as before. We also have for this example

$$\frac{\partial^2 f}{\partial \xi_\alpha^2} = 2 \cos^2 \alpha, \qquad \frac{\partial^3 f}{\partial \xi_\alpha^3} = 0$$

8.4 THE GRADIENT

For a fixed point (a, b), let us determine the direction ξ_α which will make $\dfrac{\partial f}{\partial \xi_\alpha}$ a maximum. Set

$$F(\alpha) = f_1(a, b) \cos \alpha + f_2(a, b) \sin \alpha$$

Then $F(\alpha)$ will have a maximum or minimum when

$$F'(\alpha) = -f_1 \sin \alpha + f_2 \cos \alpha = 0$$

If f_1 and f_2 are not both zero, this equation will have just two distinct solutions α_1 and α_2 between 0 and 2π determined by the equations

(3)
$$\sin \alpha_1 = \frac{f_2}{\sqrt{f_1^2 + f_2^2}}, \qquad \cos \alpha_1 = \frac{f_1}{\sqrt{f_1^2 + f_2^2}}$$

$$\sin \alpha_2 = -\frac{f_2}{\sqrt{f_1^2 + f_2^2}}, \qquad \cos \alpha_2 = -\frac{f_1}{\sqrt{f_1^2 + f_2^2}}$$

For these directions we have

$$\frac{\partial f}{\partial \xi_{\alpha_1}} = \sqrt{f_1^2 + f_2^2}, \qquad \frac{\partial f}{\partial \xi_{\alpha_2}} = -\sqrt{f_1^2 + f_2^2}$$

Hence, $\dfrac{\partial f}{\partial \xi_\alpha}$ is maximum in the direction ξ_{α_1}, and is minimum in the direction ξ_{α_2}. Of course, α_1 and α_2 differ by π. If $f_1 = f_2 = 0$, the maximum and minimum values of $\dfrac{\partial f}{\partial \xi_\alpha}$ are both zero, since the directional derivative is constantly zero.

Definition 9. *The gradient of $f(x, y)$ at a point (a, b),*

$$\text{Grad } f(x, y)|_{(a,b)}$$

is a vector of magnitude $(f_1(a, b)^2 + f_2(a, b)^2)^{1/2}$ in the direction ξ_{α_1} defined by equations (3).

EXAMPLE E.　$f(x, y) = x^2 - xy + y^2$
　　Grad $f(x, y)|_{(1,3)}$ is a vector of magnitude $\sqrt{26}$ in the direction ξ_{α_1} defined by the equations

$$\sin \alpha_1 = \frac{5}{\sqrt{26}}, \qquad \cos \alpha_1 = \frac{-1}{\sqrt{26}}$$

We have proved the following result.

Theorem 10.　1. $f(x, y) \in C^1$
　　　　　　　2. $f_1(a, b)^2 + f_2(a, b)^2 \neq 0$

$$\Rightarrow \quad \underset{0 \leq \alpha \leq 2\pi}{\text{Max}} \frac{\partial f}{\partial \xi_\alpha}\bigg|_{(a,b)} = (f_1(a, b)^2 + f_2(a, b)^2)^{1/2} = \frac{\partial f}{\partial \xi_{\alpha_1}}\bigg|_{(a,b)}$$

where ξ_{α_1} is the direction of

$$\text{Grad } f(x, y)|_{(a,b)}$$

defined by equations (3).

EXERCISES   (8)

1. Find the maximum and minimum values of

$$F(\alpha) = -3 \sin \alpha + 4 \cos \alpha.$$

For what α are these values attained, respectively?   Check by equations (3).

2. If $f(x, y) = \tan^{-1} xy$, compute $f(0.9, -1.2)$ in radians by use of tables. Then compute it approximately by use of differentials and observe that tables are not needed.

3. Define the differential of a function of three variables.

4. If $u = F(f(x, y), g(x, y), h(x, y))$, show that

$$du = F_1 \, df + F_2 \, dg + F_3 \, dh.$$

5. Show that

$$dF(f(x, y, z)) = F' \, df$$
$$dF(f(t), g(t)) = F_1 \, df + F_2 \, dg$$

6. In a 3, 4, 5 triangle the short leg is decreased, the large leg is increased by 1%. What happens to the hypotenuse, the area, and the smaller acute angle? Obtain the approximate and the exact changes.

7. If $f(x, y) = xy + x \log y$, find $\left. \dfrac{\partial f}{\partial \xi_\alpha} \right|_{(2,1)}$, $\operatorname{Grad} f|_{(2,1)}$.

8. If r, θ are polar coordinates, show that

$$\left. \frac{\partial f}{\partial \xi_\theta} \right|_{(r,\theta)} = f_1(r, \theta), \qquad \left. \frac{\partial f}{\partial \xi_{\theta + \pi/2}} \right|_{(r,\theta)} = \frac{1}{r} f_2(r, \theta)$$

9. Show that

$$\left. \frac{\partial f}{\partial \xi_{\theta + \psi}} \right|_{(r,\theta)} = f_1 \cos \psi + f_2 \frac{\sin \psi}{r}$$

10. Find the gradient of $f(r, \theta)$.

11. Show that $\dfrac{\partial f}{\partial \xi_{\alpha_1 + \psi}} = (\cos \psi) |\operatorname{Grad} f|$, where $|\operatorname{Grad} f|$ means the magnitude of the vector $\operatorname{Grad} f$ and ξ_{α_1} is the direction of that vector.

12. If $u = \sqrt{x^2 + y^2}$, ξ_α is the direction of the interior normal to the circle

$$(x - 1)^2 + (y - 3)^2 = 25$$

at the point (4, 7) and γ is the angle measured from the interior normal to the line directed from (4, 7) to (0, 0), show that

$$\left. \frac{\partial u}{\partial \xi_\alpha} \right|_{(4,7)} = -\cos \gamma$$

13. In Exercise 12, replace ξ_α by an arbitrary direction and (4, 7) by an arbitrary point (a, b) and prove the same result. Here γ is the angle measured from the direction ξ_α to the line directed from (a, b) to $(0, 0)$.

14. If

$$f_1(x, y) = g_2(x, y)$$
$$f_2(x, y) = -g_1(x, y)$$

show that

$$\frac{\partial f}{\partial \xi_\alpha} = \frac{\partial g}{\partial \xi_{\alpha+\pi/2}}$$

15. Find $\dfrac{\partial^2 f}{\partial \xi_\alpha^2}$ if $f = e^{xy}$.

16. Find $\dfrac{\partial^n f}{\partial \xi_\alpha^n}$ if $f = (x + y)^n$.

17. If

$$f(x, y) = \frac{x^3}{3} - y^2 x$$

prove

$$|\operatorname{Grad}|\operatorname{Grad} f\,||^2 = 4|\operatorname{Grad} f|$$

The absolute value signs were explained in Exercise 11.

§9. *Taylor's Theorem*

It is assumed that the student is familiar with Taylor's series with remainder for a function of one variable. However, by way of introducing the "exact" remainder, which is less generally used than the Lagrange form, we give a brief derivation of the formula.

9.1 FUNCTIONS OF A SINGLE VARIABLE

Theorem 11.   1. $f(x) \in C^{n+1}$ $\hspace{4cm}$ $|x - a| \leqq h$

$$(1) \quad \Rightarrow \quad f(x) = \sum_{k=0}^{n} \frac{f^{(k)}(a)}{k!}(x - a)^k + \int_a^x f^{(n+1)}(t)\frac{(x - t)^n}{n!}\,dt \quad |x - a| \leqq h$$

To prove this apply integration by parts to the integral appearing in equation (1):

$$R_n = \int_a^x f^{(n+1)}(t)\frac{(x - t)^n}{n!}\,dt = -f^{(n)}(a)\frac{(x - a)^n}{n!} + R_{n-1}$$

Repeated use of this equation, each time reducing the subscript of R by 1, leads finally to R_0 on the right-hand side. But

$$R_0 = \int_a^x f'(t)\,dt = f(x) - f(a)$$

Eliminating all R's except R_n, we obtain equation (1).

To obtain the familiar Lagrange or Cauchy remainders from this, we use the "first mean-value theorem" for integrals, which we prove in passing. The result will appear as a corollary to a more general theorem in §4, Chapter 5.

Theorem. 1. $f(x), g(x) \in C$ $\qquad\qquad\qquad\qquad a \leqq x \leqq b$

  2. $g(x) \geqq 0$ $\qquad\qquad\qquad\qquad\qquad a \leqq x \leqq b$

(2) $\Rightarrow \qquad \displaystyle\int_a^b f(x)g(x)\,dx = f(X)\int_a^b g(x)\,dx \qquad a < X < b$

Let M and m be the largest and smallest values of $f(x)$ in (a, b). Then

$$mg(x) \leqq f(x)g(x) \leqq Mg(x)$$

$$m \leqq \frac{\displaystyle\int_a^b f(x)g(x)\,dx}{\displaystyle\int_a^b g(x)\,dx} \leqq M$$

provided the denominator is not zero. The continuous function $f(x)$ takes on every value between m and M somewhere between a and b.* In particular, it must take on the above quotient of integrals at some point $x = X$. Hence, equation (2) holds. If the above denominator is zero, equation (2) reduces to $0 = 0$ for an arbitrary X. Note that hypothesis 2 might be replaced by $g(x) \leqq 0$. To see that X need never be a nor b see Exercise 9, p. 165.

Lagrange remainder. Take $g(t) = \dfrac{(x - t)^n}{n!}$. Then

$$g(t) \geqq 0, \qquad a \leqq t \leqq x, \qquad \text{if} \qquad x > a$$
$$(-1)^n g(t) \geqq 0, \qquad x \leqq t \leqq a, \qquad \text{if} \qquad x < a$$
$$R_n = f^{(n+1)}(X)\int_a^x \frac{(x - t)^n}{n!}\,dt = f^{(n+1)}(X)\frac{(x - a)^{n+1}}{(n + 1)!}$$

Cauchy remainder. Take $g(t) = 1$.

$$R_n = f^{(n+1)}(X)\frac{(x - X)^n}{n!}(x - a)$$

In both cases, X is between a and x.

9.2 FUNCTIONS OF TWO VARIABLES

In the proof of the next theorem we shall have to find the successive derivatives of the function

$$F(t) = f(a + ht, b + kt)$$

---

* See Exercise 11 of §6, Chapter 5.

We have
$$F'(0) = h \frac{\partial}{\partial a} f(a, b) + k \frac{\partial}{\partial b} f(a, b)$$

It is easy to show by induction that

$$F^{(n)}(0) = \sum_{j=0}^{n} \binom{n}{j} h^j k^{n-j} \frac{\partial^n f(a, b)}{\partial a^j \partial b^{n-j}}$$

$$\binom{n}{j} = \frac{n!}{j!(n-j)!} \qquad j = 0, 1, 2, \ldots, n$$

On account of the similarity of this sum to a binomial expansion, we introduce the following symbolic notation:

$$F^{(n)}(0) = \left(h \frac{\partial}{\partial a} + k \frac{\partial}{\partial b} \right)^n f(a, b)$$

Theorem 12.   1. $f(x, y) \in C^{n+1}$ $\qquad$ $|x - a| \leq |h|, \quad |y - b| \leq |k|$

(3) $\Rightarrow$ $\quad f(a + h, b + k) = \sum_{j=0}^{n} \frac{1}{j!} \left(h \frac{\partial}{\partial a} + k \frac{\partial}{\partial b} \right)^j f(a, b) + R_n$

$$R_n = \int_0^1 \frac{(1-t)^n}{n!} \left(h \frac{\partial}{\partial a} + k \frac{\partial}{\partial b} \right)^{n+1} f(a + ht, b + kt) \, dt$$

$$= \frac{1}{(n+1)!} \left(h \frac{\partial}{\partial a} + k \frac{\partial}{\partial b} \right)^{n+1} f(a + \theta h, b + \theta k) \qquad 0 < \theta < 1$$

To prove this, we have only to expand $F(t)$ in Taylor's series:

$$F(1) = \sum_{j=0}^{n} \frac{F^{(j)}(0)}{j!} + \int_0^1 \frac{(1-t)^n}{n!} F^{(n+1)}(t) \, dt$$

$$= \sum_{j=0}^{n} \frac{F^{(j)}(0)}{j!} + \frac{F^{(n+1)}(\theta)}{(n+1)!} \qquad 0 < \theta < 1$$

The result now follows by introducing the symbolic notation for $F^{(j)}(0)$.

Another useful form of equation (3) is obtained by replacing $a + h$ by x, $b + k$ by y:

$$f(x, y) = \sum_{j=0}^{n} \frac{1}{j!} \left(\overline{x - a} \frac{\partial}{\partial a} + \overline{y - b} \frac{\partial}{\partial b} \right)^j f(a, b) + R_n$$

$$R_n = \frac{1}{(n+1)!} \left(\overline{x - a} \frac{\partial}{\partial r} + \overline{y - b} \frac{\partial}{\partial s} \right)^{n+1} f(r, s)$$

where we set
$$r = a + \theta(x - a), \qquad s = b + \theta(y - b) \qquad 0 < \theta < 1$$
after the differentiation.

A particular case of the theorem of interest is obtained by taking $n = 0$:

(4) $f(a + h, b + k) - f(a, b) = f_1(a + \theta h, b + \theta k)h$
$$+ f_2(a + \theta h, b + \theta k)k \qquad 0 < \theta < 1$$

Note the resemblance between this equation and equation (2) of §3. Observe that we have replaced θ_1 and θ_2 by a single θ, which now occurs symmetrically. Equation (4) is known as *the law of the mean for functions of two variables*. It could not have been introduced in place of Theorem 3 since we could not have computed $F'(0)$ at that stage.

EXAMPLE A. $f(x, y) = x^2 + xy - y^2, \qquad a = 1, \qquad b = -2$
$$f(1, -2) = -5, \qquad f_1(1, -2) = 0, \qquad f_2(1, -2) = 5$$
$$f_{11} = 2, \qquad f_{12} = 1, \qquad f_{22} = -2$$
$$x^2 + xy - y^2 = -5 + 5(y + 2)$$
$$+ \tfrac{1}{2}[2(x - 1)^2 + 2(x - 1)(y + 2) - 2(y + 2)^2]$$

This can be checked algebraically.

EXAMPLE B. $f(x, y) \in C^1, \qquad g(x, y) \in C^1, \qquad f(0, 0) = g(0, 0) = 0$
$$g_1(0, 0) + \lambda g_2(0, 0) \neq 0$$

Find $\lim \dfrac{f(x, y)}{g(x, y)}$ as (x, y) approaches $(0, 0)$ along the line $y = \lambda x$. By Theorem 12,

$$\frac{f(x, y)}{g(x, y)} = \frac{f_1(\theta x, \theta y)x + f_2(\theta x, \theta y)y}{g_1(\theta_1 x, \theta_1 y)x + g_2(\theta_1 x, \theta_1 y)y} \qquad 0 < \theta, \theta_1 < 1$$

$$\lim \frac{f(x, y)}{g(x, y)} = \frac{f_1(0, 0) + \lambda f_2(0, 0)}{g_1(0, 0) + \lambda g_2(0, 0)} \qquad g_1(0, 0) + \lambda g_2(0, 0) \neq 0$$

EXERCISES (9)

1. Write out explicitly all terms of the symbolic expressions

   (a) $\left(2 - \dfrac{d}{dx}\right)^4 \sin x \big|_{x = \pi/2}$

   (b) $\left(h \dfrac{\partial}{\partial x} + k \dfrac{\partial}{\partial y}\right)^3 x^2 y$

   (c) $\left[(x - 1)\dfrac{\partial}{\partial a} + y \dfrac{\partial}{\partial b}\right]^3 f(a + t, b)$

2. Expand $x^3 - 2xy^2$ in Taylor's series $(a = 1, b = -1)$ and check by algebra.

3. In the previous problem stop the expansion after first degree terms and write the remainder using second partial derivatives. Do not compute θ.

4. If $f(x, y) = x^2 - 3xy + 2y^2$, use equation (4) to express the difference $f(1, 2) - f(2, -1)$ by partial derivatives. Compute θ and check that it is between 0 and 1.

5. Expand $x^2y + \sin y + e^x$ in powers of $(x - 1)$ and $(y - \pi)$ through quadratic terms and write the remainder. Do not compute θ.

6. Expand $(1 - 3x + 2y)^3$ in powers of x and y and check by algebra.

7. Expand $(1 - 3x + 2y)^3$ in powers of $x - 1$ and $y + 1$ and check.

8. Expand e^{xy} in powers of x and y. Show first that

$$\left.\frac{\partial^n e^{xy}}{\partial x^m\, \partial y^{n-m}}\right|_{(0,0)} \begin{array}{ll} = 0 & 2m \neq n \\ = m! & 2m = n \end{array} \qquad (m = 0, 1, 2, \ldots, n)$$

Check by use of the Maclaurin series for e^x. It is not required to show the convergence of the series to the function.

9. In Example B, when will the limit be independent of λ? Give an example.

10. In Example B, let the first partial derivatives of f and g be zero at $(0, 0)$. Obtain the limit under further conditions which you are to impose.

11. Let (x, y) approach $(0, 0)$ along the line $y = -x$. Find

$$\lim \frac{\sin xy + xe^x - y}{x \cos y + \sin 2y}$$

12. Same problem for

$$\lim \frac{e^{xy} - 1}{\sin x \log (1 + y)}$$

13. Extend Taylor's theorem with remainder to functions of three variables.

14. If $f(0, 0) = g(0, 0) = 0$, find

$$\lim_{x \to 0} \frac{f(x, x^2)}{g(x^2, x)}$$

What properties are you assuming for f and g?

15. If $f(0, 1, 1) = f_1(0, 1, 1) = f_3(0, 1, 1) = 0$, find

$$\lim_{x \to 0} \frac{f(x, \cos x, \cosh x)}{f(x^2, \cosh x, e^x)}$$

What assumptions are you making about $f(x, y, z)$?

16. If $f(a, b) = f_1(a, b) = f_2(a, b) = 0$, prove

$$\lim_{h \to 0} \frac{f(a + h \cos \alpha, b + h \sin \alpha)}{h^2} = \frac{1}{2} \left.\frac{\partial^2 f}{\partial \xi_\alpha^2}\right|_{(a,b)}$$

§10. *Jacobians*

We discuss here certain further results concerning Jacobians.   They are found to be useful in the problems of change of variable.   A criterion for the functional dependence of several functions can also be given in terms of Jacobians.   This latter result will be given in §12.

10.1 IMPLICIT FUNCTIONS

We have already used Jacobians in differentiating functions defined implicitly.   We now give a more general case.   Let

$$f(u, v, w, x, y, z) = 0$$
$$g(u, v, w, x, y, z) = 0$$
$$h(u, v, w, x, y, z) = 0$$

the equations being assumed to define three functions u, v, w of the variables x, y, z.   Then

$$f_1 \frac{\partial u}{\partial x} + f_2 \frac{\partial v}{\partial x} + f_3 \frac{\partial w}{\partial x} = -f_4$$

$$g_1 \frac{\partial u}{\partial x} + g_2 \frac{\partial v}{\partial x} + g_3 \frac{\partial w}{\partial x} = -g_4$$

$$h_1 \frac{\partial u}{\partial x} + h_2 \frac{\partial v}{\partial x} + h_3 \frac{\partial w}{\partial x} = -h_4$$

Solving these linear equations, we obtain

$$\frac{\partial u}{\partial x} = -\frac{\partial(f, g, h)}{\partial(x, v, w)} \frac{1}{\Delta}, \qquad \frac{\partial v}{\partial x} = -\frac{\partial(f, g, h)}{\partial(u, x, w)} \frac{1}{\Delta}, \qquad \frac{\partial w}{\partial x} = -\frac{\partial(f, g, h)}{\partial(u, v, x)} \frac{1}{\Delta}$$

where

$$\Delta = \frac{\partial(f, g, h)}{\partial(u, v, w)} \neq 0$$

10.2 THE INVERSE OF A TRANSFORMATION

Let the transformation

(1)
$$u = f(x, y)$$
$$v = g(x, y)$$

with Jacobian

$$J = \frac{\partial(u, v)}{\partial(x, y)} \neq 0$$

have an inverse with Jacobian

$$j = \frac{\partial(x, y)}{\partial(u, v)}$$

Let us investigate the relation between these two Jacobians. Computing the derivatives in question, we have

$$\frac{\partial x}{\partial u} = \frac{g_2}{J}, \qquad \frac{\partial y}{\partial u} = -\frac{g_1}{J}$$

$$\frac{\partial x}{\partial v} = -\frac{f_2}{J}, \qquad \frac{\partial y}{\partial v} = \frac{f_1}{J}$$

so that

$$j = \begin{vmatrix} g_2 & -g_1 \\ -f_2 & f_1 \end{vmatrix} \frac{1}{J^2} = \frac{1}{J}$$

Hence $Jj = 1$, or

$$\frac{\partial(u, v)}{\partial(x, y)} \frac{\partial(x, y)}{\partial(u, v)} = 1$$

Note the useful aid to memory obtained by canceling symbols.

Let us generalize to three functions,

$$\begin{aligned} u &= f(x, y, z), \\ v &= g(x, y, z), \\ w &= h(x, y, z), \end{aligned} \qquad J = \frac{\partial(u, v, w)}{\partial(x, y, z)}, \qquad j = \frac{\partial(x, y, z)}{\partial(u, v, w)}$$

For the determinant

$$J = \begin{vmatrix} f_1 & f_2 & f_3 \\ g_1 & g_2 & g_3 \\ h_1 & h_2 & h_3 \end{vmatrix}$$

write the determinant of cofactors

$$K = \begin{vmatrix} F_1 & F_2 & F_3 \\ G_1 & G_2 & G_3 \\ H_1 & H_2 & H_3 \end{vmatrix}$$

For example, the cofactor of g_3 is G_3,

$$G_3 = - \begin{vmatrix} f_1 & f_2 \\ h_1 & h_2 \end{vmatrix}$$

Then

$$\frac{\partial x}{\partial u} = \frac{F_1}{J}, \qquad \frac{\partial y}{\partial u} = \frac{F_2}{J}, \qquad \frac{\partial z}{\partial u} = \frac{F_3}{J}$$

with similar equations for the derivatives with respect to v and w. Then

$$j = \frac{K}{J^3}$$

But

$$JK = \begin{vmatrix} f_1 & f_2 & f_3 \\ g_1 & g_2 & g_3 \\ h_1 & h_2 & h_3 \end{vmatrix} \cdot \begin{vmatrix} F_1 & F_2 & F_3 \\ G_1 & G_2 & G_3 \\ H_1 & H_2 & H_3 \end{vmatrix} = \begin{vmatrix} J & 0 & 0 \\ 0 & J & 0 \\ 0 & 0 & J \end{vmatrix} = J^3$$

so that $Jj = 1$, as before.

10.3 CHANGE OF VARIABLE

If
$$u = f(x, y), \qquad x = \varphi(r, s)$$
$$v = g(x, y), \qquad y = \psi(r, s)$$

then u and v may be regarded as functions of r and s. Let us compute the Jacobian

$$\frac{\partial(u, v)}{\partial(r, s)}$$

Direct computation gives

$$(2) \qquad \frac{\partial(u, v)}{\partial(r, s)} = \begin{vmatrix} f_1\varphi_1 + f_2\psi_1 & g_1\varphi_1 + g_2\psi_1 \\ f_1\varphi_2 + f_2\psi_2 & g_1\varphi_2 + g_2\psi_2 \end{vmatrix} = \frac{\partial(u, v)}{\partial(x, y)}\frac{\partial(x, y)}{\partial(r, s)}$$

Note the analogy of this result with the formula for the differentiation of a composite function of one variable. It generalizes easily to functions of more variables.

EXERCISES (10)

1. Check equation (2) in the special case
$$u = x^2 + y^2, \qquad x = r\cos s$$
$$v = y, \qquad y = r\sin s$$
by eliminating x and y.

2. If
$$u = f(x), \qquad x = \varphi(r, s)$$
$$v = g(y), \qquad y = \psi(r, s)$$
prove
$$\frac{\partial(u, v)}{\partial(r, s)} = f'g'\frac{\partial(x, y)}{\partial(r, s)}$$
directly and by using equation (2).

3. If f, g, h are functions of x, y, z, prove
$$\frac{\partial(f, g, h)}{\partial(x, y, z)} = \frac{\partial f}{\partial x}\frac{\partial(g, h)}{\partial(y, z)} + \frac{\partial g}{\partial x}\frac{\partial(h, f)}{\partial(y, z)} + \frac{\partial h}{\partial x}\frac{\partial(f, g)}{\partial(y, z)}$$

4. In the previous exercise show that
$$\frac{\partial f}{\partial y}\frac{\partial(g, h)}{\partial(x, y)} + \frac{\partial g}{\partial y}\frac{\partial(h, f)}{\partial(x, y)} + \frac{\partial h}{\partial y}\frac{\partial(f, g)}{\partial(x, y)} = 0$$

5. Generalize equation (2) to include functions of three variables, three functions in each set.

6. If
$$u = 3x + 2y - z$$
$$v = x - y + z$$
$$w = x + 2y - z$$

find the explicit equations for the inverse transformation. Then compute J and j and show that $Jj = 1$.

7. If
$$u = 2xy, \qquad x = r \cos \theta$$
$$v = x^2 - y^2, \qquad y = r \sin \theta$$

eliminate x, y and thus compute the Jacobian $\dfrac{\partial(u, v)}{\partial(r, \theta)}$. Then verify the result by use of equation (2).

8. If f, g, h are functions of x, y, z and if
$$x = F(u, v), \qquad y = G(u, v), \qquad z = H(u, v)$$
show that

$$\frac{\partial(f, g)}{\partial(u, v)} = \frac{\partial(f, g)}{\partial(y, z)}\frac{\partial(y, z)}{\partial(u, v)} + \frac{\partial(f, g)}{\partial(z, x)}\frac{\partial(z, x)}{\partial(u, v)} + \frac{\partial(f, g)}{\partial(x, y)}\frac{\partial(x, y)}{\partial(u, v)}$$

9. In the previous example, compute $\dfrac{\partial(g, h)}{\partial(u, v)}$.

10. If
$$x = f(r, s)$$
$$y = g(r, s)$$
$$J = \frac{\partial(f, g)}{\partial(r, s)} \neq 0$$

and if h is a function of x, y, show that

$$\frac{\partial h}{\partial x} = \frac{\partial(h, g)}{\partial(r, s)}\frac{1}{J}, \qquad \frac{\partial h}{\partial y} = \frac{\partial(f, h)}{\partial(r, s)}\frac{1}{J}$$

11. If
$$f(u, v, x, y) = 0$$
$$g(u, v, x, y) = 0$$
$$\frac{\partial(f, g)}{\partial(u, v)} \neq 0$$
prove that

$$\frac{\partial(u, v)}{\partial(x, y)} = \frac{\partial(f, g)}{\partial(x, y)} \Big/ \frac{\partial(f, g)}{\partial(u, v)}$$

Illustrate by equations (1).

12. In the previous problem show that

$$\frac{\partial(u, v)}{\partial(x, y)}\frac{\partial(x, y)}{\partial(u, v)} = 1$$

13. As a special case of Exercise 11 take $f = u^2 + v^2 - x^2, g = (v/u) - \tan y$, and show that

$$\frac{\partial(u, v)}{\partial(x, y)} = x$$

Check by first solving for u and v.

§11. *Equality of Cross Derivatives*

We stated earlier that under certain very general conditions $f_{12}(x, y) = f_{21}(x, y)$. In all cases thus far encountered this has been true. We have usually been able to verify it by direct computation of the two derivatives. We shall show here that the result is true for all functions of class C^2, and we shall give an example of a function for which the cross derivatives are not equal.

11.1 A PRELIMINARY RESULT

Let us define two operators Δ_x and Δ_y on a function $f(x, y)$ as follows:

$$\Delta_x f(x_0, y_0) = f(x_0 + \Delta x, y_0) - f(x_0, y_0)$$
$$\Delta_y f(x_0, y_0) = f(x_0, y_0 + \Delta y) - f(x_0, y_0)$$

Lemma 13. *For any function* $f(x, y)$

$$\Delta_x \Delta_y f(x_0, y_0) = \Delta_y \Delta_x f(x_0, y_0)$$

For,

$$\Delta_x \Delta_y f(x_0, y_0) = \Delta_x f(x_0, y_0 + \Delta y) - \Delta_x f(x_0, y_0)$$
$$= f(x_0 + \Delta x, y_0 + \Delta y) - f(x_0, y_0 + \Delta y) - f(x_0 + \Delta x, y_0) + f(x_0, y_0)$$
$$\Delta_y \Delta_x f(x_0, y_0) = \Delta_y f(x_0 + \Delta x, y_0) - \Delta_y f(x_0, y_0)$$
$$= f(x_0 + \Delta x, y_0 + \Delta y) - f(x_0 + \Delta x, y_0) - f(x_0, y_0 + \Delta y) + f(x_0, y_0)$$

11.2 THE PRINCIPAL RESULT

Theorem 13.   1. $f(x, y) \in C^2$

$\Rightarrow$ $f_{12}(x, y) = f_{21}(x, y)$

Let (x_0, y_0) be an arbitrary point in the domain where $f \in C^2$. Then by Lemma 13 we have

(1) $\Delta_x \Delta_y f(x_0, y_0) = \Delta_y \Delta_x f(x_0, y_0)$

Set $\varphi(y) = f(x_0 + \Delta x, y) - f(x_0, y)$

Then

$$\Delta_y \Delta_x f(x_0, y_0) = \Delta_y \varphi(y_0) = \varphi(y_0 + \Delta y) - \varphi(y_0)$$
$$= \varphi'(y_0 + \theta_1 \Delta y) \Delta y \qquad\qquad 0 < \theta_1 < 1$$

(2) $= f_2(x_0 + \Delta x, y_0 + \theta_1 \Delta y) \Delta y - f_2(x_0, y_0 + \theta_1 \Delta y) \Delta y$

Set $\psi(x) = f(x, y_0 + \Delta y) - f(x, y_0)$

so that

$$\Delta_x \Delta_y f(x_0, y_0) = \Delta_x \psi(x_0) = \psi(x_0 + \Delta x) - \psi(x_0)$$
$$= \psi'(x_0 + \theta_2 \Delta x) \Delta x \qquad\qquad 0 < \theta_2 < 1$$

(3) $= f_1(x_0 + \theta_2 \Delta x, y_0 + \Delta y) \Delta x - f_1(x_0 + \theta_2 \Delta x, y_0) \Delta x$

Now apply the law of the mean to the right-hand sides of equations (2) and (3) and use equation (1). Then

(4) $f_{21}(x_0 + \theta_3 \Delta x, y_0 + \theta_1 \Delta y) \Delta y \Delta x = f_{12}(x_0 + \theta_2 \Delta x, y_0 + \theta_4 \Delta y) \Delta x \Delta y$

where $0 < \theta_3 < 1, 0 < \theta_4 < 1$. Now cancel Δx and Δy and let both approach zero. This gives the desired equality at (x_0, y_0). Observe where the hypothesis $f \in C^2$ enters into the proof. Some less restrictive hypothesis would clearly be sufficient for the various applications of the law of the mean, but the full force of the hypothesis insofar as it concerns f_{12} and f_{21} is used in the final limiting process. The conclusion of the theorem is, in fact, true under weaker hypotheses.*

11.3 AN EXAMPLE

We have already seen many examples of the theorem. The following example is one for which $f_{12} \neq f_{21}$.

Set
$$f(x, y) = 2xy \frac{x^2 - y^2}{x^2 + y^2} \qquad x^2 + y^2 \neq 0$$
$$f(0, 0) = 0$$

It would be easy to show by the formal rules of differentiation that $f_{12} = f_{21}$ when (x, y) is not the origin. These rules are not applicable at the origin, however, since the denominator of the fraction is zero there. Hence we revert to the definition of the partial derivatives.

$$f_1(0, 0) = \lim_{\Delta x \to 0} \frac{f(\Delta x, 0) - f(0, 0)}{\Delta x} = \lim_{\Delta x \to 0} \frac{0}{\Delta x} = 0$$

$$f_2(0, 0) = \lim_{\Delta y \to 0} \frac{f(0, \Delta y) - f(0, 0)}{\Delta y} = \lim_{\Delta x \to 0} \frac{0}{\Delta y} = 0$$

$$f_1(x, y) = 2y \frac{x^2 - y^2}{x^2 + y^2} + 2xy \frac{4xy^2}{(x^2 + y^2)^2} \qquad x^2 + y^2 \neq 0$$

$$f_2(x, y) = 2x \frac{x^2 - y^2}{x^2 + y^2} - 2xy \frac{4x^2 y}{(x^2 + y^2)^2} \qquad x^2 + y^2 \neq 0$$

$$f_{21}(0, 0) = \lim_{\Delta x \to 0} \frac{f_2(\Delta x, 0) - f_2(0, 0)}{\Delta x} = \lim \frac{2 \Delta x}{\Delta x} = 2$$

$$f_{12}(0, 0) = \lim_{\Delta y \to 0} \frac{f_1(0, \Delta y) - f_1(0, 0)}{\Delta y} = \lim - \frac{2 \Delta y}{\Delta y} = -2$$

* See, for example, *Differential and Integral Calculus*, by R. Courant (1936), Vol. 2. pp. 55–58.

It is important here to distinguish carefully between $f_{12}(0, 0)$ and

$$\lim_{\substack{x \to 0 \\ y \to 0}} f_{12}(x, y)$$

for of course $f_{12}(x, y)$ is not continuous at $(0, 0)$.

EXERCISES (11)

1. If $f(x, y) = x^3 - 2xy^2 + 2y^2$, compute

(a) $\Delta_x f(0, 0)$ (b) $\Delta_y \Delta_x f(1, -2)$

2. For any function $f(x, y)$ show that

$$\Delta_x \Delta_y \Delta_x f(x_0, y_0) = \Delta_y \Delta_x \Delta_x f(x_0, y_0)$$

3. Prove by use of Theorem 13 that

$$\frac{\partial^3 f}{\partial x \, \partial y \cdot \partial x} = \frac{\partial^3 f}{\partial y \, \partial x^2}$$

What assumptions on $f(x, y)$ are you making?

4. Show that

$$\frac{\partial^4 f(x, y, z)}{\partial x \, \partial y \, \partial z^2} = \frac{\partial^4 f(x, y, z)}{\partial x \, \partial z \, \partial y \, \partial z}$$

5. If $f(x, y) = xy$ when $|y| \leq |x|$ and $= -xy$ when $|y| > |x|$, show directly from Definition 5 that

$$f_1(0, b) = -b \qquad\qquad -\infty < b < \infty$$
$$f_2(a, 0) = a \qquad\qquad -\infty < a < \infty$$

Note that the cases $b = 0$ and $a = 0$ must be treated separately.

6. In Exercise 5 compute $f_1(a, 0)$ and $f_2(0, b)$.

7. In Exercise 5 show that $f_{12}(0, 0) = -1$ and $f_{21}(0, 0) = 1$.

8. In Exercise 5 compute $f_{11}(0, 0)$ and $f_{22}(0, 0)$.

9. If $f(x, y) = xy^2$ when $y > 0$ and $= -xy^2$ when $y \leq 0$, does $f_{12}(0, 0) = f_{21}(0, 0)$?

10. In Exercise 9 show that $f_{11}(0, 0) = f_{22}(0, 0) = 0$.

11. If $f(x, y) = xy$ when $y > 0$ and $= -xy^2$ when $y \leq 0$, which of the four second-order derivatives exist at the origin?

12. If

$$f(x, y) = x^2 \tan^{-1} \frac{y}{x} - y^2 \tan^{-1} \frac{x}{y} \qquad\qquad xy \neq 0$$

$$f(x, 0) = f(0, y) = 0$$

prove $f_{12}(0, 0) \neq f_{21}(0, 0)$.

13. If

$$\varphi(x) = f(x, y_0 + \Delta y) - f(x, y_0 - \Delta y)$$
$$\psi(y) = f(x_0 + \Delta x, y) - f(x_0 - \Delta x, y)$$

show that

$$\varphi(x_0 + \Delta x) - \varphi(x_0 - \Delta x) = \psi(y_0 + \Delta y) - \psi(y_0 - \Delta y)$$

14. Apply the law of the mean to the previous equation and thus give a new proof of Theorem 13.

15. If $f(x, y) \in C^2$, find

$$\lim_{h \to 0} \frac{1}{h^2} [f(a + h, b + h) - f(a, b + h) - f(a + h, b) + f(a, b)]$$

Hint: Take $\Delta x = \Delta y = h$ in §§11.1 and 11.2.

§12. *Implicit Functions*

We have hitherto assumed the existence of a function $y = f(x)$ that would satisfy an equation

$$(1) \qquad\qquad F(x, y) = 0$$

We give in this section a sufficient condition that this should be the case. It is easy to see that certain equations (1) do not define y as a single-valued function of x. Consider

$$(2) \qquad\qquad F(x, y) = x^2 + y^2 + 1$$
$$(3) \qquad\qquad F(x, y) = x^2 + y^2$$
$$(4) \qquad\qquad F(x, y) = x^2 + y^2 - 1$$

In the first case (2), equation (1) is not satisfied for any point. In the second case (3), equation (1) is satisfied for $x = y = 0$ only, so that $f(x)$ is defined at only one point. In the last case (4), equation (1) does define the two functions

$$y = \sqrt{1 - x^2}, \qquad y = -\sqrt{1 - x^2}$$

But even in this case the functions are not defined in a two-sided neighborhood of $x = 1$, or of $x = -1$. Note that in this case

$$F_2(1, 0) = 0, \qquad F_2(-1, 0) = 0$$

12.1 THE EXISTENCE THEOREM

We shall show that if

$$F(x_0, y_0) = 0, \qquad F_2(x_0, y_0) \neq 0$$

then equation (1) can be solved for y when x is in a two-sided neighborhood of x_0.

Theorem 14.   1. $F(x, y) \in C^1$ $|x - x_0| \leq \delta, \quad |y - y_0| \leq \delta$

  2. $F(x_0, y_0) = 0$

  3. $F_2(x_0, y_0) \neq 0$

$\Rightarrow$ *There exists a unique function $f(x)$ and a positive number η such that*

A. $y_0 = f(x_0)$

B. $F(x, f(x)) = 0$ $|x - x_0| < \eta$

C. $f(x) \in C^1$ $|x - x_0| < \eta$

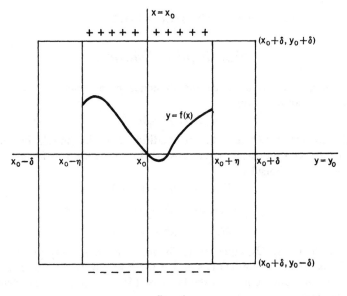

Fig. 4.

It is no restriction to suppose $F_2(x_0, y_0) > 0$. By continuity, $F_2(x, y) > F_2(x_0, y_0)/2$ in a whole neighborhood of (x_0, y_0), which we assume to be the original δ-neighborhood. (See Exercise 3 below.)   Clearly $F(x_0, y)$ is a strictly increasing function for $|y - y_0| \leq \delta$. Hence

$$F(x_0, y_0 + \delta) > F(x_0, y_0) = 0, \qquad F(x_0, y_0 - \delta) < F(x_0, y_0) = 0$$

By the continuity of $F(x, y_0 + \delta)$ and of $F(x, y_0 - \delta)$ there exists a positive number η such that (Figure 4)

$$F(x, y_0 + \delta) > 0, \qquad F(x, y_0 - \delta) < 0 \qquad |x - x_0| < \eta$$

A continuous function passing from positive to negative values must pass through zero.* Hence, for each x in the interval $x_0 - \eta < x < x_0 + \eta$, there is just one value of y, which we call $f(x)$, between $y_0 - \delta$ and $y_0 + \delta$ where $F(x, y) = 0$. If there were two such values of y, F_2 would be zero by Rolle's theorem, contrary to assumption.

We have thus established the unique existence of $f(x)$. Conclusions A and B follow from the manner of definition of $f(x)$. To prove C,

set $\qquad y_1 = f(x_1), \qquad\qquad x_0 - \eta < x_1 < x_0 + \eta$

$$y_1 + \Delta y = f(x_1 + \Delta x), \qquad x_0 - \eta < x_1 + \Delta x < x_0 + \eta$$

Then by the law of the mean for functions of two variables,

$$F(x_1 + \Delta x, y_1 + \Delta y) = 0$$
$$= F_1(x_1 + \theta\,\Delta x, y_1 + \theta\,\Delta y)\,\Delta x + F_2(x_1 + \theta\,\Delta x, y_1 + \theta\,\Delta y)\,\Delta y$$
$$0 < \theta < 1$$

This equation shows first that $\Delta y \to 0$ as $\Delta x \to 0$. For, the first term, and hence the second, approaches zero as Δx does. But since the first factor of the second term is greater than $F_2(x_0, y_0)/2$, that term cannot approach zero unless Δy does. Secondly, the above equation enables us to compute $\Delta y/\Delta x$. Finally, using the continuity of F_1 and F_2, we obtain

$$f'(x_1) = \lim_{\Delta x \to 0} \frac{\Delta y}{\Delta x} = -\frac{F_1(x_1, y_1)}{F_2(x_1, y_1)}$$

This quotient is a continuous function of x_1, $[y_1 = f(x_1)]$, so that $f \in C^1$. This completes the proof of the theorem.

The theorem can easily be generalized to include functions of more than two variables. For example, the equation

$$F(x, y, z) = 0$$

can be solved for z when (x, y) is near (x_0, y_0) if

$$F(x_0, y_0, z_0) = 0, \qquad F_3(x_0, y_0, z_0) \neq 0$$

12.2 FUNCTIONAL DEPENDENCE

Two functions $f(x, y)$ and $g(x, y)$ may be *functionally dependent*. For example, if

$$f(x, y) = \sin(x^2 + y^2), \qquad g(x, y) = \cos(x^2 + y^2)$$

there exists a function of a single variable $F(z)$ such that

$$(5) \qquad\qquad g(x, y) = F(f(x, y))$$

---

* See §6, Chapter 5, for an analytic proof.

In fact, in the present case

$$F(z) = \cos(\sin^{-1} z)$$

Observe that the Jacobian

$$\frac{\partial(f, g)}{\partial(x, y)} = \begin{vmatrix} 2x \cos(x^2 + y^2) & 2y \cos(x^2 + y^2) \\ -2x \sin(x^2 + y^2) & -2y \sin(x^2 + y^2) \end{vmatrix}$$

is identically zero. We shall see that the vanishing of this Jacobian is a characteristic of functional dependence.

12.3 A CRITERION FOR FUNCTIONAL DEPENDENCE

Note that if f and g are functionally dependent, say by virtue of equation (5), then their Jacobian is identically zero:

$$g_1 = f_1 F', \qquad g_2 = f_2 F', \qquad \frac{\partial(f, g)}{\partial(x, y)} = \begin{vmatrix} f_1 & f_1 F' \\ f_2 & f_2 F' \end{vmatrix} = 0$$

We shall now show, conversely, that under certain conditions the vanishing of this Jacobian implies the functional dependence of f and g.

Theorem 15.   1. $f(x, y), g(x, y) \in C^1$ $|x - x_0| < \delta, \quad |y - y_0| < \delta$

2. $\dfrac{\partial(f, g)}{\partial(x, y)} = 0$ $|x - x_0| < \delta, \quad |y - y_0| < \delta$

3. $f_2(x_0, y_0) \neq 0$

$\Rightarrow$ *There exists a function $F(z)$ and a number η such that*

$$g(x, y) = F(f(x, y)) \qquad |x - x_0| < \eta, \quad |y - y_0| < \eta$$

Set $z_0 = f(x_0, y_0)$. Then by the generalization of Theorem 14 to functions of three variables mentioned above, the equation

$$(6) \qquad\qquad f(x, y) - z = 0$$

can be solved for y. That is, there exists a function $\varphi(x, z)$ such that the equation

$$(7) \qquad\qquad y = \varphi(x, z)$$

is equivalent to (6) in an η-neighborhood of the point (x_0, y_0, z_0). Moreover $\varphi_1(x, z)$ can be computed in terms of f by the usual rule:

$$(8) \qquad \varphi_1(x, z) = -f_1(x, y)/f_2(x, y) \qquad\qquad y = \varphi(x, z)$$

Now compute the derivative of $g(x, \varphi(x, z))$ with respect to x, using (8) and hypothesis 2:

$$\frac{\partial}{\partial x} g(x, \varphi(x, z)) = g_1 - \frac{f_1 g_2}{f_2}$$

$$= -\frac{1}{f_2} \frac{\partial(f, g)}{\partial(x, y)} = 0 \qquad |x - x_0| < \eta, \quad |y - y_0| < \eta$$

Integrating this equation we obtain

$$g(x, \varphi(x, z)) = F(z) \qquad |x - x_0| < \eta, \quad |z - z_0| < \eta$$

for some function $F(z)$. Finally, set $z = f(x, y)$ in this equation. Since equations (6) and (7) are equivalent near (x_0, y_0, z_0) we have

$$\varphi(x, f(x, y)) = y$$

$$g(x, y) = F(f(x, y)) \qquad |x - x_0| < \eta, \quad |y - y_0| < \eta$$

This completes the proof. Note that hypothesis 3 could be replaced by $f_1(x_0, y_0) \neq 0$. Then equation (6) would have to be solved for x, but the final conclusion would be unchanged. On the other hand, if it is g_1 or g_2 that is known to be different from zero, we would show that $f(x, y) = G(g(x, y))$ for some $G(z)$.

12.4 SIMULTANEOUS EQUATIONS

Let us refer to a set of four numbers (u_0, v_0, x_0, y_0) as a point in four dimensions and to the set of values (u, v, x, y) for which

$$|u - u_0| < \delta, \quad |v - v_0| < \delta, \quad |x - x_0| < \delta, \quad |y - y_0| < \delta$$

as a δ-neighborhood, $N_\delta(u_0, v_0, x_0, y_0)$, of that point.

Theorem 16.   1. $F(u, v, x, y), G(u, v, x, y) \in C^1$ in $N_\delta(u_0, v_0, x_0, y_0)$

2. $F(u_0, v_0, x_0, y_0) = G(u_0, v_0, x_0, y_0) = 0$

3. $\dfrac{\partial(F, G)}{\partial(u, v)} \neq 0$ at (u_0, v_0, x_0, y_0)

$\Rightarrow$ *There exists a unique pair of functions* $f(x, y), g(x, y)$ *and a positive number* η *such that*

A. $f(x, y), g(x, y) \in C^1$ $\qquad\qquad |x - x_0| < \eta, \quad |y - y_0| < \eta$

B. $f(x_0, y_0) = u_0, \quad g(x_0, y_0) = v_0$

C. $F(f, g, x, y) = G(f, g, x, y) = 0 \qquad |x - x_0| < \eta, \quad |y - y_0| < \eta$

By hypothesis 3, not both F_u and F_v are zero at (u_0, v_0, x_0, y_0). Assume $F_u \neq 0$ there. Then by a generalization of Theorem 14, there exists a unique function $h(v, x, y)$ such that $h(v_0, x_0, y_0) = u_0$ and

$$F(h, v, x, y) = 0$$

in some neighborhood of (v_0, x_0, y_0). From this equation, $h_v = -F_v/F_u$. We have now to solve the equation

(9) $G(h(v, x, y), v, x, y) = 0$

for v. This is possible if the derivative of the function on the left with respect to v is different from zero at the point in question.   This derivative is

$$G_u h_v + G_v = G_v - \frac{F_v}{F_u} G_u = \frac{1}{F_u} \frac{\partial(F, G)}{\partial(u, v)}$$

This is different from zero at (v_0, x_0, y_0).   Hence, there exists a unique function $g(x, y)$, equal to v_0 at (x_0, y_0), which makes equation (9) an identity near (x_0, y_0) when it is substituted for v.   Now set $f(x, y) = h(g, x, y)$.   It is easy to see that all three conclusions of the theorem are satisfied.   A similar proof holds if $F_v \neq 0$.

EXERCISES (12)

1. Let $F(x, y) = x^2 - y^2$.   Apply Theorem 14 at the points $(1, 1)$ and $(1, -1)$, finding $f(x)$ explicitly in each case.   Discuss the situation at $(0, 0)$.   What fails there: hypothesis, conclusion, or both?

2. If $f(x) \in C$ at $x_0, f(x_0) > 0$, show that $f(x) > 0$ in a δ-neighborhood of x_0.   *Hint:* Write $\lim_{x \to x_0} f(x) = f(x_0)$ in ϵ, δ-form, choosing $\epsilon = f(x_0)/2$.   Then by use of the inequality $A - B \leq |A - B|$ show

$$f(x) \geq \frac{f(x_0)}{2} \qquad\qquad |x - x_0| < \delta$$

3. If $f(x, y) \in C$ at $(x_0, y_0), f(x_0, y_0) > 0$, show that

$$f(x, y) > f(x_0, y_0)/2$$

in a δ-neighborhood of (x_0, y_0).

4. If $f(x, y)$ and $g(x, y)$ reduce to the following functions of one variable $f(x, y) = e^{x^2}, g(x, y) = x^2 + 2x$, find the functions F and G of Theorem 15 explicitly.

5. If $f_1(x, y) = f_2(x, y) = 0$ for all $(x, y,)$, show that $f(x, y)$ is constant.   *Hint:* Use the law of the mean for functions of two variables.

6. If $f(x, y) \in C^1$ and if $f_1(x, y) = 0$ for all (x, y), show that $f(x, y) = \varphi(y)$.   *Hint:* Use the law of the mean for functions of one variable.   Show, in fact, that $\varphi(y) = f(0, y)$.

7. What does Theorem 14 become if $F = g(y) - x$?

8. Show that the functions

$$f = e^{x-y}, \qquad g = \sqrt{x^2 - 2xy + y^2 - 2x + 2y}$$

are functionally dependent.   Find $F(z)$ explicitly.

9. State and prove an implicit function theorem for three simultaneous equations in three unknowns.

10. Show by Theorem 16 that under the transformation

$$u = \frac{x}{x^2 + y^2} \qquad v = \frac{y}{x^2 + y^2} \qquad\qquad x^2 + y^2 > 0$$

for every pair of values (u, v) near $(\frac{1}{2}, \frac{1}{2})$, there is just one pair of values (x, y) near $(1, 1)$.

11. Same problem for

$$u = x^2 - y^2, \qquad v = 2xy$$

where the corresponding values are $(u_0, v_0) = (0, 2)$, $(x_0, y_0) = (1, 1)$. But show algebraically that for positive values of (u, v) near $(0, 0)$ there are two values of (x, y) near $(0, 0)$. Why does Theorem 16 fail?

12. Same problem for

$$u = x + y + z, \qquad v = x^2 + y^2 + z^2, \qquad w = x^3 + y^3 + z^3$$

where the corresponding values are $(u_0, v_0, w_0) = (0, 2, 0)$, $(x_0, y_0, z_0) = (-1, 0, 1)$. Is the implicit function theorem applicable to the corresponding values

$$(u_0, v_0, w_0) = (2, 4, 8), \qquad (x_0, y_0, z_0) = (0, 0, 2)?$$

13. Prove that hypothesis 3 of Theorem 15 can be replaced by $f_1(x_0, y_0) \neq 0$.

14. If hypothesis 3 of Theorem 15 is replaced by $g_1(x_0, y_0) \neq 0$, show that $f(x, y) = G(g(x, y))$.

15. State and prove the analogue of Theorem 14 for a function of three variables.

2

Vectors

§1. *Introduction*

The student is assumed to be at least partially familiar with three-dimensional analytic geometry. The present chapter may be regarded as a brief review of that subject, the results being here stated in vector notation. It will be evident that the use of vectors makes most of the formulas more compact.

1.1 DEFINITION OF A VECTOR

By a *vector* we mean a directed line segment. We say that two vectors are equal if the line segments defining them are parallel or coincident and their lengths and directions are the same. For example, the vector directed from the point whose coordinates are $(2, -1, 3)$ to the point whose coordinates are $(0, 1, -1)$ is the same as the vector directed from $(1, 3, 0)$ to $(-1, 5, -4)$. Each of these vectors is equal to one directed from $(0, 0, 0)$ to $(-2, 2, -4)$. The coordinates of this latter point are the differences of the respective coordinates of the terminal and initial points of either of the original vectors. For any set of equal vectors, it is clearly these differences that are common to the whole set. Consequently we shall identify a vector with the triple of numbers obtained by subtracting the three coordinates of the initial point from the respective coordinates of the terminal point. The magnitudes of these three numbers, called the *components*, represent the lengths of the projections of the vector on the three axes. The sign of a component is plus or minus, according as the directed projection is the same as or opposite to the positive sense on the corresponding axis. We now give our formal definition.

Definition 1. *A vector $\vec{r}$ is a triple of numbers (r_1, r_2, r_3). Its length $\vec{r}$ is*

$$|\vec{r}| = |r| = (r_1^2 + r_2^2 + r_3^2)^{1/2}$$

The direction cosines of the vector are $\dfrac{r_1}{|r|}, \dfrac{r_2}{|r|}, \dfrac{r_3}{|r|}$. *Its components are* r_1, r_2, r_3.

It is clear that a vector is completely determined by its length and its direction cosines. If $|r| = 0$, the vector is a *null* vector and its direction is undefined. An ordinary real number is referred to as a *scalar* when it is to be distinguished from a vector. To distinguish vectors from scalars we shall use an arrow over the letter representing a vector, as in Definition 1, at least when there is danger of ambiguity.

1.2 ALGEBRA OF VECTORS

Various operations on vectors will now be defined. To avoid too frequent use of the arrow let us agree that throughout the rest of §1 the letters $r, s, t, u, v, w, \alpha, \beta, \gamma$ shall be used for vectors; the letters k, l, m, for scalars. We adopt the following definitions:

(a) $\qquad\qquad r = 0 \Leftrightarrow r_i = 0 \qquad\qquad i = 1, 2, 3$

(b) $\qquad\qquad r = s \Leftrightarrow r_i = s_i \qquad\qquad i = 1, 2, 3$

(c) $\qquad\qquad s = kr \Leftrightarrow s_i = kr_i \qquad\qquad i = 1, 2, 3$

(d) $\qquad\qquad t = r + s \Leftrightarrow t_i = r_i + s_i \qquad\qquad i = 1, 2, 3$

(e) $\qquad\qquad r \cdot s = r_1 s_1 + r_2 s_2 + r_3 s_3$

(f) $\qquad\qquad t = r \times s \Leftrightarrow$

$$t_1 = \begin{vmatrix} r_2 & r_3 \\ s_2 & s_3 \end{vmatrix}, \qquad t_2 = \begin{vmatrix} r_3 & r_1 \\ s_3 & s_1 \end{vmatrix}, \qquad t_3 = \begin{vmatrix} r_1 & r_2 \\ s_1 & s_2 \end{vmatrix}$$

A vector equation may be changed into three scalar equations by placing the subscript i on each letter representing a vector and giving i the values 1, 2, 3, successively. The symbol 0 is used both for zero and for the null vector. Note that $r \cdot s$ is a scalar. It is called the *inner product*, the *scalar product*, or the *dot product*. On the other hand, $r \times s$ is a vector, called the *outer product*, the *vector product*, or the *cross product*. For the latter product, order of multiplication is important, since

(1) $\qquad\qquad r \times s = -s \times r$

From this, or directly from the definition, it follows that

$$r \times r = 0$$

Thus the vector product may vanish when neither factor of the product is null.

We shall abbreviate the determinant

$$\begin{vmatrix} r_1 & s_1 & t_1 \\ r_2 & s_2 & t_2 \\ r_3 & s_3 & t_3 \end{vmatrix}$$

by the symbol (rst). Expanding by the minors of a given column, we have

(2) $$(rst) = r \cdot (s \times t) = s \cdot (t \times r) = t \cdot (r \times s)$$

For, the cofactors of the elements r_1, r_2, r_3, for example, are the components of the vector $s \times t$, respectively, in the above determinant.

The following relation, known as the *Lagrange identity*, is of frequent use:

(3) $$(r \times s) \cdot (t \times u) = (r \cdot t)(s \cdot u) - (r \cdot u)(s \cdot t)$$

This may be proved by direct reference to the definitions.

1.3 PROPERTIES OF THE OPERATIONS

The following linear relations are easily verified:

(g) $$(r + s) \cdot t = (r \cdot t) + (s \cdot t)$$

(h) $$(r + s) \times t = (r \times t) + (s \times t)$$

(i) $$([r + s]tu) = (rtu) + (stu)$$

We have seen in equation (1) that the commutative law does not hold for vector multiplication. Neither does the associative law. We shall see presently that

(4) $$(r \times s) \times t = (r \cdot t)s - (s \cdot t)r$$

Hence, $$r \times (s \times t) = -(s \cdot r)t + (t \cdot r)s \neq (r \times s) \times t$$

1.4 SAMPLE VECTOR CALCULATIONS

In vector calculations it is sometimes useful to introduce artificially an *arbitrary* vector. That is, we operate on the given vectors of a problem with a new vector, say, $\vec{w}$, of our own choosing, obtaining results which will be valid for *all* $\vec{w}$. This device has the effect of converting vector equations into scalar equations. The latter are sometimes more manageable. Finally, by choosing $\vec{w}$ successively as $(1, 0, 0)$, $(0, 1, 0)$, $(0, 0, 1)$, the scalar equation is again reduced to a vector equation. For example, $\vec{r} = \vec{s}$ if, and only if,

(5) $$\vec{r} \cdot \vec{w} = \vec{s} \cdot \vec{w}$$

for every vector $\vec{w}$. If $\vec{w} = (1, 0, 0)$, equation (5) becomes $r_1 = s_1$; if $w = (0, 1, 0)$, it becomes $r_2 = s_2$, etc.

EXAMPLE A.   Prove equation (4).   For an arbitrary vector w we have from (2) that

$$([r \times s] \times t) \cdot w = ([r \times s]tw) = [r \times s] \cdot [t \times w]$$

By (3)

$$[r \times s] \cdot [t \times w] = (r \cdot t)(s \cdot w) - (s \cdot t)(r \cdot w)$$

Now converting to a scalar equation as explained above we obtain (4).

EXAMPLE B.   Let α, β, γ be such that

$$\alpha \cdot \beta = \beta \cdot \gamma = \gamma \cdot \alpha = 0$$
$$\alpha \cdot \alpha = \beta \cdot \beta = \gamma \cdot \gamma = 1 \qquad\qquad (\alpha\beta\gamma) = 1$$

Compute γ in terms of α.   By the rule for multiplying determinants we have

$$(\beta \times \gamma) \cdot w = (\beta\gamma w)(\alpha\beta\gamma) = \begin{vmatrix} \alpha \cdot \beta & \beta \cdot \beta & \gamma \cdot \beta \\ \alpha \cdot \gamma & \beta \cdot \gamma & \gamma \cdot \gamma \\ \alpha \cdot w & \beta \cdot w & \gamma \cdot w \end{vmatrix} = \alpha \cdot w$$

Hence

$$\beta \times \gamma = \alpha$$

EXERCISES (1)

1. Let r, s, t, u be the vectors $(2, 1, -1)$, $(1, -1, 2)$, $(1, 0, -1)$, $(1, 2, -3)$, respectively, and let $k = -2$.   Compute

$$kr, r + s, r - s, r \cdot s, r \times s, (rst)$$

2. For the special vectors and scalar of Exercise 1, prove (g), (h), (i), and verify equations (3) and (4).

3. Prove (g), (h), (i) in general.

4. Prove the Lagrange identity.

5. Prove:

$$(kr + ls) \cdot t = k(r \cdot t) + l(s \cdot t)$$
$$(kr + ls) \cdot (kr + ls) = k^2(r \cdot r) + 2kl(r \cdot s) + l^2(s \cdot s)$$
$$(kr + ls) \times (mt) = ?$$

6. Solve for k:

$$(r + ks) \cdot r = 0$$

Does k always exist?   Is it always unique?

7. If w is arbitrary and $r \times w = 0$, show that $r = 0$.

8. If $r \times s = 0$ and $r \neq 0$, show that $s = kr$.

9. In Example B show that $\beta = \gamma \times \alpha$ and $\gamma = \alpha \times \beta$.

10. In Example B drop the hypothesis about $(\alpha\beta\gamma)$. Show that its value must be ± 1 in any case.

11. Prove $(|s|r + |r|s) \cdot (|s|r - |r|s) = 0$.

12. Show that if k is chosen to make $|r + ks|$ minimum, then

$$|ks|^2 + |r + ks|^2 = |r|^2$$

13. Prove:

$$(r \times s \quad t \times u \quad v) = (rsu)(t \cdot v) - (rst)(u \cdot v)$$

14. If r and s are vectors whose components are functions of the scalar y, show that

$$\frac{d}{dy}(r \cdot s) = \left(\frac{dr}{dy} \cdot s\right) + \left(r \cdot \frac{ds}{dy}\right)$$

$$\frac{d}{dy}(r \times s) = \left(\frac{dr}{dy} \times s\right) + \left(r \times \frac{ds}{dy}\right)$$

15. Find

$$\frac{d^n}{dy^n}(r \cdot s), \qquad \frac{d^n}{dy^n}(r \times s)$$

§2. *Solid Analytic Geometry*

The vector notation is ideal for the formulas of solid analytic geometry. We adopt a right-handed system of rectangular coordinates, Figure 5. Denote the coordinates of a point P by (x_1, x_2, x_3). This is of course a vector $\vec{x}$, directed from the origin O to P. The usual formulas for directed line segments may now be used for vectors. We list the main formulas below in syllabus form. The angle between two vectors is defined uniquely as the angle θ, $0 \leq \theta \leq \pi$, between the corresponding directed line segments.

Fig. 5

2.1 SYLLABUS FOR SOLID GEOMETRY

(a) The length of a vector r is $\sqrt{r \cdot r}$.

(b) The vector directed from point r to point s is $s - r$.

(c) The direction components of a line segment from point r to point s are the components of the vector $s - r$.

(d) The direction cosines of a line segment directed from point r to point s are the components of

$$\frac{s - r}{|s - r|} = \frac{1}{|s - r|}(s - r)$$

(ε) The angle θ between nonzero vectors r and s is given by

$$\cos \theta = \frac{r \cdot s}{|r|\,|s|}$$

(f) $r \perp s \Leftrightarrow r \cdot s = 0$.

(g) $r \parallel s \Leftrightarrow r = ks$.

   $r \parallel s \Leftrightarrow r \times s = 0$.

(h) The common $\perp$ to r and s is $r \times s$ $[r \times s \neq 0]$.

(i) r, s, t are $\parallel$ to a plane $\Leftrightarrow (rst) = 0$.

(j) The plane through point r with direction $\vec{a}$ for the normal has equation $(x - r) \cdot a = 0$.

(k) The equation

$$a \cdot x = k \qquad\qquad a \cdot a \neq 0$$

represents a plane whose normal has direction $\vec{a}$.

(l) The distance D from point s to plane (j) is

$$D = \frac{|(s - r) \cdot a|}{|a|}$$

(m) The line through point $\vec{r}$ with direction $\vec{a}$ has equation

$$\vec{x} - \vec{r} = t\vec{a}$$

Here t is a scalar parameter. Another form is

$$(x - r) \times a = 0$$

(n) The distance D from point s to line (m) is

$$D = \frac{|(s - r) \times a|}{|a|}$$

2.2 COMMENTS ON THE SYLLABUS

Any three numbers r: (r_1, r_2, r_3), not all zero, may be the direction components of a line. They may be direction cosines $\Leftrightarrow r \cdot r = 1$. Direction components r may be converted into direction cosines: $\dfrac{\pm r}{|r|}$. The two signs correspond to the two possible senses for a given line. Direction components are used for undirected lines; direction cosines, for directed lines.

Let us prove formula (ε). Consider the triangle with vertices at points O, r, s. By the law of cosines,

$$\overline{Or}^2 + \overline{Os}^2 - 2\cos\theta\,\overline{Or}\,\overline{Os} = \overline{rs}^2$$

$$(r \cdot r) + (s \cdot s) - 2\cos\theta\sqrt{r \cdot r}\sqrt{s \cdot s} = |r - s|^2$$

$$= (r \cdot r) - 2(r \cdot s) + (s \cdot s)$$

This latter equation is equivalent to (e). We have used r as the name of a point and as the vector joining O to the point.

The equivalence of the two forms of (g) is worthy of comment. From the first form, we have

$$r \times s = (ks) \times s = k(s \times s)$$

Hence $r \times s = 0$. Conversely, $r \times s = 0$ implies

$$|r \times s|^2 = (r \cdot r)^2 (s \cdot s)^2 - (r \cdot s)^2 = 0$$

$$r \cdot s = \pm |r||s|$$

By formula (e) this latter equation means that $\cos \theta = \pm 1$ or that r is parallel to s. In (f) and (g) it is tacitly assumed that neither r nor s is null.

That $r \times s$ is perpendicular to r and s follows from

$$(r \times s) \cdot r = (rsr) = 0$$

$$(r \times s)^{\cdot} s = (rss) = 0$$

Of course $s \times r$ has the same property. The vectors r, s, $r \times s$ have the same disposition as the axes Ox_1, Ox_2, Ox_3. That is, the rotation of r into s, about a common point, through the angle between them would advance a right-handed screw, with head in the plane of r and s, in the direction of $r \times s$. Note that

$$(r \, s \, r \times s) = |r \times s|^2 \geqq 0$$

$$(r \, s \, s \times r) = -|r \times s|^2 \leqq 0$$

The sign of the determinant of three vectors thus shows their mutual disposition.

We can now interpret the meanings of $r \cdot s$ and $r \times s$. By (e) we have

$$r \cdot s = (|r| \cos \theta)|s|$$

That is, $r \cdot s$ is the product of the length of one vector by the length of the projection of the other on it. If r and s are not parallel, $r \times s$ is a common perpendicular to r and s in the sense described above. Its length is the area of the parallelogram, two of whose adjacent sides are r and s. For this area is

$$|r| \, |s| \sin \theta = |r| \, |s| \sqrt{1 - \cos^2 \theta}$$

By use of (e) and Lagrange's identity, the area reduces to

$$\sqrt{(r \cdot r)(s \cdot s) - (r \cdot s)^2} = |r \times s|$$

Equation (j) states that the vector from the variable point x to the fixed point r of a plane is always perpendicular to the normal vector. Equations (m) state that the vector from the variable point x to the fixed point r is parallel to a fixed vector a.

2.3 VECTOR APPLICATIONS

Many problems of analytic geometry are done most easily by use of vector methods. We illustrate by examples.

EXAMPLE A. If n vectors are joined end to end in succession, the terminal point of each vector being the initial point of the next, so as to form a closed polygon, their sum is zero. This is essentially the geometric interpretation of vector addition. The result will be evident if we prove it for $n = 5$. Denote the vertices by $\vec{a}, \vec{b}, \vec{c}, \vec{d}, \vec{e}$. By (b) of the syllabus the five vectors forming the sides of the pentagon are $b - a, c - b, d - c, e - d, a - e$. Their sum is clearly zero. Note that the polygon is not necessarily plane nor convex.

EXAMPLE B. Prove the Pythagorean theorem. Take three vectors r, s, t as the sides of a right triangle, s and t being the legs. By Example A, $r = -s - t$ and by (e) of the syllabus $s \cdot t = 0$. Then

$$|r|^2 = r \cdot r = (-s - t) \cdot (-s - t) = s \cdot s + t \cdot t = |s|^2 + |t|^2$$

Since the Pythagorean theorem was assumed in the proof of formula (e), this example should be regarded only as a check of the methods.

EXAMPLE C. Prove that a plane $\vec{x} \cdot \vec{a} = k$ cuts a sphere $|\vec{x} - \vec{b}|^2 = \rho^2$ orthogonally if, and only if, it passes through the center $\vec{b}$ of the sphere. Let $\vec{r}$ be a point common to plane and sphere. The vector $\vec{r} - \vec{b}$ is normal to the sphere there, and the vector $\vec{a}$ is normal to the plane. For orthogonality we must have $(\vec{r} - \vec{b}) \cdot \vec{a} = 0$, or $\vec{b} \cdot \vec{a} = k$ (since $\vec{r} \cdot \vec{a} = k$). That is, $\vec{b}$ lies in the plane, as desired.

EXERCISES (2)

1. Find the area of a parallelogram determined by the vectors $(1, 3, -1)$ and $(2, -1, 3)$.

2. Find a point midway between points r and s.

3. Write the formula for dividing a line in arbitrary ratio in vector form.

4. Prove that points r, s, t lie on a line if, and only if, there exists a scalar k such that

$$(1 - k)r + ks - t = 0$$

5. Find the center of gravity of three masses k, l, m situated at points r, s, t, respectively.

6. Prove (i).

7. Prove (n). Treat the problem as a minimum problem of the calculus.

8. Prove (1).   Let t be the foot of the perpendicular from s to the plane.   Then show that
$$t = s + ka, \quad (r - t) \cdot a = 0, \quad D = |s - t| = k|a|$$
and eliminate k.

9. Show that
$$\sin \theta = \frac{|r \times s|}{|r| \, |s|}$$

10. Prove the law of sines by vectors.

11. Show that the volume of a parallelepiped determined by the vectors r, s, t is $|r \cdot (s \times t)| = |(rst)|$.

12. $(r \times s \quad t \times u \quad v \times w) = ?$

13. If r, s, t are three points, show that a point $\frac{2}{3}$ of the way from r to the mid-point of the segment from s to t is $(r + s + t)/3$.   Hence, show that the medians of a triangle intersect in a point.

14. If u is the centroid of the triangle with vertices at r, s, t, show that the sum of the vectors $r - u$, $s - u$, $t - u$ is zero.

15. Prove that the sum of the squares of the diagonals of any quadrilateral (not necessarily plane) is twice the sum of the squares of the line segments joining the mid-points of the opposite sides.

16. Show that the mid-points of the sides of a quadrilateral (not necessarily plane) are the vertices of a parallelogram.

17. Obtain the usual formula for the area of a triangle in terms of the plane coordinates of the vertices by vector considerations.

18. Show that the area of a convex polygon with vertices at the points (x_i, y_i), $i = 1, 2, \ldots, n$, is
$$\frac{1}{2} \sum_{i=1}^{n} \begin{vmatrix} x_i & x_{i+1} \\ y_i & y_{i+1} \end{vmatrix}$$
where $x_{n+1} = x_1$, $y_{n+1} = y_n$.

§3. *Space Curves*

There are several ways of representing a space curve analytically.   We may consider the curve as the intersection of two surfaces when its equations will be
$$F(x_1, x_2, x_3) = 0$$
$$G(x_1, x_2, x_3) = 0$$
Or it may be the intersection of two cylinders,
$$x_3 = f(x_1)$$
$$x_3 = g(x_2)$$

In this case we are determining the curve by its projections on the x_1x_3-plane and on the x_2x_3-plane. But the most important representation for our purposes is the parametric one,

$$x_1 = x_1(t)$$
$$x_2 = x_2(t)$$
$$x_3 = x_3(t)$$

Here t is an arbitrary parameter. In particular, t may be the arc length s. We may write these equations in vector form

(1) $$\vec{x} = \vec{x}(t)$$

3.1 EXAMPLES OF CURVES

EXAMPLE A. A circle of radius ρ, center at $(0, 0, 0)$ lying in the plane

$$x_2 = \frac{\sqrt{3}}{3} x_1$$

See Fig. 6. Choose the central angle as the parameter t. Then

$$x_1 = \frac{\sqrt{3}}{2} \rho \sin t, \qquad x_2 = \tfrac{1}{2}\rho \sin t, \qquad x_3 = \rho \cos t$$

If the arc s is chosen as the parameter, replace t by s/ρ in the above equation.

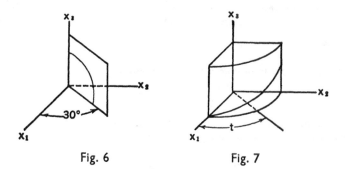

Fig. 6   Fig. 7

EXAMPLE B. *A circular helix.* This is a curve lying on a circular cylinder of radius ρ which rises at a rate proportional to the amount of turning.

Choose t as the angle indicated in Figure 7. Then if the factor of proportionality is k,

$$x_1 = \rho \cos t, \qquad x_2 = \rho \sin t, \qquad x_3 = kt$$

The helix is called *right-handed* or *left-handed* according as it resembles the threads of a right-handed or left-handed screw. If the head of a right-handed screw is placed in the x_1x_2-plane with its axis along the positive side of the x_3-axis, then rotation of the positive x_1-axis toward the positive x_2-axis makes the screw advance in the direction of the positive x_3-axis. If $k > 0$, as in Figure 7, the helix is right-handed.

EXAMPLE C. *The twisted cubic.* This is the curve whose equations are

$$x_1 = at, \qquad x_2 = bt^2, \qquad x_3 = ct^3 \qquad\qquad abc \neq 0$$

3.2 SPECIALIZED CURVES

Without further statement, let us assume throughout that the three functions $x_1(t)$, $x_2(t)$, $x_3(t)$ of equation (1) are at least of class C^3. Let us investigate what the vector equation (1) may represent.

Theorem 1. *Equation* (1) *represents a point* $\Leftrightarrow x'(t) \equiv 0$.

For, the condition is equivalent to $x = r$, where r is a constant vector.

Theorem 2.   1. $\vec{x}'(t) \times \vec{x}''(t) \equiv 0$ $\qquad\qquad\qquad\qquad a \leq t \leq b$

$\qquad\qquad$ 2. $\vec{x}'(t) \neq 0$ $\qquad\qquad\qquad\qquad\qquad\qquad a \leq t \leq b$

$\Rightarrow$ *Equation* (1) *represents a line segment.*

Hypothesis 2 means that $x'(t)$ is not the null vector for any t in (a, b). Hypothesis 1 indicates that for each t in (a, b), $x''(t)$ is either the null vector or is parallel to $x'(t)$. Hence there exists a scalar function $k(t)$ such that

$$\vec{x}''(t) = k(t)\vec{x}'(t)$$

In fact $k(t) \in C$ by hypothesis 2. The function

$$\varphi(t) = e^{-\int k(t)\,dt}$$

is an integrating factor for each of the three linear differential equations which this vector equation represents. Thus

$$\frac{d}{dt}\big[\varphi(t)x_i'(t)\big] \equiv 0 \qquad\qquad\qquad i = 1, 2, 3$$

$$x_i'(t) = a_i/\varphi(t)$$

$$x_i(t) = a_i u(t) + r_i$$

$$u(t) = \int \frac{dt}{\varphi(t)}$$

Note that a is not the null vector, for if it were, $x'(t)$ would be identically zero, contrary to hypothesis. We may clearly replace the scalar function $u(t)$ by a new variable u, which then becomes the parameter of equation (1),

$$\vec{x} = \vec{a}u + \vec{r}$$

This represents a line segment. It will not, in general, represent the entire line since the range of u will generally be limited.

Theorem 3.   1. $(x'x''x''') \equiv 0$ $a \leq t \leq b$

2. $x'(t) \times x''(t) \neq 0$ $a \leq t \leq b$

$\Rightarrow$ *Equation* (1) *represents a plane curve.*

Set $y = x' \times x''$

Differentiating, we have

$$y' = x' \times x'''$$

From equation (4) of §1,

$$y \times y' = (x'x'x''')x'' - (x''x'x''')x' \equiv 0$$

As in the previous proof,

$$\vec{y}' = k(t)\vec{y}$$

$$y_i = a_i/\varphi(t) \qquad\qquad i = 1, 2, 3$$

$$\vec{y} = \vec{a}/\varphi(t)$$

The vector $\vec{a}$ is not null by hypothesis 2.   Hence

$$x' \cdot y \equiv (x'x'x'') \equiv 0 \equiv (x' \cdot a)/\varphi(t) \equiv 0$$

Integrating we obtain

$$\vec{x}(t) \cdot \vec{a} \equiv m \qquad\qquad a \leq t \leq b$$

That is, the curve (1) lies in the plane

$$a_1 x_1 + a_2 x_2 + a_3 x_3 = m$$

E X E R C I S E S   (3)

1. Find a parametric representation for a line through two given points.

2. Find parametric equations for a circular helix that lies on the cylinder $x_1^2 + x_2^2 = 4$ and passes through the points $(2, 0, 0)$ and $(\sqrt{2}, \sqrt{2}, \sqrt{2})$. Can there be more than one such helix?

3. Find parametric equations for an ellipse that lies in the plane

$$x_2 = \frac{\sqrt{3}}{3} x_1$$

and that has its major axis in the $x_1 x_2$-plane, its minor axis in the x_3-axis.

4. Show that the twisted cubic with $a = b = c = 1$ is the intersection of the cylinders

$$x_2 = x_1^2, \qquad x_3 = x_1^3$$

5. Find a parametric representation of the curve

$$x_2^2 = x_1, \qquad x_3^2 = 1 - x_1$$

Obtain, by use of trigonometric functions, equations that do not involve radicals.

6. Solve the same problem as in Exercise 5 for the curve

$$x_1^2 + x_2^2 = \rho^2, \qquad x_1^2 + x_3^2 = \rho^2$$

What are the curves?

7. Find a parametric representation involving no radicals for the curve

$$x_1 x_2 x_3 = 1, \qquad x_2^2 = x_1$$

8. Does the twisted cubic of Exercise 4 intersect the line

$$x_1 = 1 + t$$
$$x_2 = -1 + 5t$$
$$x_3 = 1 + 7t?$$

9. Find all intersections of the curve

$$x_1 = t^2, \qquad x_2 = t^3, \qquad x_3 = t^4$$

and the surface

$$x_3^2 = x_1 + 2x_2 - 2$$

Hint: Show that the solutions are found from the roots of an eighth-degree equation, one factor of which is

$$t^4(t + 1)^2 + 2t^2(t + 1)^2 + 3t^2 + 4t + 2$$

10. What is the curve

$$x_1 = 1 + \sin t, \qquad x_2 = -1 - \sin t, \qquad x_3 = 2 \sin t?$$

Note that hypothesis 2 of Theorem 2 fails at $t = \pi/2$.

11. Is the curve

$$x_1 = \cos e^t, \qquad x_2 = \sin e^t, \qquad x_3 = \sin e^t$$

a straight line? a plane curve?

12. Show that $(x'x''x''') \equiv 0$ if equation (1) represents a plane curve.

13. What is the curve (1) if

$$\vec{x}(t) \neq 0 \quad \text{and} \quad \vec{x}(t) \times \vec{x}'(t) \equiv 0?$$

14. Show that a straight line can always have the equation (1) in such a way that the hypotheses of Theorem 2 hold.

15. Determine all functions $f(t)$ of class C^3 that will make the curve

$$x_1 = \cos t, \qquad x_2 = \sin t, \qquad x_3 = f(t)$$

plane.

16. Let $f(t) = t^4$ when $t > 0$ and $= 0$ when $t \leq 0$. Show that the curve $x_1 = f(t)$, $x_2 = f(-t)$, $x_3 = 0$ is a broken line and that hypothesis 1 of Theorem 2 is satisfied.

17. If $f(t)$ is defined as in Exercise 16, show that the curve $x_1 = f(t)$, $x_2 = f(-t)$, $x_3 = t^8$ lies in *two* planes and is not plane. Show that hypothesis 1 of Theorem 3 is satisfied.

§4. *Surfaces*

There are several ways of representing a surface. One familiar way is by a single equation of the form

$$F(x_1, x_2, x_3) = 0$$

Or this equation may be solved for one of the variables:

$$x_3 = f(x_1, x_2)$$

Perhaps the most useful representation is the parametric one:

$$x_1 = x_1(u, v), \qquad x_2 = x_2(u, v), \qquad x_3 = x_3(u, v)$$

Here there are two parameters, u and v, corresponding to the two degrees of freedom on a surface. In vector form, these equations become

(1) $$\vec{x} = \vec{x}(u, v)$$

4.1 EXAMPLES OF SURFACES

EXAMPLE A. A sphere with center at $(0, 0, 0)$ and radius ρ has the equation

$$F(x_1, x_2, x_3) = x_1^2 + x_2^2 + x_3^2 - \rho^2 = 0$$

The upper half of this sphere has the equation

$$x_3 = \sqrt{\rho^2 - x_1^2 - x_2^2}$$

Finally, a parametric representation of the sphere is

$$x_1 = \rho \cos v \cos u$$

$$x_2 = \rho \cos v \sin u$$

$$x_3 = \rho \sin v$$

Here u and v may be thought of as longitude and latitude on the sphere with Greenwich in the x_1x_3-plane.   The position of a point on the sphere is completely determined by the pair of numbers u, v.

EXAMPLE B.   A plane has equation

$$a_1x_1 + a_2x_2 + a_3x_3 + a_4 = 0, \qquad a_1^2 + a_2^2 + a_3^2 \neq 0$$

A parametric representation, if $a_3 \neq 0$, is

$$x_1 = u$$
$$x_2 = v$$
$$x_3 = \frac{a_1u + a_2v + a_4}{-a_3}$$

EXAMPLE C.   A cylinder of radius ρ and axis coinciding with the x_2-axis is

$$x_1 = \rho \cos u$$
$$x_2 = v$$
$$x_3 = \rho \sin u$$

EXAMPLE D.   A cone with vertex at $(0, h, 0)$ and axis coinciding with the x_2-axis is

$$x_1 = \frac{a}{h}(h - u) \cos v$$
$$x_2 = u$$
$$x_3 = \frac{a}{h}(h - u) \sin v$$

A single equation for this surface is

$$h^2(x_1^2 + x_3^2) = a^2(h - x_2)^2$$

EXAMPLE E.   A torus with axis along the x_3-axis and generated by the rotation of a circle of radius a, the center of which is constantly at distance ρ ($\rho > a$) from the axis is

$$x_1 = (\rho + a \cos u) \cos v$$
$$x_2 = (\rho + a \cos u) \sin v$$
$$x_3 = a \sin u$$

4.2 SPECIALIZED SURFACES

We assume throughout this section that the functions $x_i(u, v)$, $i = 1, 2, 3$, are of Class C^1.   We investigate what equation (1) may represent.

Theorem 4. *Equation* (1) *represents a point*

<⟩ $$\frac{\partial x}{\partial u} \equiv \frac{\partial x}{\partial v} \equiv 0$$

For these conditions are equivalent to $\vec{x} = \vec{r}$, where $\vec{r}$ is constant.
Equation (1) may also represent a curve. For example, the equations

$$x_1 = uv, \qquad x_2 = -uv, \qquad x_3 = u^2 v^2$$

represent the curve intersected by the plane $x_1 + x_2 = 0$ with the parabolic cylinder $x_3 = x_2^2$. If (1) represents the curve $\vec{x} = \vec{x}(t)$, then t must be a function of u and v, $t = t(u, v)$. Then

$$\vec{x} = \vec{x}(t(u, v))$$

$$\vec{x}_u = \vec{x}' t_u, \qquad \vec{x}_v = \vec{x}' t_v$$

(2) $$\vec{x}_u \times \vec{x}_v = t_u t_v \vec{x}' \times \vec{x}' \equiv 0$$

That is, if equation (1) represents a curve, (2) must hold. We shall show, conversely, that under certain conditions equation (2) implies that equation (1) represents a curve.

Theorem 5. 1. $\vec{x}_u \times \vec{x}_v \equiv 0$ *in a neighborhood of* (u_0, v_0)

     2. $\vec{x}_u(u_0, v_0) \neq 0$ *or* $\vec{x}_v(u_0, v_0) \neq 0$.

⇒ *Equation* (1) *represents a curve when* (u, v) *is in a sufficiently small neighborhood of* (u_0, v_0).

By hypothesis 2 some component of $\vec{x}_u$ or $\vec{x}_v$ is not zero at (u_0, v_0). For definiteness suppose $\partial x_2 / \partial v \neq 0$. By hypothesis (1)

$$\frac{\partial(x_1, x_2)}{\partial(u, v)} \equiv 0, \qquad \frac{\partial(x_2, x_3)}{\partial(u, v)} \equiv 0$$

By Theorem 15 of Chapter 1 there exist functions F and G such that

$$x_1(u, v) = F(x_2(u, v))$$

$$x_3(u, v) = G(x_2(u, v))$$

in a suitable neighborhood of (u_0, v_0). This means that the projections of (1) on the $x_1 x_2$-plane and on the $x_2 x_3$-plane are the curves $x_1 = F(x_2)$ and $x_3 = G(x_2)$, respectively. That is, equation (1) represents a curve. If it is another of the components of $\vec{x}_u$ or $\vec{x}_v$ which is assumed not zero, the proof may be altered in an obvious way.

Theorem 6. $\vec{x}_u \times \vec{x}_v \not\equiv 0 \Rightarrow$ *equation* (1) *represents a surface.*

Let us assume, for example, that the third component of the given cross product is different from zero, say at the point (u_0, v_0):

(3)
$$\begin{vmatrix} \dfrac{\partial x_1}{\partial u} & \dfrac{\partial x_2}{\partial u} \\[2mm] \dfrac{\partial x_1}{\partial v} & \dfrac{\partial x_2}{\partial v} \end{vmatrix} \neq 0 \quad \text{at} \quad (u_0, v_0)$$

Set $a = x_1(u_0, v_0)$, $b = x_2(u_0, v_0)$. It will be sufficient to show that

(4)
$$x_3 = f(x_1, x_2)$$

where f is defined in some two-dimensional region of the x_1x_2-plane. By Theorem 16 of Chapter 1, inequality (3) assures us that the equations

$$x_1 = x_1(u, v)$$
$$x_2 = x_2(u, v)$$

can be solved for u and v in some η-neighborhood of (a, b):

$$u = u(x_1, x_2)$$
$$v = v(x_1, x_2) \qquad |x_1 - a| < \eta, \quad |x_2 - b| < \eta$$

If these two functions are substituted in the third of equations (1), $x_3 = x_3(u, v)$, there results an equation of the form (4). If it is some other of the three Jacobians of the hypothesis that is not identically zero, the proof would be altered by permuting the letters in an obvious way.

We shall see later that $\vec{x}_u \times \vec{x}_v$, when it is not null, is the normal vector to the surface (1). A point (u, v) of (1) where $\vec{x}_u \times \vec{x}_v \neq 0$ is called *regular*; a point where $\vec{x}_u \times \vec{x}_v = 0$ is *singular*. In Example D, the vertex $u = h$ is singular; all other points are regular. In Example A, the north pole $v = \pi/2$ and the south pole $v = -\pi/2$ are singular; all other points are regular. But the poles are singular through no peculiarity of the points but only on account of the particular representation chosen. If the letters x_1, x_2, x_3 are cyclically permuted, the same sphere is represented. But it is now the points $(\pm \rho, 0, 0)$ instead of $(0, 0, \pm \rho)$ that are singular.

EXERCISES (4)

1. Find a parametric representation for an ellipsoid of revolution.

2. Find a parametric representation for an arbitrary surface of revolution and apply it to sphere, cylinder, and cone.

3. On the plane $x_2 = 2x_1$ the position of a point is determined by two parameters u, v representing, respectively, its algebraic distance to the x_3-axis and its algebraic distance to the x_1x_2-plane. Find a parametric representation of the plane with u and v as parameters. Describe explicitly the meaning of your parameters with special attention to sign.

4. Solve the same problem as in Exercise 3 if u and v are polar coordinates in the plane. Specify precisely what u and v are.

5. What surface do the following equations represent:

$$x_1 = a \sin u \sin v, \qquad x_2 = b \cos u, \qquad x_3 = a \sin u \cos v?$$

6. Find the singular points of the surface of Exercise 5. Are they singular because of a peculiarity of the surface or because of the special representation?

7. Obtain a parametric representation for a surface whose equation is

$$(x_1 - 1)^2 = x_2^2 + x_3^2$$

Test your representation for singular points.

8. If a surface $x = x(u, v)$ is plane, so that

$$a \cdot x(u, v) \equiv k$$

show that

$$(x_{uu}x_ux_v) \equiv (x_{uv}x_ux_v) \equiv (x_{vv}x_ux_v) \equiv 0$$

9. Show that the three determinants of Exercise 8 vanish identically if

$$x_1 = e^u + uv, \qquad x_2 = 3 \sin \dot{v} - e^u + 3, \qquad x_3 = 2uv + 6 \sin v - 7$$

What plane do these equations represent?

10. Apply Theorem 5 to

$$x_1 = e^u - 3v - 3, \quad x_2 = e^{2u} - 6ve^u + 9v^2, \quad x_3 = (e^u - 3v)(e^u - 3v + 1)$$

11. Has the torus of Example E any singular points?

§5. *A Symbolic Vector*

We now introduce a symbolic vector $\vec{\nabla}$ ("del"). It is an operator and acquires meaning only when operating on a scalar or vector function. The usefulness of the symbol lies chiefly in the fact that it makes many physical formulas more compact.

5.1 DEFINITION OF $\vec{\nabla}$

The operator $\vec{\nabla}$ is a symbolic vector with components $\dfrac{\partial}{\partial x_1}, \dfrac{\partial}{\partial x_2}, \dfrac{\partial}{\partial x_3}$. It may be applied to a scalar function $F(x_1, x_2, x_3)$ or to a vector function

$\vec{y}(x_1, x_2, x_3)$ with components $y_i(x_1, x_2, x_3)$, $i = 1, 2, 3$. In the latter case, we have either the scalar product $\vec{\nabla} \cdot \vec{y}$ or the vector product $\vec{\nabla} \times \vec{y}$. Finally, we may have the scalar product $\vec{\nabla} \cdot \vec{\nabla}$, a symbolic operator which may be applied to a scalar function.

Definition 2. $\vec{\nabla} F(x_1, x_2, x_3)$ *is a vector function with components*

$$\frac{\partial F}{\partial x_1}, \frac{\partial F}{\partial x_2}, \frac{\partial F}{\partial x_3}$$

It is called the gradient of F:

$$\mathrm{Grad}\ F = \vec{\nabla} F$$

Definition 3. $\vec{\nabla} \cdot \vec{y} = \dfrac{\partial y_1}{\partial x_1} + \dfrac{\partial y_2}{\partial x_2} + \dfrac{\partial y_3}{\partial x_3}$. *This scalar function is called the divergence of the vector function $\vec{y}$:*

$$\mathrm{Div}\ \vec{y} = \vec{\nabla} \cdot \vec{y}$$

Definition 4. $\vec{\nabla} \times \vec{y}$ *is a vector function with components*

$$\frac{\partial y_3}{\partial x_2} - \frac{\partial y_2}{\partial x_3}, \quad \frac{\partial y_1}{\partial x_3} - \frac{\partial y_3}{\partial x_1}, \quad \frac{\partial y_2}{\partial x_1} - \frac{\partial y_1}{\partial x_2}$$

This vector function is called the curl of the vector function $\vec{y}$:

$$\mathrm{Curl}\ \vec{y} = \vec{\nabla} \times \vec{y}.$$

Definition 5. $\vec{\nabla} \cdot \vec{\nabla} F = \dfrac{\partial^2 F}{\partial x_1^2} + \dfrac{\partial^2 F}{\partial x_2^2} + \dfrac{\partial^2 F}{\partial x_3^2}$. *This scalar function is called the Laplacian of F. The equation*

$$\Delta F = \vec{\nabla} \cdot \vec{\nabla} F = 0$$

is Laplace's differential equation. Any solution of class C^2 is a harmonic function.

EXAMPLE A. $F = x_1^2 - x_2^2 + 2x_2 x_3$. Then

$$\mathrm{Grad}\ F = \vec{\nabla} F: \quad 2x_1, \quad -2x_2 + 2x_3, \quad 2x_2$$

$$\vec{\nabla} \cdot \vec{\nabla} F = 2 - 2 = 0$$

F is harmonic.

EXAMPLE B. $\vec{y}$: $\quad x_1^2 + x_2 x_3,\ x_1 e^{x_2 + x_3},\ x_1 x_2 \sin x_3$. Then

$$\mathrm{Div}\ \vec{y} = 2x_1 + x_1 e^{x_2 + x_3} + x_1 x_2 \cos x_3$$

$$\mathrm{Curl}\ \vec{y} = x_1 \sin x_3 - x_1 e^{x_2 + x_3}, \quad x_2 - x_2 \sin x_3, \quad e^{x_2 + x_3} - x_3$$

5.2 DIRECTIONAL DERIVATIVES

We now define directional derivatives for functions of three variables. Let $\vec{a}$ be a given vector and $\vec{r}$ a given point. We shall refer to the direction of the vector as the direction ξ_a. Let its direction cosines be $\cos \alpha_1$, $\cos \alpha_2$, $\cos \alpha_3$. The notation for the directional derivative of the function $F(x_1, x_2, x_3)$ at the point $\vec{r}$ in the direction ξ_a will be

$$\frac{\partial F}{\partial \xi_a}\bigg|_{(r_1, r_2, r_3)} = \frac{\partial F}{\partial \xi_a}(r_1, r_2, r_3)$$

Definition 6. $\dfrac{\partial F}{\partial \xi_a}(r_1, r_2, r_3)$

$$= \lim_{\Delta s \to 0} \frac{F(r_1 + \Delta s \cos \alpha_1, r_2 + \Delta s \cos \alpha_2, r_3 + \Delta s \cos \alpha_3) - F(r_1, r_2, r_3)}{\Delta s}$$

For example, if $\vec{a}$ is taken successively as $(1, 0, 0)$, $(0, 1, 0)$, $(0, 0, 1)$, then $\dfrac{\partial F}{\partial \xi_a}$ is successively the partial derivatives $\dfrac{\partial F}{\partial x_1}, \dfrac{\partial F}{\partial x_2}, \dfrac{\partial F}{\partial x_3}$. Just as in two dimensions, the general directional derivative can be expressed in terms of these partial derivatives. From the very definition of $\dfrac{\partial F}{\partial \xi_a}$, we see that it is equal to the rate of change of F in the direction ξ_a.

Theorem 7. 1. $F(x_1, x_2, x_3) \in C^1$

$$\Rightarrow \qquad \frac{\partial F}{\partial \xi_a} = \frac{\partial F}{\partial x_1} \cos \alpha_1 + \frac{\partial F}{\partial x_2} \cos \alpha_2 + \frac{\partial F}{\partial x_3} \cos \alpha_3$$

The proof of this is analogous to that of Theorem 9, Chapter 1, and is omitted.

EXAMPLE C. Take F, as in Example A, $\vec{r}$: $(1, 1, -1)$, $\vec{a}$: $(1, 0, -2)$. Then

$$\frac{\partial F}{\partial \xi_a}(1, 1, -1) = \frac{2}{\sqrt{5}} + 0 - \frac{4}{\sqrt{5}} = -\frac{2}{\sqrt{5}}$$

That is, F is decreasing at a rate $2/\sqrt{5}$ in the direction ξ_a.

5.3 MEANING OF THE GRADIENT

We shall show that Grad F is a vector whose direction is the direction of maximum increase of F and whose length

$$L = |\vec{\nabla} F|$$

is the magnitude of that maximum rate of increase. The direction cosines of the direction of the vector Grad F are

$$\frac{1}{L}\frac{\partial F}{\partial x_1}, \quad \frac{1}{L}\frac{\partial F}{\partial x_2}, \quad \frac{1}{L}\frac{\partial F}{\partial x_3}$$

If ξ_a is an arbitrary direction with direction cosines $\cos \alpha_1$, $\cos \alpha_2$, $\cos \alpha_3$ and makes an angle θ with the vector Grad F, then, by formula (e) of section 2,

$$L \cos \theta = \frac{\partial F}{\partial x_1} \cos \alpha_1 + \frac{\partial F}{\partial x_2} \cos \alpha_2 + \frac{\partial F}{\partial x_3} \cos \alpha_3$$

But this is $\dfrac{\partial F}{\partial \xi_a}$ by Theorem 7. Since $|\cos \theta| \leq 1$,

$$\left| \frac{\partial F}{\partial \xi_a} \right| \leq L$$

Moreover, $\dfrac{\partial F}{\partial \xi_a}$ is equal to L when ξ_a coincides with the direction of Grad F and is consequently maximum in that direction.

EXAMPLE D. Define F, r, ξ_a as in Example C. Then

$$\text{Grad } F|_{(1,1,-1)}: \quad 2, \quad -4, \quad 2$$

$$L = 2\sqrt{6}$$

$$\cos \theta = \frac{2-4}{2\sqrt{6}\sqrt{5}} = -\frac{1}{\sqrt{30}}$$

$$L \cos \theta = -\frac{2}{\sqrt{5}} = \frac{\partial F}{\partial \xi_a} (1, 1, -1)$$

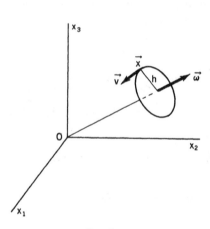

Fig. 8.

EXAMPLE E. The rotation of a body with angular velocity w (say, in radians per second) is conveniently represented by a vector $\vec{\omega}$ along the axis of rotation with $|\vec{\omega}| = w$. Its sense is determined by the *right-hand rule*: if the fingers of the right hand are curled around the axis in the direction of rotation the thumb points in the direction of $\vec{\omega}$, (Figure 8). Let the origin lie on the axis and let $\vec{x}$ be a point a distance h from it. The instantaneous velocity of the point $\vec{x}$ is given by the vector $\vec{v} = \vec{\omega} \times \vec{x}$. For $\vec{v}$ is perpendicular to $\vec{\omega}$ and to $\vec{x}$ and hence is in the direction of the motion of $\vec{x}$ (the direction of the fingers above). Moreover $|\vec{v}| = wh$, the area of the parallelogram determined by $\vec{\omega}$ and $\vec{x}$. Thus the speed of $\vec{x}$ is the angular velocity w multiplied by the radius h of the circle in which $\vec{x}$ is moving. Simple computation gives

$$\vec{\omega} \times \vec{x} = (\omega_2 x_3 - \omega_3 x_2, \omega_3 x_1 - \omega_1 x_3, \omega_1 x_2 - \omega_2 x_1)$$

$$\text{Curl } \vec{\omega} \times \vec{x} = (\omega_1 + \omega_1, \omega_2 + \omega_2, \omega_3 + \omega_3) = 2\vec{\omega}$$

EXERCISES (5)

1. Find $\vec{\nabla}F$ and $(\vec{\nabla} \cdot \vec{\nabla})F$ if

$$F = \log (x_1^2 + x_2^2 + x_3^2)$$

2. Find the divergence and curl of the vector y: x_1, x_2, x_3.

3. Find the directional derivative of the function F of Exercise 1 at an arbitrary point in an arbitrary direction.

4. Same question if the point is $(1, 2, -1)$ and the direction is from that point to the origin.

5. Prove Theorem 7.

6. Find the gradient of $F = x_1 x_2 x_3$. Compute $\dfrac{\partial F}{\partial \xi_a}(1, -1, 2)$ in the direction a: $2, -1, 1$ in two ways, first by Theorem 7 and then by use of the gradient.

7. Find the divergence and curl of the vector:

$$y: \quad \frac{x_1}{r}, \quad \frac{x_2}{r}, \quad \frac{x_3}{r}, \qquad r = \sqrt{x_1^2 + x_2^2 + x_3^2}$$

8. Show that F is harmonic if

$$F = \frac{1}{\sqrt{x_1^2 + x_2^2 + x_3^2}}$$

9. If F is defined as in Exercise 8, show that $\dfrac{\partial F}{\partial \xi_a}(x_1, x_2, x_3)$ is the component of the attraction between unit particles at $(0, 0, 0)$ and (x_1, x_2, x_3) in the direction ξ_a. *Hint:* Represent the attraction as a vector directed from (x_1, x_2, x_3) to $(0, 0, 0)$ and of length $1/(x_1^2 + x_2^2 + x_3^2)$. Then resolve it in the direction ξ_a.

10. If

$$F = \sqrt{x_1^2 + x_2^2 + x_3^2}$$

and if θ is the angle between the vector x_1, x_2, x_3 and the direction ξ_a, show that

$$\frac{\partial F}{\partial \xi_a} = \cos \theta$$

11. Prove Curl $(\vec{a} - \vec{b}) \times (\vec{x} - \vec{b}) = 2\vec{a} - 2\vec{b}$.

12. Prove Curl Grad $F = 0$.

13. Prove Div Curl $\vec{y} = 0$.

14. Prove Curl Curl $\vec{y}$ = Grad Div $\vec{y}$ provided each component of $\vec{y}$ is harmonic.

15. Prove Div $(\vec{y} \times \vec{z}) = \vec{z} \cdot$ Curl $\vec{y} - \vec{y}$ Curl $\vec{z}$.

§6. *Invariants*

The great usefulness of vectors is in large measure due to the fact that certain operations upon them are invariant under rigid motions. Formulas involving such operations will consequently be the same, no matter what system of rectangular coordinates is chosen. For this reason, vectors are particularly useful to represent physical quantities such as force, velocity, acceleration, etc. that are intrinsic in the physical situation and hence independent of a coordinate system. We shall show that scalar and vector products are invariants.

6.1 CHANGE OF AXES

There are two types of change of coordinates corresponding, respectively, to translation and rotation. For the first we have

$$(1) \qquad\qquad x_i' = x_i + a_i \qquad\qquad i = 1, 2, 3$$

This is a transformation from the coordinates (x_1, x_2, x_3) to the coordinates (x_1', x_2', x_3'). Here vectors themselves are invariant, as one sees from their very definition. Analytically, the vector from the point r to the point s in the x-coordinates is transformed to the vector from the point r' to the point s' in the x'-coordinates, where

$$r' = r + a$$

$$s' = s + a$$

The components of the vector are actually the same in each system since

$$s' - r' = s - r$$

Let us determine a rotation about the origin O by three mutually perpendicular unit vectors α, β, γ:

$$(2) \qquad\qquad \alpha \cdot \alpha = \beta \cdot \beta = \gamma \cdot \gamma = 1$$

$$\alpha = \beta \times \gamma, \qquad \beta = \gamma \times \alpha, \qquad \gamma = \alpha \times \beta.$$

Let the new axes x_1', x_2', x_3' have the directions of α, β, γ, respectively. For example, a point one unit distance from O in the positive x_1'-direction has coordinates $(\alpha_1, \alpha_2, \alpha_3)$ in the x-system of coordinates. Let P be an arbitrary point with coordinates (x_1, x_2, x_3) and (x_1', x_2', x_3') in the two systems.

Denote the angle between the vector from O to P and the positive x_i'-axis by θ_i. Then

$$x_i' = L \cos \theta_i, \qquad i = 1, 2, 3$$

where L is the length of OP.

But

$$\cos \theta_1 = \frac{x \cdot \alpha}{\sqrt{x \cdot x}}, \qquad \cos \theta_2 = \frac{x \cdot \beta}{\sqrt{x \cdot x}}, \qquad \cos \theta_3 = \frac{x \cdot \gamma}{\sqrt{x \cdot x}}$$

so that the equations of the transformation become

(3)
$$\begin{aligned}
x_1' &= x \cdot \alpha \\
x_2' &= x \cdot \beta \\
x_3' &= x \cdot \gamma
\end{aligned}$$

6.2 INVARIANCE OF INNER PRODUCT

Let r and s be two arbitrary vectors from O to points r and s, respectively. We shall show that $r \cdot s$ is invariant under the transformation (3). That is, if points r and s transform into r' and s', respectively, by equations (3), then

$$r \cdot s = r' \cdot s'$$

This is obvious geometrically, since

$$r \cdot s = |r| \, |s| \cos \theta$$

where θ is the angle between r and s. Clearly, length and angle must be invariant under a rigid motion. But we shall give an analytic proof.

We have by equations (3)

(4)
$$\begin{aligned}
r_1' &= (r \cdot \alpha), & s_1' &= (s \cdot \alpha) \\
r_2' &= (r \cdot \beta), & s_2' &= (s \cdot \beta) \\
r_3' &= (r \cdot \gamma), & s_3' &= (s \cdot \gamma)
\end{aligned}$$

We must show that

$$(r \cdot \alpha)(s \cdot \alpha) + (r \cdot \beta)(s \cdot \beta) + (r \cdot \gamma)(s \cdot \gamma) = (r \cdot s)$$

or by equations (2) that

(5)
$$(r \cdot \alpha)(s\beta\gamma) + (r \cdot \beta)(s\gamma\alpha) + (r \cdot \gamma)(s\alpha\beta) = (r \cdot s)$$

This follows from the identity

$$\begin{vmatrix}
s_i & \alpha_i & \beta_i & \gamma_i \\
s_1 & \alpha_1 & \beta_1 & \gamma_1 \\
s_2 & \alpha_2 & \beta_2 & \gamma_2 \\
s_3 & \alpha_3 & \beta_3 & \gamma_3
\end{vmatrix} = 0 \qquad\qquad i = 1, 2, 3$$

$$s_i(\alpha\beta\gamma) - \alpha_i(s\beta\gamma) + \beta_i(s\alpha\gamma) - \gamma_i(s\alpha\beta) = 0$$

Now take the inner product of the vector on the left with the vector r:

$$(r \cdot s) - (r \cdot \alpha)(s\beta\gamma) + (r \cdot \beta)(s\alpha\gamma) - (r \cdot \gamma)(s\alpha\beta) = 0$$

This is clearly equivalent to equation (5).

In particular, when $r = s$ this gives an analytic proof that length $\sqrt{r \cdot r}$, is invariant.

6.3 INVARIANCE OF OUTER PRODUCT

That $r \times s$ is also invariant follows from its geometric meaning. By the invariance of this operation we mean that if r and s are transformed to r' and s', respectively, by equations (4) and if the vector $r \times s$ is also transformed by the transformation (3) to a new vector t',

$$t'_1 = (rs\alpha)$$
$$t'_2 = (rs\beta)$$
$$t'_3 = (rs\gamma)$$

then $t' = r' \times s'$:

$$(rs\alpha) = (r \cdot \beta)(s \cdot \gamma) - (r \cdot \gamma)(s \cdot \beta) = (r \times s) \cdot (\beta \times \gamma)$$
$$(rs\beta) = (r \cdot \gamma)(s \cdot \alpha) - (r \cdot \alpha)(s \cdot \gamma) = (r \times s) \cdot (\gamma \times \alpha)$$
$$(rs\gamma) = (r \cdot \alpha)(s \cdot \beta) - (r \cdot \beta)(s \cdot \alpha) = (r \times s) \cdot (\alpha \times \beta)$$

But these equations are true by virtue of the relations (2).

EXERCISES (6)

1. Solve equations (3) for x in terms of x'.

2. Show that the transformation

$$3x'_1 = x_1 - 2x_2 + 2x_3$$
$$3x'_2 = 2x_1 + 2x_2 + x_3$$
$$3x'_3 = -2x_1 + x_2 + 2x_3$$

is a rotation about the origin. Find α, β, γ.

3. Take r: 1, -1, 1, s: 1, 2, 1. Under the transformation of Exercise 2 show

$$r' \cdot s' = r \cdot s$$
$$(r \times s)' = r' \times s'$$

4. Are the results of Exercise 3 true for the transformation

$$3x'_1 = x_1 - 2x_2 + 2x_3$$
$$3x' = -2x_1 + x_2 + 2x_3$$
$$3x'_3 = 2x_1 + 2x_2 + x_3?$$

Show that this is not a rotation.

5. Find the fixed points [$x' = x$] of Exercise 2 and thus find the axis of rotation. Find the angle of rotation about the axis.

6. Find the fixed points of the transformation of Exercise 4. Interpret the transformation.

7. Show analytically that the area of the triangle with vertices O, r, s is invariant under a rotation.

8. Same problem for a triangle with vertices r, s, t.

9. Show analytically that the volume of a tetrahedron with vertices O, r, s, t is invariant under a rotation.

10. Show that the gradient of a scalar function is invariant under a rotation. First state carefully what is meant.

3

Differential
Geometry

§1. *Arc Length of a Space Curve*

Let a curve be given parametrically by the vector equation

$$(1) \qquad \vec{x} = \vec{x}(t)$$

The arc length between two points $t = a$ and $t = b$ of the curve is defined as follows. Consider a *subdivision* of the interval (a, b),

$$a = t_0 < t_1 < \ldots < t_n = b$$

of *norm* δ,

$$\delta = \max (t_1 - t_0, t_2 - t_1, \ldots, t_n - t_{n-1})$$

The length L of the arc is defined as

$$(2) \qquad L = \lim_{\delta \to 0} \sum_{i=1}^{n} |\vec{x}(t_i) - \vec{x}(t_{i-1})|$$

whenever this limit exists. The curve is then said to be *rectifiable*. The sum (2) is clearly the length of a broken line inscribed in the curve.

I.I AN INTEGRAL FORMULA FOR ARC LENGTH

If $x_1(t)$, $x_2(t)$, $x_3(t) \in C^1$, then

$$L = \int_a^b |\vec{x}'(t)| \, dt$$

For, by the law of the mean

$$L = \lim_{\delta \to 0} \sum_{i=1}^n \sqrt{x_1'(\xi_i)^2 + x_2'(\eta_i)^2 + x_3'(\zeta_i)^2}(t_i - t_{i-1})$$

where ξ_i, η_i, ζ_i all lie in the interval $t_{i-1} < t < t_i$. Then by Duhamel's theorem (§6.5, Chapter 5) we have

$$L = \int_a^b \sqrt{x_1'(t)^2 + x_2'(t)^2 + x_3'(t)^2} \, dt$$

$$L = \int_a^b |\vec{x}'(t)| \, dt$$

Let s be the arc length measured from a fixed point t_0 to a variable point t, s being taken as positive when $t > t_0$ and negative when $t < t_0$. Then

(3) $$s = \int_{t_0}^t \sqrt{x' \cdot x'} \, dt$$

(4) $$\frac{ds}{dt} = |\vec{x}'|$$

$$ds^2 = dx_1^2 + dx_2^2 + dx_3^2 = |d\vec{x}|^2$$

Note that s increases as t increases. The direction of increasing s is called the *positive sense* of the curve.

EXAMPLE A.   Consider the circular helix

$$x_1 = \cos t, \qquad x_2 = \sin t, \qquad x_3 = t$$

Choose $t_0 = 0$. Then

$$x_1' = -\sin t, \qquad x_2' = \cos t, \qquad x_3' = 1$$

$$|\vec{x}'| = \sqrt{2}, \qquad s = \int_0^t \sqrt{2} \, dt = \sqrt{2}t$$

Introduce s as the parameter:

$$x_1 = \cos(s/\sqrt{2}), \qquad x_2 = \sin(s/\sqrt{2}), \qquad x_3 = s/\sqrt{2}$$

The positive sense is that which makes the x_3-coordinate increase.

Theorem 1. *Let $x_i(t) \in C^1$, $i = 1, 2, 3$. Then the parameter t is the arc length of the curve* (1)

$$(5) \quad \Leftrightarrow \qquad\qquad |\vec{x}'(t)| \equiv 1$$

First suppose t is the arc, $t - t_0 = s$. Then $ds = dt$. Now equation (4) gives (5). Conversely, if (5) holds, equation (3) gives $s = t - t_0$, so that t is the arc measured from a suitable point.

EXAMPLE B.   For the curve

$$x_1 = \frac{\sin t}{\sqrt{2}}, \qquad x_2 = \frac{\sin t}{\sqrt{2}}, \qquad x_3 = \cos t$$

we have $|\vec{x}'(t)| \equiv 1$, so that the parameter t is the arc length. The curve is, in fact, a circle in the plane $x_1 = x_2$ with center at the origin and of radius unity.

1.2 TANGENT TO A CURVE

Let us now assume that the parameter is the arc length

$$(6) \qquad\qquad\qquad x = x(s)$$

A tangent line is defined, as for plane curves, as the limit of the secant. The *positive direction* of the tangent corresponds with the positive sense of the curve.

Definition 1. *The tangent vector to the curve* (6) *at a point $s = s_0$ is a unit vector α from the point s_0 in the positive direction of the tangent at s_0.*

Theorem 2. *The tangent vector to the curve* (6) *is $\alpha = x'(s)$.*

Let s_0 be an arbitrary point of the curve (6) and $s_0 + \Delta s$ a neighboring point of the curve. If $\Delta s > 0$, the vector directed from the first to the second point,

$$x(s_0 + \Delta s) - x(s_0)$$

has a direction which corresponds to the positive sense of the curve. If we divide this vector by the positive scalar Δs, we do not change the direction of the vector, but merely alter its length. But as Δs approaches zero, the vector approaches the vector $x'(s_0)$. This is a unit vector since the arc is the parameter.

If the parameter is not the arc, the direction components of the tangent are still the components of the vector $x'(t)$, and

$$(7) \qquad\qquad\qquad \alpha = \frac{dx}{ds} = \frac{dx}{dt}\frac{dt}{ds} = \frac{x'(t)}{|x'(t)|}$$

This assumes, of course, that the denominator is different from zero. Points where $|x'(t)| = 0$ are called *singular points* and are excluded from discussion.

We may now write the equations of the tangent line and normal plane to the curve (6) at a point $x(s_0)$.

Tangent line: $X = x(s_0) + tx'(s_0)$

Normal plane: $[X - x(s_0)] \cdot x'(s_0) = 0$

We use the letters X_1, X_2, X_3 for the running coordinates. By (7) we see that the tangent line and normal plane to the curve (1) at the point $\bar{x}(t_0)$ are obtained by replacing s_0 by t_0 in the above equations.

EXAMPLE C.   The tangent vector to the circle of Example B at the point $t = \pi/4$ or $(\frac{1}{2}, \frac{1}{2}, \sqrt{2}/2)$ is $\alpha = (\frac{1}{2}, \frac{1}{2}, -\sqrt{2}/2)$. Note that it is perpendicular to the vector from $(0, 0, 0)$ to $(\frac{1}{2}, \frac{1}{2}, \sqrt{2}/2)$, a radius of the circle.

Tangent line: $X_1 = \dfrac{1}{2} + \dfrac{t}{2}$, $X_2 = \dfrac{1}{2} + \dfrac{t}{2}$, $X_3 = \dfrac{\sqrt{2}}{2} - \dfrac{\sqrt{2}t}{2}$

Normal plane: $X_1 + X_2 - \sqrt{2}X_3 = 0$

Observe that the normal plane passes through the center of the circle and that the tangent line intersects the x_3-axis.

EXERCISES (1)

1. Give a parametric representation of the circle

$$\bar{x} \cdot \bar{x} = 4a^2, \qquad x_1 = a$$

and find its total length by integration.   Check by elementary geometry.

2. Find the arc length from $t = 0$ to $t = 1$ of the curve

$$x_1 = 6t, \qquad x_2 = 3t^2, \qquad x_3 = t^3$$

3. Introduce the arc as the parameter for the curve

$$x_1 = e^t, \qquad x_2 = e^{-t}, \qquad x_3 = \sqrt{2}t$$

4. Find the equations of the tangent line and normal plane at an arbitrary point of the curve of Exercise 2.

5. Find the equation of the tangent line and normal plane of the helix of Example A at an arbitrary point P.   If the normal plane cuts the x_3-axis in a point Q, show that the line PQ is parallel to the x_1x_2-plane.

6. Find the angle between the curves

$$\begin{cases} x_2^2 = x_1 \\ x_3^2 = 2 - x_1 \end{cases} \qquad \begin{cases} x_1 = t \\ x_2 = t^2 \\ x_3 = t^3 \end{cases}$$

at the point $(1, 1, 1)$.

7. Find the equations of the tangent line and normal plane to the curve (1).

8. Same problem for a curve given as the intersection of two cylinders

$$x_2 = f(x_1), \qquad x_3 = g(x_1)$$

9. Same problem for the cylinders

$$F(x_1, x_2) = 0, \qquad G(x_1, x_3) = 0$$

What are you assuming about the functions F and G?

10. Illustrate Exercise 9 by the first curve of Exercise 6 at the point $(1, 1, 1)$.

11. Find the arc length of the curve of Exercise 8.

12. Show that the components of the tangent vector to the curve

$$F(x_1, x_2, x_3) = 0$$

$$G(x_1, x_2, x_3) = 0$$

are proportional to

$$\frac{\partial(F, G)}{\partial(x_2, x_3)}, \qquad \frac{\partial(F, G)}{\partial(x_3, x_1)}, \qquad \frac{\partial(F, G)}{\partial(x_1, x_2)}$$

if these are not all zero. *Hint:* Assume that the given equations can be solved for two of the variables and use Exercise 9.

13. Illustrate Exercise 12 by the curve

$$x_1^2 + 3x_2^2 + 2x_3^2 = 9$$

$$x_1^2 + x_2^2 + x_3^2 = 6$$

at the point $(2, 1, 1)$.

14. The curve

$$x_1 = t^2, \qquad x_2 = t^3, \qquad x_3 = t^4$$

has a singular point at the origin. Find the direction of the tangent line there.

15. Is the parameter t the arc for the curve

$$x_1 = \frac{\sqrt{t^2 + 4} + t}{2}, \quad x_2 = \frac{\sqrt{t^2 + 4} - t}{2}, \quad x_3 = \sqrt{2} \log \frac{\sqrt{t^2 + 4} + t}{2}$$

§2. *Osculating Plane*

A tangent plane to a space curve at a point is any plane containing the tangent line at the point. In general, there is one of these planes that is closer to the curve than any other. It is called the *osculating plane*. We proceed to make these ideas precise.

2.1 ZEROS. ORDER OF CONTACT

Let $\varphi(t) \in C^n$.

Definition 2. $\varphi(t)$ *has a zero of order n at* $t = t_0$

$$\Leftrightarrow \qquad \varphi^{(k)}(t_0) = 0 \qquad\qquad k = 0, 1, \ldots, n - 1$$

$$\varphi^{(n)}(t_0) \neq 0.$$

For example, $\sin t$ has a zero of order 1, $(1 - \cos t)$ has a zero of order 2 at $t = 0$. By use of Taylor's theorem with the Lagrange remainder, it may be shown that $\varphi(t)$ has a zero of order n at $t = t_0 \Leftrightarrow$

$$\lim_{t \to t_0} \frac{\varphi(t)}{(t - t_0)^n} = A$$

where A is a constant not zero.

Definition 3.   *A curve* $x = x(t)$ *and a plane* $(X - a) \cdot \gamma = 0$ *have contact of order n at a common point* $a = x(t_0) \Leftrightarrow$ *the distance* $\varphi(t)$ *from a point* t *of the curve to the plane has a zero of order* $n + 1$ *at* $t = t_0$.

We say that the contact is of order greater than n if, and only if,

$$\varphi^{(k)}(t_0) = 0 \qquad\qquad k = 0, 1, 2, \ldots, n + 1$$

Here we do not determine the precise order of contact by determining the exact order of the first nonvanishing derivative. In fact, if $\varphi(t) = t^{5/3}$, then $\phi'(0) = 0$, $\varphi''(0) = \infty$, and there is no precise order of contact. We may say, however, that the order of contact is greater than zero. We do not define fractional orders of contact.

EXAMPLE A.   Find the order of contact between the helix

$$x_1 = \cos (s/\sqrt{2}), \qquad x_2 = \sin (s/\sqrt{2}), \qquad x_3 = s/\sqrt{2}$$

and the plane $x_2 = x_3$. They intersect at $(1, 0, 0)$, where $s = 0$. But

$$\varphi(s) = \frac{s}{2} - \frac{1}{\sqrt{2}} \sin \frac{s}{\sqrt{2}}$$

$$\varphi(0) = \varphi'(0) = \varphi''(0) = 0$$

$$\varphi'''(0) \neq 0$$

The zero is of order three; the contact, of order two.

2.2 EQUATION OF THE OSCULATING PLANE

Definition 4.   *A tangent plane to a curve which has contact of order greater than unity with the curve is called an osculating plane.*

Let us now determine the equation of the osculating plane to the curve

(1) $$\vec{x} = \vec{x}(t)$$

in a case in which it is uniquely determined.

Theorem 3.   1. $x_i(t) \in C^2$ $i = 1, 2, 3$

2. $\vec{x}' \times \vec{x}'' \neq 0$ at $t = t_0$

$\Rightarrow$ *The curve* (1) *has a unique osculating plane at the point* $t = t_0$. *Its equation is*

(2) $$(X - x \quad x' \quad x'') = 0$$

The vectors $\vec{x}$, $\vec{x}'$, $\vec{x}''$ are formed at $t = t_0$, so that equation (2) can also be written as

$$\begin{vmatrix} X_1 - x_1(t_0) & x_1'(t_0) & x_1''(t_0) \\ X_2 - x_2(t_0) & x_2'(t_0) & x_2''(t_0) \\ X_3 - x_3(t_0) & x_3'(t_0) & x_3''(t_0) \end{vmatrix} = 0$$

The distance from the point t of the curve to the plane

$$(X - x(t_0)) \cdot \gamma = 0$$

is

$$\varphi(t) = \pm \frac{(x(t) - (x(t_0)) \cdot \gamma}{\sqrt{\gamma \cdot \gamma}}$$

Hence

$$\pm \sqrt{\gamma \cdot \gamma} \varphi'(t_0) = x'(t_0) \cdot \gamma, \qquad \pm \sqrt{\gamma \cdot \gamma} \varphi''(t_0) = x''(t_0) \cdot \gamma$$

Now if

(3) $$x'(t_0) \cdot \gamma = x''(t_0) \cdot \gamma = 0$$

$\varphi(t)$ will have a zero of order greater than 2 at $t = t_0$.   But the direction of the normal, γ, is uniquely determined by equations (3).   It is the common perpendicular to the vectors $x'(t_0)$ and $x''(t_0)$.   Thus there is a unique tangent plane having contact of order greater than unity with the curve at $t = t_0$.   It is the osculating plane and its equation is (2).

EXAMPLE B.   The helix of Example A has the osculating plane at $(1, 0, 0)$ with equation

$$\begin{vmatrix} X_1 - 1 & X_2 & X_3 \\ 0 & 1/\sqrt{2} & 1/\sqrt{2} \\ -\frac{1}{2} & 0 & 0 \end{vmatrix} = 0$$

or $$X_2 = X_3$$

We saw in Example A that this plane has contact of order 2 with the curve.

2.3 TRIHEDRAL AT A POINT

With each point of the curve (1) are associated three mutually perpendicular unit vectors α, β, γ. We define them first for the curve

$$x = x(s)$$

where the parameter s is the arc length. They are

$$\alpha = x'(s), \qquad \beta = \frac{x''(s)}{|x''(s)|}, \qquad \gamma = \frac{x'(s) \times x''(s)}{|x''(s)|}$$

Direct computation shows that

$$\alpha \cdot \beta = \beta \cdot \gamma = \gamma \cdot \alpha = 0, \qquad \alpha \cdot \alpha = \beta \cdot \beta = \gamma \cdot \gamma = 1$$

$$\alpha = \beta \times \gamma, \qquad \beta = \gamma \times \alpha, \qquad \gamma = \alpha \times \beta, \qquad (\alpha\beta\gamma) = 1$$

One has only to make use of the identities

$$x' \cdot x' \equiv 1, \qquad \frac{d}{ds}(x' \cdot x') \equiv 2(x' \cdot x'') \equiv 0$$

The vectors α, β, γ, in that order, have the same disposition as the axes x_1, x_2, x_3. They are called the *tangent vector*, *principal normal vector*, and the *binormal vector*, respectively. The corresponding indefinite straight lines through the point are the *tangent*, the *principal normal*, and the *binormal*, respectively. The principal normal lies in the osculating plane, the binormal is perpendicular to it.

The faces of the trihedral are the normal plane, the *rectifying plane*, and the osculating plane. The rectifying plane is a tangent plane containing the binormal. (See Figure 9.)

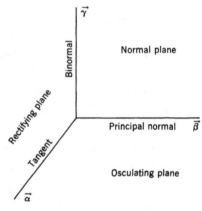

Fig. 9.

EXAMPLE C. For the helix of the previous example we have at $(1, 0, 0)$

Tangent vector α:	$0, 1/\sqrt{2}, 1/\sqrt{2}$
Principal normal vector β:	$-1, 0, 0$
Binormal vector γ:	$0, -1/\sqrt{2}, 1/\sqrt{2}$
Normal plane:	$X_2 + X_3 = 0$
Osculating plane:	$X_2 - X_3 = 0$
Rectifying plane:	$X_1 = 1$

For the curve $x = x(t)$, where t is no longer the arc length it may be shown by use of the equation

$$\frac{ds}{dt} = |\vec{x}'(t)|$$

that

(4) $\alpha = \dfrac{x'}{|x'|}, \qquad \beta = \dfrac{(x' \cdot x')x'' - (x' \cdot x'')x'}{|x'| \, |x' \times x''|}, \qquad \gamma = \dfrac{x' \times x''}{|x' \times x''|}$

EXERCISES (2)

1. Show that the osculating plane to the curve
$$x_1 = t, \qquad x_2 = 1 - t, \qquad x_3 = t + t^2$$
at the point $(1, 0, 2)$ is parallel to the x_3-axis.

2. Find the osculating plane to the twisted cubic at an arbitrary point.

3. Same problem for the curve
$$x_1 = \cos t, \qquad x_2 = \sin t, \qquad x_3 = \sin t$$

4. Same problem for the curve
$$x_1 = 2 \sin^2 t, \qquad x_2 = \sin 2t, \qquad x_3 = 2 \cos t$$

5. Same problem for the curve
$$x_1^2 + x_2^2 = a^2, \qquad 2x_1 x_2 = ax_3$$

6. Find the order of contact of the twisted cubic
$$x_1 = t, \qquad x_2 = t^2, \qquad x_3 = t^3$$
with each of the three coordinate planes.

7. Find the order of contact of the curve
$$x_3 = x_2^2, \qquad x_1^2 = 1 - x_3$$
with its osculating plane at the point $(1, 0, 0)$.

8. Show that the osculating plane of a plane curve is the plane of the curve.

9. Prove that a curve, all of whose osculating planes are parallel to a fixed plane, is a plane curve.

10. Find α, β, γ for the curves of Exercises 3 and 4.

11. Find α, β, γ for the curves of Exercises 5 and 6.

12. Prove formulas (4).

13. Find the order of contact of the curve
$$x_1 = 4(t - 1)^3, \qquad x_2 = -6(t + 2 \cos t), \qquad x_3 = 3(1 - e^{-2t})$$
with the plane
$$x_1 + x_2 - x_3 + 16 = 0$$
at the point $(-4, -12, 0)$. Is the plane an osculating plane to the curve?

14. If $\vec{x}' \times \vec{x}'' = 0$ but $\vec{x}' \times \vec{x}''' \neq 0$ at a point of the curve (1), show that every tangent plane is osculating but that there is still a unique plane of closest contact. Find its equation and show that the order of contact will be greater than 2.

15. Generalize the previous exercise.

§3. Curvature and Torsion

The notion of the curvature of a plane curve is familiar. It is essentially the rate at which the tangent line is turning. It is natural to replace this single quantity by two others for a space curve. The first will be called the *curvature* and is a measure of the rate at which the curve is turning away from its tangent line at a point; the second is called the *torsion* and is a measure of the rate at which the curve is twisting out of its osculating plane at a point. For a plane curve the torsion is zero, and the new notion of curvature reduces to the old.

3.1 CURVATURE

Let us consider a space curve with equation

(1) $$\vec{x} = \vec{x}(s)$$

the parameter s being the arc length.

Definition 5. *The curvature of the curve* (1) *at the point* s_0 *is*

$$\frac{1}{R} = \lim_{\Delta s \to 0} \left| \frac{\Delta \theta}{\Delta s} \right|$$

where $\Delta\theta$ *is the angle between the tangents at the points* s_0 *and* $s_0 + \Delta s$.

EXAMPLE A. Find the curvature of the circle

$$x_1 = \frac{a}{\sqrt{2}} \sin \frac{s}{a}, \qquad x_2 = \frac{a}{\sqrt{2}} \sin \frac{s}{a}, \qquad x_3 = a \cos \frac{s}{a}$$

The tangent vectors at points s_0 and $s_0 + \Delta s$ are

$$\frac{1}{\sqrt{2}} \cos \frac{s_0}{a}, \qquad \frac{1}{\sqrt{2}} \cos \frac{s_0}{a}, \qquad -\sin \frac{s_0}{a}$$

$$\frac{1}{\sqrt{2}} \cos \frac{s_0 + \Delta s}{a}, \qquad \frac{1}{\sqrt{2}} \cos \frac{s_0 + \Delta s}{a}, \qquad -\sin \frac{s_0 + \Delta s}{a}$$

so that

$$\cos \Delta \theta = \cos \frac{s_0}{a} \cos \frac{s_0 + \Delta s}{a} + \sin \frac{s_0}{a} \sin \frac{s_0 + \Delta s}{a} = \cos \frac{\Delta s}{a}$$

Hence,
$$\Delta\theta = \frac{\Delta s}{a}, \qquad \frac{1}{R} = \frac{1}{a}$$

and the curvature is constantly equal to the reciprocal of the radius.

Theorem 4. 1. $x_i(s) \in C^2$ $\qquad\qquad i = 1, 2, 3$
  The curvature of the curve (1) *at the point* s_0 *is*

$$\frac{1}{R} = |x''(s_0)|$$

For, the tangent vectors at s_0 and $s_0 + \Delta s$ are

$$\alpha = x'(s_0), \qquad \alpha + \Delta\alpha = x'(s_0 + \Delta s)$$

so that
$$\cos\Delta\theta = \alpha \cdot (\alpha + \Delta\alpha) = 1 + (\alpha \cdot \Delta\alpha)$$

Since the parameter is the arc, α and $\alpha + \Delta\alpha$ are unit vectors. Hence,

$$(\alpha + \Delta\alpha) \cdot (\alpha + \Delta\alpha) = 1 + 2(\alpha \cdot \Delta\alpha) + (\Delta\alpha \cdot \Delta\alpha) = 1$$

Consequently,
$$\frac{2(1 - \cos\Delta\theta)}{\Delta s^2} = \frac{\Delta\alpha}{\Delta s} \cdot \frac{\Delta\alpha}{\Delta s}$$

$$\frac{1}{R^2} = \frac{d\alpha}{ds} \cdot \frac{d\alpha}{ds} = |x''(s_0)|^2$$

and the proof is complete. Observe that the principal normal vector may now be written $\beta = Rx''$ and that $\alpha' = x'' = \beta/R$.

In Example A, we have for the vector x'' the components

$$-\frac{1}{a\sqrt{2}}\sin\frac{s}{a}, \qquad -\frac{1}{a\sqrt{2}}\sin\frac{s}{a}, \qquad -\frac{1}{a}\cos\frac{s}{a}$$

Hence,
$$\frac{1}{R} = |x''| = \frac{1}{a}$$

3.2 TORSION

Definition 6. *The torsion of the curve* (1) *at the point* s_0 *is*

$$\frac{1}{T} = \pm\lim_{\Delta s \to 0} \frac{\Delta\varphi}{\Delta s}$$

where $\Delta\varphi$ is the angle between the osculating planes at the points s_0 and $s_0 + \Delta s$.

The sign will be determined so that the torsion of a right-handed helix is positive.

EXAMPLE B. Find the torsion of the helix,

$$x_1 = \cos \frac{s}{\sqrt{2}}, \qquad x_2 = \sin \frac{s}{\sqrt{2}}, \qquad x_3 = \frac{s}{\sqrt{2}}$$

at the point $s = 0$. The components of the vector γ, normal to the osculating plane, are

$$\frac{1}{\sqrt{2}} \sin \frac{s}{\sqrt{2}}, \qquad -\frac{1}{\sqrt{2}} \cos \frac{s}{\sqrt{2}}, \qquad \frac{1}{\sqrt{2}}$$

The angle between this vector at $s = 0$ and the same vector at Δs is given by

$$\cos \Delta\varphi = \frac{1}{2} \cos \frac{\Delta s}{\sqrt{2}} + \frac{1}{2}$$

$$\frac{2(1 - \cos \Delta\varphi)}{\Delta s^2} = \frac{1 - \cos \dfrac{\Delta s}{\sqrt{2}}}{\Delta s^2}$$

$$\frac{1}{T^2} = \lim_{\Delta s \to 0} \frac{1 - \cos \dfrac{\Delta s}{\sqrt{2}}}{\Delta s^2} = \frac{1}{4}$$

$$\frac{1}{T} = \pm \frac{1}{2}$$

We must now choose the positive sign since the present helix is right-handed (see §3.1 of Chapter 2).

Theorem 5.   1. $x_i(s) \in C^3$ $i = 1, 2, 3$

2. $|x''(s_0)| \neq 0$

$\Rightarrow$ *The torsion of the curve* (1) *at the point* s_0 *is*

$$\frac{1}{T} = \frac{(x'x''x''')_{s_0}}{(x'' \cdot x'')_{s_0}}$$

As in the proof of Theorem 4, we obtain at once

(2) $$\frac{1}{T^2} = (\gamma' \cdot \gamma')$$

We shall show that

(3) $$\gamma' \cdot \gamma' = (\gamma \cdot \beta')^2$$

For,

$$\beta \succ \gamma' = (\gamma \times \alpha) \times \gamma' = (\gamma \cdot \gamma')\alpha - (\alpha \cdot \gamma')\gamma = (\alpha' \cdot \gamma)\gamma = \frac{1}{R}(\beta \cdot \gamma)\gamma = 0$$

Here we have used the relations

$$(\gamma \cdot \gamma) = 1, \qquad (\gamma \cdot \gamma') = 0, \qquad (\alpha \cdot \gamma) = 0$$

$$(\alpha \cdot \gamma') + (\alpha' \cdot \gamma) = 0, \qquad \alpha' = \frac{\beta}{R}$$

Since the vector $\beta \times \gamma'$ is a null vector, its length is zero:

$$(\beta \times \gamma') \cdot (\beta \times \gamma') = (\gamma' \cdot \gamma') - (\beta \cdot \gamma')^2 = (\gamma' \cdot \gamma') - (\beta' \cdot \gamma)^2 = 0$$

Here we have used the fact that

$$(\beta \cdot \gamma) = 0, \qquad (\beta' \cdot \gamma) + (\beta \cdot \gamma') = 0$$

Since $\qquad \gamma = R(x' \times x''), \qquad \beta = Rx'', \qquad \beta' = R'x'' + Rx'''$

we have from equations (2) and (3)

$$\frac{1}{T} = \pm R(x' \times x'') \cdot (R'x'' + Rx''')$$

(4)

$$= \pm R^2 (x'x''x''') = \pm (x'x''x''')/(x'' \cdot x'')$$

Finally, we must choose the positive sign in equation (4), as the following example will show. Note that this choice of sign gives $1/T = \gamma \cdot \beta'$.

To compute the torsion of the helix of Example B by Theorem 5, we have

$$\frac{1}{T} = 4 \begin{vmatrix} 0 & \dfrac{1}{\sqrt{2}} & \dfrac{1}{\sqrt{2}} \\ -\dfrac{1}{2} & 0 & 0 \\ 0 & -\dfrac{1}{2\sqrt{2}} & 0 \end{vmatrix} = \frac{1}{2}$$

Our choice of sign in (4) must have been correct since it has produced a positive torsion for this right-handed helix.

EXERCISES (3)

1. Compute $1/R$ for the helix of Example B at $s = \pi/\sqrt{2}$ directly from the definition. Check by Theorem 4.

2. Solve the same problem for $1/T$, checking by Theorem 5.

3. Show that the curvature of a circular helix is constant.

4. Solve the same problem for the torsion.

5. For the curve $x = x(t)$, show that

$$\frac{1}{R} = \frac{|x' \times x''|}{|x'|^3}$$

6. For the curve $x = x(t)$, show that

$$\frac{1}{T} = \frac{(x'x''x''')}{|x' \times x''|^2}$$

7. Show that the torsion of a plane curve is zero. Take the axes in an arbitrary position.

8. Reconcile the present formula for curvature with the familiar one for a plane curve.

9. Find $1/R$ for the twisted cubic at an arbitrary point.

10. Work out the same problem for $1/T$.

11. Show that a curve is a straight line $\Leftrightarrow 1/R \equiv 0$.

12. Show that a curve for which $\vec{x}'(s)$ and $\vec{x}''(s)$ are never null is plane $\Leftrightarrow 1/T \equiv 0$.

13. Show that for the curve (1)

$$\beta' = \frac{d}{ds} \frac{x''}{|x''|} = \frac{(x'' \times x''') \times x''}{|x''|^3}$$

14. Show that for the curve (1)

$$\gamma' = \alpha \times \beta' = -\frac{(x'x''x''')x''}{|x''|^3}$$

15. Show that for the curve (1)

$$|\gamma'|^2 = \frac{(x'x''x''')^2}{|x''|^4}$$

Exercises 13, 14, 15 together give a new proof of Theorem 5 without any sleight of hand.

16. Prove the final result of Exercise 14 by differentiating $(x' \times x'')/|x''|$.

§4. *Frenet-Serret Formulas*

The Frenet-Serret formulas are three equations expressing the derivatives of the vectors α, β, γ with respect to s as linear combinations of α, β, γ. They are of fundamental importance in the theory of space curves. Of course any vector is a linear combination of α, β, γ:

$$\vec{\eta} = k\vec{\alpha} + l\vec{\beta} + m\vec{\gamma}$$

In fact,

$$k = \alpha \cdot \eta, \qquad l = \beta \cdot \eta, \qquad m = \gamma \cdot \eta$$

When η is α', β', or γ' the scalars k, l, m can be very simply expressed in terms of the curvature and torsion. The resulting equations are known in differential geometry as the *Frenet-Serret formulas*.

4.1 DERIVATION OF THE FORMULAS

Two of the formulas we have essentially obtained in the previous section. From the definitions of α and β and from the formula for the curvature, we have

$$\alpha = x'(s), \qquad \frac{d\alpha}{ds} = x''(s)$$

$$\beta = \frac{x''(s)}{\sqrt{x'' \cdot x''}}, \qquad \frac{d\alpha}{ds} = \beta\sqrt{x'' \cdot x''} = \frac{\beta}{R}$$

Since $(\gamma' \cdot \gamma) = (\gamma' \cdot \alpha) = 0$, the vector γ' is parallel to β, $\gamma' = k\beta$. Since $\frac{1}{T} = (\gamma \cdot \beta') = -(\gamma' \cdot \beta)$, the scalar k must be $-\frac{1}{T}$, and we have

$$\frac{d\gamma}{ds} = -\frac{\beta}{T}$$

Finally, $\qquad \dfrac{d\beta}{ds} = \dfrac{d}{ds}\gamma \times \alpha = \gamma \times \dfrac{\beta}{R} - \dfrac{\beta}{T} \times \alpha = -\dfrac{\alpha}{R} + \dfrac{\gamma}{T}.$

These three formulas are more easily remembered when put in the following arrangement:

$$\frac{d\alpha}{ds} = \quad * \quad \frac{\beta}{R} \quad *$$

$$\frac{d\beta}{ds} = -\frac{\alpha}{R} \quad * \quad +\frac{\gamma}{T}$$

$$\frac{d\gamma}{ds} = \quad * \quad -\frac{\beta}{T} \quad *$$

4.2 AN APPLICATION

By use of the Frenet-Serret formulas, one may obtain the Taylor expansion of the vector $x(s)$, the coefficients in the series being expressed in terms of $\frac{1}{R}, \frac{1}{T}$ and their successive derivatives with respect to s and in terms of α, β, γ. For,

$$x' = \alpha, \qquad x'' = \frac{\beta}{R}, \qquad x''' = -\frac{\alpha}{R^2} + \left(\frac{1}{R}\right)'\beta + \frac{\gamma}{RT}, \qquad \cdots$$

Thus, for the development about $s = 0$, we have

$$(1) \qquad x(s) = x(0) + \alpha s + \frac{\beta}{R}\frac{s^2}{2!} + \left(-\frac{\alpha}{R^2} + \left(\frac{1}{R}\right)'\beta + \frac{\gamma}{RT}\right)\frac{s^3}{3!} + \cdots$$

Here $\alpha, \beta, \gamma, \dfrac{1}{R}, \dfrac{1}{T}, \left(\dfrac{1}{R}\right)' \cdots$ are all formed for $s = 0$.

To study the usual form of a curve at an arbitrary point P, let us choose our system of coordinates with origin at P and with the positive x_1-, x_2-, and x_3-axes coinciding, respectively, with the vectors α, β, γ formed at P. Then

$$\alpha: \quad 1, 0, 0; \qquad \beta: \quad 0, 1, 0; \qquad \gamma: \quad 0, 0, 1$$

The vector equation (1) becomes the three equations

$$x_1(s) = s \ + * \qquad\quad - \frac{s^3}{6R^2} \qquad\quad + \cdots$$

(2)
$$x_2(s) = * \ + \frac{s^2}{2R} \ + \left(\frac{1}{R}\right)' \frac{s^3}{6} \ + \cdots$$

$$x_3(s) = * \ + * \qquad + \frac{s^3}{6RT} \qquad + \cdots$$

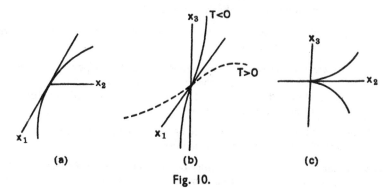

Fig. 10.

If neither curvature nor torsion is zero at $s = 0$, we can determine the behavior of the projections of the curve on the three coordinate planes. For the behavior very near the origin, we may neglect all but the first terms in the above series. The projections then are approximately:

(a) $\qquad x_2 = \dfrac{x_1^2}{2R}$ $\qquad$ in the osculating plane

(b) $\qquad x_3 = \dfrac{x_1^3}{6RT}$ $\qquad$ in the rectifying plane

(c) $\qquad x_3^2 = \dfrac{2R}{9T^2}\, x_2^3$ $\qquad$ in the normal plane

We graph these curves in character in Figure 10.

We can now give another interpretation of the effect of our choice of sign in the formula (4) §3 for the torsion. As a point moves in the direction of increasing s, the curve cuts through the osculating plane in the direction of the vectors γ or $-\gamma$ according as $T > 0$ or $T < 0$.

EXERCISES   (4)

1. For the helix

$$x_1 = \cos \frac{s}{\sqrt{2}}, \qquad x_2 = \sin \frac{s}{\sqrt{2}}, \qquad x_3 = \frac{s}{\sqrt{2}}$$

compute α, β, γ, α', β', γ', $1/R$, $1/T$ at an arbitrary point without use of the Frenet-Serret formulas. Then verify the formulas for this special curve.

2. If we had chosen the negative sign in formula (4) of §3, what would the Frenet-Serret formulas have been?

3. Express $x^{(4)}(s)$ and $x^{(5)}(s)$ as linear combinations of α, β, γ.

4. In equations (2) suppose that $1/(RT) \neq 0$. Find the order of contact of the curve with the three coordinate planes.

5. By use of the Frenet-Serret formulas show that a curve whose curvature is identically zero is a line.

6. The center of curvature of the curve $x = x(s)$ at a point s_0 is the point $x(s_0) + \beta_0 R_0$, where $1/R_0$ is the curvature and β_0 is the principal normal vector at the point s_0. Show that the locus of the centers of curvatures of the helix of Exercise 1 is another helix.

7. Show that the center of curvature of the space curve (2) is the same as the center of curvature of the parabola (a).

8. Reconcile the definition of center of curvature given in Exercise 6 with the familiar coordinates for the center of curvature for the plane curve $y = f(x)$.

$$X = x - \frac{f'[1 + (f')^2]}{f''}, \qquad Y = y + \frac{1 + f'^2}{f''}$$

9. Write the equations of the six elements of the trihedral at $t = 0$ for the curve

$$x_1 = 1 + \sin t, \qquad x_2 = te^t - 1, \qquad x_3 = \log(1 + t)$$

10. Define order of contact of a curve with a sphere. Show that of all spheres through a point P of a curve and with centers on the binormal to the curve at P that one with the center at the center of curvature has closest contact with the curve.

11. For the curve $x = x(s)$ show that

$$\alpha' \cdot \gamma' = -\frac{1}{R}\frac{1}{T}$$

12. For the curve $x = x(s)$ show that

$$x' \cdot x''' = -\frac{1}{R^2}$$

13. Using Exercise 14 of §3 prove the last of the Frenet-Serret formulas.

14. Using the expression for β' from Exercise 13 of §3 and the definitions of α, β, γ, show that

$$\alpha \cdot \beta' = -\frac{1}{R}, \qquad \dot{\gamma} \cdot \beta' = \frac{1}{T}$$

thus proving again the second of the Frenet-Serret formulas.

15. For the curve $x = x(s)$ show that

$$x' \cdot x'''' = -\frac{3}{R}\left(\frac{1}{R}\right)'$$

§5. Surface Theory

We give next an introduction to surface theory. There are three important ways of representing a surface:

(1) $$x_3 = f(x_1, x_2)$$

(2) $$F(x_1, x_2, x_3) = 0$$

(3) $$x = x(u, v)$$

Equation (3) is, of course, a vector equation, and u and v are parameters corresponding to the two degrees of freedom on a surface. For example, a sphere of radius ρ with center at the origin may be represented in each of the three ways:

$$x_3 = \sqrt{\rho^2 - x_1^2 - x_2^2}$$

(4) $$x_1^2 + x_2^2 + x_3^2 - \rho^2 = 0$$

$$x_1 = \rho \cos u \cos v, \qquad x_2 = \rho \cos u \sin v, \qquad x_3 = \rho \sin u$$

The first equation represents only half of the sphere. In the parametric representation the parameters u and v are latitude and longitude, respectively.

5.1 THE NORMAL VECTOR

Let us find the normal vector to the surface (3) at a point (u_0, v_0). The equation

$$x = x(u, v_0)$$

represents a curve on the surface through the point. (See Figure 11.) In the above example of the sphere, it is a meridian. The tangent to this curve at the point has direction components equal to the components of the vector

$x_u(u_0, v_0)$, where the subscript u indicates partial differentiation with respect to u. Also the tangent to the curve

$$x = x(u_0, v)$$

at the point (u_0, v_0) has direction components equal to the components of the vector $x_v(u_0, v_0)$. The normal to the surface is the common perpendicular to these two tangents. We define the normal vector $\vec{\zeta}$ as a unit vector along

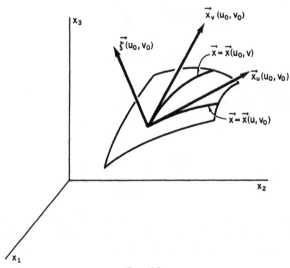

Fig. 11.

along this normal with such a sense that the three vectors x_u, x_v, ζ will have the same disposition as the axes. We thus have

$$\zeta = \frac{x_u \times x_v}{|x_u \times x_v|}$$

whenever the denominator is different from zero. It cannot be identically zero for a bona fide surface and can only vanish at singular points of the surface. (See §4.2 of Chapter 2.) We state our result as a theorem.

Theorem 6. 1. $x_i(u, v) \in C^1$ $\qquad\qquad\qquad\qquad\qquad i = 1, 2, 3$

$\qquad\qquad$ 2. $|x_u \times x_v|_{(u_0, v_0)} \neq 0$

$\Rightarrow$ *The normal vector to the surface* (3) *at the point* (u_0, v_0) *is*

$$\zeta = \frac{x_u \times x_v}{|x_u \times x_v|}\bigg|_{(u_0, v_0)}$$

If the surface has equation (1), then we have

$$x_3 = f(x_1, x_2), \qquad x_1 = x_1, \qquad x_2 = x_2$$

so that $u = x_1$, $v = x_2$. Then

$$x_u: \quad 1, \quad 0, \quad f_1$$
$$x_v: \quad 0, \quad 1, \quad f_2$$

$$\zeta: \quad -\frac{f_1}{D}, \quad -\frac{f_2}{D}, \quad \frac{1}{D} \qquad D = \sqrt{1 + f_1^2 + f_2^2}$$

If the surface has equation (2), then

$$\zeta: \quad \frac{F_1}{D}, \quad \frac{F_2}{D}, \quad \frac{F_3}{D} \qquad D = \sqrt{F_1^2 + F_2^2 + F_3^2}$$

5.2 TANGENT PLANE

It is now easy to write down the equation of the tangent plane to a surface. We have for the surface (1) at a point $(a, b, f(a, b))$

$$x_3 - f(a, b) = f_1(a, b)(x_1 - a) + f_2(a, b)(x_2 - b)$$

At a point (a_1, a_2, a_3) of surface (2) the tangent plane is

$$(x - a) \cdot \nabla F = 0$$

where ∇F is the gradient of F at the point,

$$\nabla F: \quad F_1(a_1, a_2, a_3), \quad F_2(a_1, a_2, a_3), \quad F_3(a_1, a_2, a_3)$$

Finally, the tangent plane to the surface (3) at (u_0, v_0) is

$$(x - a \quad x_u \quad x_v) = 0$$

where $a = x(u_0, v_0)$, $x_u = x_u(u_0, v_0)$, $x_v = x_v(u_0, v_0)$.

5.3 NORMAL LINE

The normal line to surface (1) at the given point is

$$x_1 = a + f_1(a, b)t$$
$$x_2 = b + f_2(a, b)t$$
$$x_3 = f(a, b) - t$$

The normal to surface (2) is

$$(x - a) \times \nabla F = 0$$

where a and ∇F are the vectors a_1, a_2, a_3 and $F_1(a_1, a_2, a_3)$, $F_2(a_1, a_2, a_3)$, $F_3(a_1, a_2, a_3)$. Finally, the normal to the surface (3) at (u_0, v_0) is

$$x = x(u_0, v_0) + t(x_u \times x_v)|_{(u_0, v_0)}$$

5.4 AN EXAMPLE

As an example of the use of some of the foregoing results, let us show that any circular cone with vertex at the origin cuts any sphere with center at the origin orthogonally all along their curve of intersection. The sphere has equations (4). The cone has equations

$$(5) \qquad x_1 = u \cos \alpha \cos v, \qquad x_2 = u \cos \alpha \sin v, \qquad x_3 = u \sin \alpha$$

If $u = \alpha$ in equations (4) and $u = \rho$ in equations (5), the two sets of equations are identical and represent the circle of intersection. At any point on this circle, we get by simple computation the normal vector for the sphere

$$\zeta: \qquad -\cos \alpha \cos v, \qquad -\cos \alpha \sin v, \qquad -\sin \alpha$$

For the cone along the same circle

$$\bar{\zeta}: \qquad -\sin \alpha \cos v, \qquad -\sin \alpha \sin v, \qquad \cos \alpha$$

Since $\zeta \cdot \bar{\zeta} = 0$, our result is established.

EXERCISES (5)

1. Find the equations of the tangent plane and normal line to the surface

$$x_2 x_3 + 2x_1^{x_3} - 3x_3^{x_2} = 0$$

   at the point $(1, 1, 1)$.

2. At what points is the tangent plane to the ellipsoid

$$2(x_1^2 + x_2^2 + x_3^2) + x_1 x_2 + x_2 x_3 + x_3 x_1 = 17$$

   parallel to the $x_1 x_2$-plane?

3. Determine all the points in which the curve

$$x_1 = 2t, \qquad x_2 = t^2, \qquad x_3 = 3t^3$$

   meets the surface

$$x_3 = x_1^2 - x_2^2$$

   and find the angle of intersection at one of them.

4. For the surface

$$x_1 = \tfrac{1}{2} \sin u \sin v, \qquad x_2 = \tfrac{1}{2}(u - \sin u \cos v), \qquad x_3 = \sin u$$

   find the equations of the normal line at $u = \pi/2$, $v = 0$. Show that for the curve $u = v$ on this surface u represents the arc. Find the principal normal vector for this curve at an arbitrary point.

5. The surface

$$x_1 = (1 + u) \cos v, \qquad x_2 = (1 - u) \sin v, \qquad x_3 = u$$

   has two tangent planes at $(1, 0, 1)$. Find both.

6. Write parametric equations for a right circular cone and show that the normal vector is the same all along the straight lines of the cone.

7. A circular cylinder has radius a and has its axis along the x_2-axis. Find the equation of the tangent plane at the point $(a/2, a\sqrt{2}/2, a\sqrt{3}/2)$, using all three forms of the surface: (1), (2), (3).

8. A curve $x_3 = f(x_1)$ in the x_1x_3-plane is rotated about the x_3-axis. Show analytically that the normal line at any point of the resulting surface of revolution either intersects the x_3-axis or is parallel to it.

9. Show analytically that the normal vector at a point of a sphere has the same direction as the radius to that point.

10. Write the equation of the normal to a torus at an arbitrary point. Show that it intersects or is parallel to the axis of the torus.

11. Show that the spheres

$$x \cdot x = 1 \qquad (x - a) \cdot (x - a) = 1$$

intersect orthogonally if, and only if, $a \cdot a = 2$. Interpret geometrically.

12. Find a condition that three surfaces in the form (2) should have a common tangent line at a common point of intersection.

13. At what angle does the curve

$$x_2^2 = x_1, \qquad x_3^2 = 1 - x_1$$

intersect the surface

$$6x_1^2 + 3x_2^2 - 2x_3^2 = 9$$

14. Find the angle between tangent planes to the surfaces (1) and (2) at a common point.

15. Solve the same problem for the surfaces (2) and (3).

§6. *Fundamental Differential Forms*

In this section we shall introduce two differential forms which are of the greatest importance in studying the characteristics of a surface and the behavior of curves on the surface. As an example of their use, we shall discuss briefly the curvature of a normal cross section of a surface.

6.1 FIRST FUNDAMENTAL FORM

Let us take the vector equation

(1) $\vec{x} = \vec{x}(u, v)$

as the representation of the surface. A curve on this surface generally is determined by a single relation between u and v,

(2) $H(u, v) = 0$

or

(3) $v = h(u)$

The direction components of the tangent to this curve,

$$x = x(u, h(u))$$

will be the components of the vector

$$\vec{x}_u + \vec{x}_v h'$$

For the arc length s of the curve, we have

$$\frac{ds^2}{du^2} = |\vec{x}_u + \vec{x}_v h'|^2 = (\vec{x}_u \cdot \vec{x}_u) + 2(\vec{x}_u \cdot \vec{x}_v)h' + (\vec{x}_v \cdot \vec{x}_v)h'^2$$

If the curve is in the form (2) with neither variable preferred, we write

(4) $$ds^2 = (x_u \cdot x_u)\, du^2 + 2(x_u \cdot x_v)\, du\, dv + (x_u \cdot x_v)\, dv^2$$

Since $$H_u\, du + H_v\, dv = 0$$

we could easily compute $\dfrac{ds}{du}$ or $\dfrac{ds}{dv}$ from equation (4).

Definition 7. *The first fundamental form of the surface* (1) *is*

$$\vec{dx} \cdot \vec{dx} = E\, du^2 + 2F\, du\, dv + G\, dv^2$$

where $$E = (x_u \cdot x_u), \qquad F = (x_u \cdot x_v), \qquad G = (x_v \cdot x_v)$$

6.2 ARC LENGTH AND ANGLE

Equation (4) permits one to compute the arc length of a curve on a surface. For example, the length of the curve (3) between points (u_0, v_0) and (u_1, v_1) is

$$\pm \int_{u_0}^{u_1} \sqrt{E + 2Fh' + G(h')^2}\, du$$

The first fundamental form also enables one to compute the angle θ between two curves on a surface. For example, if the curves are (3) and

$$v = k(u)$$

then

(5) $$\cos \theta = \frac{(\vec{x}_u + \vec{x}_v h') \cdot (\vec{x}_u + \vec{x}_v k')}{|\vec{x}_u + \vec{x}_v h'|\, |\vec{x}_u + \vec{x}_v k'|}$$

$$= \frac{E + Fh' + Fk' + Gh'k'}{\sqrt{E + 2Fh' + G(h')^2}\, \sqrt{E + 2Fk' + G(k')^2}}$$

EXAMPLE A.   Let u and v be longitude and latitude on a sphere.   Find the angle at which the curve

$$v = u$$

cuts the equator. Choose units so that the radius is unity. Then the equations of the sphere are

$$x_1 = \cos v \cos u, \qquad x_2 = \cos v \sin u, \qquad x_3 = \sin v$$

Hence, $$E = \cos^2 v, \qquad F = 0, \qquad G = 1$$

$$ds^2 = \cos^2 v \, du^2 + dv^2$$

For the angle θ at the point $u = v = 0$, we have by formula (5)

$$h(u) = u, \qquad k(u) = 0$$
$$\cos \theta = 1/\sqrt{2}, \qquad \theta = \pi/4$$

6.3 SECOND FUNDAMENTAL FORM

Definition 8. *The second fundamental form of the surface* (1) *is*

$$-\overrightarrow{dx} \cdot \overrightarrow{d\zeta} = e \, du^2 + 2f \, du \, dv + g \, dv^2$$

where

$$e = -(x_u \cdot \zeta_u), \qquad f = -(x_u \cdot \zeta_v) = -(x_v \cdot \zeta_u), \qquad g = -(x_v \cdot \zeta_v)$$

Here $\vec{\zeta}$ is the normal vector defined in §5.1, and

$$\overrightarrow{d\zeta} = \vec{\zeta}_u \, du + \vec{\zeta}_v \, dv$$

From the relations

$$(\zeta \cdot x_u) = 0, \qquad (\zeta \cdot x_v) = 0$$

we obtain by differentiation

$$(\zeta_u \cdot x_u) = -(\zeta \cdot x_{uu}), \qquad (\zeta_v \cdot x_u) = -(\zeta \cdot x_{uv})$$
$$(\zeta_u \cdot x_v) = -(\zeta \cdot x_{uv}), \qquad (\zeta_v \cdot x_v) = -(\zeta \cdot x_{vv})$$

The two expressions given for f in Definition 8 are seen in this way to be equal. We have also obtained new expressions for e, f, g in terms of the vector $x(u, v)$ and its derivatives:

$$e = \frac{1}{D}(x_{uu}x_ux_v), \qquad f = \frac{1}{D}(x_{uv}x_ux_v), \qquad g = \frac{1}{D}(x_{vv}x_ux_v)$$

$$D = |x_u \times x_v| = \sqrt{EG - F^2}$$

6.4 CURVATURE OF A NORMAL SECTION OF A SURFACE

At a point of the surface (1), draw a normal plane. It will cut the surface in a normal section whose equation we assume to be

$$v = \varphi(u)$$

The tangent vector of this curve is $\dfrac{dx}{ds}$, and since this is orthogonal to the normal vector ζ, we have

$$\frac{dx}{ds} \cdot \zeta = \alpha \cdot \zeta = 0$$

Differentiating with respect to s and using the Frenet-Serret formulas, we obtain

$$\frac{1}{R}(\beta \cdot \zeta) + \left(\frac{dx}{ds} \cdot \frac{d\zeta}{ds}\right) = 0$$

where $1/R$ is the curvature of the normal section. Since the principal normal β lies in the osculating plane (here the plane of section), $\beta = \pm \zeta$ and

$$(6) \qquad \frac{1}{R} = \pm \frac{(dx \cdot d\zeta)}{(dx \cdot dx)} = \pm \frac{e + 2f\varphi' + g(\varphi')^2}{E + 2F\varphi' + G(\varphi')^2}$$

The derivative $\varphi'(u)$ might be regarded as a generalized "slope" defining the direction at which the curve leaves the point in question. The curvature of the various normal sections of a given surface at a fixed point depends on this slope, as is indicated in formula (6). Replace φ' by λ and choose arbitrarily the positive sign in equation (6). The resulting quantity is called the *normal curvature* $1/r$ of the surface at the point in question in the direction λ:

$$\frac{1}{r} = \frac{e + 2f\lambda + g\lambda^2}{E + 2F\lambda + G\lambda^2}$$

EXAMPLE B. Find the normal curvature of the paraboloid of revolution

$$x_1 = r \cos \theta, \qquad x_2 = r \sin \theta, \qquad x_3 = 1 - r^2$$

at the point $\theta = \pi/4$, $r = 1$.

$$x_r: \quad \cos \theta, \quad \sin \theta, \quad -2r \qquad\qquad \frac{1}{\sqrt{2}}, \quad \frac{1}{\sqrt{2}}, \quad -2$$

$$x_\theta: \quad -r \sin \theta, \quad r \cos \theta, \quad 0 \qquad\qquad -\frac{1}{\sqrt{2}}, \quad \frac{1}{\sqrt{2}}, \quad 0$$

$$x_{rr}: \quad 0, \quad 0, \quad -2 \qquad\qquad 0, \quad 0, \quad -2$$

$$x_{r\theta}: \quad -\sin \theta, \quad \cos \theta, \quad 0 \qquad\qquad -\frac{1}{\sqrt{2}}, \quad \frac{1}{\sqrt{2}}, \quad 0$$

$$x_{\theta\theta}: \quad -r \cos \theta, \quad -r \sin \theta, \quad 0 \qquad\qquad -\frac{1}{\sqrt{2}}, \quad -\frac{1}{\sqrt{2}}, \quad 0$$

$$E = 5, \qquad F = 0, \qquad G = 1, \qquad D = \sqrt{5}$$

$$e = \frac{-2}{\sqrt{5}}, \qquad f = 0, \qquad g = \frac{-2}{\sqrt{5}}$$

Hence,

$$\frac{1}{r} = \frac{-2 - 2\lambda^2}{\sqrt{5}(5 + \lambda^2)}$$

It is easy to determine the maximum and the minimum values of the normal curvature. In the present case, they are found to be $-2/5^{3/2}$ and $-2/\sqrt{5}$, corresponding to $\lambda = 0$ and $\lambda = \infty$, respectively. The product of these two, 4/25, is called the *total* or *Gaussian curvature* at the point. The sum of the maximum and minimum values of $1/r$, here $-12/5^{3/2}$, is called the *mean curvature*.

EXERCISES (6)

1. Rewrite equation (5) for the curve (2) and show that on a sphere a circle of latitude cuts a meridian orthogonally.

2. Show that the curves $u = u_0$ and $v = v_0$ on the surface (1) intersect orthogonally if, and only if, $F = 0$.

3. Find e, f, g for the sphere of Example A.

4. Show that the normal curvature of a sphere is constant.

5. What are the mean and total curvatures of a sphere?

6. In Example B, show that the normal curvature of the paraboloid at the point $\theta = \pi/4$, $r = 1$ in the direction of the curve $\theta = \pi/4$ is numerically equal to the curvature of the generating parabola at the appropriate point.

7. If for a surface $f = F = 0$, show that the total curvature is eg/EG.

8. Find E, F, G, e, f, g for the hyperbolic paraboloid
$$x_1 = a(u + v), \qquad x_2 = b(u - v), \qquad x_3 = uv$$

9. Find the Gaussian curvature of the hyperbolic paraboloid.

§7. Mercator Maps

By way of illustrating the uses of surface theory, we will discuss briefly the geometry of curves on spheres and cylinders. We shall compare analytic geometry on a sphere with plane analytic geometry by placing corresponding formulas side by side in parallel columns. Finally, we shall set up a Mercator map. This is a method of representing the points of a sphere on a cylinder of equal radius in such a way that the angle between intersecting curves is preserved. Since a plane map of a cylinder can easily be made by cutting the cylinder along a ruling and unrolling, one obtains then a conformal (angle preserving) plane map of the sphere.

7.1 CURVES ON A SPHERE

Let φ, θ be latitude and longitude on a sphere of radius a, the meridian of Greenwich, $\theta = 0$, lying in the x_1x_3-plane. A parametric representation of the sphere is

(1) $x_1 = a \cos \varphi \cos \theta, \qquad x_2 = a \cos \varphi \sin \theta, \qquad x_3 = a \sin \varphi$

We now make a table of corresponding formulas for plane and sphere.

	Plane	*Sphere*

Coordinates:

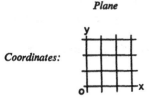

Fig. 12.

Parametric curves:

Plane	Sphere
$x = x_0$ is a parallel to the y-axis.	$\theta = \theta_0$ is a meridian.
$y = y_0$ is a parallel to the x-axis.	$\varphi = \varphi_0$ is a circle of latitude.

Arbitrary curves:

$$y = f(x) \quad \text{or} \quad F(x, y) = 0 \qquad\qquad \varphi = f(\theta) \quad \text{or} \quad F(\theta, \varphi) = 0$$

Arc length:

$$ds^2 = dx^2 + dy^2 \qquad\qquad ds^2 = a^2\, d\varphi^2 + a^2 \cos^2 \varphi\, d\theta^2$$

Slope:

$$\tan w = \frac{dy}{dx} = f'(x) = -\frac{F_1}{F_2} \qquad \tan w = \frac{f'(\theta)}{\cos f(\theta)} = \frac{1}{\cos \varphi}\frac{d\varphi}{d\theta}$$

$$\cos w = \frac{1}{\sqrt{1 + (f')^2}} = \frac{dx}{ds} \qquad \cos w = \frac{\cos f}{\sqrt{\cos^2 f + (f')^2}} = a \cos \varphi \frac{d\theta}{ds}$$

Rhumb lines. A *rhumb line* or *loxodrome* on a sphere is a curve that cuts all meridians at equal angles. We can easily obtain its equation by solving a differential equation

$$\tan w_0 = \lambda = \frac{dy}{dx} \qquad \tan w_0 = \lambda = \frac{1}{\cos \varphi}\frac{d\varphi}{d\theta}$$

$$y = \lambda x + C \qquad \log \tan\left(\frac{\varphi}{2} + \frac{\pi}{4}\right) = \lambda\theta + C$$

We make several remarks about the above formulas. The first differential form for the sphere was obtained in the previous section. It may be obtained directly from equation (1) by use of the identity

$$ds^2 = \vec{dx} \cdot \vec{dx}$$

Along a meridian, $s = a\varphi$.

To obtain the angle w between the curve $\varphi = f(\theta)$ and a circle of latitude, we find the tangent vectors of the two curves:

$$x_1 = a \cos f(\theta) \cos \theta, \qquad x_2 = a \cos f(\theta) \sin \theta, \qquad x_3 = a \sin f(\theta)$$

$$y_1 = a \cos \varphi_0 \cos\theta, \qquad y_2 = a \cos \varphi_0 \sin \theta, \qquad y_3 = a \sin \varphi_0$$

Then

$$\cos w = \frac{x' \cdot y'}{|x'|\, |y'|}$$

Observe that on the equator the formulas for ds and for w are the same in the two columns. The analogy between the equation of a loxodrome on a sphere and a straight line in a plane may be seen by noting that the first term in the Maclaurin expansion of $\log \tan (\varphi/2 + \pi/4)$ is precisely φ.

EXAMPLE A. A ship sails from equator to pole, always keeping latitude equal to longitude. How far does it go? The result is

$$s = \int ds = a \int_0^{\pi/2} \sqrt{1 + \cos^2 \varphi} \, d\varphi$$

$$= a\sqrt{2} \int_0^{\pi/2} \sqrt{1 - \tfrac{1}{2} \sin^2 \varphi} \, d\varphi$$

This integral cannot be evaluated in terms of the elementary functions. But it is a well-known "elliptic integral."* We find

$$s = a\sqrt{2}E(k), \qquad k = \frac{1}{\sqrt{2}}, \qquad \sin^{-1} k = 45°$$

$$s = a\sqrt{2}(1.3506) = 1.91a$$

7.2 CURVES ON A CYLINDER

Consider a circular cylinder which is tangent to the sphere (1) along the equator. Choose as parameters the same angle θ as for the sphere and $z = x_3$, the absolute value of which is the distance from the point in question to the plane of the equator. The parametric equations are

(2) $x_1 = a \cos \theta, \qquad x_2 = a \sin \theta, \qquad x_3 = z$

Parametric curves

Here $\theta = \theta_0$ is a ruling or generating line of the cylinder. And $z = z_0$ is a circle whose plane is parallel to the plane of the equator.

Arbitrary curves:

$$z = f(\theta) \qquad \text{or} \qquad F(\theta, z) = 0$$

Arc length: $ds^2 = a^2 \, d\theta^2 + dz^2$

Slope. $\tan w = \dfrac{f'(\theta)}{a} = \dfrac{1}{a} \dfrac{dz}{d\theta}$

$$\cos w = \frac{a}{\sqrt{a^2 + (f')^2}} = \frac{a \, d\theta}{\sqrt{dz^2 + a^2 \, d\theta^2}}$$

---

* See formula 544, and page 133 of *Short Table of Integrals* by B. O. Peirce and R. M. Foster, New York: Ginn and Company, 1956.

7.3 MERCATOR MAPS

Let us make an arbitrary point P, (θ, φ), of the sphere (1) correspond to the point Q, (θ, z), of the cylinder (2) in such a way that the angle between two arbitrary curves on the sphere will be the same as the angle between corresponding curves on the cylinder, Figure 13. Since θ has the same meaning in the two representations, we have only to determine z as a function of φ, $z = g(\varphi)$. Since meridian $\theta = \theta_0$ is transformed into ruling $\theta = \theta_0$,

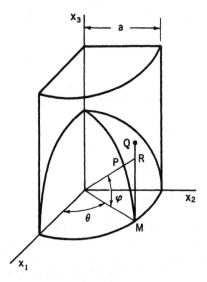

Fig. 13.

it will be sufficient to consider the angle between a single curve on a sphere and a meridian. Thus the slope must be preserved from sphere to cylinder. That is,

$$\frac{1}{\cos \varphi} \frac{d\varphi}{d\theta} = \frac{1}{a} \frac{dz}{d\theta}$$

This differential equation can now be solved for the unknown function $g(\varphi)$. We find

$$z = a \log \tan \left(\frac{\varphi}{2} + \frac{\pi}{4}\right) + C$$

If we make the equator correspond to the circle $z = 0$, we have $C = 0$. Thus the map is completely determined. For example, a point on the sphere with longitude 45° and latitude 30° will be transformed to the point $\theta = \pi/4$, $z = (a/2) \log 3 = 0.549a$. To obtain the plane map, we must now unroll the cylinder. We then obtain a plane, and we may choose coordinates x, y so that

$$x = a\theta$$

$$y = z = a \log \tan \left(\frac{\varphi}{2} + \frac{\pi}{4} \right)$$

In this plane, the above point will have coordinates $(0.785a, 0.549a)$. The plane can, of course, be reduced in scale by dividing all coordinates by a, for example.

EXERCISES (7)

1. Find integral formulas for the lengths of curves $F(\theta, \varphi) = 0$ and $F(\theta, z) = 0$ on the sphere and cylinder, respectively.

2. Find the "slopes" ($\tan w$) for the curves of Exercise 1.

3. Prove the formula for $\cos w$ given in the text for sphere and cylinder.

4. Solve the same problem for $\tan w$.

5. Find the length of a spherical loxodrome from pole to pole.

6. Develop a theory of curves on a cone, as was done in the text for a sphere.

7. Define a loxodrome on a cylinder and find the length of one revolution.

8. Solve the same problem for a cone. Specify which revolution you are considering.

9. A curve $\theta = \varphi$ is transformed by Mercator projection. Find the equation of the transformed curves on the cylinder (θ, z) and on the plane (x, y).

10. A ship sails steadily north-northeast from the point $\theta = 45°$, $\varphi = 15°$. Find the equation (in x and y) of its path on the Mercator map.

11. Find the angle between the curves $\theta = \varphi$, $\varphi = 2 - \theta^2$ on a sphere at points of intersection. Find the angle between the corresponding curves on the Mercator map at points of intersection.

12. *Central projection* is defined as a transformation which carries a point P on the sphere (1) into a point R on the cylinder (2) in such a way that the line PR passes through the center of the sphere (Figure 13). Show that the transformation is not conformal by consideration of a curve on the sphere which passes northeast from the point $\theta = 0$, $\varphi = \pi/4$.

4

Applications of
Partial
Differentiation

§1. Maxima and Minima

In this chapter we shall discuss certain applications of partial differentiation. We shall be concerned principally with the determination of the maximum and minimum values of functions of several variables. We shall revert to the more familiar notation (x, y, z) for the rectangular coordinates of a point in three dimensional space. In the present section we shall review the facts about extreme values of a function of one variable in order to have a basis for generalization to higher dimensions.

1.1 NECESSARY CONDITIONS

We recall first that the vanishing of the first derivative of a function of class C^1 is a necessary condition for a maximum or a minimum of the function. If the existence of one or the other is known independently, this condition is usually the practical one for the applications. We state the result in the following form.

Theorem A.   1. $f(x) \in C^1$ $a \leq x \leq b$

2. $f(a) < f(c), f(b) < f(c)$ *for some c between a and b*

$\Rightarrow$ *There exists at least one number X $(a < X < b)$ such that*

A. $f(x) \leq f(X)$ $a \leq x \leq b$

B. $f'(X) = 0$

Condition 2 insures that the graph of $f(x)$ is "low at the sides and high in the middle." Hence, the function must have a maximum value, taken on at one or more places between a and b. The result is thus obvious geometrically. The analytic proof is similar to that of Rolle's theorem, and is not repeated here. Observe that the function $1 - |x|$ has the maximum value 1, taken on at $x = 0$. But Theorem A does not apply since the function is not of class C^1.

EXAMPLE A. $f(x) = 12x^2 - 4x^3 - 3x^4$ $-3 \leq x \leq 2$

$f(-3) = -27 < f(0) = 0$

$f(2) = -32 < f(0) = 0$

$f'(x) = 12x(1 - x)(2 + x)$

Hence, X must be one of the numbers 0, 1, −2. Since

$$f(0) = 0, \qquad f(1) = 5, \qquad f(-2) = 32$$

it is clear that $X = -2$ and

$$f(x) \leq 32 \qquad\qquad -3 \leq x \leq 2$$

In fact, it is easy to see that this relation holds for all x.

1.2 SUFFICIENT CONDITIONS

By use of the derivatives of higher order, we may obtain sufficient conditions for a relative extremum.

Theorem B.   1. $f(x) \in C^{2n}$ $a \leq x \leq b$

2. $f^{(k)}(X) = 0$ $k = 1, \ldots, 2n - 1;$ $a < X < b$

3. $f^{(2n)}(X) < 0$

$\Rightarrow$ *There exists a positive number ϵ such that*

$$f(x) < f(X) \qquad\qquad 0 < |x - X| < \epsilon$$

The conclusion is that $f(x)$ has a relative maximum at X. By Taylor's expansion, we have

$$(1) \qquad f(x) - f(X) = \frac{(x - X)^{2n}}{(2n)!} f^{(2n)}(\xi)$$

where ξ is between x and X. As x approaches X, ξ does also, and $f^{(2n)}(\xi)$ approaches a negative number by condition 3. Hence, for all $x \neq X$ sufficiently near X the right-hand side of equation (1) is negative, and the result is established. The theorem is easily modified for relative minima.

This theorem is particularly useful for $n = 1$, for then two derivatives need only be computed. In Example A above,

$$f''(x) = 12(2 - 2x - 3x^2)$$

$$f''(1) = -36, \qquad f''(-2) = -72, \qquad f''(0) = 24$$

Hence, $f(x)$ has relative maxima at $x = 1$ and $x = -2$. It has a relative minimum at $x = 0$. It is easy to see that the absolute maximum is at $x = -2$. The absolute minimum in the interval $-3 \leq x \leq 2$ is $f(2) = -32$. There is no absolute minimum in the infinite interval $-\infty < x < \infty$.

EXAMPLE B. $\quad f(x) = 1 - x^6$

$$X = 0, \qquad f^{(6)}(0) = -6!$$

The relative (and absolute) maximum is $f(0) = 1$.

EXAMPLE C. $\quad f(x) = x^4 - x^5$ $\hfill -2 \leq x \leq 2$

This function has a relative minimum $f(0) = 0$, a relative maximum $f(4/5) = 4^4/5^5$, an absolute minimum $f(2) = -16$, and an absolute maximum $f(-2) = 48$.

1.3 POINTS OF INFLECTION

A *point of inflection of a curve* is a point where the curve crosses its tangent at the point. We can obtain a derivative condition for such a point.

Theorem C. 1. $f(x) \in C^{2n+1}$ $\hfill a \leq x \leq b$

2. $f^{(k)}(X) = 0$ $\hfill k = 2, \ldots, 2n; \quad a < X < b$

3. $f^{(2n+1)}(X) \neq 0$

$\Rightarrow$ *The graph of $f(x)$ has a point of inflection at X.*

For, as in the previous proof,

$$f(x) - f(X) - f'(X)(x - X) = \frac{(x - X)^{2n+1}}{(2n + 1)!} f^{(2n+1)}(\xi)$$

The left-hand side is the difference between corresponding ordinates of the curve $y = f(x)$ and its tangent; the right-hand side changes sign as x passes through X. Hence, the theorem is proved.

EXAMPLE D. $f(x) = x^5 + x + 1$

$$f^{(k)}(0) = 0 \qquad\qquad k = 2, 3, 4$$
$$f^{(5)}(0) = 5!$$

The graph of $f(x)$ has a point of inflection at $x = 0$.

EXERCISES (1)

In Exercises 1–6 find the relative and the absolute maxima and minima. State which, if any, of the theorems of the text you are using.

1. $x^4 - 4x^3 + 1$ $-1 \leqq x \leqq 1$

2. $(4 - x^2)^{-1/2}$ $-1 \leqq x \leqq 1$

3. x^x $0.1 \leqq x \leqq 1$

4. $\displaystyle\int_0^x (t^3 - t)^3 \, dt$ $-1 \leqq x \leqq 1$

5. $x^{2/3}$ $-1 \leqq x \leqq 1$

6. $x^m(1 - x)^n$ (m, n are positive integers) $-\infty < x < \infty$

In Exercises 7–9 determine if $x = 0$ is a maximum, minimum, or point of inflection.

7. $x(1 - e^x) \sin x$

8. $x \sin x - \sin^2 x$

9. $x \tan^{-1} x - x^2$

10. A man can walk twice as fast as he can swim. To get from a point on the edge of a circular pool to a point diametrically opposite he may walk around the edge, swim straight across, or walk part way around and swim the rest of the way in a straight line. How shall he proceed if he is to make the trip in the least time? greatest time?

11. Weary Willie, sleeping $\frac{1}{4}$ mile from the tracks, is awakened just as the locomotive is nearest him. The freight train is $\frac{1}{3}$ mile long and is traveling 20 mph. If he starts running immediately and cuts across a field in a straight line, how slowly can he run and still catch the freight? In what direction must he head?

12. Find the minimum distance from the ellipse
$$9x^2 + 25y^2 = 225$$
to the point $(4, 0)$. Solve the problem first with x and then with y as independent variable.

§2. *Functions of Two Variables*

In this section, we shall prove a result for functions of two variables analogous to Theorem A of the previous section. It will provide a sufficient condition for the existence of an absolute maximum or minimum at an interior point of the region of definition. A further conclusion, the vanishing of the two first-order derivatives at such points, will provide a means of determining their positions.

2.1 ABSOLUTE MAXIMUM OR MINIMUM

Definition 1. *A function $f(x, y)$ has an absolute maximum at a point (X, Y) of a region $R \Leftrightarrow$*

$$f(X, Y) \geq f(x, y)$$

for all (x, y) in R.

Definition 2. *A function $f(x, y)$ has a relative maximum at a point (X, Y) of a region $R \Leftrightarrow$ there exists a positive number δ such that*

$$f(X, Y) > f(x, y)$$

for all (x, y) of R at which $0 < (x - X)^2 + (y - Y)^2 < \delta$.

Obvious modifications of the inequalities are necessary for the definition of absolute or relative minima.

For example, consider the surface $z = f(x, y)$ obtained by rotating the curve $z = x^4 - 2x^2$ about the z-axis. In the circle $x^2 + y^2 \leq 4$ the function $f(x, y)$ has the absolute maximum value 8, realized on the circumference of the circle. It has a relative maximum equal to zero at $(0, 0)$ and absolute minima equal to -1 at all points of the circle $x^2 + y^2 = 1$. There is no relative minimum in the strict sense of Definition 2.

Theorem 1. 1. $f(x, y) \in C^1$ in a bounded region R consisting of a domain D and a boundary curve Γ.

2. $f(a, b) > f(x, y)$ for some $(a, b) \in D$ and all $(x, y) \in \Gamma$

$\Rightarrow$ *There exists a point $(X, Y) \in D$ such that*

A. $f(x, y) \leq f(X, Y)$ for all $(x, y) \in R$

B. $f_1(X, Y) = f_2(X, Y) = 0$

Since $f(x, y)$ is continuous in the closed region R, it has a maximum*

---

* Compare §6.5, Chapter 5. The proof for functions of two variables is similar to the one given there.

there, say, at (X, Y), which must be in D by virtue of hypothesis 2. Then

$$\frac{f(X + \Delta x, Y) - f(X, Y)}{\Delta x} \geqq 0 \qquad \Delta x < 0$$

$$\leqq 0 \qquad \Delta x > 0$$

Allowing Δx to approach zero, we obtain in the two cases, respectively,

$$f_1(X, Y) \geqq 0$$
$$f_1(X, Y) \leqq 0$$

Hence, $f_1(X, Y)$ is zero. A similar argument shows that $f_2(X, Y) = 0$.

2.2 ILLUSTRATIVE EXAMPLES

EXAMPLE A. $f(x, y) = \sqrt{4 - x^2 - y^2}$ $\qquad\qquad x^2 + y^2 \leqq 1$

Choose $a = b = 0$. Then

$$f(0, 0) = 2 > f(x, y)|_{x^2 + y^2 = 1} = \sqrt{3}$$

Hence, the absolute maximum exists at an interior point (X, Y). To find it, we have

$$f_1(X, Y) = \frac{-X}{\sqrt{4 - X^2 - Y^2}} = 0$$

$$f_2(X, Y) = \frac{-Y}{\sqrt{4 - X^2 - Y^2}} = 0$$

Hence, the absolute maximum for $f(x, y)$ occurs at the origin where $f(x, y)$ has the value 2. The result is also obvious by inspection.

EXAMPLE B. $f(x, y) = 1 - \sqrt{x^2 + y^2}$ $\qquad\qquad x^2 + y^2 \leqq 1$

Theorem 1 is not applicable since $f(x, y) \notin C^1$. But one sees by inspection that the function has the absolute maximum value 1 at $(0, 0)$.

EXAMPLE C. $f(x, y) = x + y$ $\qquad\qquad x^2 + y^2 \leqq 1$

Here hypothesis 2 fails. The function has an absolute maximum at $\left(\dfrac{1}{\sqrt{2}}, \dfrac{1}{\sqrt{2}}\right)$. But, of course, the partial derivatives of first order vanish nowhere.

EXAMPLE D. $f(x, y) = x^4 + y^4 - x^2 - y^2 + 1$ $\qquad\qquad x^2 + y^2 < \infty$

To establish hypothesis 2, introduce polar coordinates:

$$f(r \cos \theta, r \sin \theta) = r^4 (\cos^4 \theta + \sin^4 \theta) - r^2 + 1$$

On the circle $r = r_0$, we see easily that the first term is at least $r_0^4/2$. Hence, on the circle $r = 2$,

$$f \geq 8 - 4 + 1 = 5$$

and $f(0, 0) = 1$, so that an absolute minimum exists. To find it, we have

$$4X^3 - 2X = 0, \qquad 4Y^3 - 2Y = 0$$

There are thus nine points where both equations hold. By trying all nine, we find that there are absolute minima at four of them $\left(\pm \dfrac{1}{\sqrt{2}}, \pm \dfrac{1}{\sqrt{2}} \right)$, where f is equal to $\frac{1}{4}$. Hence, $f \geq \frac{1}{4}$ in all the plane.

2.3 CRITICAL TREATMENT OF AN ELEMENTARY PROBLEM

A familiar problem requires the rectangular parallelepiped of given surface area and maximum volume. Let us examine it carefully in the light of Theorem 1.

Let x, y, z be the lengths of the three sides. Then we must maximize the function xyz subject to such a condition as

$$(1) \qquad xy + yz + zx = 1$$

Eliminating z, we consider the function

$$f(x, y) = xy \left(\frac{1 - xy}{x + y} \right) \qquad x \geq 0, y \geq 0, xy \leq 1$$

It would be natural to choose the region R as defined by the above inequalities, for $f(x, y)$ is zero on its boundary. But the region is not bounded. Moreover, $f(x, y)$ is not defined at the origin.

Let us choose R as the region in the first quadrant between the following six curves:

$$x = 0, \quad x = h, \quad y = 0, \quad y = h, \quad x + y = d, \quad xy = 1$$

where d is to be chosen small and h large. We have taken the infinite region between the axes and a branch of the hyperbola $xy = 1$ and made it finite by cutting off the two infinite "tails." Moreover we have excluded the origin from R. Choose $a = b = \frac{1}{2}$, $f(a, b) = \frac{3}{16}$. Since $x + y \geq x$, $x + y \geq y$, $1 - xy \leq 1$ in R, we see that

$$(2) \qquad f(x, y) \leq \frac{xy}{x} = y, \qquad f(x, y) \leq \frac{xy}{y} = x$$

On the segment of the line $x = h$ belonging to R we use the first inequality (2) and find $f(h, y) \leq 1/h$ there. Also $f(x, h) \leq 1/h$ on the corresponding segment $y = h$ by the second inequality (2). On the segment $x + y = d$ we have $f(x, y) \leq d$ by either inequality (2). Thus if we choose $h = 16$

and $d = \frac{1}{16}$, for example, we have $f(x, y) \leq \frac{1}{16}$ on the whole boundary of R. All hypotheses of the theorem are satisfied, and an absolute maximum exists in R. To find it, we have

$$1 - X^2 - 2XY = 0$$
$$1 - Y^2 - 2XY = 0$$

from which
$$X = Y = \frac{1}{\sqrt{3}}$$

It can be seen (Exercise 14) that $f(x, y) \leq \frac{1}{16}$ also in the three excluded regions. Hence the absolute maximum applies to the whole region of definition of $f(x, y)$. The desired solid is a cube of volume $\sqrt{3}/9$.

Observe that if the existence of the maximum is assumed, it is unnecessary to eliminate the variable z. We would have from equation (1)

$$\frac{\partial z}{\partial x} = -\frac{y + z}{x + y}, \qquad \frac{\partial z}{\partial y} = -\frac{x + z}{x + y}$$

Hence,
$$\frac{\partial}{\partial x}(xyz) = yz + xy\frac{\partial z}{\partial x} = yz - \frac{xy(y + z)}{x + y}$$

$$\frac{\partial}{\partial y}(xyz) = xz + xy\frac{\partial z}{\partial y} = xz - \frac{xy(x + z)}{x + y}$$

Equating these two functions to zero, we obtain $x = y = z$. Then from equation (1) we again see that all three dimensions must be $1/\sqrt{3}$. To obtain relations between the variables at an extremum, it is often simpler to use the implicit method. The explicit method may be shorter if the actual values of the variables at an extremum are desired.

EXERCISES (2)

1. The sum of three positive numbers is unity. What is the maximum value of their product?

2. Determine the rectangular parallelepiped of maximum surface area which can be inscribed in a sphere.

3. Find the rectangular parallelepiped of minimum surface area for a given volume. Show the existence of the absolute minimum.

4. Same problem for a rectangular tank open at the top.

5. Show that the function
$$x^4 + y^4 - 2x^2 + 8y^2 + 4$$
has an absolute minimum.

6. Find the minimum value of the function of the previous exercise. At how many points does it occur?

7. Examine the function
$$x^4 - y^3 + x^2 - y + 1$$
for absolute maxima and minima.

8. Same problem for
$$Ax^2 + 2Bxy + Cy^2 + Dx + Ey + F \qquad B^2 - AC < 0$$
Treat all cases. If an extremum exists, find its position.

In the following problems, the existence of the extremum may be assumed.

9. Find the volume of the greatest rectangular parallelepiped inscribed in an ellipsoid, the axes of the ellipsoid being perpendicular to the faces of the parallelepiped.

10. Find the best dimensions of a tent with horizontal ridge pole. Assume the two ends closed by isosceles triangles. There is no floor.

11. A cylinder is capped at its ends by equal cones. Find the maximum volume for a given surface area.

12. Find the triangle of maximum area for a given perimeter.

13. Write the equation of a torus obtained by rotating a circle in the xz-plane about the z-axis. Show analytically that the surface has infinitely many "highest" and infinitely many "lowest" points.

14. Prove the facts stated about the function f of equation (2). How can we be sure that the desired maximum (§2.3) does not lie in part of the first quadrant outside of the region R?

15. Find the triangle of maximum area that can be inscribed in a circle. *Hint:* Take the polar coordinates of the vertices as $(a, 0)$, (a, θ), (a, φ). The area of the triangle can now be computed from the areas of three isosceles triangles.

§3. Sufficient Conditions

In this section we shall obtain sufficient conditions for relative maxima and minima. They will involve derivatives of the second order at a point, the theorem being analogous to Theorem B ($n = 1$) of §1.

We shall also prove an analogue of Theorem C.

3.1 RELATIVE EXTREMA

Theorem 2. 1. $f(x, y) \in C^2$
      2. $f_1 = f_2 = 0$ at (X, Y)
      3. $f_{12}^2 - f_{11}f_{22} < 0$ at (X, Y)
      4. $f_{11} < 0$ at (X, Y)
$\Rightarrow$ $f(x, y)$ has a relative maximum at (X, Y).

We use Taylor's theorem with remainder to obtain the equation

(1) $\quad \Delta f = f(X + h, Y + k) - f(X, Y) = \frac{1}{2}[Ah^2 + 2Bhk + Ck^2]$

(2) $\quad A = f_{11}(X + \theta h, Y + \theta k), \qquad B = f_{12}(X + \theta h, Y + \theta k)$

$$C = f_{22}(X + \theta h, Y + \theta k)$$

where $0 < \theta < 1$. By hypothesis 1, inequalities 3 and 4 will hold also in some circle of radius δ and center at (X, Y). This circle will contain in its interior the point $(X + \theta h, Y + \theta k)$ if $h^2 + k^2 < \delta^2$, and hence $A < 0$, $AC - B^2 > 0$. Consequently,

$$\Delta f = \frac{1}{2A}\left[(Ah + Bk)^2 + (AC - B^2)k^2\right] < 0$$

That is, $f(x, y)$ has a relative maximum at (X, Y), and the proof is complete.

To apply the theorem for a minimum, one has only to reverse the inequality in hypothesis 4.

EXAMPLE A. $\quad f(x, y) = x^4 + y^4 - x^2 - y^2 + 1$

We saw in Example D of §2 that hypothesis 2 holds at the points

$$(0, 0), \quad \left(\pm\frac{1}{\sqrt{2}}, \pm\frac{1}{\sqrt{2}}\right), \quad \left(0, \pm\frac{1}{\sqrt{2}}\right), \quad \left(\pm\frac{1}{\sqrt{2}}, 0\right)$$

Furthermore,

$f_{11} = -2,$	$f_{12} = 0,$	$f_{22} = -2$	at $(0, 0)$
$= 4,$	$= 0,$	$= 4$	at $\left(\pm\dfrac{1}{\sqrt{2}}, \pm\dfrac{1}{\sqrt{2}}\right)$
$= -2,$	$= 0,$	$= 4$	at $\left(0, \pm\dfrac{1}{\sqrt{2}}\right)$
$= 4,$	$= 0,$	$= -2$	at $\left(\pm\dfrac{1}{\sqrt{2}}, 0\right)$
$f_{12}^2 - f_{11}f_{22} = -4$			at $(0, 0)$
$= -16$			at $\left(\pm\dfrac{1}{\sqrt{2}}, \pm\dfrac{1}{\sqrt{2}}\right)$
$= 8$			at $\left(0, \pm\dfrac{1}{\sqrt{2}}\right)$
$= 8$			at $\left(\pm\dfrac{1}{\sqrt{2}}, 0\right)$

Hence, there is a relative maximum at $(0, 0)$. There are relative minima at the four points $(\pm 1/\sqrt{2}, \pm 1/\sqrt{2})$. (We saw earlier that the minima are actually absolute.) Finally, the theorem is not applicable for the remaining four points, since hypothesis 3 fails there.

3.2 SADDLE-POINTS

A function $f(x, y)$ has a *saddle-point* at (X, Y) if $f_1(X, Y) = f_2(X, Y) = 0$ and if the difference Δf, defined by equation (1), has both positive and negative values in every neighborhood of (X, Y). For example, the function xy has

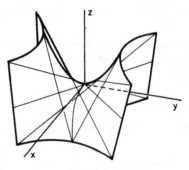

Fig. 14.

such a point at the origin, since it is positive in quadrants one and three and is negative in quadrants two and four. The surface $z = xy$ is the familiar hyperbolic paraboloid. The reason for the term saddle-point is clear from the appearance of this surface. (See Figure 14.)

Theorem 3.   1. $f(x, y) \in C^2$

2. $f_1 = f_2 = 0$ at (X, Y)

3. $f_{12}^2 - f_{11}f_{22} > 0$ at (X, Y)

$\Rightarrow$ $f(x, y)$ has a saddle-point at (X, Y).

Define A, B, C by equations (2), and set

$$a = f_{11}(X, Y), \qquad b = f_{12}(X, Y), \qquad c = f_{22}(X, Y)$$

so that A, B, C approach a, b, c, respectively, as h and k approach zero. We treat three cases.

CASE I. $a \neq 0$.   First set $h = \lambda$, $k = 0$.   Then

$$\lim_{\lambda \to 0} \frac{\Delta f}{\lambda^2} = \lim_{\lambda \to 0} \frac{A}{2} = \frac{a}{2}$$

Next set $h = -\lambda b$, $k = \lambda a$. Then

$$\lim_{\lambda \to 0} \frac{\Delta f}{\lambda^2} = \lim_{\lambda \to 0} \frac{1}{2}[Ab^2 - 2Bab + Ca^2] = \frac{a}{2}(ac - b^2)$$

By hypothesis 3, these two limits have opposite signs.   Hence, Δf will have opposite signs for small λ in the two cases by hypothesis 1.

CASE II. $c \neq 0$.   This case is treated like Case I.

CASE III. $a = c = 0$.   Then $b \neq 0$.   First set $h = k = \lambda$, when

$$\lim_{\lambda \to 0} \frac{\Delta f}{\lambda^2} = b$$

and then set $h = -k = \lambda$, when

$$\lim_{\lambda \to 0} \frac{\Delta f}{\lambda^2} = -b$$

The desired conclusion now follows as in Case I.

EXAMPLE B. $f(x, y) = xy$

$$f_{12}^2 - f_{11}f_{12} = 1 > 0$$

The origin is a saddle-point.

We can now see that the four points $(0, \pm 1/\sqrt{2})$, $(\pm 1/\sqrt{2}, 0)$ of Example A, which we were unable to test by Theorem 2, are saddle-points.

We point out that if $f_{12}^2 - f_{11}f_{12} = 0$ at a point where $f_1 = f_2 = 0$, the point may be a maximum, a minimum, or a saddle-point.   The function

$$f = y^2 - x^3$$

clearly has a saddle-point at the origin.   But the function

$$f = y^2 + x^4 + y^4$$

has a minimum there.   The first of these two functions serves to illustrate an important point of the theory.   In section 1 we saw that we could distinguish between maximum, minimum, and point of inflection by looking merely at the first nonvanishing term of the Taylor expansion.   The situation is very different here; the above two functions have the same term of the second degree.   The reason for the difference becomes clear if we consider an approach to the origin along the parabola $y = x^2$.   Along this curve, the two functions become, respectively,

$$f = -x^3 + x^4$$

$$f = 2x^4 + x^8$$

In the first, the cubic term has now become dominant in the neighborhood of the origin. For the existence of a minimum, it is not sufficient that the homogeneous polynomial of lowest degree (>1) in the Taylor expansion should be always positive.

3.3 LEAST SQUARES

As a further example, let it be required to pass a line

$$y = ax + b$$

"through" the points $(x_1, y_1), (x_2, y_2), \ldots, (x_n, y_n)$ by the method of *least squares*. That is, one must determine constants a and b so that

$$f(a, b) = \sum_{i=1}^{n} (ax_i + b - y_i)^2$$

should be minimum.

We have

(3) $$f_1(a, b) = 2 \sum_{i=1}^{n} (ax_i + b - y_i)x_i = 0$$

(4) $$f_2(a, b) = 2 \sum_{i=1}^{n} (ax_i + b - y_i) = 0$$

$$f_{11}(a, b) = 2 \sum_{i=1}^{n} x_i^2$$

$$f_{12}(a, b) = 2 \sum_{i=1}^{n} x_i$$

$$f_{22}(a, b) = 2 \sum_{i=1}^{n} 1 = 2n$$

Since

(5) $$\left(\sum_{i=1}^{n} 1 \right) \left(\sum_{i=1}^{n} x_i^2 \right) - \left(\sum_{i=1}^{n} x_i \right)^2 = \tfrac{1}{2} \sum_{i=1}^{n} \sum_{j=1}^{n} (x_i - x_j)^2 > 0$$

Theorem 2 is applicable. The unique relative minimum thus assured must also be an absolute minimum since Theorem 1 is also applicable. If $n = 3$, equation (5) follows from Lagrange's identity, §1.2 of Chapter 2. It is easily proved generally.

We have now only to solve equations (3) and (4) for a and b and to substitute these values in the equation of the line. We obtain

$$\begin{vmatrix} x & y & 1 \\ \sum_{i=1}^{n} x_i & \sum_{i=1}^{n} y_i & \sum_{i=1}^{n} 1 \\ \sum_{i=1}^{n} x_i^2 & \sum_{i=1}^{n} x_i y_i & \sum_{i=1}^{n} x_i \end{vmatrix} = 0$$

EXAMPLE C. The line "through" the points $(1, 2)$, $(0, 0)$, $(2, 2)$ is

$$\begin{vmatrix} x & y & 1 \\ 3 & 4 & 3 \\ 5 & 6 & 3 \end{vmatrix} = -6x + 6y - 2 = 0$$

EXERCISES (3)

Test the functions of Exercises 1–5 for relative maxima, relative minima, and saddle-points.

1. $x^2 + 2xy + 2y^2 + 4x$.

2. $x^3 - y^3 + 3x^2 + 3y^2 - 9x$.

3. $x^2 - xy + y^4$.

4. $(x + y)^3 + (x - y)^2 - 12(x + y)$.

5. $(x - 2)^n + (x + 1)^n + (y - 3)^n + (y + 1)^n$ (n is a positive integer).

6. Is the origin a relative maximum, relative minimum, or a saddle-point for the function

$$ax^3 + by^3 + cx^2 + dxy + ey^2 \qquad\qquad d^2 - 4ce \neq 0$$

7. Test the functions z defined by the equation

$$x^2 + 2y^2 + 3z^2 - 2xy - 2yz = 2$$

for maxima and minima by use of Theorem 2.

8. Find the shortest distance from a point to a plane.

9. Find the shortest distance from the line $x = y = z$ to the line

$$x = 1, \qquad y = 0$$

10. Same problem for the lines $y = 2x$, $z = 3x$ and $y = x - 3$, $z = x$.

11. Find the triangle of largest area which can be inscribed in a circle. Use the second derivative test of Theorem 2.

12. Prove equation (5).

13. Pass a line "through" the following points by least squares:

$$(-2, 0), (-1, 0), (0, 1), (1, 3), (2, 2)$$

Plot the line and the given points.

14. Discuss the problem of least squares of §3.3 with the roles of x and y interchanged. That is, you are to minimize the sum of the squares of the errors in the abscissas.

15. Apply the result of Exercise 14 to the points of Exercise 13. Plot.

§4. *Functions of Three Variables*

A theorem analogous to Theorem 1 for functions of three variables is easily developed. We omit it since the proof is practically the same as that of Theorem 1. To obtain a result analogous to Theorem 2, we must first develop briefly the theory of definite quadratic forms in three variables.

4.1 QUADRATIC FORMS

By a *quadratic form* in three variables we mean

$$(1) \qquad F(x_1, x_2, x_3) = \sum_{i=1}^{3} \sum_{j=1}^{3} a_{ij} x_i x_j \qquad\qquad a_{ij} = a_{ji}$$

$$\begin{aligned}
= \; & a_{11}x_1^2 && + a_{12}x_1x_2 && + a_{13}x_1x_3 \\
+ \; & a_{21}x_2x_1 && + a_{22}x_2^2 && + a_{23}x_2x_3 \\
+ \; & a_{31}x_3x_1 && + a_{32}x_3x_2 && + a_{33}x_3^2
\end{aligned}$$

It is *positive definite* if, and only if, $F(x_1, x_2, x_3) > 0$, except when $x_1 = x_2 = x_3 = 0$. Clearly $F(0, 0, 0) = 0$. It is *positive semidefinite* if, and only if, $F \geqq 0$, the equality holding for certain values of x_1, x_2, x_3, not all zero. For example, if

$$F = x_1^2 + x_2^2 + x_3^2, \qquad G = x_1^2 + x_2^2$$

F is positive definite and G is positive semidefinite. Note that $G \geqq 0$, but $G(0, 0, 1) = 0$. If G were being considered as a form in two variables x_1, x_2, it would be positive definite. Negative definite forms may be defined, *mutatis mutandis*. It is a familiar fact that the form in two variables $Ax_1^2 + 2Bx_1x_2 + Cx_2^2$ is positive definite if, and only if,

$$(2) \qquad A > 0, \qquad \begin{vmatrix} A & B \\ B & C \end{vmatrix} > 0$$

We now develop a similar result for the form (1).

Lemma 4. *The form* (1) *is positive definite* $\Leftrightarrow$

$$(3) \qquad a_{11} > 0, \qquad \begin{vmatrix} a_{11} & a_{12} \\ a_{21} & a_{22} \end{vmatrix} > 0, \qquad \begin{vmatrix} a_{11} & a_{12} & a_{13} \\ a_{21} & a_{22} & a_{23} \\ a_{31} & a_{32} & a_{33} \end{vmatrix} > 0$$

We prove only the sufficiency of condition (3). Denote the three-rowed determinant (3) by Δ and the cofactor of its element a_{ij} by A_{ij}. By use of the formula for the product of two determinants, we have

$$\Delta \begin{vmatrix} 1 & 0 & 0 \\ 0 & A_{22} & A_{23} \\ 0 & A_{32} & A_{33} \end{vmatrix} = \begin{vmatrix} a_{11} & a_{21} & a_{31} \\ -a_{11}A_{21} & \Delta - a_{21}A_{21} & -a_{31}A_{21} \\ -a_{11}A_{31} & -a_{21}A_{31} & \Delta - a_{31}A_{31} \end{vmatrix}$$

$$= \begin{vmatrix} a_{11} & a_{21} & a_{31} \\ 0 & \Delta & 0 \\ 0 & 0 & \Delta \end{vmatrix} = a_{11}\Delta^2$$

Hence,

$$\begin{vmatrix} A_{22} & A_{23} \\ A_{32} & A_{33} \end{vmatrix} = a_{11}\Delta$$

Now collect terms in x_1^2 and in x_1 in the form (1) as follows:

$$F = Ax_1^2 + 2Bx_1 + C$$

$$A = a_{11}, \quad B = a_{12}x_2 + a_{13}x_3, \quad C = a_{22}x_2^2 + 2a_{23}x_2x_3 + a_{33}x_3^2$$

We shall show that $AC - B^2 > 0$ unless $x_2 = x_3 = 0$, and this will prove $F > 0$ by (2). If $x_2 = x_3 = 0$, $F = a_{11}x_1^2$, and this is positive unless x_1 is also zero, so that F is positive definite.

In $AC - B^2$ collect terms in x_2^2, in x_2x_3, and in x_3^2 as follows:

$$AC - B^2 = A_{33}x_2^2 - 2A_{23}x_2x_3 + A_{22}x_3^2$$

To show that this is always positive, unless $x_2 = x_3 = 0$, we again use (2). We need

$$A_{33} = \begin{vmatrix} a_{11} & a_{12} \\ a_{21} & a_{22} \end{vmatrix} > 0$$

$$\begin{vmatrix} A_{22} & A_{23} \\ A_{32} & A_{33} \end{vmatrix} = a_{11} \begin{vmatrix} a_{11} & a_{12} & a_{13} \\ a_{21} & a_{22} & a_{23} \\ a_{31} & a_{32} & a_{33} \end{vmatrix} > 0$$

But these facts follow at once by hypothesis.

We observe in passing one very important distinction between forms in two variables and forms in more than two variables. The former is positive semidefinite if, and only if, the sign $>$ is replaced by $\geqq$ in (2) (not both $>$). If a corresponding change is made in inequalities (3), a necessary but not a sufficient condition for (1) to be positive semidefinite is obtained. For, suppose all $a_{ij} = 1$, except $a_{33} = 0$. Then

$$a_{11} > 0, \quad \begin{vmatrix} a_{11} & a_{12} \\ a_{21} & a_{22} \end{vmatrix} = 0, \quad \begin{vmatrix} a_{11} & a_{12} & a_{13} \\ a_{21} & a_{22} & a_{23} \\ a_{31} & a_{32} & a_{33} \end{vmatrix} = 0$$

$$F = (x_1 + x_2 + x_3)^2 - x_3^2$$

$$F(1, 1, -2) = -4 < 0$$

4.2 RELATIVE EXTREMA

We can now establish our main result.

Theorem 4. 1. $f(x, y, z) \in C^2$

2. $f_1 = f_2 = f_3 = 0$ at (X, Y, Z)

3. $f_{11} > 0$, $\begin{vmatrix} f_{11} & f_{12} \\ f_{21} & f_{22} \end{vmatrix} > 0$, $\begin{vmatrix} f_{11} & f_{12} & f_{13} \\ f_{21} & f_{22} & f_{23} \\ f_{31} & f_{32} & f_{33} \end{vmatrix} > 0$

at (X, Y, Z)

$\Rightarrow$ $f(x, y, z)$ has a relative minimum at (X, Y, Z).

By Taylor's theorem, we have

$$\Delta f = f(X + h_1, Y + h_2, Z + h_3) - f(X, Y, Z)$$
$$= \tfrac{1}{2} \sum_{i=1}^{3} \sum_{j=1}^{3} f_{ij}(X + \theta h_1, Y + \theta h_2, Z + \theta h_3) h_i h_j$$

where $0 < \theta < 1$. By hypothesis 1 it is clear that inequalities 3 also hold in some neighborhood of (X, Y, Z). If the point $(X + h_1, Y + h_2, Z + h_3)$ is in this neighborhood, the coefficients of the quadratic form Δf will satisfy the conditions of Lemma 4, so that $\Delta f > 0$ throughout the neighborhood, except at $h_1 = h_2 = h_3 = 0$, where $\Delta f = 0$. Hence, f has a relative minimum at (X, Y, Z).

For a relative maximum the first and third of the inequalities of hypothesis 3 must be reversed.

EXAMPLE A. $f(x, y, z) = x^2 + y^2 + 3z^2 - xy + 2xz + yz$

$$f_1 = 2x - y + 2z$$
$$f_2 = -x + 2y + z$$
$$f_3 = 2x + y + 6z$$

Conditions 3 become for $X = Y = Z = 0$

$$2 > 0, \quad \begin{vmatrix} 2 & -1 \\ -1 & 2 \end{vmatrix} = 3 > 0, \quad \begin{vmatrix} 2 & -1 & 2 \\ -1 & 2 & 1 \\ 2 & 1 & 6 \end{vmatrix} = 4 > 0$$

Hence, $f(x, y, z) \geq f(0, 0, 0) = 0$

EXERCISES (4)

1. Discuss the behavior of the function

$$f(x, y, z) = x^4 - y^2 z^2 + xyz - x^2 - 2y^2 - z^2$$

at the origin.

2. Same problem if the sign of the term in x^2 is changed to plus. *Hint:* Consider the function of two variables $f(x, y, 0)$.

3. Find the distance from the point (a, b, c, d) in four dimensions to the hyperplane
$$Ax + By + Cz + Du + E = 0$$

4. Pass a curve
$$y = a + bx + cx^2$$
"through" n given points (x_i, y_i) by the method of least squares. Illustrate by the points $(-1, 1)$, $(0, 0)$, $(1, 1)$, $(3, 2)$. Plot.

5. Find the best shape of a wall tent. The two ends are closed by vertical rectangles capped on top by vertical isosceles triangles. There is no bottom. The existence of the extremum may be assumed.

6. Maximize the function $xyzw$ subject to the conditions
$$xyz + xyw + xzw + yzw = 1 \qquad x > 0, \quad y > 0, \quad z > 0, \quad w > 0$$

7. State without proof a sufficient condition for $f(x, y, z, t)$ to have a relative maximum (minimum).

8. Check your guess in the previous exercise by use of the two functions $\pm(x^2 + y^2 + z^2 + t^2)$, one of which clearly has a minimum, the other a maximum at $(0, 0, 0, 0)$.

9. Show by inspection that the function $(x - y)^2 + z^2$ has a whole "ridge" of minima along the line $y = x$, $z = 0$. Since none of these are relative minima in the sense of Definition 2, some hypothesis of Theorem 4 must be violated at each. Which?

10. Prove the necessity of conditions (3) in Lemma 4. *Hint:* Choose the variables successively as $(1, 0, 0)$, $(-a_{21}, a_{11}, 0)$, (A_{31}, A_{32}, A_{33}).

§5. *Lagrange's Multipliers*

If the variables of a function which is to be maximized are not independent but are connected by one or more relations, no new theory is needed. The derivatives which are to be equated to zero can be computed by the methods of Chapter 1. However, the formal procedure can be freed of any consideration of which variables are to be regarded as independent by the introduction of extraneous parameters, known as *Lagrange's multipliers*. We shall illustrate the method in several cases, from which the general procedure may be inferred.

5.1 ONE RELATION BETWEEN TWO VARIABLES

A typical problem of elementary calculus is to maximize a function
$$(1) \qquad\qquad u = f(x, y)$$

where x and y are connected by an equation

(2) $$g(x, y) = 0$$

Let us suppose

$$f, g \in C^1, \qquad g_1^2 + g_2^2 > 0$$

in a region of the xy-plane. If g_2 is not zero, we may solve equation (2) for y and substitute in equation (1), thus regarding x as the independent variable. A necessary condition for a maximum (or minimum) is thus seen to be

$$\frac{du}{dx} = f_1 - f_2 \frac{g_1}{g_2} = 0$$

The points desired will then be included among the simultaneous solutions of the equations

(3) $$\frac{\partial(f, g)}{\partial(x, y)} = 0, \qquad g(x, y) = 0$$

On the other hand, if g_1 is not zero, we take y as the independent variable. But in this case we are led to the same pair of equations (3).

To solve the same problem by the method of Lagrange, introduce the Lagrange multiplier λ, forming the function

$$V = f(x, y) + \lambda g(x, y)$$

We now proceed as if x and y were independent variables and set

(4) $$\frac{\partial V}{\partial x} = f_1 + \lambda g_1 = 0$$

(5) $$\frac{\partial V}{\partial y} = f_2 + \lambda g_2 = 0$$

We can now solve at least one of these equations for λ and substitute in the other equation. Combining the result with equation (2) we arrive anew at equations (3). Thus, instead of solving the two equations (1), (2) for x and y, we must now solve the three equations (2), (4), (5) for x, y, and λ. We arrive at the same pairs (x, y). As we mentioned above, the advantage of the Lagrange method is only that it does not require us to discuss which variable is independent. We state our results as a theorem.

Theorem 5. 1. $f(x, y), g(x, y) \in C^1$ *in a domain* D

2. $g_1^2 + g_2^2 > 0$ *in* D

$\Rightarrow$ *The set of points* (x, y) *on the curve* $g(x, y) = 0$, *where* $f(x, y)$ *has maxima or minima, is included in the set of simultaneous solutions* (x, y, λ) *of the equations*

$$f_1(x, y) + \lambda g_1(x, y) = 0$$
$$f_2(x, y) + \lambda g_2(x, y) = 0$$
$$g(x, y) = 0$$

Observe that a domain includes no boundary points. Hence, we are excluding from consideration the type of extremum that can occur on the boundary of a region and for which the derivatives in question need not vanish.

EXAMPLE A. Find the rectangle of perimeter l which has maximum area. If the lengths of the sides are x and y, then

$$V = xy + \lambda(2x + 2y - l)$$

Equations (4) and (5) become

$$y + 2\lambda = 0$$
$$x + 2\lambda = 0$$

The solution of equations (2), (4), (5) is $x = y = l/4$, $\lambda = -l/8$, so that the rectangle of maximum area is a square.

EXAMPLE B. An instructive example is that of finding the shortest distance from the point $(1, 0)$ to the parabola $y^2 = 4x$. We must minimize the function

$$u = (x - 1)^2 + y^2$$

where

$$y^2 = 4x$$

If we eliminate y and set du/dx equal to zero, we find $x = -1$, an absurd result since the parabola has no real point with negative abscissa. The valid range is $x \geqq 0$, and the minimum occurs at $x = 0$, where the derivative du/dx is not zero.

The method of Lagrange is applicable, however. Take the domain D as the entire xy-plane. Then

$$V = (x - 1)^2 + y^2 + \lambda(y^2 - 4x)$$

and we must solve the system

$$2(x - 1) - 4\lambda = 0$$
$$2y + 2\lambda y = 0$$
$$y^2 - 4x = 0$$

From the second equation either $y = 0$ or $\lambda = -1$. The latter must be rejected since it would lead to $x = -1$. Hence, the only real solution is $x = 0$, $y = 0$, $\lambda = -1/2$, and the required distance is unity. Note that we could not eliminate λ from the system by solving the second equation thereof for λ. For, $g_2(x, y) = 0$ at the very point which yields the minimum. This is, of course, mirrored in the fact, observed above, that x is not a suitable independent variable. The strength of the Lagrange method in not singling out any variable as independent is thus brought forcefully to our attention.

5.2 ONE RELATION AMONG THREE VARIABLES

We next consider the case

$$u = f(x, y, z)$$

$$g(x, y, z) = 0$$

$$g_1^2 + g_2^2 + g_3^2 > 0$$

It is easily seen by elimination that the desired extrema will lie among the simultaneous solutions of one of the three systems:

$$\begin{cases} g = 0 \\ \dfrac{\partial(f, g)}{\partial(x, y)} = 0 \\ \dfrac{\partial(f, g)}{\partial(x, z)} = 0 \end{cases} \qquad \begin{cases} g = 0 \\ \dfrac{\partial(f, g)}{\partial(y, x)} = 0 \\ \dfrac{\partial(f, g)}{\partial(y, z)} = 0 \end{cases} \qquad \begin{cases} g = 0 \\ \dfrac{\partial(f, g)}{\partial(z, x)} = 0 \\ \dfrac{\partial(f, g)}{\partial(z, y)} = 0 \end{cases}$$

according as it is g_1, g_2, or g_3 which is different from zero.

If we look for extrema of the following function of three variables x, y, z,

$$V = f(x, y, z) + \lambda g(x, y, z)$$

we are led to the system

$$g = 0$$
$$f_1 + \lambda g_1 = 0$$
$$f_2 + \lambda g_2 = 0$$
$$f_3 + \lambda g_3 = 0$$

We can solve at least one of these for λ and thus arrive at one of the above systems.

EXAMPLE C.   Find the rectangular parallelepiped of surface area a^2 and maximum volume.   We have

$$V = xyz + \lambda(2xy + 2yz + 2zx - a^2)$$
$$yz + \lambda(2y + 2z) = 0$$
$$xz + \lambda(2x + 2z) = 0$$
$$xy + \lambda(2x + 2y) = 0$$

Since the variables x, y, z must all be positive, no coefficient of λ is zero, so that

$$\frac{x}{y} = \frac{x + z}{y + z}, \qquad \frac{y}{z} = \frac{x + y}{x + z}$$

whence $x = y = z = \dfrac{a}{\sqrt{6}}$, $\lambda = -\dfrac{a}{4\sqrt{6}}$.   The desired solid is a cube.

5.3 TWO RELATIONS AMONG THREE VARIABLES

The next case to be considered is

$$u = f(x, y, z)$$

$$g(x, y, z) = 0$$

$$h(x, y, z) = 0$$

(6)
$$\left[\frac{\partial(g, h)}{\partial(x, y)}\right]^2 + \left[\frac{\partial(g, h)}{\partial(y, z)}\right]^2 + \left[\frac{\partial(g, h)}{\partial(z, x)}\right]^2 > 0$$

There is now a single independent variable which must be chosen in accordance with the Jacobian which is not zero. All three cases lead to the system

(7)
$$g = h = \frac{\partial(f, g, h)}{\partial(x, y, z)} = 0$$

The Lagrange method introduces two parameters λ and μ and leads to the system of five equations in x, y, z, λ, μ,

$$f_1 + \lambda g_1 + \mu h_1 = 0$$
$$f_2 + \lambda g_2 + \mu h_2 = 0$$
$$f_3 + \lambda g_3 + \mu h_3 = 0$$
$$g = 0$$
$$h = 0$$

Under conditions (6) this system is easily seen to reduce to the system (7) when λ and μ are eliminated.

EXAMPLE D. Show that the shortest distance from a point to a line in space is the perpendicular distance. In vector notation, we have as equations of the given line

$$\vec{a} \cdot \vec{x} = k$$
$$\vec{b} \cdot \vec{x} = l$$

$$\vec{a} \times \vec{b} \neq 0$$

Let c: (c_1, c_2, c_3) be a point off the given line. The letters k and l represent scalars.

$$V = (x - c) \cdot (x - c) + \lambda[(a \cdot x) - k] + \mu[(b \cdot x) - l]$$

The system to be solved for $x_1, x_2, x_3, \lambda, \mu$ is

$$2(x_i - c_i) + \lambda a_i + \mu b_i = 0 \qquad\qquad i = 1, 2, 3$$

$$(a \cdot x) = k$$

$$(b \cdot x) = l$$

Eliminating λ and μ from the first three equations, which we may do since the vectors a and b are not parallel, we get

$$(a \quad b \quad x - c) = 0 = (a \times b) \cdot (x - c)$$

That is, the vector $x - c$ is perpendicular to the vector $a \times b$, as we wished to prove.

EXERCISES (5)

In the following problems use the Lagrange method. No discussion of the existence of the maximum or minimum is expected unless expressly required.

1. Prove that the length of the shortest line from the point $(0, 0, 25/9)$ to the surface $z = xy$ is $\sqrt{41}/3$. What angle does this shortest line make with the z-axis?

2. Find the point on the parabola $y^2 = 2x$, $z = 0$ which is nearest the plane $z = x + 2y + 8$. Show that this minimum distance is $\sqrt{6}$.

3. Show that the minimum of the sum of the squares of the distances of a point from the four faces of a tetrahedron is nine times the square of the volume of the tetrahedron divided by the sum of the squares of the areas of its faces. Why is this problem equivalent to Exercise 3 of §4 with $a = b = c = d = 0$?

4. Derive the plane formula for distance from point to line.

5. Find the direction of the axes of the ellipse

$$5x^2 - 6xy + 5y^2 - 4x - 4y - 4 = 0$$

by maximizing (minimizing) the distance to the center.

6. Same problem for the general ellipse.

7. Find the largest and the smallest distances from $(0, 0, 0)$ to the ellipsoid

$$\frac{x^2}{a^2} + \frac{y^2}{b^2} + \frac{z^2}{c^2} = 1 \qquad\qquad a < b < c$$

8. Express the number 12 as the sum of three parts x, y, z so that xy^2z^2 shall be a maximum.

9. Discuss the Lagrange method for a function of four variables bound by two conditions.

10. Same problem with three conditions.

11. Find the rectangular parallelepiped of maximum volume, the sum of the lengths of all the edges being given. (Show the existence of the maximum.)

12. A function z is defined implicitly by the equations

$$f(x, y, z) = 0$$
$$g(x, y, z) = 0$$

Obtain a necessary condition that z should have a maximum or a minimum.

13. By the result of Exercise 12, find the highest and lowest points of the circle
$$x^2 + y^2 + z^2 = 16$$
$$(x + 1)^2 + (y + 1)^2 + (z + 1)^2 = 27$$

14. Find the minimum distance from the origin to the surface
$$(x - y)^2 - z^2 = 1$$

15. In Exercise 10 of §2 denote the length of the equal legs of the triangle by x, of the base by $2y$, of the ridge pole by z. Find the maximum volume V for a fixed surface area 2:
$$V = yz\sqrt{x^2 - y^2}, \qquad xz + y\sqrt{x^2 - y^2} = 1$$
Solve by eliminating y. *Hint:* The meaningful range of the variables x, z is:
$$0 \leqq z, \quad 0 \leqq x, \qquad 2(1 - xz) \leqq x^2$$

16. In the previous exercise denote the surface area by $F(x, y, z)$ and show that $F_2(x, y, z)$ vanishes at the point (x, y, z) which makes V maximum.

§6. Families of Plane Curves

By a *family* of curves is usually meant an infinite set of curves. In most cases the curves are all of the same type; for example, all circles or all parabolas, the individuals of the family differing only in size or position. If each individual of a family of plane curves has attached to it a number α, we may represent the whole family by the single equation

(1) $f(x, y, \alpha) = 0$

If we set α equal to the value corresponding to a given curve, equation (1) is to reduce to the equation of that curve.

An example of a family of lines is the set of all tangent lines to the unit circle, center at the origin. Let us take as the parameter α attached to a given line the angle which the normal through the origin makes with the positive x-axis. Then equation (1) becomes

(2) $x \cos \alpha + y \sin \alpha = 1$

By the *envelope* of a family is meant a curve touched by all members of the family. In the above example, the unit circle itself is an envelope of the family (2). Any curve is the envelope of all its tangents. We shall discuss here methods of finding envelopes of given families of curves.

6.1 ENVELOPES

We begin with a more precise definition of an envelope.

Definition 3. *The family of curves* (1) *has an envelope*

(3) $x = g(\alpha), \qquad y = h(\alpha)$

if, and only if, for each $\alpha = \alpha_0$ *the point* $(g(\alpha_0), h(\alpha_0))$ *of the curve* (3) *lies on the curve* $f(x, y, \alpha_0) = 0$ *and both curves have the same tangent line there.*

EXAMPLE A.   The family (2) has the unit circle

$$(4) \qquad\qquad x = \cos \alpha, \qquad y = \sin \alpha$$

as an envelope. For each α the point $(\cos \alpha, \sin \alpha)$ lies on the curve (2).
The slope of the line (2) for a given α is $-\cot \alpha$, and the slope of the unit
circle for the same α has the same value. The tangents are vertical if $\alpha = 0$
or if $\alpha = \pi$.

Theorem 6.   1. $f(x, y, \alpha), g(\alpha), h(\alpha) \in C^1$

2. $f_1^2 + f_2^2 \neq 0$

3. $(g')^2 + (h')^2 \neq 0$

4. $f(g(\alpha), h(\alpha), \alpha) \equiv 0$

5. $f_3(g(\alpha), h(\alpha), \alpha) \equiv 0$

$\Rightarrow$ *The family* (1) *has the curve* (3) *as an envelope.*

For each α the point $(g(\alpha), h(\alpha))$ lies on the curve (1) by hypothesis 4.
For each α the slope of the curve (1) is

$$\frac{dy}{dx} = -\frac{f_1}{f_2} \qquad\qquad f_2 \neq 0$$
$$= \infty \qquad\qquad f_2 = 0$$

By hypotheses 4 and 5,

$$f_1 g' + f_2 h' + f_3 = f_1 g' + f_2 h' = 0$$

Hence,

$$(5) \qquad\qquad -\frac{f_1}{f_2} = \frac{h'}{g'} \qquad\qquad g' \neq 0$$
$$= \infty \qquad\qquad g' = 0$$

Since the right-hand side of (5) is precisely the slope of the curve (3), the
proof is complete. It is clear that when f_2 vanishes f_1 does not and that then
g' must also vanish. Both slopes are then infinite.

This theorem provides a simple method of determining the functions g
and h. We have only to solve the equations

$$(6) \qquad\qquad f(x, y, \alpha) = 0, \qquad f_3(x, y, \alpha) = 0$$

as simultaneous equations in x and y. In Example A, these equations
become

$$x \cos \alpha + y \sin \alpha = 1$$
$$-x \sin \alpha + y \cos \alpha = 0$$

The solution of the system is given precisely by equations (4).

We observe that the conditions of the theorem are not necessary. The family of curves

$$f = x - \sqrt{y} + \alpha = 0$$

obtained by translating half a parabola parallel to the x-axis, clearly has the x-axis as an envelope. But the theorem is not applicable, since $f \notin C^1$. Moreover, no simultaneous solution of equations (6) exists, since $f_3 \equiv 1$. If the entire parabola is translated, the method is applicable:

$$f = (x + \alpha)^2 - y = 0$$
$$\frac{\partial f}{\partial \alpha} = 2(x + \alpha) = 0$$

We clearly obtain the x-axis as an envelope.

EXAMPLE B. Find the envelope of the family of lines, the sum of the squares of whose intercepts is unity. The family has equations

(7) $$\frac{x}{a} + \frac{y}{b} = 1$$

(8) $$a^2 + b^2 = 1$$

Here it is convenient to retain b as an auxiliary parameter. Then

$$-\frac{x}{a^2} - \frac{y}{b^2}\frac{db}{da} = -\frac{x}{a^2} + \frac{ay}{b^3} = 0$$

(9) $$\frac{x^{1/3}}{a} = \frac{y^{1/3}}{b}$$

Solving equations (7) and (9) for a and b, we obtain

$$a = x^{1/3}(x^{2/3} + y^{2/3}), \qquad b = y^{1/3}(x^{2/3} + y^{2/3})$$

Substituting these values in equation (8), we find the equation of the locus,

$$x^{2/3} + y^{2/3} = 1$$

6.2 CURVE AS ENVELOPE OF ITS TANGENTS

Let a plane curve, not a straight line, be given in the form

(10) $$y = f(x), \qquad f \in C^2$$

The family of its tangents is

$$y - f(\alpha) - f'(\alpha)(x - \alpha) = 0$$

To obtain the envelope of the family, we must solve this equation with the equation

$$-f''(\alpha)(x - \alpha) = 0$$

for x and y. Since $f''(\alpha)$ is not identically zero, the solution must be

$$x = \alpha, \qquad y = f(\alpha)$$

a pair of equations that represents the given curve.

6.3 EVOLUTE AS ENVELOPE OF NORMALS

Consider the family of normals to the curve (10),

$$x - \alpha + f'(\alpha)(y - f(\alpha)) = 0$$

Differentiate with respect to α,

$$-1 + f''(\alpha)(y - f(\alpha)) - f'(\alpha)^2 = 0$$

and solve for x, y:

$$x = \alpha - \frac{f'(1 + f'^2)}{f''}$$

$$y = f + \frac{1 + f'^2}{f''}$$

But these are the parametric equations of the evolute of the given curve. In elementary calculus, the evolute is defined as the locus of the centers of curvature. It is shown that the normal to the curve is tangent to the evolute. That result has now been verified by the present methods.

EXERCISES (6)

In the first five problems, find the envelopes of the families of lines described. Plot several of the lines.

1. $\alpha y = \alpha^2 x + 1$.

2. $2\alpha y = 2x + \alpha^2$.

3. The sum of the intercepts is constant.

4. The sum of the intercepts is equal to their product.

5. The area of the triangle made with the axes is constant.

6. Show that the curve (3) is the envelope of its tangents.

7. Find the evolute of the curve (3).

8. Find the evolute of the ellipse given in parametric form.

9. State and prove a result analogous to Theorem 6 for a two-parameter family of surfaces,

$$f(x, y, z, \alpha, \beta) = 0$$

10. Prove that a surface

$$z = f(x, y)$$

is the envelope of its tangent planes. You may assume that the surface is not developable; that is,

$$f_{11}f_{22} \neq f_{12}^2$$

11. Find the envelope of the family of spheres

$$(x - a)^2 + (y - \beta)^2 + (z - \beta + 2)^2 = 2$$

12. Solve the same problem for

$$(x - \alpha)^2 + y^2 + (z - \beta)^2 = 2\alpha + 2\beta$$

§7. Families of Surfaces

We shall discuss in this section one-parameter families of surfaces in a manner analogous to that used in the previous section for families of curves. An example of such a family is the set of all tangent planes to a cone. Since it will be convenient to employ vector analysis, we shall revert to the notation of Chapter 3. Consider the cone

$$x_1^2 + x_2^2 = x_3^2$$

which may be written in parametric form

$$x_1 = u \cos v, \qquad x_2 = u \sin v, \qquad x_3 = u$$

The tangent plane at a point (u, v) of the surface is

$$x_1 \cos v + x_2 \sin v = x_3$$

This does not depend on u. The equation represents a one-parameter family of planes. The general one-parameter family of surfaces will have the form

(1) $$f(x_1, x_2, x_3, t) = 0$$

the parameter being t.

7.1 ENVELOPES OF FAMILIES OF SURFACES

We begin with a definition.

Definition 4. *The family of surfaces* (1) *has an envelope*

(2) $$\vec{x} = \vec{g}(t, u)$$

if, and only if, for each $t = t_0$ *the curve* $\vec{x} = \vec{g}(t_0, u)$ *lies on the surface* $f(x_1, x_2, x_3, t_0) = 0$ *and if along that curve the surface* (1) *(with* $t = t_0$) *and the surface* (2) *have the same tangent planes.* *The curve* $\vec{x} = \vec{g}(t_0, u)$ *is called the characteristic curve of the surface* $f(x_1, x_2, x_3, t_0) = 0$.

In the above example, the characteristic curve of the plane

$$x_1 \cos v_0 + x_2 \sin v_0 = x_3$$

will be the straight line

$$x_1 = u \cos v_0, \qquad x_2 = u \sin v_0, \qquad x_3 = u$$

For, this line lies in the plane, and its locus, when v_0 varies, is precisely the original cone, which is the envelope of the family of planes.

Theorem 7. 1. $f(x_1, x_2, x_3, t)$, $g_i(t, u) \in C^1$ $i = 1, 2, 3$

2. $\vec{\nabla} f \neq 0$

3. $\vec{g}_t \times \vec{g}_u \neq 0$

4. $f(g_1, g_2, g_3, t) \equiv 0$

5. $f_4(g_1, g_2, g_3, t) \equiv 0$

$\Rightarrow$ *the family* (1) *has the surface* (2) *as an envelope.*

Hypothesis 2 means that the partial derivatives of f, with respect to the first three variables, do not vanish simultaneously for any t. In 4 and 5, the identities are in both t and u. By 4, it is clear that the curve $x = g(t_0, u)$ lies on the surface $f(x_1, x_2, x_3, t_0) = 0$. Differentiating identity 4 first with respect to t, using 5, and then with respect to u, we have

$$(\vec{\nabla} f \cdot \vec{g}_t) = 0, \qquad (\vec{\nabla} f \cdot \vec{g}_u) = 0$$

Hence, the normal to the surface $f(x_1, x_2, x_3, t_0) = 0$ has the direction of the vector

$$\vec{g}_t \times \vec{g}_u|_{(t_0, u)}$$

But this is also the direction of the normal to the surface (2) along the curve where $t = t_0$, and the proof is complete.

This theorem provides the means of finding envelopes. The pair of equations

$$f(x_1, x_2, x_3, t) = 0, \qquad f_4(x_1, x_2, x_3, t) = 0$$

determines the characteristic curve for any fixed t as the intersection of two surfaces. The locus of these curves, as t varies, is the desired envelope. Its equation is obtained by eliminating t between the two equations. In the example of the cone, the equations are

$$x_1 \cos v + x_2 \sin v = x_3, \qquad -x_1 \sin v + x_2 \cos v = 0$$

By squaring both sides of these equations and adding, we get

$$x_1^2 + x_2^2 = x_3^2$$

the equation of the envelope.

7.2 DEVELOPABLE SURFACES

The envelope of a one-parameter family of planes is called a *developable surface*. Thus, a cone is a developable surface, as we saw above. Clearly, a cylinder has the same property. It can be shown that any developable surface, like the cylinder and cone, can be cut along a straight line of the surface and then rolled out onto a plane without tearing or stretching. We shall now illustrate a third type of developable surface.

Definition 5. *The locus of the tangent lines to a space curve is called the tangent surface to the curve.*

EXAMPLE A.   Consider the twisted cubic

$$x_1 = t, \qquad x_2 = t^2, \qquad x_3 = t^3$$

The tangent has the direction of the vector whose components are 1, $2t$, $3t^2$. For $t = t_0$, the tangent line has equations

$$x_1 = t_0 + u, \qquad x_2 = t_0^2 + 2t_0 u, \qquad x_3 = t_0^3 + 3t_0^2 u$$

If now we allow t_0 to vary, these equations represent the tangent surface to the twisted cubic.

We shall show that the envelope of the osculating planes of a given curve is the tangent to the curve. Let the curve be given by the equation

$$x = x(s)$$

where the parameter is the arc length. The family of osculating planes is

$$(X - x(s)) \cdot \gamma = 0, \qquad \gamma = \frac{x' \times x''}{|x''|}$$

Differentiating with respect to s, we have, by use of the Frenet-Serret formulas,

$$-\frac{1}{T}(X - x(s)) \cdot \beta - x'(s) \cdot \gamma = 0$$

Since $x'(s) = \alpha$ and $(\alpha \cdot \gamma) = 0$, this becomes

$$(X - x(s)) \cdot \beta = 0$$

whence

$$X - x(s) = u\beta \times \gamma$$

$$X = x(s) + u\alpha$$

But this is precisely the vector equation of the tangent surface to the given curve.

There are two other important developable surfaces connected with a curve. The envelope of the family of rectifying planes is called the *rectifying developable.* It can be shown that the original curve lies on it and that, if the surface is rolled out onto a plane, the curve becomes a straight line. It is from this property that the word rectifying derives. Finally, the envelope of the normal planes is called the *polar developable.*

EXERCISES (7)

Find the characteristic lines and the envelopes of the families described in the first five problems.

1. $(x_1 - t)^2 + x_2^2 + x_3^2 = 1$.

2. $x_1^2 + x_2^2 + (x_3 - t)^2 = t^2/2$.

3. $x_1 \cos \theta + x_2 \cos \theta + x_3 \sin \theta = \sqrt{2}$.

4. $x_1 \sin \theta - x_2 \cos \theta + x_3 = 0$.

5. $x_1^2 + x_2^2 + x_3^2 + 2tx_1 + kt^2 = 0$.

6. Find the equation of the rectifying developable of a curve. Show that the given curve lies on it.

7. Find the equation of the polar developable of a curve.

8. Show that the polar developable of a plane curve is a cylinder (not necessarily circular).

9. Find the envelope of the family

$$t^3 - 3t^2x_1 + 3tx_2 - x_3 = 0$$

10. Does the family of spheres

$$(x_1 - t)^2 + x_2^2 + x_3^2 = t^2$$

have an envelope? Is Theorem 7 applicable?

5

Stieltjes Integral

§1. Introduction

The student is assumed to be familiar with the ordinary theory of the definite integral. The Stieltjes integral is, however, only a slight generalization of that familiar integral, so that what follows may be used as a review or solidification of the classical theory. The student has only to replace the *integrator function* $\alpha(x)$ of the present chapter by the function x in order to revert to the *Riemann integral*, which is referred to in elementary texts as the *integral as the limit of a sum*. By way of introduction we recall here the definition of and two basic theorems about the Riemann integral.

Definition. *Divide the interval* $a \leqq x \leqq b$ *into* n *subintervals by points*

$$a = x_0 < x_1 < x_2 < \ldots < x_n = b$$

and set

$$\delta = \max (x_1 - x_0, x_2 - x_1, \ldots, x_n - x_{n-1})$$

Choose points ξ_k, $x_{k-1} \leqq \xi_k \leqq x_k$, $k = 1, 2, \ldots, n$. *Then the Riemann integral of* $f(x)$ *with respect to* x *from* a *to* b *is*

$$(1) \qquad \int_a^b f(x)\, dx = \lim_{\delta \to 0} \sum_{k=1}^{n} f(\xi_k)(x_k - x_{k-1})$$

if this limit exists.

149

Theorem A. $f(x) \in C, a \leqq x \leqq b \Rightarrow \int_a^b f(x)\, dx$ exists.

Proof of this theorem is often omitted in introductory courses because of its dependence on the notion of *uniform continuity*. The student who wishes to postpone study of the Stieltjes integral may now turn to §6 and §7, replacing throughout $\alpha(x)$ by x. He will thus obtain the fundamental properties of continuous functions and a simple proof of Theorem A.

Theorem B.   1. $f(x) \in C$ $a \leqq x \leqq b$

           2. $F'(x) = f(x)$ $a \leqq x \leqq b$

$$\Rightarrow \qquad \int_a^b f(x)\, dx = F(b) - F(a)$$

This result is known as the *fundamental theorem of integral calculus*. We include a proof for purposes of review. By Theorem A the limit (1) exists uniquely, *no matter how* the points ξ_k are chosen. For our present convenience we choose them in a special way, namely, so that

$$F(x_k) - F(x_{k-1}) = F'(\xi_k)(x_k - x_{k-1}) \qquad x_{k-1} < \xi_k < x_k$$

This is possible by Theorem 2, Chapter 1. By hypothesis 2 the sum (1) becomes

$$\sum_{k=1}^n [F(x_k) - F(x_{k-1})] = F(b) - F(a)$$

That is, it does not change as $n \to \infty$, $\delta \to 0$, and hence has the stated limit.

For the definition of the Stieltjes integral (Definition 3 below) the factor $(x_k - x_{k-1})$ in (1) is replaced by $[\alpha(x_k) - \alpha(x_{k-1})]$, and the basic existence theorem states that the integral exists if $f(x) \in C$, $\alpha(x)$ is nondecreasing. Since the function x is increasing, it is clear why Theorem A is a special case of Theorem 1 below.

Although the Stieltjes definition differs little from the Riemann definition, the change is very important. The Stieltjes integral is an ideal tool in physical applications. It is a familiar fact that the ordinary integral enables one to define physical concepts involving continuous distribution of mass by analogy with corresponding concepts for a distribution of particles. For example, the formula for the moment of inertia of n particles on a line is

$$I = \sum_{k=1}^n m_k x_k^2$$

the moment of inertia of a continuous distribution of mass is

$$I = \int_a^b m(x) x^2 \, dx$$

But the two situations, one discrete and the other continuous, must be treated separately, sums being used in one case and integrals in the other.

However, the relation between the sign Σ and the sign $\int$ is more than analogy. By use of the Stieltjes integral, the two cases may be treated by a single formula. In fact, we may even use this generalized integral to take care of distributions of mass which are partly discrete, partly continuous. The integral is even more important in theoretical mathematics, chiefly because of this capacity for including both sums and limits of sums.

1.1 DEFINITIONS

As in the definition of a Riemann integral, we begin by dividing the interval of integration into subintervals. To simplify the writing we introduce certain terms and notations.

Definition 1. *A subdivision Δ of an interval (a, b) is a set of numbers $\{x_k\}_0^n$, or points, such that*

$$a = x_0 < x_1 < \ldots < x_n = b$$

A subdivision involving $n + 1$ points divides the interval into n adjoining subintervals (x_0, x_1), (x_1, x_2), $\ldots$, (x_{n-1}, x_n).

Definition 2. *The norm $\|\Delta\|$ of a subdivision Δ is*

$$\|\Delta\| = \max(x_1 - x_0, x_2 - x_1, \ldots, x_n - x_{n-1})$$

In other words, it is the length of the largest of the subintervals.

Definition 3. *The Stieltjes integral of $f(x)$ with respect to $\alpha(x)$ from a to b is*

$$(2) \qquad \int_a^b f(x)\, d\alpha(x) = \lim_{\|\Delta\|\to 0} \sum_{k=1}^n f(\xi_k)[\alpha(x_k) - \alpha(x_{k-1})]$$

where $\qquad x_{k-1} \leqq \xi_k \leqq x_k \qquad\qquad k = 1, 2, \ldots, n$

The left-hand side of equation (2) is the notation employed for the Stieltjes integral. It reduces to the usual notation for the classical integral if $\alpha(x) = x$, as it should in view of the right-hand side of equation (2). The notion of the limit (2) needs amplification. The norm $\|\Delta\|$ may indeed be regarded as an independent variable. But

$$\sigma_\Delta = \sum_{k=1}^n f(\xi_k)[\alpha(x_k) - \alpha(x_{k-1})]$$

is not a single-valued function of $\|\Delta\|$. For, there are clearly many different subdivisions all having the same norm. And even with a given Δ there are usually infinitely many values of σ_Δ corresponding to the infinitely many choices of the points ξ_k. When we say that

$$\lim_{\|\Delta\|\to 0} \sigma_\Delta = I$$

we mean that to an arbitrary positive number ϵ there corresponds a number δ such that

$$|\sigma_\Delta - I| < \epsilon$$

for *all* values of σ_Δ corresponding to *any* Δ whose norm is less than δ. It should be clearly understood that the limit (2) may or may not exist, depending on what functions $f(x)$ and $\alpha(x)$ are used. It is only when the limit exists that the integral is defined.

EXAMPLE A. $a = 0, b = 2$

$$f(x) = \alpha(x) = 0 \qquad\qquad 0 \leq x \leq 1$$
$$= 1 \qquad\qquad 1 < x \leq 2$$

Here the limit (2) does not exist, and the Stieltjes integral is undefined. For, let Δ be an arbitrary subdivision of $(0, 2)$. There is just one of the differences

$$\alpha(x_k) - \alpha(x_{k-1}) \qquad\qquad k = 1, 2, \ldots, n$$

that is different from zero. This difference has the value 1, say for $k = m$. Accordingly, σ_Δ can have the two values 0 or 1, depending on the way in which ξ_m is chosen. Clearly, σ_Δ, always having *two* values differing by 1, cannot approach any limit as $\|\Delta\|$ approaches zero.

EXAMPLE B.   Let a, b, and $\alpha(x)$ have the same definition as in Example A and let $f(x)$ be identically 1. In this case,

$$\int_0^2 f(x)\, d\alpha(x) = 1$$

For, the only nonvanishing term in any σ_Δ must have the unique value 1, regardless of the choice of the ξ_k.

1.2 EXISTENCE OF THE INTEGRAL

We now state a condition in which the limit (2) exists. We use the symbols $\uparrow$ and $\downarrow$ to indicate the classes of nondecreasing and nonincreasing functions, respectively.

Theorem 1.   1. $f(x) \in C$ $\qquad\qquad\qquad\qquad a \leq x \leq b$

2. $\alpha(x) \in \uparrow$ $\qquad\qquad\qquad\qquad a \leq x \leq b$

$\Rightarrow \qquad\qquad\qquad \int_a^b f(x)\, d\alpha(x)$ *exists*

The proof of this theorem depends on some of the more delicate properties of continuous functions and will be deferred until later. The meaning of the result is entirely clear without the proof.   Obviously, hypothesis 2 may be replaced by $\alpha(x) \in \downarrow$.

EXAMPLE C. $a = 0,\quad b = 1,\quad f(x) = x,\quad \alpha(x) = x^2$
Since $x \in C$ and $x^2 \in \uparrow$ in $0 \leq x \leq 1$, we know by Theorem 1 that

$$\int_0^1 x \, dx^2$$

exists.   Let us find its value.   Since the limit (2) exists independently of the manner of subdivision and of the choice of the points ξ_k, we may make our choice in any convenient way.   Let us choose the subintervals all equal and choose $\xi_k = x_k$.   Then

(3) $$\int_0^1 f(x) \, d\alpha(x) = \lim_{n \to \infty} \sum_{k=1}^n f\left(\frac{k}{n}\right)\left[\alpha\left(\frac{k}{n}\right) - \alpha\left(\frac{k-1}{n}\right)\right]$$

$$\int_0^1 x \, dx^2 = \lim_{n \to \infty} \sum_{k=1}^n \frac{k}{n}\frac{2k-1}{n}\frac{1}{n}$$

But

(4) $$\sum_{k=1}^n k^2 = \frac{n(n+1)(2n+1)}{6}, \qquad \sum_{k=1}^n k = \frac{n(n+1)}{2}$$

Hence,

$$\int_0^1 x \, dx^2 = \lim_{n \to \infty} \frac{1}{n^3}\left[\frac{n(n+1)(2n+1)}{3} - \frac{n(n+1)}{2}\right] = \frac{2}{3}$$

EXERCISES   (1)

1. In Example B change $f(x)$ to any continuous function.   Show that the integral still exists and has the value $f(1)$.

2. Let $f(x) \in C$ in $a \leq x \leq b$; $\alpha(a) = c$, $\alpha(x) = c + h$ in $a < x \leq b$.   Show that

$$\int_a^b f(x) \, d\alpha(x) = f(a)h$$

3. Change $\alpha(x)$ in Exercise 2 to a step-function with a single jump at b.

4. Change $\alpha(x)$ in Exercise 2 to a step-function with a single jump at an interior point.

5. Let $\alpha(x)$ and $f(x)$ both be step-functions, both having a single jump at a common point c, $a \leq c \leq b$.   Show that the limit (2) cannot exist.

6. Evaluate the limit (3) of Example C by applying the law of the mean to the difference $\alpha(k/n) - \alpha([k-1]/n)$.   Then, using Theorem 9 below, show that the limit is an ordinary integral.   Evaluate it by Theorem B.

7. Prove equation (4) by induction.

8. Prove that

$$\sum_{k=1}^{n} k(k+1) = \frac{n(n+1)(n+2)}{3}$$

by use of the relation

$$k(k+1) = \frac{[k(k+1)(k+2) - (k-1)k(k+1)]}{3}$$

9. Prove equation (4) by use of Exercise 8 and the relation

$$k^2 = k(k+1) - k$$

10. Prove

$$\sum_{k=1}^{n} k^3 = \frac{n^2(n+1)^2}{4}$$

11. Find

$$\sum_{k=1}^{n} k^4$$

Hint: Write

$$k^4 = k(k+1)(k+2)(k+3) + ak(k+1)(k+2) + bk(k+1) + ck$$

and use the method of Exercise 8.

12. Evaluate

$$\int_0^1 x^2 \, dx^2$$

by the method of Example C.

13. Solve the same problem for

$$\int_0^2 x \, dx^2$$

14. Solve the same problem for

$$\int_0^3 x \, dx^2$$

15. Solve the same problem for

$$\int_0^1 x \, dx^3$$

16. Verify the answer of Exercise 12 by the method of Exercise 6.

17. Solve the same problem for Exercise 15.

18. If $\alpha_1(x) = \alpha(x) + k$ in (a, b) except that $\alpha_1(a) = \alpha(a)$, compare the two integrals

$$\int_a^b f(x) \, d\alpha_1(x), \qquad \int_a^b f(x) \, d\alpha(x)$$

§2. *Properties of the Integral*

We collect here some of the elementary properties of the Stieltjes integral. The proofs of these are almost identical with the corresponding ones for the Riemann integral and are omitted. We show how the Stieltjes integral may reduce to a sum or to a Riemann integral under certain circumstances.

2.1 A TABLE OF PROPERTIES

In the following list k is a constant, the functions $f(x)$ and $\alpha(x)$, with or without subscripts, are, respectively, continuous and nondecreasing in $a \leqq x \leqq b$.

$$\text{I. } \int_a^b d\alpha(x) = \alpha(b) - \alpha(a).$$

$$\text{II. } \int_a^b f(x)\, d[\alpha(x) + k] = \int_a^b f(x)\, d\alpha(x).$$

$$\text{III. } \int_a^b kf(x)\, d\alpha(x) = k\int_a^b f(x)\, d\alpha(x).$$

$$\text{IV. } \int_a^b [f_1(x) + f_2(x)]\, d\alpha(x) = \int_a^b f_1(x)\, d\alpha(x) + \int_a^b f_2(x)\, d\alpha(x).$$

$$\text{V. } \int_a^b f(x)\, d[\alpha_1(x) + \alpha_2(x)] = \int_a^b f(x)\, d\alpha_1(x) + \int_a^b f(x)\, d\alpha_2(x).$$

$$\text{VI. } \int_a^b f(x)\, d\alpha(x) = \int_a^c f(x)\, d\alpha(x) + \int_c^b f(x)\, d\alpha(x) \qquad a < c < b$$

$$\text{VII. } f_1(x) \leqq f_2(x) \Rightarrow \int_a^b f_1(x)\, d\alpha(x) \leqq \int_a^b f_2(x)\, d\alpha(x).$$

$$\text{VIII. } \left| \int_a^b f(x)\, d\alpha(x) \right| \leqq \int_a^b |f(x)|\, d\alpha(x).$$

$$\text{IX. } \left| \int_a^b f(x)\, d\alpha(x) \right| \leqq [\alpha(b) - \alpha(a)] \max_{a \leqq x \leqq b} |f(x)|.$$

Properties I-VII follow from Definition 3 and Theorem 1. For example, IV and V result from the fact that the limit of a sum is a sum of limits. The same fact is used for VI, but here the subdivision Δ must be chosen to include the point c. This is permissible, for the very meaning of Theorem 1 is that the defining limit exists *independently* of the choice Δ (provided only that $\|\Delta\| \to 0$).

Property VIII follows from VII by use of the inequalities

$$-|f(x)| \leqq f(x) \leqq |f(x)| \qquad a \leqq x \leqq b$$

Property IX is proved by use of VIII, VII, III, and I.   Observe that all these properties could be used, with slight modification, in the case in which the functions $\alpha(x) \in \downarrow$.

2.2 SUMS

Let $\alpha(x)$ be a step-function with jumps at the points c_k of amounts h_k, where

$$a < c_1 < c_2 < \ldots < c_n < b$$

That is, $\alpha(x)$ is constant in the subintervals created by the introduction of the points c_k and

$$\alpha(c_k+) - \alpha(c_k-) = h_k \qquad\qquad k = 1, 2, \ldots, n$$

Let $f(x) \in C$ in $a \leqq x \leqq b$.   Then

$$(1) \qquad\qquad \int_a^b f(x)\, d\alpha(x) = \sum_{k=1}^n h_k f(c_k)$$

This can be proved by use of properties V or VI combined with Exercise 4 of §1.

2.3 RIEMANN INTEGRALS

Let $f(x) \in C$ and $\alpha(x) \in C^1$ in $a \leqq x \leqq b$.   Then

$$(2) \qquad\qquad \int_a^b f(x)\, d\alpha(x) = \int_a^b f(x)\alpha'(x)\, dx$$

The integral on the right is an ordinary Riemann integral.   To prove equation (2) we have by the law of the mean

$$\sigma_\Delta = \sum_{k=1}^n f(\xi_k)\alpha'(\eta_k)(x_k - x_{k-1}) \qquad\qquad x_{k-1} < \eta_k < x_k$$

The result is now immediate by use of Duhamel's theorem (see Theorem 9).

2.4 EXTENSIONS

In the table of §2.1 it was assumed that the functions $f(x) \in C$ and that the functions $\alpha(x) \in \uparrow$.   Under these conditions all integrals appearing exist by Theorem 1.   Properties I to VI still hold, as one can easily prove, without these conditions, provided only that all integrals appearing are known to exist.   In fact, a property like V still holds if only two of the integrals appearing are assumed to exist, for then the third does also by virtue of the theorem concerning the limit of a sum of two variables.   As a consequence, we see that, if $f(x) \in C$ and $\alpha(x) = \alpha_1(x) + \alpha_2(x)$, where $\alpha_1(x) \in \uparrow$, $\alpha_2(x) \in \downarrow$, then $\int_a^b f(x) \, d\alpha(x)$ exists and

(3) $$\int_a^b f(x) \, d\alpha(x) = \int_a^b f(x) \, d\alpha_1(x) + \int_a^b f(x) \, d\alpha_2(x)$$

Definition 4. *A function $\alpha(x)$ is of bounded variation in an interval* $a \leq x \leq b \Leftrightarrow$

$$\alpha(x) = \alpha_1(x) + \alpha_2(x)$$

where $\alpha_1(x) \in \uparrow$ and $\alpha_2(x) \in \downarrow$ in $a \leq x \leq b$.

EXAMPLE A. The function $\sin x$ is of bounded variation in $0 \leq x \leq \pi$. For we may take

$$\alpha_1(x) = \sin x \qquad\qquad 0 \leq x \leq \pi/2$$
$$= 1 \qquad\qquad\qquad \pi/2 \leq x \leq \pi$$

$$\alpha_2(x) = 0 \qquad\qquad\quad 0 \leq x \leq \pi/2$$
$$= \sin x - 1 \qquad\quad \pi/2 \leq x \leq \pi$$

We might equally well have defined $\alpha_1(x)$ to be 2 and $\alpha_2(x)$ to be $\sin x - 2$ in $\pi/2 \leq x \leq \pi$. Clearly, there are infinitely many possible definitions of $\alpha_1(x)$ and $\alpha_2(x)$.

In accordance with the above remarks, it is clear that, if $f(x) \in C$ and $\alpha(x)$ is of bounded variation in $a \leq x \leq b$, then the integral on the left of equation (3) exists and has the value given by the right-hand side.

EXAMPLE B. $f(x) \in C$ in $0 \leq x \leq 2$; $\alpha(x) = 1$ except that $\alpha(1) = 0$. Take

$$\alpha_1(x) = 0 \qquad\qquad 0 \leq x \leq 1$$
$$= 1 \qquad\qquad\quad 1 < x \leq 2$$

$$\alpha_2(x) = 1 \qquad\qquad 0 \leq x < 1$$
$$= 0 \qquad\qquad\quad 1 \leq x \leq 2$$

Then

$$\int_0^2 f(x) \, d\alpha_1(x) = f(1), \qquad \int_0^2 f(x) \, d\alpha_2(x) = -f(1)$$

$$\int_0^2 f(x) \, d\alpha(x) = f(1) - f(1) = 0$$

EXERCISES (2)

1. Under the assumptions of §2.1 prove properties I, II, III.

2. Same problem for properties IV and V.

3. Same problem for property VI. Explain what to do about a subdivision of (a, b), no point of which coincides with c.

4. Prove properties VII, VIII, and IX.

5. State and prove properties VII, VIII, and IX, if $\alpha(x) \in \downarrow$.

In the next three problems use the method of §2.3.

6. $\displaystyle\int_0^\pi x \, d\sin x = ?$

7. $\displaystyle\int_0^\pi \cos x \, d\sin x = ?$

8. $\displaystyle\int_{-\pi}^\pi e^{|x|} \, d\cos x = ?$

9. Prove equation (1) by both methods suggested in the text.

10. If $\alpha(x) = 2$ except in the interval $(-2, 2)$ where $\alpha(x) = x$, find

$$\int_{-4}^9 x^3 \, d\alpha(x)$$

11. Define $\alpha(x)$ so that

$$\int_0^{10} f(x) \, d\alpha(x) = f(0) - f(1) + 2f(5) - 3.7f(3.7) + 4f(10)$$

where $\qquad f(x) \in C \qquad$ in $\qquad 0 \leq x \leq 10$

12. In Example A, find

$$\int_0^\pi x \, d\alpha_1(x), \quad \int_0^\pi x \, d\alpha_2(x), \quad \int_0^\pi x \, d\alpha(x)$$

and verify equation (3). All properties of the table and equation (2) may be used whenever applicable.

13. In Example B find

$$\int_0^1 f \, d\alpha, \quad \int_1^2 f \, d\alpha$$

and check by property VI.

14. In Example B, find

$$\int_0^2 f(x) \, d\alpha(x)$$

directly from Definition 3.

15. Show that $\sin 3x$ is of bounded variation in $(0, \pi)$.

16. Same problem for $|\sin 3x|$.

17. If $\alpha(x)$ is of bounded variation in (a, b) and the points $\{x_k\}_0^n$ form a subdivision Δ of (a, b), show that there exists a constant M such that

$$\sum_{i=1}^n |\alpha(x_i) - \alpha(x_{i-1})| < M$$

for all Δ. This property can be shown to be equivalent to the defining property and is usually taken as the definition.

18. Define $\alpha(0) = 0$, $\alpha(x) = 0$ in the interval $1/(2k + 1) < x \leqq 1/(2k)$ and $\alpha(x) = 1$ in $1/(2k) < x \leqq 1/(2k - 1)$ for $k = 1, 2, \ldots$. Show how to choose a subdivision Δ of $0 \leqq x \leqq 1$ so that the sum of Exercise 17 is equal to n or $n - 1$. Thus, $\alpha(x)$ is not of bounded variation in $(0, 1)$.

§3. Integration by Parts

One of the most useful processes used in the theory of Stieltjes integrals is integration by parts. We develop the formula in the present section.

3.1 PARTIAL SUMMATION

Let Δ be an arbitrary subdivision of (a, b) and let σ_Δ be defined for a function $f(x)$ as in Definition 3,

$$(1) \qquad \sigma_\Delta = \sum_{k=1}^{n} f(\xi_k)[\alpha(x_k) - \alpha(x_{k-1})]$$

By rearranging the terms in this sum, we have

$$(2) \qquad \sigma_\Delta = \sum_{k=1}^{n-1} \alpha(x_k)[f(\xi_k) - f(\xi_{k+1})] - f(\xi_1)\alpha(x_0) + f(\xi_n)\alpha(x_n)$$

The process employed in the rearrangement is of frequent occurrence in mathematics and is called *partial summation*. The similarity of equation (2) to the familiar formula for integration by parts of Riemann integrals is evident.

Note that the sum (2) resembles closely the sum (1), the functions f and α being interchanged and the points ξ_k being replaced by x_k. But there is one important difference. The points $\{\xi_k\}_1^n$ do not form a subdivision of (a, b) since ξ_1 need not be a and ξ_n may differ from b. We can remedy this difficulty by defining $\xi_0 = a$ and $\xi_{n+1} = b$ and by adding and subtracting the terms $\alpha(x_0)f(\xi_0) = \alpha(a)f(a)$ and $\alpha(x_n)f(\xi_{n+1}) = \alpha(b)f(b)$ to the right-hand side of equation (2). We obtain

$$(3) \qquad \sigma_\Delta = \sum_{k=0}^{n} \alpha(x_k)[f(\xi_k) - f(\xi_{k+1})] + \alpha(b)f(b) - \alpha(a)f(a)$$

Now the points $\{\xi_k\}_0^{n+1}$ form a subdivision of (a, b), except for the fact that in certain cases two successive ξ_k may coincide. If this happens, the corresponding term of the sum (3) disappears. Hence, the sum (3) is always of the type appearing in Definition 3. Since the ξ_k and x_k occur alternately on the line, in so far as they do not coincide, it is clear that as $\|\Delta\| \to 0$ the norm of the subdivision formed by the points $\{\xi_k\}_0^{n+1}$ also $\to 0$.

3.2 THE FORMULA

We state the main result as a theorem.

Theorem 2.   1. $f(x) \in \uparrow$ $\qquad\qquad\qquad\qquad$ $a \leqq x \leqq b$

$\qquad\qquad$ 2. $\alpha(x) \in C$ $\qquad\qquad\qquad\qquad\qquad$ $a \leqq x \leqq b$

(4) $\Rightarrow \quad \displaystyle\int_a^b f(x)\, d\alpha(x) + \int_a^b \alpha(x)\, df(x) = \alpha(b)f(b) - \alpha(a)f(a)$

To prove this, let Δ be an arbitrary subdivision of (a, b).   Form the sum
(1) as prescribed by Definition 3.   Rewrite it in the form (3).   Let $\|\Delta\| \to 0$.
Then, by Theorem 1, the right-hand side of equation (3) approaches a limit.
Hence, the left-hand side does also, and we obtain equation (4).   Observe
that we have proved that a monotonic function is integrable with respect to
a continuous function.   Also, since $f(x)$ and $\alpha(x)$ appear symmetrically in
formula (4), it is clear that the hypotheses may be reversed to read $f(x) \in C$
and $\alpha(x) \in \uparrow$.   As in §2, we may replace monotonic functions by functions
of bounded variation.

Let us do Example B of §1 by the present method.   Clearly $f(x) \in C$
and $\alpha(x) \in \uparrow$.   By equation (4)

$$\int_0^2 f(x)\, d\alpha(x) = 1 - \int_0^2 \alpha(x)\, df(x)$$

The integral on the right is zero by Property II, §2, since $f(x)$ is constant.
The result may also be obtained immediately from property I.

EXAMPLE A.   Find $\displaystyle\int_{-1}^1 x\, d|x|$ by two methods.

By formula (4)

$$\int_{-1}^1 x\, d|x| = x|x| \Big|_{-1}^1 - \int_{-1}^1 |x|\, dx = 2 - 1 = 1$$

By Property VI

$$\int_{-1}^1 x\, d|x| = \int_{-1}^0 x\, d(-x) + \int_0^1 x\, dx = \tfrac{1}{2} + \tfrac{1}{2} = 1$$

EXERCISES (3)

*In the following exercises $[x]$ means the largest integer $\leqq x$.   For example,
$[\pi] = 3$ and $[-3/2] = -2$.*

1. $\displaystyle\int_0^1 x\, de^{2x} = ?$

2. $\displaystyle\int_{\pi/6}^{\pi/4} x\, d\tan x = ?$

3. $\displaystyle\int_0^5 (x^2 + 1)\, d[x] = ?$

4. $\displaystyle\int_0^5 e^x\, d\{x + [x]\} = ?$

5. $\displaystyle\int_{-2}^3 [|x|]\, d|x| = ?$

6. $\displaystyle\int_0^6 (x^2 + [x])\, d|3 - x| = ?$

7. $\displaystyle\int_{-1}^1 [|x|]\, d\,\frac{1}{1 + e^{-1/x}} = ?$

8. Compute the same integral from -2 to 2.

9. $\displaystyle\int_{1/4}^{3/4} [x]\, d[2x] = ?$

10. Does $\displaystyle\int_{1/4}^{5/4} [2x]\, d[x]$ exist?

11. Show that if $f(x) \in C^1$ and $\alpha(x) \in C$ in $a \leqq x \leqq b$ then

$$\int_a^b f(x)\, d\alpha(x)$$

exists.

12. If $f(x)$ and $g(x) \in C$ in $a \leqq x \leqq b$ and

$$\alpha(x) = \int_a^x g(t)\, dt \qquad\qquad a \leqq x \leqq b$$

prove that

$$\int_a^b f(x)\, d\alpha(x) = \int_a^b f(x)g(x)\, dx$$

§4. Laws of the Mean

As in the ordinary theory of integration, there are two very useful mean-value theorems for the Stieltjes integral. We shall prove them here. As corollaries, we shall obtain the familiar laws of the mean for Riemann integrals. This method of treating the Riemann integral as a special case of the Stieltjes integral is particularly useful in the proof of the second mean-value theorem since it avoids the partial summation necessary in the usual proof. That process is now subsumed, once for all, in the process of integration by parts.

4.1 FIRST MEAN-VALUE THEOREM

Theorem 3.   1. $f(x) \in C$ $a \leq x \leq b$

2. $\alpha(x) \in \uparrow$ $a \leq x \leq b$

(1) $\Rightarrow$ $\displaystyle\int_a^b f(x)\, d\alpha(x) = f(\xi) \int_a^b d\alpha(x)$ $a \leq \xi \leq b$

Set

$$M = \max_{a \leq x \leq b} f(x), \qquad m = \min_{a \leq x \leq b} f(x)$$

Then by Property VII, §2.1, we have

$$m \leq f(x) \leq M \qquad\qquad a \leq x \leq b$$

(2) $\displaystyle m[\alpha(b) - \alpha(a)] \leq \int_a^b f(x)\, d\alpha(x) \leq M[\alpha(b) - \alpha(a)]$

If $\alpha(b) = \alpha(a)$, then $\alpha(x)$ is constant and both sides of equation (1) are zero, no matter what value of ξ is chosen. Since the continuous function $f(x)$ takes on every value* between m and M in the interval (a, b), there is certainly one point ξ where it takes on the value

$$[\alpha(b) - \alpha(a)]^{-1} \int_a^b f(x)\, d\alpha(x)$$

which does lie between m and M, if $\alpha(b) \neq \alpha(a)$, by inequalities (2). That is,

$$f(\xi) = [\alpha(b) - \alpha(a)]^{-1} \int_a^b f(x)\, d\alpha(x) \qquad a \leq \xi \leq b$$

This completes the proof of the theorem.

Corollary 3.   1. $f(x), g(x) \in C$ $a \leq x \leq b$

2. $g(x) \geq 0$ $a \leq x \leq b$

(3) $\Rightarrow$ $\displaystyle\int_a^b f(x)g(x)\, dx = f(\xi) \int_a^b g(x)\, dx$ $a \leq \xi \leq b$

Set

(4) $\displaystyle \alpha(x) = \int_a^x g(t)\, dt$ $a \leq x \leq b$

Then† by equation (2), §2.3, equations (1) and (3) are equivalent. Clearly $\alpha(x)$, as defined by equation (4), is nondecreasing by virtue of hypothesis 2. It could be shown that ξ may always be chosen different from a and b in equation (3). The same is not true of equation (1).

---

* See Exercise 11, §6.
† Corollary 3 was proved directly in §9.1, Chapter 1.

4.2 SECOND MEAN-VALUE THEOREM

Theorem 4.   1. $f(x) \in \uparrow$ $a \leq x \leq b$

2. $\alpha(x) \in C$ $a \leq x \leq b$

(5) $\Rightarrow$ $\displaystyle\int_a^b f(x)\, d\alpha(x) = f(a) \int_a^\xi d\alpha(x) + f(b) \int_\xi^b d\alpha(x)$ $a \leq \xi \leq b$

By Theorem 2 and Theorem 3,

$$\int_a^b f(x)\, d\alpha(x) = f(b)\alpha(b) - f(a)\alpha(a) - \int_a^b \alpha(x)\, df(x)$$

$$= f(b)\alpha(b) - f(a)\alpha(a) - \alpha(\xi) \int_a^b df(x) \quad a \leq \xi \leq b$$

Rearrangement of terms in the latter equation gives equation (5).

Corollary 4.1.   1. $g(x) \in C$ $a \leq x \leq b$

2. $f(x) \in \uparrow$ $a \leq x \leq b$

(6) $\Rightarrow$ $\displaystyle\int_a^b f(x)g(x)\, dx = f(a) \int_a^\xi g(x)\, dx + f(b) \int_\xi^b g(x)\, dx$ $a \leq \xi \leq b$

This follows at once from Theorem 4, if $\alpha(x)$ is defined by equation (4). Equation (6) is known as the *Weierstrass form of Bonnet's theorem*.

Corollary 4.2.   1. $g(x) \in C$ $a \leq x \leq b$

2. $f(x) \in \uparrow$ $a \leq x \leq b$

3. $f(x) \geq 0$ $a \leq x \leq b$

(7) $\Rightarrow$ $\displaystyle\int_a^b f(x)g(x)\, dx = f(b) \int_\xi^b g(x)\, dx$ $a \leq \xi \leq b$

Let us alter the definition of $f(x)$ so that $f(a) = 0$. Then $f(x)$ remains nondecreasing and nonnegative. Equation (6) is still valid, but the first term on the right of the equation now disappears. Moreover, the Riemann integral on the left-hand side of equation (7) is unaltered by changing the definition of $f(x)$ at $x = a$.

EXAMPLE A.   Show that the integral

(8) $\displaystyle\int_0^\infty f(x)g(x)\, dx$

converges if $f(x) \in \downarrow$ and tends to zero as $x \to \infty$ and if there exists a constant M such that

$$\left| \int_0^R g(x)\, dx \right| < M$$

for all positive R. We are assuming that $f(x),\, g(x) \in C$ in $0 \leq x < \infty$.

Let ϵ be an arbitrary positive number. It will be sufficient* to show the existence of a number R_0 such that for all numbers R', R'' greater than R_0

$$\left| \int_{R'}^{R''} f(x)g(x)\,dx \right| < \epsilon$$

Under the present hypotheses, it is the second term on the right of (6) which may be made to disappear. Hence,

$$\left| \int_{R'}^{R''} f(x)g(x)\,dx \right| \leq f(R') \left| \int_{R'}^{\xi} g(x)\,dx \right| \qquad R' \leq \xi \leq R''$$
$$\leq 2Mf(R')$$

Since $f(R)$ tends to zero with $1/R$, the existence of R_0 such that

$$2Mf(R') < \epsilon$$

for all $R' > R_0$ is evident, and the proof is complete.

EXERCISES (4)

1. By use of Corollary 3 show that, if $f(x) \in C^1$ in $a \leq x \leq b$,
$$f(b) - f(a) = f'(\xi)(b - a) \qquad\qquad a \leq \xi \leq b$$

2. Give an example to show that the conclusion of Theorem 3 may not be altered to read $a < \xi < b$. *Hint:* Choose $\alpha(x)$ as a step-function with jump at a or at b.

3. Prove Theorem 3 if $\uparrow$ is replaced by $\downarrow$.

4. Solve the same problem for Theorem 4.

5. State and prove two results like Corollary 4.2 with $f(x) \leq 0$ and monotonic.

6. Prove Corollary 4.1 from Corollary 4.2. *Hint:* Consider the nonnegative function $f(x) - f(a)$.

7. Under the hypotheses of Corollary 3, show that, if
$$\int_a^b g(x)\,dx = 0$$
then $g(x)$ is identically zero in $a \leq x \leq b$.

8. Under the hypotheses of Corollary 3, show that the integral on the left of equation (3) cannot equal either of the values
$$m \int_a^b g(x)\,dx, \qquad M \int_a^b g(x)\,dx$$
if $f(x)$ is equal to m or M at a or b only and $g(x)$ is not identically zero. *Hint:* Consider, for example, the integral
$$\int_a^b [f(x) - m]g(x)\,dx$$

---

* Compare Theorem 7, Chapter 8. Cauchy's criterion applies equally well to a continuous variable like the present R.

9. Show that the conclusion of Corollary 3 may be altered to read $a < \xi < b$. *Hint:* Use Exercise 8.

10. By use of Example A prove that the integral

$$\int_0^\infty \frac{\sin x}{x}\, dx$$

converges.

11. Same problem for

$$\int_0^\infty \frac{\sin x}{x^p}\, dx \qquad\qquad 0 < p < 1$$

12. Show that the integral (8) converges if

1. $f(x), g(x) \in C$ $0 \leqq x < \infty$

2. $f(x) \in \downarrow , \geqq 0$ $0 \leqq x < \infty$

3. $\displaystyle\int_0^\infty g(x)\, dx$ converges.

13. Illustrate Exercise 12 by an example in which $f(\infty) \neq 0$.

14. Under the conditions of Theorem 3, find the limit

$$\lim_{x \to a+} \int_a^x f(t)\, d\alpha(t)$$

15. $\displaystyle\lim_{x \to 1+} \int_1^x e^t\, d(t + [t]) = ?$

16. If $f(x) \in C$ in $a \leqq x \leqq b$ and if $g(x) = f(x)$ except that $g(a) = f(a) + h$, $h \neq 0$, show that

$$\int_a^b f(x)\, dx = \int_a^b g(x)\, dx$$

§5. *Physical Applications*

In §1 we pointed out that the Stieltjes integral is useful in the definition of certain physical concepts which involve a combination of discrete distributions and continuous distributions of mass. We illustrate here by a few of the many possible examples.

5.1 MASS OF A MATERIAL WIRE

Let us take the physical notion of mass as undefined in our mathematical system. Of course, the mathematical situation we are about to describe can be closely approximated by a physical one in which mass is well defined. A particle can be approximated by a small pellet of matter, and a curve with a mass distribution can be nearly realized by a fine wire of heavy material. The masses of these physical objects can be determined by the process of weighing.

Let us consider a plane curve which can be given parametrically, the arc s being the parameter:

(1) $$x = x(s), \qquad y = y(s)$$

Assume that $x(s)$, $y(s) \in C$ in $0 \leqq s \leqq l$, where l is the total length of the curve. The position of a point on the curve can be determined by a single coordinate s. A particle on the curve is to be thought of as a quantity of mass situated at a geometrical point of the curve. We may define it mathematically as follows.

Definition 5. *A particle of mass m at a point s of the curve (1) is the number pair (s, m).*

Definition 6. *A distribution of mass on the curve (1) is a function $M(s)$ such that*

$$M(0) = 0, \qquad M(s) \in \uparrow \qquad\qquad 0 \leqq s \leqq l$$

The mass of the segment of the curve between any two points $s = a$ and $s = b$ $(0 \leqq a < b \leqq l)$ is

(2) $$M(b) - M(a)$$

If, for example, the distribution consists entirely of the n particles

(3) $$(s_k, m_k) \qquad\qquad k = 1, 2, \dots, n$$

where $0 < s_1 < s_2 < \dots < s_n \leqq l$, then

$$
\begin{aligned}
M(s) &= 0 & 0 \leqq s < s_1 \\
&= m_1 & s_1 \leqq s < s_2 \\
(4) \qquad &= m_1 + \dots + m_{n-1} & s_{n-1} \leqq s < s_n \\
&= m_1 + \dots + m_n & s_n \leqq s \leqq l
\end{aligned}
$$

That is, $M(s)$ is a step-function with jump m_k at the point s_k. We make the convention that a particle situated at the point b of Definition 6 is to belong to the segment (a, b) and a particle at a is not to belong. With this understanding, the mass of the segment (a, b) is given by (2), when $M(s)$ is described by equations (4). The total mass of the wire is $M(l)$. The mass of the particle at s_k is

$$m_k = M(s_k) - M(s_k-) \qquad\qquad k = 1, 2, \dots, n$$

Definition 7. *A distribution of mass $M(s)$ is continuous $\Leftrightarrow M(s) \in C^1$.*

Definition 8. *The density of a continuous distribution $M(s)$ at a point a is $M'(a)$.*

This definition conforms with our intuitive notion of density. Average density of a wire is thought of as mass per unit length. The average density of the arc (a, b) of Definition 6 is

$$\frac{M(b) - M(a)}{b - a}$$

and the limit of this is $M'(a)$ as b approaches a. For a continuous distribution, the total mass is the integral of the density

$$M(l) = \int_0^l M'(s)\, ds$$

For an arbitrary distribution, we have a similar formula using the Stieltjes integral

$$M(l) = \int_0^l dM(s)$$

5.2 MOMENT OF INERTIA

Assume as known the formula for the moment of inertia about an axis of a set of particles. For the set (3) it is

(5) $$I = \sum_{k=1}^n m_k r_k^2$$

where r_k is the distance of the particle (s_k, m_k) from the axis. Let us observe the following facts about this formula.

A. *If a total mass is divided into several parts, the moment of inertia of the whole is the sum of the moments of inertia of the parts.*

B. *If new mass is added, the moment of inertia is increased.*

C. *If mass is moved farther from the axis, the moment of inertia is increased.*

It is implicit in B that, if mass is removed, I is decreased. Likewise, it is to be understood in C that, if mass is moved nearer to the axis, I is decreased and that, if it is moved parallel to the axis, I is unchanged.

Let us now assume that the moment of inertia of any distribution is to satisfy these three properties and is to be given by formula (5) if the mass is concentrated in the set of particles (3). We shall show that under these assumptions the moment of inertia of an arbitrary distribution is a uniquely determined number equal to a certain Stieltjes integral.

Let us find the moment of inertia about the x-axis of a distribution $M(s)$ on the curve (1). Let the points $\{s_k\}_0^n$ be a subdivision Δ of the interval $0 \leqq s \leqq l$. By property A, the moment of inertia desired will be

$$I = \sum_{k=1}^n I_k$$

where I_k is the moment of inertia of the arc (s_{k-1}, s_k).   Set

$$y(s_k'') = \max_{s_{k-1} \leq s \leq s_k} |y(s)|$$

$$y(s_k') = \min_{s_{k-1} \leq s \leq s_k} |y(s)|$$

The mass of the arc (s_{k-1}, s_k) is $M(s_k) - M(s_{k-1})$.   If this mass were concentrated in a particle at s_k' or at s_k'', mass would have been moved nearer to or farther from the x-axis, respectively.   Hence, by property C we have

$$\sum_{k=1}^n y^2(s_k')[M(s_k) - M(s_{k-1})] \leq \sum_{k=1}^n I_k \leq \sum_{k=1}^n y^2(s_k'')[M(s_k) - M(s_{k-1})]$$

By Theorem 1 both extremes of these inequalities approach the same limit as $\|\Delta\| \to 0$.   Hence

(6) $$I = \int_0^l y^2(s)\, dM(s)$$

Observe that we have not used property B.

EXAMPLE A.   Let the curve (1) be the straight line

$$y = \frac{s}{\sqrt{2}}, \qquad x = 1 - \frac{s}{\sqrt{2}} \qquad\qquad 0 \leq s \leq \sqrt{2}$$

Let the distribution be a combination of a continuous one in which the density is proportional to the distance from the end point $s = 0$ and a discrete one consisting of the two particles $\left(\frac{1}{\sqrt{2}}, 2\right)$, $(\sqrt{2}, 4)$.   More explicitly,

$$M(s) = M_1(s) + M_2(s)$$

where
$$M_1(s) = \int_0^s t\, dt = \frac{s^2}{2}$$

$$\begin{aligned} M_2(s) &= 0 & 0 \leq s < 1/\sqrt{2} \\ &= 2 & 1/\sqrt{2} \leq s < \sqrt{2} \\ &= 6 & s = \sqrt{2} \end{aligned}$$

Then
$$I = \int_0^{\sqrt{2}} \frac{s^2}{2}\, dM_1(s) + \int_0^{\sqrt{2}} \frac{s^2}{2}\, dM_2(s)$$

$$= \int_0^{\sqrt{2}} \frac{s^3}{2}\, ds + \frac{1}{2}\left(\frac{1}{\sqrt{2}}\right)^2 2 + \frac{1}{2}(\sqrt{2})^2 4 = 5$$

EXAMPLE B.   A plane lamina is bounded by the four curves

$$x = a, \qquad x = b, \qquad y = 0, \qquad y = f(x) > 0$$

where $f(x) \in C$ in $a \leq x \leq b$. The density is constant along vertical lines, so that the distribution can be described by a function of one variable $M(x)$. The mass of a narrow vertical strip varies as its height.   Find the moment of inertia of the lamina about the y-axis.

Let the points $\{x_k\}_0^n$ be a subdivision Δ of (a, b).   Erecting ordinates at the points x_k divides the lamina into n vertical strips.   Let m_k and I_k be the mass and the moment of inertia, respectively, of the kth vertical strip.   Let $f(x_k')$ and $f(x_k'')$ be the minimum and maximum ordinates of the kth strip. By Property B, we have

$$f(x_k')[M(x_k) - M(x_{k-1})] \leq m_k \leq f(x_k'')[M(x_k) - M(x_{k-1})]$$

By properties A and C, we see that

$$\sum_{k=1}^n x_{k-1}^2 f(x_k')[M(x_k) - M(x_{k-1})] \leq I \leq \sum_{k=1}^n x_k^2 f(x_k'')[M(x_k) - M(x_{k-1})]$$

Now by use of Duhamel's theorem, §6.5, we have

(7) $$I = \int_a^b x^2 f(x) \, dM(x)$$

We have assumed that $a > 0$, but this assumption was not essential to the final result.

EXERCISES   (5)

1. In equation (1) take $x(s) = \cos s$, $y(s) = \sin s$.   Find $M(s)$ if the density at a point is equal to its distance from $(1, 0)$.

2. Same problem if density at a point is equal to its shortest distance to $(1, 0)$, measured along the arc.

3. Do exercise 1 if in addition to the continuous distribution there are unit particles at $(0, 1)$ and $(0, -1)$.

4. Find the moment of inertia about the y-axis of the wire of §5.2.

5. Illustrate Exercise 4 by the wire of Example A.

6. Find the moment of inertia about an axis perpendicular to the coordinate plane at the point (x_0, y_0) of the wire of §5.2.

7. Illustrate Exercise 6 by the wire of Example A with $x_0 = y_0 = 0$.

8. Alter the definitions (4) of $M(s)$ if $s_1 = 0$.

9. Find I in equation (6) if the curve (1) is the circular arc

$$x = \sin s, \qquad y = \cos s \qquad\qquad 0 \leq s \leq \pi/2$$

and the distribution consists of a continuous part with density $D(s) = s$ and of three particles of masses 1, 2, 3 at the points $s = 0$, $\pi/4$, $\pi/2$, respectively.

10. Check the result of Example A by finding the explicit definition of $M(s)$ in $0 \leqq s \leqq 1/\sqrt{2}$ and in $1/\sqrt{2} < s \leqq \sqrt{2}$ and by integrating by parts.

11. Derive formula (7) when $a < b < 0$ and also when $a < 0$, $b > 0$.

12. Illustrate Example B with $a = 0$, $b = 2$, $f(x) = e^x$,
$$M(x) = e^x \qquad\qquad 0 \leqq x \leqq 1$$
$$= e^{2x} \qquad\qquad 1 < x \leqq 2$$

13. Write a set of properties like A, B, C for the moment of mass, and thus discuss the center of gravity of the wire of §5.2.

14. Find the coordinates of the center of gravity of the wire of Example A.

15. Find the coordinates of the center of gravity of the lamina of Example B.

§6. Continuous Functions

We shall prove here a few of the important properties of continuous functions. Some of them we have already used in view of the fact that they appear self-evident to most students. To investigate the more delicate aspects of continuity, we need to base our study firmly on the definition, in terms of limits, of continuity.

6.1 THE HEINE-BOREL THEOREM

We prove first a result discovered independently by two mathematicians and hence referred to by the hyphenated name. Probably it will seem obvious to most students without proof. Let there correspond to each point c of the closed interval $a \leqq x \leqq b$ a number δ_c and an interval I_c of length $2\delta_c$ with c the center point,

$$(1) \qquad\qquad I_c: \quad c - \delta_c < x < c + \delta_c$$

The Heine-Borel theorem states that a finite number of the intervals (1) can be chosen which will cover the whole interval $a \leqq x \leqq b$. That is, every point of $a \leqq x \leqq b$ will be in at least one of the above mentioned finite number of intervals (1). In order to emphasize the need for proving this result, let us give an example to show it false if the interval (a, b) were open instead of closed.

Let $a = 0$, $b = 1$, and define I_c, for $0 < c < 1$, as

$$(2) \qquad\qquad I_c: \quad c/2 < x < 3c/2$$

That is, $\delta_c = c/2$. No finite set of the intervals (2) will cover the interval $0 < x < 1$. For, consider such a set, $I_{c_1}, I_{c_2}, \dots, I_{c_n}$, where $0 < c_1 < c_2 \dots < c_n$. Of these, the interval I_{c_1} reaches farthest to the left. Hence, no point to the left of $c_1/2$ is covered by the set.

Theorem 5. *To each c, $a \leqq c \leqq b$ corresponds an interval* (1)

$\Rightarrow$ *There exist points $c_1, c_2, \ldots, c_n$ of $a \leqq x \leqq b$ such that every point of the interval $a \leqq x \leqq b$ is in at least one of the intervals $I_{c_1}, I_{c_2}, \ldots, I_{c_n}$.*

Call a point A of the interval I, $a \leqq x \leqq b$, *accessible* if the interval $a \leqq x \leqq A$ can be covered by a finite sequence of the intervals I_c. Clearly, if A is accessible, every point of I to its left is also. Hence, there must either be a point B of I dividing accessible points from inaccessible ones, or else all points of I are accessible. (Some points are accessible, since all points of I in I_a are covered by the single interval I_a.) But the existence of the dividing point B, $B < b$, is impossible. For, if $I_{c_1}, \ldots, I_{c_n}$ is a set of intervals covering $(a, B - \delta)$, $\delta = \delta_B/2$, then $I_{c_1}, \ldots, I_{c_n}, I_B$ covers $a \leqq x \leqq B + \delta$, so that there are accessible points to the right of B. This is a contradiction, so that b must be accessible.

6.2 BOUNDS OF CONTINUOUS FUNCTIONS

We show next that, if $f(x) \in C$ in $a \leqq x \leqq b$, then $f(x)$ is bounded there. This result would be false in an open interval. For, $1/x \in C$ in $0 < x \leqq 1$, but $1/x$ is certainly not bounded in $0 < x \leqq 1$.

Theorem 6.   1. $f(x) \in C$, $a \leqq x \leqq b$

$\Rightarrow$ *There exists a number M such that*

$$|f(x)| \leqq M \qquad\qquad\qquad a \leqq x \leqq b$$

Define $f(x)$ outside (a, b) to be $f(a)$ for $x < a$ and $f(b)$ for $x > b$, so that $f(x) \in C$ in $-\infty < x < \infty$. This is done so that the end points a and b may be *interior* points of intervals (1) in applying the Heine-Borel theorem. Let $a \leqq c \leqq b$. Since $f(x) \in C$ at $x = c$, a number δ_c corresponds to $\epsilon = 1$ such that

$$|f(x) - f(c)| \leqq 1 \qquad\qquad\qquad |x - c| < \delta_c$$

whence

(3) $|f(x)| \leqq |f(c)| + 1 = M_c$ $x \in I_c$

Now, by Theorem 5, we choose $I_{c_1}, I_{c_2}, \ldots, I_{c_n}$ covering $a \leqq x \leqq b$. Set

(4) $M = \max (M_{c_1}, M_{c_2}, \ldots, M_{c_n})$

Since an arbitrary point x of $a \leqq x \leqq b$ is in some I_{e_k}, $k = 1, 2, \ldots, n$, we have by (3) and (4)

$$|f(x)| \leqq M_{c_k} \leqq M \qquad\qquad\qquad a \leqq x \leqq b$$

and the proof is complete.

6.3 MAXIMA AND MINIMA OF CONTINUOUS FUNCTIONS

We shall prove next that if $f(x) \in C$ in $a \leqq x \leqq b$, then $f(x)$ has a maximum M in $a \leqq x \leqq b$ and there is a point c such that $f(c) = M$,

$a \leqq c \leqq b$. This result would also be false in an open interval. The function $f(x) = x$ has no maximum in the interval $0 < x < 1$.

Theorem 7. 1. $f(x) \in C$, $a \leqq x \leqq b$

⟹ *There exist numbers* m, M, c_1, c_2 *such that*

$$m \leqq f(x) \leqq M \qquad\qquad a \leqq x \leqq b$$

$$f(c_1) = m, \qquad f(c_2) = M \quad a \leqq c_1 \leqq b, \ a \leqq c_2 \leqq b$$

It will be sufficient to prove the part of the theorem which concerns c_2 and M, for this result can then be applied to $-f(x)$ to prove the rest.

By Theorem 6, $f(x)$ has an upper bound in $a \leqq x \leqq b$. Let M be the least upper bound. Then $f(x) \leqq M$, and we must show that the equality holds for at least one value of x in the interval $a \leqq x \leqq b$. Suppose the contrary. Then $M - f(x) > 0$ and $[M - f(x)]^{-1} \in C$ in $a \leqq x \leqq b$. By Theorem 6, $[M - f(x)]^{-1}$ is bounded. But this is impossible since $f(x)$ becomes arbitrarily near its least upper bound M in (a, b). Hence, the existence of the desired number c_2 is assured.

6.4 UNIFORM CONTINUITY

Definition 9. *The function* $f(x)$ *is uniformly continuous in* $a \leqq x \leqq b$ ⟺ *to an arbitrary* $\epsilon > 0$ *corresponds a number* δ *such that for all points* x' *and* x'' *of* $a \leqq x \leqq b$ *with* $|x' - x''| < \delta$ *we have*

$$|f(x') - f(x'')| < \epsilon$$

We shall show that $f(x) \in C$ in $a \leqq x \leqq b$ implies that $f(x)$ is uniformly continuous there. This result would not hold for an open interval, as the example $f(x) = x^{-1}$ in $0 < x < 1$ shows. Here the difference

$$f(x') - f(x' + \delta) = \frac{\delta}{x'(x' + \delta)}$$

for any fixed $\delta > 0$, can be made arbitrarily large by choosing x' near zero.

Theorem 8. 1. $f(x) \in C$ $a \leqq x \leqq b$

⟹ $f(x)$ *is uniformly continuous in* $a \leqq x \leqq b$.

Extend the definition of $f(x)$ to $(-\infty, \infty)$ as in the proof of Theorem 6. To an arbitrary $\epsilon > 0$ corresponds for each c, $a \leqq c \leqq b$, a number δ_c such that

$$|f(x) - f(c)| < \epsilon/2$$

for all x in the interval I_c: $c - \delta_c < x < c + \delta_c$. By Theorem 5 there exist intervals, finite in number,

(5) $I_{c_1}, I_{c_2}, \dots, I_{c_n}$

covering (a, b). Those end points of the intervals (5) which lie in (a, b)

form with the points a and b themselves a subdivision of (a, b). Choose the number δ of Definition 9 as the length of the smallest of the subintervals into which the interval (a, b) is divided by the subdivision. Now let x' and x'' be points of (a, b) for which $|x' - x''| < \delta$. Then it is clear that there can be at most one of the points of subdivision between x' and x''. In fact, both x' and x'' lie in a single one of the intervals (5), say in I_{c_k}. Consequently,

$$|f(x') - f(x'')| \leq |f(x') - f(c_k)| + |f(x'') - f(c_k)| < \epsilon/2 + \epsilon/2 = \epsilon$$

This completes the proof of the theorem.

The notion of uniform continuity extends in an obvious way to functions of more variables. For example, $f(x, y, z)$ is uniformly continuous in the region R, $a \leq x \leq b$, $c \leq y \leq d$, $e \leq z \leq g$,$\Leftrightarrow$for every $\epsilon > 0$ there is a positive δ such that when (x, y, z) and (x', y', z') are any points of R so close that

$$|x - x'| < \delta, \qquad |y - y'| < \delta, \quad |z - z'| < \delta,$$

then $\qquad\qquad |f(x, y, z) - f(x', y', z')| < \epsilon$

The analogue of Theorem 8 is true; that is, ordinary continuity in a closed bounded region implies uniform continuity there. We omit the proof.

6.5 DUHAMEL'S THEOREM

As a first application of uniform continuity let us prove for Stieltjes integrals a result analogous to one form of the familiar Duhamel's theorem for Riemann integrals.

Theorem 9. 1. $f(x), g(x) \in C$ $\qquad\qquad\qquad\qquad\qquad a \leq x \leq b$

2. $\alpha(x) \in \ \uparrow$

3. $\{x_k\}_0^n$ is a subdivision Δ of (a, b)

4. $x_{k-1} \leq \xi_k \leq x_k,\quad x_{k-1} \leq \eta_k \leq x_k \qquad\qquad k = 1, 2, \ldots, n$

(6) $\Rightarrow \displaystyle\lim_{\|\Delta\| \to 0} \sum_{k=1}^{n} f(\xi_k)g(\eta_k)[\alpha(x_k) - \alpha(x_{k-1})] = \int_a^b f(x)g(x)\, d\alpha(x)$

Set σ'_Δ equal to the sum on the left of equation (6), and σ_Δ equal to the same sum in which $\eta_k = \xi_k$, $k = 1, 2, \ldots, n$. Let ϵ be an arbitrary positive number and let δ be the number which corresponds to it, according to Definition 9. Then if $\|\Delta\| < \delta$, we have, by the uniform continuity of $f(x)$,

$$|\sigma_\Delta - \sigma'_\Delta| \leq \epsilon \sum_{k=1}^{n} |g(\eta_k)|[\alpha(x_k) - \alpha(x_{k-1})]$$

$$\leq \epsilon[\alpha(b) - \alpha(a)] \max_{a \leq x \leq b} |g(x)|$$

Hence,
$$\lim_{||\Delta||\to 0} \sigma_\Delta' = \lim_{||\Delta||\to 0} \sigma_\Delta = \int_a^b f(x)g(x)\,d\alpha(x)$$

This completes the proof. Of course, if $\alpha(x) = x$, we have a conventional form of Duhamel's theorem, sometimes called the theorem of G. A. Bliss.

The integrand (6) was a product of two functions, but it might as well have been a more complicated function. We state a more general result.

Theorem 10. 　1. $f(x, y, z) \in C$ 　　　　　　　　　　　$a \leqq x, y, z \leqq b$

　　　　　　　2. $\alpha(x) \in \uparrow$

　　　　　　　3. $\{x_k\}_0^n$ is a subdivision Δ of (a, b),

　　　　　　　4. $x_{k-1} \leqq \xi_k, \eta_k, \zeta_k \leqq x_k$ 　　　　　　　$k = 1, 2, \ldots, n$

$$\Rightarrow \lim_{||\Delta||\to 0} \sum_{k=1}^n f(\xi_k, \eta_k, \zeta_k)[\alpha(x_k) - \alpha(x_{k-1})] = \int_a^b f(x, x, x)\,d\alpha(x)$$

The proof follows from the uniform continuity of $f(x, y, z)$ in the given closed cube. With obvious notations,

$$|\sigma_\Delta - \sigma_\Delta'| \leqq \sum_{k=1}^n |f(\xi_k, \xi_k, \xi_k) - f(\xi_k, \eta_k, \zeta_k)|[\alpha(x_k) - \alpha(x_{k-1})]$$

$$\leqq \epsilon \sum_{k=1}^n [\alpha(x_k) - \alpha(x_{k-1})] = \epsilon[\alpha(b) - \alpha(a)]$$

$$\lim \sigma_\Delta = \lim \sigma_\Delta' = \int_a^b f(x, x, x)\,d\alpha(x)$$

EXAMPLE A. Compute the limit L of §1.1, Chapter 3. Here $\alpha(x) = x$ and

$$f(x, y, z) = \sqrt{x_1'(x)^2 + x_2'(y)^2 + x_3'(z)^2} \in C$$
$$a \leqq x, y, z \leqq b$$

The continuity follows since $\sqrt{w} \in C$, $0 \leqq w < \infty$ and $\vec{x}(t)$ was assumed of class C^1. By Theorem 10

$$L = \int_a^b f(x, x, x)\,dx = \int_a^b |\vec{x}'(t)|\,dt$$

6.6 ANOTHER PROPERTY OF CONTINUOUS FUNCTION

As a further application of uniform continuity, let us prove that, if $f(x) \in C$ in $a \leqq x \leqq b$ and if $f(a)f(b) < 0$, then $f(X) = 0$ for some X, $a < X < b$. Suppose the contrary. Then $1/f(x) \in C$ in $a \leqq x \leqq b$ and is bounded there. Let ϵ be an arbitrary positive number. By uniform continuity, there is a sequence of points between a and b such that the variation of $f(x)$ between any consecutive pair is less than ϵ. Now $f(x)$ must change sign between two consecutive points of the set, so that $|f(x)| < \epsilon$ at each. Hence, $|1/f(x)| > 1/\epsilon$, and $1/f(x)$ is not bounded. The contradiction is evident.

6.7 CRITICAL REMARKS

For a thorough understanding of the proofs of Theorems 5 and 7 a fuller appreciation of the structure of the set of all real numbers is needed than we have hitherto assumed. For example, in §6.3 we used the phrase least upper bound. Is it entirely evident that such a number exists? The fact that every bounded set of numbers has a least upper bound and a greatest lower bound is, in fact, a property of the real number system. The property states essentially that there are no "holes" in the system. For, consider a system which had no number zero, for example. What would then be the least upper bound of all negative numbers? Every positive number would be an upper bound; no negative number could be. Yet there is no smallest positive number, so that a least upper bound would not exist. Again in §6.1 the existence of the point of division B again depends on the existence of a least upper bound for the set of all accessible points.

The following may be taken as a characterization of the least upper bound A of a set E of real numbers. To an arbitrary positive number ϵ corresponds at least one number a of E such that $a > A - \epsilon$. Moreover, all points of E are less than or equal to A. We make the convention that the least upper bound of a set which is unbounded above is $+\infty$, that the greatest lower bound of a set which is unbounded below is $-\infty$.

EXERCISES (6)

1. To the intervals (2) add two more: the interval $|x| < .1$ corresponding to the point $x = 0$ and the interval $|x - 1| < .1$ corresponding to the point $x = 1$. Describe explicitly a finite number of these intervals which cover the interval $0 \leq x \leq 1$. Give also a second set involving a larger number of intervals.

2. For the intervals (1) take $\delta_c = 1 - c$ when $0 \leq c < 1$ and define I_1 as the interval $|x - 1| < 10^{-10}$. From this set, find a finite set of covering intervals for the interval $0 \leq x \leq 1$. What is the smallest number of covering intervals that can be used?

3. Give an example of a function defined on $a \leq x \leq b$ which has no maximum value.

4. In Definition 9 choose $f(x) = x^2$ on $0 \leq x \leq 1$. If $\epsilon = .1$, find the least upper bound of numbers δ corresponding.

5. In the proof of Theorem 8 give details to show that x' and x'' both lie in a single interval (5).

6. Prove the Heine-Borel theorem, modified so that the intervals (1) include their end points.

7. Prove the Heine-Borel theorem if c is a point not the mid-point of the *open* interval I_c corresponding.

8. If $M(s)$ is the function defined by equations (4), §5.1, describe those of the four following functions which are well defined:

$$f(x) = \max_{0 \leq s \leq x} M(s), \qquad h(x) = \max_{0 \leq s < x} M(s)$$
$$g(x) = \min_{0 \leq s \leq x} M(s), \qquad k(x) = \min_{0 \leq s < x} M(s)$$

9. Same problem if $M(s)$ is replaced by $s + M(s)$.

10. State and prove a form of Duhamel's theorem involving the product of three continuous functions.

11. Prove that a continuous function takes on every value between its maximum and its minimum. State the result in precise theorem form.

12. Find the least upper bound (lub) and the greatest lower bound (glb) for all the values of the following functions:

$$x, \ e^{-x^2}, \ x^2 e^{-x^2}, \ e^{-x}, \ (x+1)(x^2+1)^{-1/2}$$

What can you say of the maxima and minima of these functions? No proofs are required.

13. Give and ϵ-characterization of a finite glb.

14. Give an ϵ-characterization of an infinite glb.

15. Give an ϵ-characterization of an infinite lub.

16. Prove glb $a = -\text{lub}\,(-a)$

17. Prove lub $[f(x) + g(x)] \leq \text{lub}\, f(x) + \text{lub}\, g(x)$.

Give examples to show that either the equality or the inequality may occur.

§7. *Existence of Stieltjes Integrals*

By use of the results about continuous functions established in the preceding section, we can now give a proof of Theorem 1.

7.1 PRELIMINARY RESULTS

For an arbitrary subdivision Δ of (a, b) let us define, in addition to the numbers σ_Δ of §1, two additional numbers, S_Δ and s_Δ. We assume throughout this section that $f(x) \in C$, $\alpha(x) \in \uparrow$ in $a \leq x \leq b$. Set

$$M_k = \max_{x_{k-1} \leq x \leq x_k} f(x), \qquad m_k = \min_{x_{k-1} \leq x \leq x_k} f(x)$$

$$S_\Delta = \sum_{k=1}^n M_k[\alpha(x_k) - (x_{k-1})], \qquad s_\Delta := \sum_{k=1}^n m_k[\alpha(x_k) - \alpha(x_{k-1})]$$

Clearly, $s_\Delta \leq S_\Delta$.

We say that a subdivision Δ of (a, b) undergoes *refinement* if new points of subdivision are interpolated among those of Δ. We can now prove some results useful in the proof of Theorem 1.

Lemma 1.1. $S_\Delta \in \downarrow$, $s_\Delta \in \uparrow$ *under refinement of* Δ

Suppose Δ_1 is a refinement of Δ obtained by introducing a single point t between x_{i-1} and x_i. Let M_i' and M_i'' be the maxima of $f(x)$ in (x_{i-1}, t) and (t, x_i), respectively, so that neither of these maxima is greater than M_i. Hence,

$$M_i'[\alpha(t) - \alpha(x_{i-1})] + M_i''[\alpha(x_i) - \alpha(t)] \leq M_i[\alpha(x_i) - \alpha(x_{i-1})]$$

and $S_{\Delta_1} \leq S_\Delta$. Since any refinement of Δ can be accomplished by successively adding a single point, one half the result is established. The other half may be proved in like manner.

Lemma 1.2. 1. Δ_1 *and* Δ_2 *are subdivisions of* (a, b)

$\Rightarrow$ 　　　　　　　　$s_{\Delta_1} \leq S_{\Delta_2}, \qquad s_{\Delta_2} \leq S_{\Delta_1}$

Let Δ_3 be a third subdivision made up of all the points of Δ_1 and Δ_2, coincident points being counted as a single point. By Lemma 1.1,

$$s_{\Delta_2} \leq s_{\Delta_3} \leq S_{\Delta_3} \leq S_{\Delta_1}$$

Since Δ_1 and Δ_2 occur symmetrically in the hypothesis, they must do so in the conclusion, so that $s_{\Delta_1} \leq S_{\Delta_2}$ also.

By Lemma 1.2, it is clear that, for all subdivisions Δ, the numbers s_Δ have a least upper bound s, the numbers S_Δ a greatest lower bound S and that $s \leq S$.

Lemma 1.3. 1. $f(x) \in C$　　　　　　　　　　$a \leq x \leq b$

$\Rightarrow$ 　　　　　　　　　　$s = S$

By Theorem 8, $f(x)$ is uniformly continuous in $a \leq x \leq b$. Let ϵ and δ be the corresponding numbers described in Definition 5. Let Δ be a subdivision of (a, b) with $\|\Delta\| < \delta$. Then

$$0 \leq S_\Delta - s_\Delta = \sum_{k=1}^n [M_k - m_k][\alpha(x_k) - \alpha(x_{k-1})] \leq \epsilon[\alpha(b) - \alpha(a)]$$

(1) $0 \leq S_\Delta - s_\Delta = (S_\Delta - S) + (S - s) + (s - s_\Delta) \leq \epsilon[\alpha(b) - \alpha(a)]$

Since each term in parentheses is nonnegative, we have

$$0 \leqq S - s \leqq \epsilon[\alpha(b) - \alpha(a)]$$

Since S and s do not depend on Δ, $S = s$, and the proof is complete.

Lemma 1.4.   1. $f(x) \in C$ $a \leqq x \leqq b$

$$\Rightarrow \qquad \lim_{\|\Delta\| \to 0} s_\Delta = \lim_{\|\Delta\| \to 0} S_\Delta = s = S$$

For, by inequalities (1) we have, when $\|\Delta\| < \delta$,

$$0 \leqq S_\Delta - S \leqq \epsilon[\alpha(b) - \alpha(a)]$$
$$0 \leqq s \;\; - s_\Delta \leqq \epsilon[\alpha(b) - \alpha(a)]$$

7.2 PROOF OF THEOREM I

By the definition of M_k and m_k, it is clear for any Δ that $s_\Delta \leqq \sigma_\Delta \leqq S_\Delta$. Since s_Δ and S_Δ both approach s or S as $\|\Delta\| \to 0$, the same must be true of σ_Δ, and the proof is complete.

7.3 THE RIEMANN INTEGRAL

By Theorem A, which now follows as a corollary of Theorem 1, the Riemann integral of a continuous functions exists. But certain discontinuous functions are also integrable. For example, if $f(x) \in C$ on $a \leqq x \leqq c$ and on $c \leqq x \leqq b$ but $f(c-) \neq f(c+)$, then it is easy to see that $f(x)$ is integrable from a to b and that

$$\int_a^b f(x)\, dx = \int_a^c f(x)\, dx + \int_c^b f(x)\, dx$$

The interpretation of each integral as an area makes this equation evident. On the other hand, many discontinuous functions are not integrable (Exercise 5). There is an elegant criterion which distinguishes between the two classes. We state it here without proof.* To make the statement simpler we first introduce the notion of point sets of *measure zero*.

Definition 10. *A set of points on a line is of measure zero if, and only if, for every positive ϵ (however small) it can be covered by a set of intervals $I_1, I_2, I_3, \ldots$ of lengths $\epsilon/2, \epsilon/2^2, \epsilon/2^3, \ldots$, respectively.*

EXAMPLE A. The set $x_0 = 0$, $x_k = 1/k$ $(k = 1, 2, 3, \ldots)$ is of measure zero. For, make the center of I_k coincide with the point x_{k-1} for $k = 1, 2, 3, \ldots$. Each point is covered. Whether the intervals overlap or not is of no concern.

---

* See, for example, *The Theory of Functions of Real Variables*, by L. M. Graves, McGraw-Hill (1946), p. 89.

EXAMPLE B. The set of all points in the interval $0 \leq x \leq 1$ is not of measure zero. For, if this unit interval were covered by the intervals I_k, certainly its length would be at most equal to ϵ, the sum of the lengths of the I_k. If $\epsilon = \frac{1}{2}$, for example, we have a contradiction.

Theorem C. *If $f(x)$ is bounded on $a \leq x \leq b$, then $\int_a^b f(x)\, dx$ exists if, and only if, the set of discontinuities of $f(x)$ is of measure zero.*

EXAMPLE C. If

$$f(x) = \sin\left[1/\sin\left(\pi/x\right)\right]$$

except at the points $x = 0$, $x = 1/k$ ($k = 1, 2, 3, \ldots$), where $f(x)$ is given any definition desired, then $\int_0^1 f(x)\, dx$ exists. For, $f(x)$ is bounded, and its only discontinuities on $(0, 1)$ are at the point set of Example A.

EXERCISES (7)

1. Compute S_Δ if $f(x) = \cos x$, $\alpha(x) = x$, $a = -\pi/2$, $b = \pi/2$,

Δ is $\left\{-\dfrac{\pi}{2} + \dfrac{k\pi}{m}\right\}_{k=0}^{m}$, $m = 2^n$.

Hint: $\frac{1}{2} + \cos x + \ldots + \cos nx = \dfrac{\sin\left(n + \frac{1}{2}\right)x}{2\sin\dfrac{x}{2}}$

2. In Exercise 1 show directly that

$$\lim_{n\to\infty} S_\Delta = \int_a^b f(x)\, d\alpha(x)$$

3. In Lemma 1.1, prove that $s_\Delta \in \uparrow$.

4. Let ϵ and δ be the numbers described in the proof of Lemma 1.3. If Δ_1 and Δ_2 are any subdivisions of (a, b) such that $\|\Delta_1\| < \delta$, $\|\Delta_2\| < \delta$, show that

$$\left|\sigma_{\Delta_1} - \sigma_{\Delta_2}\right| \leq 2\epsilon[\alpha(b) - \alpha(a)]$$

5. Let $f(x)$ be zero for rational x and unity for irrational x. Let $\alpha(x) \in \uparrow$ in $\alpha \leq x \leq b$ with $\alpha(b) > \alpha(a)$. Show that the Stieltjes integral

$$\int_a^b f(x)\, d\alpha(x)$$

cannot exist.

6. If $[x]$ is the largest integer $\leq x$, use Theorem C to show that

$$\int_1^7 [x]\, dx$$

exists. Find its value.

7. In the definition of the Riemann integral in §1, choose $f(x) = x^{-2}$ when $x \neq 0$ and $f(0) = 0$; choose $a = 0$, $b = 1$, $x_k = \xi_k = k/n$. Thus show that the integral (1) of §1 does not exist. Show that Theorem C is false if the hypothesis of boundedness is omitted.

8. If the bounded function $f(x) \in C$ on $a \leq x \leq b$ except at a *countable* set of points (that is, a set that can be arranged in a sequence $x_1, x_2, x_3, \ldots$), show that $\int_a^b f(x)\,dx$ exists.

9. In Example A, show that the intervals I_k, $(k = 2, 3, 4, \ldots)$ do not overlap if $\epsilon < 16/9$.

10. In Example A, show that I_{k+1} will overlap I_1 when $k(1 + 2^{-k}) > 4/\epsilon$.

6

Multiple Integrals

§1. *Introduction*

In this chapter, we shall discuss double and triple integrals. We shall follow as closely as possible the analogy with the theory of simple integrals developed in the previous chapter.

1.1 REGIONS

We have already discussed in Chapter 1 regions of the plane. Let us collect here the notations which will be needed in the present chapter.

A *domain* D is an open connected set of points. That is, every point of D is the center of some δ-neighborhood, all of whose points are points of D; and any two points of D can be joined by a broken line having a finite number of segments, all of whose points are points of D. A domain is *bounded* if all its points lie inside some square.

A *region* R is a closed point set consisting of a bounded domain plus its boundary points. We shall assume further that the boundary of R consists of a finite number of closed curves that do not cross themselves nor each other. Note that the regions here defined and designated by the letter R are special cases of the more general ones of §3.1, Chapter 1. In practical problems, R will usually be given in terms of its boundary curves. For example, R might be the set of points between two concentric circumferences plus the points on the circumferences. More frequently, we shall meet regions that can be most simply described by use of functions. Accordingly, we shall have a special notation for these.

Let $\varphi(x)$ and $\psi(x) \in C$ in $a \leq x \leq b$ and $\varphi(x) < \psi(x)$ in $a < x < b$. Then the *region* R_x, or $R[a, b, \varphi(x), \psi(x)]$, is the region bounded by the curves

$$x = a, \quad x = b, \quad y = \varphi(x), \quad y = \psi(x)$$

If (x_1, y_1) is a point of R_x, then $a \leq x_1 \leq b$ and $\varphi(x_1) \leq y_1 \leq \psi(x_1)$. A line $x = x_1, a < x_1 < b$ cuts the boundary of R_x in just two points. For example, the region $R[-1, 1, -\sqrt{1 - x^2}, \sqrt{1 - x^2}]$ is the circle $x^2 + y^2 \leq 1$. We could define in an obvious way a region R_y. The region R described above as lying between two concentric circles is neither an R_x nor an R_y. It could be divided into four regions R_x, for example, by two vertical lines tangent to the inner circle. These vertical lines would be counted twice, as the boundary of adjoining regions.

A region R is *simply connected* if its boundary consists of a single closed curve. The concept of the area of a region R will be assumed known. Of course, the area of R_x is known from elementary calculus, and the area of R could be defined by use of a limiting process (§10.4). The *diameter* of a region R is the length of the longest line segment that joins two points of R. In the case of a circle this coincides with the elementary notion of diameter. Observe that, if a region R varies so that its diameter approaches zero, then its area also approaches zero. The converse is not true.

1.2 DEFINITIONS

We begin by dividing a given region R of area A into subregions.

As in the case of simple integrals, we introduce certain simplifying notations.

Fig. 15.

Definition 1. *A subdivision* Δ *of a region* R *is a set of closed curves** $\{C_k\}_1^n$ *which divides* R *into* n *subregions* R_k *of area* $\Delta S_k, k = 1, 2, \ldots, n$.

For example, in the adjoining figure, R is divided into 12 subregions. In all but two of the subregions the boundary curve C_k is composed partly of the boundary of R. One convenient method of making a subdivision is to draw equally spaced lines parallel to the axes. If the distance between

---

* It is sufficiently general for all practical purposes to suppose that these curves are of such a nature that two subdivisions superimposed make a new subdivision (forming a *finite* number of subregions). This is true for example if the subdivisions are made by use of broken lines with a finite number of segments.

the lines is small compared with the diameter of R, most of the subregions will be squares.

Definition 2. *The norm $\|\Delta\|$ of a subdivision Δ is the maximum diameter of the subregions produced by the subdivision.*

Definition 3. *The double integral of $f(x, y)$ over the region R is*

$$(1) \qquad \iint_R f(x, y)\, dS = \lim_{\|\Delta\| \to 0} \sum_{k=1}^{n} f(\xi_k, \eta_k)\, \Delta S_k$$

where (ξ_k, η_k) is a point of R_k.

For clarification of the meaning of the limit (1) see remarks following Definition 3, Chapter 5.

1.3 EXISTENCE OF THE INTEGRAL

The limit (1) may or may not exist depending on the function $f(x, y)$. We give at once a sufficient condition for existence.

Theorem 1.   1. $f(x, y) \in C$ $(x, y) \in R$

$$=> \qquad \iint_R f(x, y)\, dS \text{ exists.}$$

The proof of this theorem will be given at the end of this chapter.

EXAMPLE A. $R_x = R[0, 1, -x, x], f(x, y) = x$.

Since $f(x, y) \in C$ in R_x, the limit (1) exists. Hence, we may choose the subregions R_k and the points (ξ_k, η_k) in a special way. Choose R_k a square (except near the lines $y = \pm x$) of side Δx. With obvious notations, we have

$$\iint_R f(x, y)\, dS = \lim_{\Delta x \to 0} \Delta x \sum_{k=0}^{n-1} x_k(x_k + x_{k+1})$$

$$= \int_0^1 2x^2\, dx = 2/3$$

Here we have collected in a single term those terms of the sum (1) coming from the regions R_k in a vertical line. Since $f(x, y) = x$ does not change on a vertical line, the sum in question reduces to a constant times the area of a certain trapezoid.

EXAMPLE B. $R_x = R[0, 1, 0, 1]$; $f(x, y) = 0$, when x and y are both rational; $f(x, y) = 1$, when either x or y is irrational. Then the sum (1) may be made to equal either 0 or 1 for an arbitrary subdivision, depending on the way in which the points (ξ_k, η_k) are chosen. Hence, the double integral of $f(x, y)$ over R does not exist.

EXERCISES (1)

1. State which of the sets of points below is a domain D and describe the sets: the points (x, y) for which

   (a) $3x - 2y + 1 > 0$

   (b) $x^2 + y^2 - 1 > 0$

   (c) $x^2 + y^2 - 1 > 0, |x| < 1, |y| < 1$

   (d) $-5 \leqq x^2 - 2x + y^2 - 4y < -4$

2. State which of the sets of points below is a region R and describe the sets:

   (a) $x + y - 1 \geqq 0, |x| \leqq 1, |y| \leqq 1$

   (b) $|x| \leqq 1, |y| \leqq 1, |x| \neq |y|$

   (c) $x^2 + y^2 \leqq 4, 1 \leqq |x| \leqq 2$

   (d) $(x^2 + y^2 + x)^2 \leqq x^2 + y^2$

3. Same problem for

   (a) $|x + y| \leqq 1$

   (b) $1 \leqq |x| + |y|, |x| \leqq 1, |y| \leqq 1$

   (c) $|x + y| + |x - y| \leqq 4$

   (d) $y^2 - x^2 + 2x \leqq 1$

4. Describe the region
   $$R_y = R[-4, 5, -\sqrt{25 - y^2}, 3y/4]$$
   Express it as the sum of several regions R_x.

5. Same problem for
   $$R_y = R[-m, m, y^2 - m^2, 4m^2 - 4y^2]$$

6. Describe the region
   $$R_x = [0, \tfrac{1}{2}, \cos^{-1}(1 - x), \cos^{-1} x]$$
   Express it as the sum of several regions R_y.

7. Decompose the set of points (x, y) for which $2 \leqq x^2 + y^2 \leqq 4$ into two regions R_x. A point of the set may be a boundary point of both regions R_x.

8. Replace R_x by R_y in Exercise 7.

9. Decompose the set (c) of Exercise 2 and the set (b) of Exercise 3 into regions R_x.

10. Show analytically that a line joining an interior point with an exterior point of a region R_x cuts the boundary. Treat all cases.

§2. Properties of Double Integrals

We shall set down in this section certain elementary properties of double integrals. The proofs of these properties are all simple and may be supplied by the student. We shall also introduce the iterated integral. It will be seen that this latter is to the theory of integration as partial differentiation is to the theory of differentiation. Finally, we shall express the volume of a solid by use of a double integral.

2.1 A TABLE OF PROPERTIES

In the following table, $f(x, y)$, $g(x, y)$ are assumed continuous in the region R over which the function is integrated; k is a constant; A is the area of R. The statement $R = R_1 + R_2$ in property IV means that R, R_1, R_2 are regions R of the type described in §1.1 and that R is composed of R_1 and R_2. That is, every point of R is a point of R_1, a point of R_2, or a boundary point of both. The regions R_1 and R_2 do not overlap and have no points in common except some of their boundary points. For example, $R = R_1 + R_2$ if $R_1 = R[-1, 0, -\sqrt{1 - x^2}, \sqrt{1 - x^2}]$, $R_2 = R[0, 1, -\sqrt{1 - x^2}, \sqrt{1 - x^2}]$, $R = [-1, 1, -\sqrt{1 - x^2}, \sqrt{1 - x^2}]$.

$$\text{I.} \quad \iint_R dS = A.$$

$$\text{II.} \quad \iint_R kf(x, y) \, dS = k \iint_R f(x, y) \, dS.$$

$$\text{III.} \quad \iint_R [f(x, y) + g(x, y)] \, dS = \iint_R f(x, y) \, dS + \iint_R g(x, y) \, dS.$$

$$\text{IV.} \quad R = R_1 + R_2 \Rightarrow \iint_R f(x, y) \, dS = \iint_{R_1} f(x, y) \, dS + \iint_{R_2} f(x, y) \, dS.$$

$$\text{V.} \quad f(x, y) \leq g(x, y) \Rightarrow \iint_R f(x, y) \, dS \leq \iint_R g(x, y) \, dS.$$

$$\text{VI.} \quad \left| \iint_R f(x, y) \, dS \right| \leq \iint_R |f(x, y)| \, dS.$$

$$\text{VII.} \quad \left| \iint_R f(x, y) \, dS \right| \leq A \max_{(x,y)\in R} |f(x, y)|.$$

2.2 ITERATED INTEGRALS

Definition 4. *An iterated integral is an integral of the form*

$$\int_a^b dx \int_{\varphi(x)}^{\psi(x)} f(x, y) \, dy$$

This means that for each fixed x between a and b the integral

(1) $$F(x) = \int_{\varphi(x)}^{\psi(x)} f(x, y)\, dy$$

is evaluated; then the integral

$$\int_a^b F(x)\, dx$$

is computed.   In view of the fact that x is held constant during the integration (1), it is clear that the computation of an iterated integral is somewhat analogous to the process of partial differentiation.

EXAMPLE A. $a = -\frac{1}{2}$, $b = 1$, $\varphi(x) = -x$, $\psi(x) = 1 + x$

$$f(x, y) = x^2 + y$$

$$\int_{-1/2}^{1} dx \int_{-x}^{1+x} (x^2 + y)\, dy = \int_{-1/2}^{1} \left[x^2 y + \frac{y^2}{2} \right]_{-x}^{1+x} dx$$

$$= \int_{-1/2}^{1} [2x^3 + x^2 + x + \tfrac{1}{2}]\, dx = \frac{63}{32}$$

2.3 VOLUME OF A SOLID

The usefulness of the double integral derives mainly from the fact that many physical quantities can be expressed in terms of it.   For example, the volume of certain solids can be so expressed.   At every point of the boundary C of a simply connected region R of the xy-plane, erect a perpendicular to the plane, thus generating a cylinder.   Let us find the volume bounded by this cylinder, the surface $z = f(x, y)$, and the plane $z = 0$.   Assume $f(x, y) \in C$ and $f(x, y) > 0$ in R.   We make two postulates about the volume of a solid:

A. *Volume is additive.   That is, if a solid A of volume V is composed of two other distinct solids A_1 and A_2 of volumes V_1 and V_2, respectively, then $V = V_1 + V_2$.*

B. *If a solid A_1 of volume V_1 is a part of solid A_2 of volume V_2, then $V_1 \leq V_2$.*

Make a subdivision of R by curves $\{C_k\}^n$ and erect cylinders on each curve C_k.   Denote the volume between the cylinder on C_k, the surface $z = f(x, y)$ and the plane $z = 0$, by ΔV_k.   If $f(x'_k, y'_k)$ and $f(x''_k, y''_k)$ are the minimum and maximum values, respectively, of $f(x, y)$ in the subregion bounded by C_k, then from the volume of a cylinder and by postulate B we have

$$f(x'_k, y'_k)\, \Delta S_k \leq \Delta V_k \leq f(x''_k, y''_k)\, \Delta S_k$$

Here ΔS_k is the area of the subregion bounded by C_k. By postulate A, we see that the required volume V lies between two sums,

$$\sum_{k=1}^{n} f(x'_k, y'_k)\,\Delta S_k \leq V \leq \sum_{k=1}^{n} f(x''_k, y''_k)\,\Delta S_k$$

If now the norm of the subdivision approaches zero, we see by Theorem 1 that

$$V = \int\int_R f(x, y)\,dS$$

EXAMPLE B. The volume of a sphere of radius a is

$$V = 2\int\int_R \sqrt{a^2 - x^2 - y^2}\,dS$$

where

$$R = R[-a, a, -\sqrt{a^2 - x^2}, \sqrt{a^2 - x^2}]$$

E X E R C I S E S (2)

1. Under the assumptions of §2.1, prove Properties I, II, III.

2. Work out the same problem for IV. Explain what to do about a subdivision of R which produces subregions lying partly in R_1 and partly in R_2.

3. Prove, V, VI, and VII.

4. $\displaystyle\int_0^{\pi/4} dx \int_0^{\sec x} y^3\,dy = ?$

5. $\displaystyle\int_{-1}^1 dx \int_0^{\sqrt{1-x^2}} \left(1 - x^2 - y^2\right) dy = ?$

6. $\displaystyle\int_{-2}^2 dx \int_{2-x}^{e^x} (x^2 + 3y^2)\,dy = ?$

In the following three problems, express the volume of the solid as a double integral, defining R as a region R_x:

7. A tetrahedron with vertices at the points $(a, 0, 0)$, $(0, b, 0)$, $(0, 0, c)$, $(0, 0, 0)$, where a, b, and c are positive numbers.

8. A general ellipsoid.

9. A solid bounded by the planes $z = x + a$, $z = -x - a$, and by the cylinder $x^2 + y^2 = a^2$.

10. Prove that the mass of a lamina of variable density $\rho = f(x, y)$ is

$$M = \int\int_R f(x, y)\,dS$$

Define average density and density at a point. State carefully what postulates about mass you are accepting.

11. Express as a double integral the mass of a right triangle whose density varies as the square of the distance from an acute-angled vertex.

12. Solve the same problem for an ellipse, the density being proportional to the square of the distance to the nearest focus.

13. Prove the law of the mean for double integrals:

$$\iint_R f(x, y)\, dS = f(\xi, \eta)A \qquad\qquad (\xi, \eta) \in R$$

You may assume that a continuous function has a maximum and a minimum in a closed region R, and that it takes on each value between the two at some point of R.

§3. Evaluation of Double Integrals

The definition of a double integral as a limit of a sum gives no clue as to a method of evaluating it. An iterated integral, on the other hand, can be evaluated by successive integrations. We shall show in this section that a double integral may be expressed in certain cases as an iterated integral.

3.1 THE FUNDAMENTAL THEOREM

Theorem 2.   1. $f(x, y) \in C$ in R_x

2. $R_x = R[a, b, \varphi(x), \psi(x)]$

$$\Rightarrow \qquad \iint_{R_x} f(x, y)\, dS = \int_a^b dx \int_{\varphi(x)}^{\psi(x)} f(x, y)\, dy$$

Recall that by the definition of R_x the functions $\varphi(x)$ and $\psi(x)$ are continuous in $a \leq x \leq b$ and $\varphi(x) < \psi(x)$ in $a < x < b$. Choose a constant A so that $A < \varphi(x)$ for $a \leq x \leq b$. If $R'_x = R[a, b, A, \varphi(x)]$ and $R''_x = R[a, b, A, \psi(x)]$ then $R' + R = R''$. Suppose that the theorem has been proved for the special case in which $\varphi(x)$ is replaced by a constant A. Extend the definition of $f(x, y)$ to the region R'' so that $f(x, y) \in C$ in R''. Then by property IV of 2.1,

$$\iint_{R''} f(x, y)\, dS = \iint_{R'} f(x, y)\, dS + \iint_R f(x, y)\, dS$$

Applying the theorem to the integrals over R'' and R', both of admissible type, we have

$$\iint_R f(x, y)\, dS = \int_a^b dx \int_A^{\psi(x)} f(x, y)\, dy - \int_a^b dx \int_A^{\varphi(x)} f(x, y)\, dy$$

$$= \int_a^b dx \int_{\varphi(x)}^{\psi(x)} f(x, y)\, dy$$

Thus, the theorem in its general form would follow from its special form. By an easy change of coordinates, we may also assume that $A = 0$.

Suppose that the minimum of $\psi(x)$ in $a \leq x \leq b$ is $m > 0$. Set $\Delta x = (b - a)/n$ and

$$F(x) = \int_0^{\psi(x)} f(x, y) \, dy$$

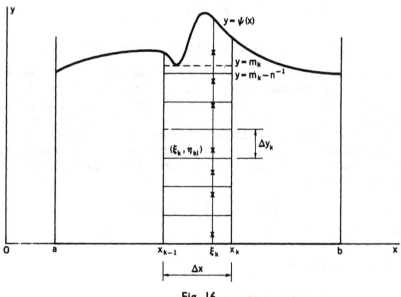

Fig. 16.

Then $F(x) \in C$ in $a \leq x \leq b$ (compare §7, Chapter 10). Hence,

(1) $$\int_a^b F(x) \, dx = \lim_{n \to \infty} \Delta x \sum_{k=1}^n F(\xi_k)$$

where

$$x_{k-1} = a + (k - 1) \Delta x \leq \xi_k \leq a + k \, \Delta x = x_k \qquad k = 1, 2, \ldots, n$$

Actually, we choose ξ_k so that (Figure 16)

$$\int_{x_{k-1}}^{x_k} \psi(x) \, dx = \psi(\xi_k) \, \Delta x$$

Let m_k be the minimum of $\psi(x)$ in $x_{k-1} \leq x \leq x_k$. Choose n_0 so large that when $n > n_0$ we will have $n^{-1} < m$. Divide the region $R[x_{k-1}, x_k, 0, \psi(x)]$ into $n + 1$ subregions by drawing the horizontal lines

$$y = i \, \Delta y_k, \qquad i = 1, 2, \ldots, n, \qquad \text{where} \qquad \Delta y_k = [m_k - n^{-1}]n^{-1}$$

All but one of these will be rectangles of area $\Delta x \, \Delta y_k$. The area of the exceptional one is

$$\int_{x_{k-1}}^{x_k} [\psi(x) - m_k + n^{-1}]\, dx = \Delta x [\psi(\xi_k) - m_k + n^{-1}]$$

Now by the law of the mean for integrals, we have

$$F(\xi_k) = \sum_{i=1}^{n} \int_{(i-1)\Delta y_k}^{i\Delta y_k} f(\xi_k, y)\, dy + \int_{m_k-(1/n)}^{\psi(\xi_k)} f(\xi_k, y)\, dy$$

(2) $$F(\xi_k) = \Delta y_k \sum_{i=1}^{n} f(\xi_k, \eta_{ki}) + [\psi(\xi_k) - m_k + n^{-1}] f(\xi_k, \eta_{k\,n+1})$$

If we multiply equations (2) by Δx and sum for $k = 1, 2, \ldots, n$, we shall have on the left the sum which appears on the right of equation (1). On the right of equation (2), we shall have a sum of $n^2 + n$ terms corresponding to the $n^2 + n$ subregions of our subdivision Δ of R. The term corresponding to a given subregion is a product of two factors, one of which is the area of the subregion, the other of which is $f(x, y)$ formed at a point of the subregion. As n becomes infinite, $\|\Delta\|$ approaches zero. Hence, by Theorem 1 and equation (1), we have

$$\int_a^b F(x)\, dx = \int\int_R f(x, y)\, dS$$

and our theorem is proved.

3.2 ILLUSTRATIONS

We illustrate by two examples.

EXAMPLE A. $R_x = R[-\tfrac{1}{2}, 1, -x, 1 + x)], f(x, y) = x^2 + y$
The region R_x is a triangle with vertices at the points

$$(-\tfrac{1}{2}, \tfrac{1}{2}), \quad (1, 2), \quad (1, -1).$$

Then

$$\int\int_{R_x} (x^2 + y)\, dS = \int_{-1/2}^{1} dx \int_{-x}^{1+x} (x^2 + y)\, dy = \frac{63}{32}$$

The iterated integral was evaluated in §2.2.

EXAMPLE B. Find the volume of a sphere of radius a. We have from Example B of §2.3

$$V = 2 \int_{-a}^{a} dx \int_{-\sqrt{a^2-x^2}}^{\sqrt{a^2-x^2}} \sqrt{a^2 - x^2 - y^2}\, dy$$

$$= \pi \int_{-a}^{a} (a^2 - x^2)\, dx = \frac{4}{3}\pi a^3$$

EXERCISES (3)

1. Find the volume of Exercise 7, §2.

2. Find the volume of Exercise 8, §2.

3. Find the volume of Exercise 9, §2.

4. Find the mass of the lamina of Exercise 11, §2.

5. Find the mass of the lamina of Exercise 12, §2.

6. Give the details of the change of coordinates in §3.1 designed to show that A may be taken equal to zero.

7. In §3.1 extend the definition of $f(x, y)$ to the region R'' by defining $f(x, y)$ to be constant on vertical lines outside the region R. Show that $f(x, y) \in C$ in R''.

8. Show that the diameter of the region $R[x_{k-1}, x_k, m_k - n^{-1}, \psi(x)]$ is not greater than $\sqrt{\Delta x^2 + h_k^2}$, where $h_k = M_k - m_k + n^{-1}$ and M_k is the maximum of $\psi(x)$ in $x_{k-1} \leq x \leq x_k$.

9. Give details of the proof that the norm of the subdivision described in §3.1 tends to zero with $1/n$. What theorem about continuous functions do you use to show that the maximum of the numbers h_k of Exercise 8 approaches zero with $1/n$?

10. State a theorem analogous to Theorem 2 for a region R_y.

11. Prove that, if $f(x, y) \in C$ in a suitable region R,

$$\int_a^b dx \int_a^x f(x, y) \, dy = \int_a^b dy \int_y^b f(x, y) \, dx$$

What is R?

12. Let $f(x) \in C$ and

$$f^{(0)}(x) = f(x), \qquad f^{(-n-1)}(x) = \int_0^x f^{(-n)}(t) \, dt$$

Prove that

$$f^{(-n-1)}(x) = \int_0^x \frac{(x-t)^n}{n!} f(t) \, dt$$

13. If $R_x = R[-1, 0, 0, x^2] + R[0, 2, 0, x^2]$, show that

$$\iint_{R_x} y^{3/2} \, dS = \frac{13}{3}$$

Iterate in both orders.

§4. Polar Coordinates

We shall obtain here a result analogous to Theorem 2 for the case in which the rectangular coordinates (x, y) of that theorem are replaced by the polar coordinates (θ, r). We shall proceed by analogy, omitting some of the more obvious details.

4.1 REGION R_θ AND R_r

We define special regions R_θ and R_r, just as in rectangular coordinates, except that we now exclude the pole from our regions. Thus for $R_\theta = R[\alpha, \beta, \varphi(\theta), \psi(\theta)]$ we demand, in addition to previous requirements, that $0 < \varphi(\theta)$ for $\alpha \leq \theta \leq \beta$. For $R_r = R[a, b, \varphi(r), \psi(r)]$ we now assume $0 < a$. In problems involving regions which include the pole a limiting process is usually effective (see Examples A and C below).

For example, the region $R_\theta = R[\alpha, \beta, a, b]$ is the region between the rays $\theta = \alpha$, $\theta = \beta$ and between the circles $r = a$ and $r = b$. Here $0 \leq \alpha < \beta \leq 2\pi$ and $0 < a < b$. By an elementary integral formula, the area of this region is

$$(1) \qquad A = \tfrac{1}{2} \int_\alpha^\beta [b^2 - a^2] \, d\theta$$

More generally, the area of the region $R_\theta = \dot{R}[\alpha, \beta, a, \psi(\theta)]$ is

$$(2) \qquad A = \tfrac{1}{2} \int_\alpha^\beta [\psi^2(\theta) - a^2] \, d\theta$$

4.2 THE FUNDAMENTAL THEOREM

Theorem 3.   1. $f(\theta, r) \in C$ in R_θ

2. $R_\theta = R[\alpha, \beta, \varphi(\theta), \psi(\theta)]$

$$\Rightarrow \qquad \iint_{R_\theta} f(\theta, r) \, dS = \int_\alpha^\beta d\theta \int_{\varphi(\theta)}^{\psi(\theta)} f(\theta, r) r \, dr$$

As in the proof of Theorem 2, it is sufficient to suppose that $\varphi(\theta) = A$. Set $\Delta\theta = (\beta - \alpha)/n$, $\theta_k = \alpha + k \, \Delta\theta$ and

$$F(\theta) = \int_0^{\psi(\theta)} f(\theta, r) r \, dr$$

Then

$$(3) \qquad \int_\alpha^\beta F(\theta) \, d\theta = \lim_{n \to \infty} \Delta\theta \sum_{k=1}^n F(\xi_k)$$

In this case, we choose ξ_k so that

$$(4) \qquad \frac{1}{2} \int_{\theta_{k-1}}^{\theta_k} \psi^2(\theta) \, d\theta = \tfrac{1}{2} \psi^2(\xi_k) \, \Delta\theta$$

Our subregions are now introduced by drawing the rays $\theta = \theta_k$ and the circles

$$r = i \, \Delta r_k, \qquad i = 1, 2, \dots, n$$

where

$$\Delta r_k = [m_k - n^{-1}] n^{-1}$$

The constant m_k is the minimum of $\psi(\theta)$ in $\theta_{k-1} \leq \theta \leq \theta_k$. Using Corollary 3 §4.1 of Chapter 5 (taking r as the positive function g), we have

$$\Delta\theta F(\xi_k) = \Delta\theta \sum_{i=1}^{n} \int_{(i-1)\Delta r_k}^{i\Delta r_k} f(\xi_k, r) r \, dr + \Delta\theta \int_{m_k-(1/n)}^{\psi(\xi_k)} f(\xi_k, r) r \, dr$$

(5)
$$= \sum_{i=1}^{n} f(\xi_k, \eta_{ki}) A_{ki} + f(\xi_k, \eta_{k\,n+1}) A_{k\,n+1}$$

where

$$A_{ki} = \Delta\theta \int_{(i-1)\Delta r_k}^{i\Delta r_k} r \, dr \qquad\qquad i = 1, 2, \ldots, n$$

$$A_{k\,n+1} = \Delta\theta \int_{m_k-(1/n)}^{\psi(\xi_k)} r \, dr$$

By formulas (1) and (2) and by the choice of ξ_k, so as to make equation (4) valid, it is clear that A_{ki} is the area of a certain subregion and that the point $\theta = \xi_k$, $r = \eta_{ki}$ is a point of that subregion. Since the norm of the subdivision tends to zero with $1/n$, we see from the definition of the double integral that the right-hand side of equation (3), with the sum expanded by use of equation (5), is equal to the double integral of f extended over R_θ. This completes the proof.

4.3 ILLUSTRATIONS

EXAMPLE A.   Find the area of a circle.   We compute the area A_ϵ of the region $R_\theta = R[\epsilon, 2\pi, \epsilon, a]$ and let $\epsilon \to 0$.

$$A_\epsilon = \int\int_{R_\theta} dS = \int_\epsilon^{2\pi} d\theta \int_\epsilon^a r \, dr = \tfrac{1}{2}(2\pi - \epsilon)(a^2 - \epsilon^2) \to \pi a^2$$

Frequently this work is abbreviated, as in Example C below, by replacing ϵ by 0 in the iterated integral. But it should be understood that Theorem 3 is not applicable to regions which include the pole without intervention of a limit process.

EXAMPLE B.   The semicircle $y = \sqrt{a^2 - (x - b)^2}$, $b > a$, is rotated in the xy-plane about the origin through $90°$.   Find the area traced out.

The equation of the curve in polar coordinates is

$$\theta = f(r) = \cos^{-1}\left(\frac{r^2 + b^2 - a^2}{2br}\right)$$

The region whose area is required is not a region R_θ. It is, however, the region $R_r = R[b - a, b + a, f(r), f(r) + \pi/2]$.   Hence, the area is

$$A = \int\int_{R_r} dS = \int_{b-a}^{b+a} r \, dr \int_{f(r)}^{f(r)+\pi/2} d\theta = \pi a b$$

EXAMPLE C. Find the mass of a circular lamina whose density is proportional to the square of the distance from a fixed diameter. Here

$$M = k \iint_R x^2 \, dS$$

where R is the region of Example A and k is a constant of proportionality. Then

$$M = k \int_0^{2\pi} d\theta \int_0^a r^3 \cos^2 \theta \, dr = \frac{k\pi a^4}{4}$$

EXERCISES (4)

1. Find the mass of a circular disc whose density is proportional to the distance from the center.

2. Solve the same problem for a square.

3. Find the area of a right triangle by the methods of the present section.

4. Solve the same problem for a general triangle.

5. Find the area of one lobe of the lemniscate

$$r^2 = a^2 \cos 2\theta$$

6. If R is the set of points (x, y) for which $x^2 + y^2 \leq 1$, find

$$\iint_R e^{x^2 + y^2} \, dS$$

7. Show by elementary calculus that the area of a region $R_r = R[A, B, C, \psi(r)]$ is

$$\int_A^B r[\psi(r) - C] \, dr$$

8. Show that for a suitable number ξ between A and B the area of Exercise 7 is equal to $[\psi(\xi) - C](B - A)D$, where D is the arithmetic mean of A and B.

9. If $F(r) \in C$ in $A \leq r \leq B$ and $r_k = A + k(B - A)n^{-1}$, $k = 0, 1, \ldots, n$, show that

$$\int_A^B F(r)r \, dr = \lim_{n \to \infty} \sum_{k=1}^n F(\xi_k)(r_k^2 - r_{k-1}^2)/2 \qquad r_{k-1} \leq \xi_k \leq r_k$$

10. If $R_r = R[A, B, 0, \psi(r)]$, prove that

$$\iint_{R_r} f(\theta, r) \, dS = \int_A^B r \, dr \int_0^{\psi(r)} f(\theta, r) \, d\theta$$

Hint: Set

$$F(r) = \int_0^{\psi(r)} f(\theta, r) \, d\theta$$

and use Exercise 9. The points ξ_k of that exercise should be chosen so that the area of the region $R[r_{k-1}, r_k, C, \psi(r)]$ is $[\psi(\xi_k) - C][r_k - r_{k-1}^2]/2$. This is possible by Exercise 8. Now express the integral defining $F(\xi_k)$ as the sum of $(n + 1)$ others and apply the law of the mean to each.

11. Find the area bounded by the curve $r = \sin \theta$, using both orders for the iterated integrals.

12. Find the area between the curve $\theta = \sin r$ and the line segment $\theta = 0$, $0 \leqq r \leqq \pi$.

13. Find the double integral of the function $\sqrt{1 - r^2}$ over the region $R_\theta = R\left[-\dfrac{\pi}{2}, \dfrac{\pi}{2}, \cos \theta, 1 \right]$. Observe that the result must be positive since the integrand is nonnegative over the whole region.

§5. *Change in Order of Integration*

In this section we shall illustrate by examples the method of changing the order of integration in an iterated integral. No new theory is involved. The method is an application of Theorems 2 and 3. The iterated integral is first expressed as a double integral and the corresponding area of integration determined. Then the double integral is again expressed as an iterated integral, but this time the integration is in the opposite order. Frequently, it will be necessary to break the area into several parts since it need not be, in the first instance, a region R_x or R_y.

5.1 RECTANGULAR COORDINATES

We have already seen in Exercise 11 of §3 that

(1) $$\int_a^b dx \int_a^x f(x, y)\, dy = \int_a^b dy \int_y^b f(x, y)\, dx$$

Both integrals are equal to the double integral of $f(x, y)$ over a triangle. Equation (1) is known as *Dirichlet's formula*.

EXAMPLE A. Invert the order of integration in

$$\int_a^b dx \int_0^x f(x, y)\, dy \qquad\qquad 0 < a < b$$

This is a double integral over $R_x = R[a, b, 0, x]$. This is a trapezoid with parallel sides vertical. It is the sum of two regions R_y, a rectangle $R_y = R[0, a, a, b]$ and a triangle $R_y = R[a, b, y, b]$. Hence, the given integral is equal to

$$\int_0^a dy \int_a^b f(x, y)\, dx + \int_a^b dy \int_y^b f(x, y)\, dx$$

EXAMPLE B.

$$\int_0^1 dy \int_{1-y}^{1+y} f(x, y)\, dx = \int_0^1 dx \int_{1-x}^1 f(x, y)\, dy + \int_1^2 dx \int_{x-1}^1 f(x, y)\, dy$$

Here the region is the triangle bounded by the lines $y = 1$, $y = 1 - x$,
$y = x - 1$. As a check take $f(x, y) = 1$. Both
sides reduce to 1, the area of the triangle.

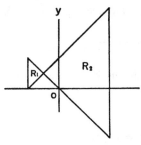

Fig. 17.

EXAMPLE C. Interchange the order of
integration in the iterated integral

$$I = \int_{-1}^1 dx \int_{-x}^{1+x} f(x, y)\, dy$$

Here the lines $x = -1$, $x = 1$, $y = -x$, $y = 1 + x$ do not bound a single region but two,
the triangles R_1 and R_2 of Figure 17. Note that
the integral I is the *difference* of the double integrals over R_2 and R_1

$$I = \iint_{R_2} f(x, y)\, dS - \iint_{R_1} f(x, y)\, dS$$

This is because $-x < 1 + x$ in the interval $-\tfrac{1}{2} < x < 1$ and $-x > 1 + x$
in the interval $-1 < x < -\tfrac{1}{2}$. Hence,

$$I = \int_{-1}^{1/2} dy \int_{-y}^1 f\, dx + \int_{1/2}^2 dy \int_{y-1}^1 f\, dx$$

$$- \int_0^{1/2} dy \int_{-1}^{y-1} f\, dx - \int_{1/2}^1 dy \int_{-1}^{-y} f\, dx$$

If $f(x, y) = 1$, $I = 2$. This is the difference between the areas of R_2
and R_1.

5.2 POLAR COORDINATES

The same procedure is applicable in polar coordinates.

EXAMPLE D. Obtain Dirichlet's formula by use of polar coordinates.
Clearly equation (1) must be unaltered if x is replaced by r, and y by θ
throughout, since x and y are "dummy" variables and do not appear in the
final result. We have to consider now the region $R_r = R[a, b, a, r]$ between
the circle $r = b$, the line $\theta = a$, and the spiral $r = \theta$. We have

$$\int_a^b dr \int_a^r f(r, \theta)\, d\theta = \iint_{R_r} \frac{f(r, \theta)}{r}\, dS$$

Introduction of the factor $1/r$ does not affect the continuity of the integrand since the origin is not in the region R_r.   Since R_r is also the region $R_\theta = R[a, b, \theta, b]$ we again get equation (1).

EXAMPLE E.   Invert the order of integration in

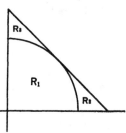

$$I = \int_0^{\pi/2} d\theta \int_0^{(\cos\theta + \sin\theta)^{-1}} f(r, \theta)\, dr$$

This is the double integral of f/r over the region $R_\theta = R[0, \pi/2, 0, (\cos\theta + \sin\theta)^{-1}]$.   This is the triangle bounded by the lines $x = 0$, $y = 0$, $x + y = 1$.   It is the sum of three regions R_r, indicated by R_1, R_2, R_3 in Figure 18.   Using the principal value of $\sin^{-1} r$ as usual, we have

Fig. 18.

$$I = \int_0^{1/\sqrt{2}} dr \int_0^{\pi/2} f\, d\theta + \int_{1/\sqrt{2}}^1 dr \int_0^{\varphi(r)} f\, d\theta + \int_{1/\sqrt{2}}^1 dr \int_{\psi(r)}^{\pi/2} f\, d\theta$$

where

$$\varphi(r) = -\frac{\pi}{4} + \sin^{-1}\frac{1}{r\sqrt{2}}$$

$$\psi(r) = \frac{3\pi}{4} - \sin^{-1}\frac{1}{r\sqrt{2}}$$

Since the region includes the pole the critical remarks in Example A, §4, are here applicable.

EXERCISES (5)

Interchange the order of integration in the following eleven integrals and sketch the region of integration for the corresponding double integrals.

1. $\displaystyle\int_0^1 dx \int_{x^2}^1 f(x, y)\, dy.$

2. $\displaystyle\int_{-2}^1 dx \int_{x^2}^1 f(x, y)\, dy.$

3. $\displaystyle\int_{-2}^1 dx \int_x^{x^2} f(x, y)\, dy.$

4. $\displaystyle\int_{1/3}^{2/3} dx \int_{x^2}^{\sqrt{x}} f(x, y)\, dy.$

5. $\displaystyle\int_a^{a\sqrt{2}} dy \int_{\cos^{-1}(a/y)}^{\pi/4} f(x, y)\, dx.$

6. $\displaystyle\int_0^{\pi/2} d\theta \int_0^{\cos\theta} rf(r, \theta)\, dr.$

7. $\displaystyle\int_{\pi/4}^{\pi/2} d\theta \int_{\sin\theta}^{2} f(r,\,\theta)\,dr.$

8. $\displaystyle\int_{0}^{\pi/2} d\theta \int_{1}^{2\cos\theta} rf(r,\,\theta)\,dr.$

9. $\displaystyle\int_{0}^{1} r\,dr \int_{-\sin^{-1}r}^{\sin^{-1}r} f(r,\,\theta)\,d\theta.$

10. $\displaystyle\int_{\sqrt{2}}^{4} dr \int_{\csc^{-1}r}^{\sec^{-1}r} f(r,\,\theta)\,d\theta.$

11. $\displaystyle\int_{1}^{4} dr \int_{\csc^{-1}r}^{\sec^{-1}r} f(r,\,\theta)\,d\theta.$

12. If R is the region of the first quadrant bounded by the curve $xy = 2$, the axes, and the lines $y = x + 1$, $y = x - 1$, express the double integral of $f(x,\,y)$ over R as an iterated integral in two ways.

13. Prove that

$$2!\int_{a}^{b} f(x)\,dx \int_{x}^{b} f(y)\,dy = \left[\int_{a}^{b} f(x)\,dx\right]^{2}$$

Hint: Apply Dirichlet's formula to the left-hand side and add the two iterated integrals.

14. Prove that

$$n!\int_{a}^{b} f(x_1)\,dx_1 \int_{x_1}^{b} f(x_2)\,dx_2 \ldots \int_{x_{n-1}}^{b} f(x_n)\,dx_n = \left[\int_{a}^{b} f(x)\,dx\right]^{n}$$

Illustrate by choosing $a = 0$, $b = \pi/2$, $f(x) = \cos x$.

§6. *Applications*

The double integral, like the simple Riemann integral, is a very useful tool in the formulation of certain physical concepts. We illustrate here by a number of examples. In many of these applications we need a result analogous to Theorem 9, Chapter 5. We refer to it as Duhamel's theorem.

6.1 DUHAMEL'S THEOREM

Theorem 4. 1. $f(x,\,y)$, $g(x,\,y) \in C$ $\hspace{4cm}(x,\,y) \in R$

$\hspace{1.8cm}$ 2. *A subdivision* Δ *divides* R *into subregions* R_k

$\hspace{9cm} k = 1,\,2,\,\ldots,\,n$

$\hspace{1.8cm}$ 3. $(x_k,\,y_k)$, $(\xi_k,\,\eta_k)$ *are points of* R_k $\hspace{1.5cm} k = 1,\,2,\,\ldots,\,n$

$$\Rightarrow \hspace{1cm} \lim_{\|\Delta\|\to 0} \sum_{k=1}^{n} f(x_k,\,y_k)g(\xi_k,\,\eta_k)\,\Delta S_k = \int\int_{R} f(x,\,y)g(x,\,y)\,dS$$

The proof is very similar to that of Theorem 9, Chapter 5. It depends, of course, on the uniform continuity of a continuous function in a closed bounded region.

6.2 CENTER OF GRAVITY OF A PLANE LAMINA

Let us illustrate the method by finding the center of gravity of a lamina, geometrically represented by a region R. Let the density at a point (x, y) of R be $f(x, y)$ and let $f(x, y) \in C$ in R. The center of gravity of n particles of masses $m_1, m_2, \ldots, m_n$ situated at points $(x_1, y_1), (x_2, y_2), \ldots, (x_n, y_n)$, respectively, is known to be at the point $(\bar{x}, \bar{y})$:

$$\bar{x} = \frac{1}{M} \sum_{k=1}^{n} x_k m_k, \qquad \bar{y} = \frac{1}{M} \sum_{k=1}^{n} y_k m_k, \qquad M = \sum_{k=1}^{n} m_k$$

The numbers $M\bar{x}$ and $M\bar{y}$ are known as the *x-moments* and the *y-moments* of mass about the origin. To understand their physical meaning, think of a weightless plane in a vertical position and let pellets of mass m_k be inserted into the plane at the points (x_k, y_k). If the plane has x-axis extending horizontally to the right and y-axis vertically upward, $M\bar{x}$ measures the tendency of the plane to turn about the origin in the clockwise sense. If the y-axis is vertical and the y-axis horizontal, $M\bar{y}$ measures the same tendency. The point $(\bar{x}, \bar{y})$ could be defined as a point such that the tendency of the plane, in either position, to turn about it will be zero.

We now formulate the following postulates about x-moments or y-moments. We state them for x-moments only.

A. *The x-moment is additive.*
B. *If mass is moved to the right (left), the x-moment is increased (decreased).*
C. *If mass to the right (left) of the origin is increased, the x-moment is increased (decreased).*

These postulates continue to have meaning if the mass in the plane has a continuous distribution instead of consisting of isolated particles. In fact, they are sufficient to define the x-moment of the lamina R described above.

Make a subdivision Δ of R, dividing it into subregions $R_k, k = 1, 2, \ldots, n$. Let (x_k', y_k') and (x_k'', y_k'') be points of R_k where the density $f(x, y)$ has its minimum and maximum values, respectively. If ΔS_k is the area of R_k and ΔM_k its mass, we have

$$f(x_k', y_k') \Delta S_k \leq \Delta M_k \leq f(x_k'', y_k'') \Delta S_k$$

By postulate A the x-moment μ_x of R is the sum of the x-moments of the subregions R_k. Let (ξ_k', η_k') be a point of R_k having minimum abscissa and (ξ_k'', η_k'') a point of maximum abscissa. By concentrating all the mass of R_k

first at one of these points and then at the other, we see by postulate B that

$$\sum_{k=1}^{n} \xi_k' f(x_k', y_k') \Delta S_k \leq \mu_x \leq \sum_{k=1}^{n} \xi_k'' f(x_k'', y_k'') \Delta S_k$$

By Theorem 4, both these sums approach the same double integral as $\|\Delta\| \to 0$, and

$$\mu = \iint_R xf(x, y) \, dS$$

Proceeding in a similar way for the y-moment, we get for the coordinates of the center of gravity

(1) $\bar{x} = \dfrac{1}{M} \iint_R xf(x, y) \, dS, \quad \bar{y} = \dfrac{1}{M} \iint_R yf(x, y) \, dS \quad M = \iint_R f(x, y) \, dS$

EXAMPLE A.   Find the center of gravity of a triangular region. The most general triangle may be so placed as to be the region $R_x = R[0, h, \lambda_1 x, \lambda_2 x]$, where $\lambda_1 < \lambda_2$. Then

$$\bar{x} = \frac{1}{A} \iint_R x \, dS = \frac{1}{A} \int_0^h x \, dx \int_{\lambda_1 x}^{\lambda_2 x} dy = \frac{1}{A} (\lambda_2 - \lambda_1) \frac{h^3}{3}$$

$$A = \iint_R dS = (\lambda_2 - \lambda_1) \frac{h^2}{2}$$

$$\bar{x} = \tfrac{2}{3}h$$

That is, the center of gravity lies on a line parallel to one base and half as far from that base as from the opposite vertex. Applying this same result, we see that $\bar{y} = (\lambda_1 + \lambda_2)h/3$, and the center of gravity is at the intersection of the medians of the triangle.

6.3 MOMENTS OF INERTIA

Using the postulates of §5.2, Chapter 5, we obtain the moment of inertia of the lamina of §6.2 about an axis perpendicular to it and passing through the point (a, b) as

(2) $$I = \iint_R [(x - a)^2 + (y - b)^2] f(x, y) \, dS$$

Similarly, if $g(x, y)$ is the square of the distance from the point (x, y) to the line L in the plane, then

(3) $$I = \iint_R g(x, y) f(x, y) \, dS$$

is the moment of inertia of the lamina about L.

EXAMPLE B. Find the moment of inertia of a circle about a diameter. Use polar coordinates and take the circle $r = a$, the diameter in question being the y-axis. Then if the density is taken as unity, the mass M is πa^2 and

$$I = \int\int_R x^2 \, dS = \int_0^{2\pi} \cos^2 \theta \, d\theta \int_0^a r^3 \, dr = \frac{\pi a^4}{4} = \frac{Ma^2}{4}$$

The *radius of gyration* k of R with respect to L is $(I/M)^{1/2}$, where M is given by equations (1) and I by (3). If all the mass of R is concentrated in a particle a distance k from L, its moment of inertia is unchanged.

EXERCISES (6)

1. Show that the center of gravity of an ellipse lies at its center.

2. Find the center of gravity of the area bounded by the curve

$$x^{2n} + y^{2n} = 1$$

where n is a positive integer.

3. The density of a circular lamina at a point (r, θ) is $(1 + r^4)^{-1}$. The center is at the pole. Find the moment of inertia about a perpendicular at the center.

4. Find the moment of inertia of a square about a perpendicular at the center if the density is proportional to the distance from the center.

5. Find the radius of gyration in the previous problem.

6. Express each coordinate of the center of gravity of the region $R_x = R[a, b, 0, \varphi(x)]$ as a simple integral.

7. Find the center of gravity of one lobe of the lemniscate $r^2 = a^2 \cos 2\theta$.

8. Find the moment of inertia of the area of the previous question about a perpendicular to the plane at the pole.

9. What is meant by the x-moment and the y-moment of a lamina about a point (a, b)? Obtain integral formulas for these numbers. Use these formulas to show that if (a, b) is determined so as to make both moments zero, then (a, b) is the center of gravity.

10. Obtain an integral formula for the moment of inertia of a solid of revoluton about the axis of revolution.

11. Find the moment of inertia of an anchor ring (solid torus) about its axis.

12. Make a subdivision Δ of R into n subregions R_k of equal area, and let (x_k, y_k) lie in R_k. Define the arithmetic average of $f(x, y)$ over R as the limit of the arithmetic averages of $f(x_1, y_1), f(x_2, y_2), \dots, f(x_n, y_n)$ as $\|\Delta\| \to 0$. Obtain an integral formula for it.

13. A man's height is the average height of a room whose ceiling is in the form of a hemisphere. At what points of the floor can he stand upright?

14. Solve the same problem for a room in the form of a right circular cone.

15. Find the center of gravity of one lobe of the curve $r = a \cos 3\theta$.

16. By using the elementary formula for volume of revolution prove the theorem of Pappus,

$$V = 2\pi A\bar{y}$$

for the volume V of a solid of revolution obtained by rotating the region $R[a, b, \varphi(x), \psi(x)]$, $\varphi(x) \geqq 0$, of area A and centroid $(\bar{x}, \bar{y})$ about the x-axis.

17. By using the theorem of Pappus find the volume of the set of points (x, y, z) for which

$$x^2 + y^2 + z^2 \leqq a^2, \qquad 4x^2 + 4y^2 \leqq (z - a)^2$$

§7. Further Applications

In this section, we apply the method of §6 to obtain several additional applications. We shall obtain an integral formula for the area of a surface and another for the attractive force between a lamina and a particle under the Newtonian law of attraction.

7.1 DEFINITION OF AREA

We begin by defining the area of a surface whose equation is $z = f(x, y)$, the function $f(x, y)$ being of class C^1 in a region R of the xy-plane. Make a subdivision Δ of R into subregions R_k, $k = 1, 2, \dots, n$. On R_k erect a cylinder with its rulings perpendicular to the xy-plane. At a point (x_k, y_k) of R_k erect a perpendicular to the xy-plane intersecting the given surface in a point P_k. At P_k draw a tangent plane to the given surface and denote by $\Delta\sigma_k$ the area of this plane cut out by the cylinder on R_k. The area of the surface $z = f(x, y)$ cut out by the cylinder on R is defined as

$$\lim_{\|\Delta\| \to 0} \sum_{k=1}^{n} \Delta\sigma_k$$

We shall show that this limit exists under the conditions assumed and that it has the value

(1) $$A = \iint_R \sqrt{1 + f_1^2 + f_2^2}\, dS$$

7.2 A PRELIMINARY RESULT

Let A be the area of a square and B the area of a rectangle, the square and rectangle lying in two different planes which make an acute angle γ with each other. Suppose further that both quadrilaterals have two sides parallel to the line of intersection of the planes and that the square is the projection of the rectangle on the plane of the square (Figure 19). Then $B = A \sec \gamma$. For, if the length of the side of the square is l, the dimensions of the rectangle are l by $l \sec \gamma$. More generally, the area of any region projects by use of the same equation. For, any area is the limit of a sum of rectangles or squares.

7.3 THE INTEGRAL FORMULA

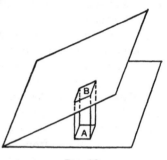

Fig. 19.

In the definition of the area of the surface $z = f(x, y)$, denote the area of R_k by ΔS_k. Then by the result of §7.2, $\Delta \sigma_k = \Delta S_k \sec \gamma_k$, where γ_k is the acute angle between the tangent plane at P_k and the xy-plane. Then

$$A = \lim_{\|\Delta\| \to 0} \sum_{k=1}^{n} \Delta S_k \sec \gamma_k = \iint_R \sec \gamma \, dS$$

provided $\sec \gamma$ is a continuous function of x, y in R. The direction components of the normal to the surface at a point $(x, y, f(x, y))$ are $f_1(x, y)$. $f_2(x, y), -1$, so that for the acute angle γ we have

$$\cos \gamma = \frac{1}{\sqrt{f_1^2 + f_2^2 + 1}}$$

and formula (1) is proved.

EXAMPLE A.   Find the area of a sphere of radius a.   Take the equation of the hemisphere as

$$z = f(x, y) = \sqrt{a^2 - x^2 - y^2}$$

Note that $f(x, y)$ does not belong to C^1 in the circle $x^2 + y^2 \leq a^2$. Let us find the area of the surface above the circle $x^2 + y^2 = b^2$, $b < a$, and then let $b \to a$. With obvious notations,

$$A_b = \iint_{R_b} \sqrt{1 + \frac{x^2}{z^2} + \frac{y^2}{z^2}} \, dS = a \int_0^{2\pi} d\theta \int_0^b \frac{r}{\sqrt{a^2 - r^2}} \, dr$$

$$= 2\pi a[a - \sqrt{a^2 - b^2}]$$

$$\lim_{b \to a-} A_b = 2\pi a^2$$

The area of the whole sphere is $4\pi a^2$.

7.4 CRITIQUE OF THE DEFINITION

In view of the student's experience with the definition of arc length of a curve as the limit of the lengths of inscribed polygons, the definition of area in §7.1 may be unexpected. It might seem more natural to consider the area as a limit of areas of inscribed polyhedra. But the latter limit need not exist, even for very simple surfaces. Let us illustrate by a right cylinder of altitude a erected on the circle $x^2 + y^2 = 1$. Its curved surface has area $2\pi a$. Let us inscribe a polyhedron whose faces consist of isosceles triangles as follows. Divide the circumference of each base into n equal arcs subtending angles $\Delta\theta = 2\pi/n$ at the centers, but let the points of subdivision of the top circumference lie midway between those of the bottom circumference. Draw a straight line from each point of its two neighbors on the same circle and to the two nearest points of subdivision on the other circle. The inscribed polyhedron thus formed has $2n$ isosceles triangles for faces. The base of each triangle is $2 \sin (\Delta\theta/2)$ and its altitude, computed by the Pythagorean theorem, is

$$c = \sqrt{a^2 + \left(1 - \cos \frac{\Delta\theta}{2}\right)^2}$$

Hence, the area of the inscribed polyhedron is $2nc \sin (\Delta\theta/2)$.

Next suppose that the number of sides of the polyhedron is increased by first dividing the cylinder into m equal cylinders by planes parallel to the base and then proceeding with each as above. In the expression for c we must replace a by a/m, and we must note that the total number of faces is now $2mn$. The total area of the inscribed polyhedron is

$$A(m, n) = 2n \sin (\pi/n)\sqrt{a^2 + m^2[1 - \cos (\pi/n)]^2}$$

Note that

$$\lim_{m \to \infty} A(m, n) = \infty, \qquad \lim_{n \to \infty} A(n, n) = 2\pi a$$

$$\lim_{n \to \infty} A(n^2, n) = 2\pi\sqrt{a^2 + (\pi^4/4)}$$

Hence, $A(m, n)$ approaches no limit as the number of faces becomes infinite.[*]

7.5 ATTRACTION

Two particles of masses m_1 and m_2 a distance r apart attract each other, according to the Newtonian law, with a force equal to

$$F = K \frac{m_1 m_2}{r^2}$$

---

[*] This example is due to H. A. Schwarz, *Gesammelte Mathematische Abhandlungen*, Vol. 2, p. 309. Berlin: Julius Springer, 1890.

where K is a constant depending upon the units employed. From this law we could set up postulates like those of §6.2 which would continue to have meaning for a continuous distribution of mass. Without giving details, let us find the attraction of a lamina R on a unit particle in the plane of the lamina but outside R. Let (r, θ) be the polar coordinates of a point of R and let the density of the lamina at that point be $f(r, \theta)$. Assume the particle to be at the pole and compute the component of the attraction F_x in the prime direction. Then with the usual notations we have

$$F_x = \lim_{\|\Delta\| \to 0} \sum_{k=1}^{n} K \frac{f(r_k, \theta_k)}{r_k^2} \cos \theta_k \, \Delta S_k$$

(2)
$$= K \int\int_R \frac{f(r, \theta) \cos \theta}{r^2} \, dS$$

EXAMPLE B. Find the attraction on a particle at the pole by a lamina $R_\theta = R[-\pi/2, \pi/2, a, b]$ of unit density. The region R lies between two concentric semicircles. By symmetry the total attraction will be equal to the x-component

$$F_x = K \int_{-\pi/2}^{\pi/2} \cos \theta \, d\theta \int_a^b \frac{dr}{r} = 2K \log \frac{b}{a}$$

If the symmetry of the present example is lacking, it is necessary to obtain the components of the attraction in two perpendicular directions. The total attraction can then be obtained by use of the parallelogram of forces.

EXERCISES (7)

1. Find the area of the surface cut out of a sphere of radius a by a cylinder of diameter a if one of the rulings of the cylinder is a diameter of the sphere.

2. Find the total area cut out of the surface

$$z = \tan^{-1} \frac{y}{x}$$

by the cylinder $x^2 + y^2 = a^2$ Describe the surface. Note that it is discontinuous where it is cut by the yz-plane.

3. Find the areas of a cone and a cylinder by the present methods.

4. Find the area cut out of a sphere of radius a by a square hole of side $2b$ ($b < a/\sqrt{2}$), the axis of the hole being a diameter of the sphere.

5. Find the area of a torus by the present methods.

6. Find the area of the surface $z = xy$ over the circle $x^2 + y^2 = a^2$.

7. A region R is bounded by the rays $\theta = \pi/2$, $\theta = 2\pi$ and by the branches of the spirals $r = \theta, r = 2\theta$ nearest the pole. Find the total force of attraction of R on a particle at the pole. Describe its direction.

8. Find the attraction of a semicircular lamina on a particle at the point of the circumference (extended) farthest from the lamina.

9. Give the postulates mentioned in §7.5.   Use them to derive formula (2).

10. Show that one component of the attraction of the surface of §7.1 (of unity density) on a particle is

$$K \int \int_R \frac{\cos \psi}{\rho^2} \sqrt{1 + f_1^2 + f_2^2} \, dS$$

   Describe ρ and ψ.

11. Find the attraction of a hemispherical shell on a particle at the point of the sphere (produced) which is farthest from the shell.

12. Give the details in the computation of the function $A(m, n)$ of §7.4.

13. Show that the faces of the polyhedra of §7.4 approach a position parallel to the base of the cylinder as $m \to \infty$ (n fixed).   What relation has this fact with the equation

$$\lim_{m \to \infty} A(m, n) = \infty$$

14. Give an example to show that $A(m, n)$ may become infinite when both m and n become infinite.

15. Obtain the elementary formula for the area of a surface of revolution

$$2\pi \int_a^b y \sqrt{1 + (y')^2} \, dx$$

   from equation (1).

§8. *Triple Integrals*

In this section we discuss integrals of functions of three variables over regions of three dimensional space.   The development is very similar to the corresponding one for double integrals, so that fewer details will be given.

8.1 DEFINITION OF THE INTEGRAL

Let $f(x, y, z)$ be defined in a closed bounded three dimensional region V having volume.   We define *subdivision* Δ and *norm* $\|\Delta\|$ in the obvious way. Suppose V is divided by Δ into subregions $V_1, V_2, \ldots, V_n$ and that (x_k, y_k, z_k) is a point of V_k, $k = 1, 2, \ldots, n$.   Then the triple integral of $f(x, y, z)$ over V is defined as

(1) $$\lim_{\|\Delta\| \to 0} \sum_{k=1}^n f(x_k, y_k, z_k) \, \Delta V_k = \int \int \int_V f(x, y, z) \, dV$$

when this limit exists.   The symbol ΔV_k denotes the volume of V_k.   A result analogous to Theorem 1 holds here: The integral (1) exists if $f \in C$ in V.   Properties I to VII of §2.1 apply here also, *mutatis mutandis*.

8.2 ITERATED INTEGRAL

The actual evaluation of a triple integral depends upon its expression as an iterated integral. This is possible for special types of regions which we shall denote, for example, by $V_{xy} = V[R, \varphi(x, y), \psi(x, y)]$. This is the region bounded by the surfaces $z = \varphi(x, y)$, $z = \psi(x, y)$ and the cylinder whose rulings are perpendicular to the xy-plane on the boundary of a region R of that plane. We suppose that φ, $\psi \in C$ in R and that $\varphi(x, y) < \psi(x, y)$ in R except perhaps on the boundary. As an illustration we may take

$$V[R, -\sqrt{a^2 - x^2 - y^2}, \sqrt{a^2 - x^2 - y^2}],$$

where the region R is

$$R_x = R[-a, a, -\sqrt{a^2 - x^2}, \sqrt{a^2 - x^2}]$$

Then V is the region bounded by the spherical surface $x^2 + y^2 + z^2 = a^2$.
The chief result here is contained in the following theorem.

Theorem 5.   1. $f(x, y, z) \in C$ in V_{xy}

2. $V_{xy} = V[R, \varphi(x, y), \psi(x, y)]$

$$\Rightarrow \qquad \iiint_{V_{xy}} f(x, y, z)\, dV = \iint_R dS \int_{\varphi(x,y)}^{\psi(x,y)} f(x, y, z)\, dz$$

By use of property IV we see as in §3.1 that it is sufficient to suppose φ identically zero. By following the proof of that section, it will be easy to fill in the details of the following sketch.
Set

$$(2) \qquad\qquad F(x, y) = \int_0^{\psi(x,y)} f(x, y, z)\, dz$$

and make a subdivision of R into n subregions R_k all of equal area ΔS, so that if the norm of this subdivision approaches zero as $n \to \infty$, we have

$$(3) \qquad \iint_R F(x, y)\, dS = \lim_{n\to\infty} \Delta S \sum_{k=1}^{n} F(x_k, y_k) \qquad (x_k, y_k) \in R_k.$$

We choose (x_k, y_k) so that

$$\iint_{R_k} \psi(x, y)\, dS = \psi(x_k, y_k)\, \Delta S$$

Set

$$m = \min_{(x,y)\in R} \psi(x, y), \qquad m_k = \min_{(x,y)\in R_k} \psi(x, y) \quad k = 1, 2, \ldots, n$$

We take n so large that $n^{-1} < m \; (m > 0)$.
Now divide the cylinder under ψ on R_k into $(n + 1)$ subregions by n equally spaced planes from $z = 0$ to $z = m_k - n^{-1}$, denoting the distance between successive planes by Δz_k. The volume of n of these subregions will

be $\Delta z_k \, \Delta S$, and the volume of the top one will be

$$\iint_{R_k} [\psi(x, y) - m_k + n^{-1}] \, dS = \Delta S[\psi(x_k, y_k) - m_k + n^{-1}]$$

Write the integral (2) formed for (x_k, y_k) as the sum of $n + 1$ integrals corresponding to the $(n + 1)$ intervals into which the interval $(0 \leq z \leq \psi(x_k, y_k))$ is divided by the n horizontal planes described above. Apply the law of the mean to each. For example,

$$\Delta S \int_{m_k - n^{-1}}^{\psi(x_k, y_k)} f(x_k, y_k, z) \, dz = f(x_k, y_k, \zeta_k) \Delta S[\psi(x_k, y_k) - m_k + n^{-1}]$$

The right-hand side is a product of f, formed at a point of the subregion at the top of the cylinder on R_k, by the volume of that subregion. Substituting the values thus obtained for $F(x_k, y_k)$ in the sum (3) we get a new sum of $n(n + 1)$ terms, each of which is in the form appearing in the sum (1). Hence, the limit is the triple integral of f over V, and the result is established. The result can also be applied to regions V_{yz} and V_{zx}.

EXAMPLE A.   Find the volume of the tetrahedron bounded by the planes $x = 0, y = 0, z = 0, a^{-1}x + b^{-1}y + c^{-1}z = 1$. Here

$$V_{xy} = V[R_x, 0, c(1 - a^{-1}x - b^{-1}y)]$$
$$R_x = R[0, a, 0, b(1 - a^{-1}x)]$$

The volume required is

$$\iiint_V dV = \iint_{R_x} dS \int_0^{c(1 - a^{-1}x - b^{-1}y)} dz$$
$$= c \int_0^a dx \int_0^{b(1 - a^{-1}x)} (1 - a^{-1}x - b^{-1}y) \, dy = \frac{abc}{6}$$

8.3 APPLICATIONS

We list here several integral formulas. A detailed discussion of them is omitted in view of their similarity to corresponding formulas in two dimensions.

I. Mass

$$M = \iiint_V f(x, y, z) \, dV$$

II. Center of gravity

$$\bar{x} = \frac{1}{M} \iiint_V x f(x, y, z) \, dV$$

III. Moment of inertia

$$I = \iiint_V r^2 f(x, y, z) \, dV$$

IV. Force of attraction

$$F_z = K \iiint_V f(r, \theta, \varphi) \frac{\cos \varphi}{r^2} \, dV$$

In all these integrals, f is the variable density of a solid V. In II only the x-coordinate of the center of gravity is given. Analogous formulas hold for y and z. The integral in III gives the moment of inertia of the solid V about an axis, and r is the distance from the point (x, y, z) to the axis. In IV (r, θ, φ) are the spherical coordinates of a point P of V:

$$x = r \sin \varphi \cos \theta, \qquad y = r \sin \varphi \sin \theta, \qquad z = r \cos \varphi$$

and F_z is the z-component of the force exerted on a unit particle at the origin by the solid V (supposed not to include the origin).

EXERCISES (8)

1. Find the moment of inertia of the solid of Example A about the x-axis. Express the triple integral involved as a triply iterated integral in the six possible ways. Evaluate two of them.

2. Find the center of gravity of the solid of Example A.

3. The density of a cube is proportional to the square of the distance from one vertex. Find the mass.

4. A column has for lower base the square with sides $\pm x \pm y = a$. The axis of the column is the z-axis. The upper base is the plane $z = h - x$. How far is the center of gravity from the axis?

5. Change the order of integration in
$$\int_0^a dx \int_0^x dy \int_0^y f(x, y, z)\, dz$$
(Five answers.)

6. Express the volume of the solid between the hemisphere
$$z = \sqrt{a^2 - x^2 - y^2}$$
and the plane $z = a/2$ as iterated integral in six ways. Do not integrate.

7. Solve the same problem for the cone with vertex at the origin and with base bounded by the circle
$$z = h, \qquad x^2 + y^2 = a^2$$

8. Describe the region V if
$$\iiint_V f(x, y, z)\, dV = \int_0^1 dy \int_{y-1}^{1-y} dx \int_{-\sqrt{(1-y)^2 - x^2}}^{\sqrt{(1-y)^2 - x^2}} f(x, y, z)\, dz$$

9. Find the volume of the region
$$x^2 + y^2 \leqq z^2, \qquad x^2 + y^2 + z^2 \leqq 1$$

10. Find the x-coordinate of the center of gravity of the solid

$$x^2 + y^2 \leqq 2ax, \qquad 0 \leqq az \leqq x^2 + y^2 \qquad (a > 0)$$

11. Prove the law of the mean for triple integrals:

$$\iiint_V f(x, y, z)\, dV = f(\xi, \eta, \zeta) U \qquad (\xi, \eta, \zeta) \in V$$

where U is the volume of V. You may assume that a continuous function has a maximum and a minimum in a closed region V and that it takes on each value between the two at some point of V.

§9. Other Coordinates

Many physical situations are more simply described in some system of coordinates not rectangular. The force of attraction given by formula IV of the previous section is a case in point. Again the position of a point near the surface of the earth might be fixed by its latitude, longitude, and distance above or below the surface of the earth. In this section we shall obtain results analogous to Theorem 5 for cylindrical and spherical coordinates.

9.1 CYLINDRICAL COORDINATES

Let the cylindrical coordinates of a point be (θ, r, z), related to the Cartesian coordinates of the point by the equations

$$x = r \cos \theta, \qquad y = r \sin \theta, \qquad z = z$$

In this system of coordinates a region $V_{\theta r} = V[R, \varphi(\theta, r), \psi(\theta, r)]$ is the region bounded by the surfaces $z = \varphi(\theta, r)$, $z = \psi(\theta, r)$ and the cylinder whose rulings are perpendicular to the plane $z = 0$ on the boundary of a region R. We suppose that φ, $\psi \in C$ in R and that $\varphi < \psi$ in R expect perhaps on the boundary.

Theorem 6. 1. $f(\theta, r, z) \in C$ in $V_{\theta r}$

2. $V_{\theta r} = V[R, \varphi(\theta, r), \psi(\theta, r)]$

$$\Rightarrow \qquad \iiint_{V_{\theta r}} f(\theta, r, z)\, dV = \iint_R dS \int_{\varphi(\theta, r)}^{\psi(\theta, r)} f(\theta, r, z)\, dz$$

The proof of this theorem is the same as that of Theorem 5. In applying it, however, one would evaluate the double integral over R by use of polar coordinates rather than rectangular.

EXAMPLE A. Find the volume of the cone $0 \leq z \leq h(a - r)/a$.

Here $V_{\theta r} = V[R, 0, h(a - r)/a]$, $R = R_\theta = R[0, 2\pi, 0, a]$. Hence, the volume is

$$\iiint_{V_{\theta r}} dV = \int_0^{2\pi} d\theta \int_0^a r\, dr \int_0^{h(a-r)/a} dz = \tfrac{1}{3}\pi a^2 h$$

9.2 SPHERICAL COORDINATES

Let us introduce spherical coordinates by means of the equations $x = r \sin \varphi \cos \theta$, $y = \sin \varphi \sin \theta$, $z = r \cos \varphi$. It will now be clear, for example, what is meant by a region $V_{\theta\varphi} = V[R, g(\theta, \varphi), h(\theta, \varphi)]$. We could describe it as the set of points (r, θ, φ) for which

(1) $$g(\theta, \varphi) \leq r \leq h(\theta, \varphi) \qquad\qquad (\theta, \varphi) \in R$$

where R is the set (θ, φ), for example, for which

(2) $$G(\theta) \leq \varphi \leq H(\theta) \qquad\qquad \alpha \leq \theta \leq \beta$$

Theorem 7.   1. $f(r, \theta, \varphi) \in C$ in $V_{\theta\varphi}$

         2. $V_{\theta\varphi} = V[R, g(\theta, \varphi), h(\theta, \varphi)]$

         3. $R = R_\theta = R[\alpha, \beta, G(\theta), H(\theta)]$

$\Rightarrow$ $$\iiint_{V_{\theta\varphi}} f(r, \theta, \varphi)\, dV = \int_\alpha^\beta d\theta \int_{G(\theta)}^{H(\theta)} \sin \varphi\, d\varphi \int_{g(\theta,\varphi)}^{h(\theta,\varphi)} f(r, \theta, \varphi) r^2\, dr$$

Make a subdivision Δ of V by the coordinate surfaces, obtained by holding each coordinate constant while the other two vary. Let us compute the volume of a typical subregion bounded, for example, by the spheres $r = r_0$, $r = r_0 + \Delta r$, the planes $\theta = \theta_0$, $\theta = \theta_0 + \Delta\theta$, and by the cones $\varphi = \varphi_0$, $\varphi = \varphi_0 + \Delta\varphi$. This region is obtained by rotating the area A of Figure 20 through an angle $\Delta\theta$ about the z-axis. By elementary calculation the required volume is

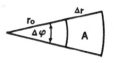

Fig. 20.

(3) $$\Delta V = \frac{\Delta\theta}{3} [\cos \varphi_0 - \cos (\varphi_0 + \Delta\varphi)][(r_0 + \Delta r)^3 - r_0^3]$$

By the law of the mean,

$$\Delta V = (r')^2 \sin \varphi' \, \Delta r \, \Delta\theta \, \Delta\varphi$$

where r' lies between r_0 and $r_0 + \Delta r$ and φ' is between φ_0 and $\varphi_0 + \Delta\varphi$. Hence,

(4) $$\iiint_V f(r, \theta, \varphi)\, dV = \lim_{\|\Delta\| \to 0} \sum_{k=1}^n f(r_k, \theta_k, \varphi_k) \Delta V_k$$

$$= \lim_{\|\Delta\| \to 0} \sum_{k=1}^n f(r_k, \theta_k, \varphi_k)(r')^2 \sin \varphi_k' \, \Delta r_k \, \Delta\theta_k \, \Delta\varphi_k$$

Next interpret r, θ, φ as rectangular coordinates. The region defined by inequalities (1) and (2) will now have a different shape. Call it V^*. By Duhamel's theorem the limit (4) will be the same if the accents are removed. It is equal then to

$$\iiint_{V^*} f(x, y, z)x^2 \sin z \, dV$$

and this may be written as an iterated integral by Theorem 5:

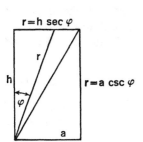

$r = h \sec \varphi$

$r = a \csc \varphi$

$$\iiint_{V^*} f(x, y, z)x^2 \sin z \, dV$$

$$= \int_\alpha^\beta dy \int_{G(y)}^{H(y)} \sin z \, dz \int_{g(y,z)}^{h(y,z)} f(x, y, z)x^2 \, dx$$

This is equivalent to the desired result when the dummy variables are renamed.

Fig. 21.

EXAMPLE B. Find the volume of a cylinder by use of spherical coordinates. Generate the cylinder by rotating a rectangle about the z-axis, Figure 21. Take the rectangle as the region R in $V_{r\varphi} = V[R, 0, 2\pi]$. Since the boundary of the rectangle cannot be given by a single equation, we must break the integral over $V_{r\varphi}$ into two parts:

$$\int_0^{\tan^{-1}(a/h)} \sin\varphi \, d\varphi \int_0^{h\sec\varphi} r^2 \, dr \int_0^{2\pi} d\theta + \int_{\tan^{-1}(a/h)}^{\pi/2} \sin\varphi \, d\varphi \int_0^{a\csc\varphi} r^2 \, dr \int_0^{2\pi} d\theta$$

$$= 2\pi \frac{h^3}{3} \int_0^{\tan^{-1}(a/h)} \sin\varphi \sec^3\varphi \, d\varphi + 2\pi \frac{a^3}{3} \int_{\tan^{-1}(a/h)}^{\pi/2} \csc^2\varphi \, d\varphi$$

$$= \frac{\pi}{3} a^2 h + \frac{2}{3}\pi a^2 h = \pi a^2 h$$

EXERCISES (9)

1. Compute the moment of inertia of a solid sphere about a diameter.

2. Solve the same problem for a solid cone about the axis.

3. Find the attraction of a solid cone of revolution on a particle at the vertex, using cylindrical coordinates.

4. Solve the same problem, using spherical coordinates.

5. Find the attraction of a pipe (a solid between two coaxial cylinders) on a particle on its axis. Discuss the limiting case in which the length of the pipe becomes infinite in one direction.

6. Find the volume of a cube using cylindrical coordinates.

7. Solve the same problem for spherical coordinates.

8. Find the attraction on a particle at the origin by a cube bounded by the planes $x = \pm h$, $y = \pm h$, $z = h$, $z = 3h$.

9. Using rectangular coordinates, express the triple integral of f over the region
$$x^2 + y^2 + z^2 \leq 2ay \leq 2a^2 \qquad (a > 0)$$
as an iterated integral.

10. Solve the same problem, using spherical coordinates.

11. Solve the same problem, using cylindrical coordinates.

12. Describe the region defined by inequalities (1) and (2) if
$$g(\theta, \varphi) = a \cos \varphi, \quad h(\theta, \varphi) = b \cos \varphi$$
$$G(\theta) = 0, \quad H(\theta) = \pi, \quad \alpha = 0, \quad \beta = 2\pi$$

13. Solve the same problem if r, θ, φ are thought of as rectangular coordinates.

14. Fill in the limits of integration in the equation
$$\int_0^\pi d\theta \int_0^{\pi/2} d\varphi \int_{a \cos \varphi}^a r^2 f(r, \theta, \varphi)\, dr = \int r^2\, dr \int d\theta \int f(r, \theta, \varphi)\, d\varphi$$

15. Express the following iterated integral in cylindrical coordinates:
$$\int_0^1 dx \int_0^{1-x} dy \int_0^{1-x-y} dz$$

16. Solve the same problem in spherical coordinates.

17. Prove equation (3). First show by simple integration that the volume of revolution obtained by rotating a circular vector (radius a and angle ω) about a radius is
$$\frac{2\pi a^3}{3} (1 - \cos \omega).$$

§10. *Existence of Double Integrals*

In this section we give a proof of the existence of the double integral of a continuous function. The proof is very similar to that given in §7 of Chapter 5, so that we shall omit some of the details. It is easy to see how the proof could be modified to apply to triple integrals.

10.I UNIFORM CONTINUITY

The existence of the double integral depends vitally on the uniform continuity of the function to be integrated.

Definition 5. *The function $f(x, y)$ is uniformly continuous in a region R $\Leftrightarrow$ to an arbitrary $\epsilon > 0$ corresponds a number δ such that for all points (x', y') and (x'', y'') of R for which $|x' - x''| < \delta$, $|y' - y''| < \delta$ we have*

$$|f(x', y') - f(x'', y'')| < \epsilon$$

As in §6.4 of Chapter 5, we could prove the following important result.

Theorem 8. 1. $f(x, y) \in C$ *in* R

$\Rightarrow$ $\qquad\qquad$ $f(x, y)$ *is uniformly continuous in* R

Recall that we defined R in §1 to be a closed bounded region. This is an essential part of the hypothesis of the theorem. By use of this result, we could now prove Theorem 4. By a corresponding result in three dimensions, we could also prove the form of Duhamel's theorem needed in §9.2.

10.2 PRELIMINARY RESULTS

For an arbitrary subdivision Δ of R into subregions R_k of area ΔS_k, $k = 1, 2, \ldots, n$, introduce the following notations:

$$M_k = \max_{(x,y)\in R_k} f(x, y), \qquad m_k = \min_{(x,y)\in R_k} f(x, y)$$

$$S_\Delta = \sum_{k=1}^{n} M_k \Delta S_k, \qquad s_\Delta = \sum_{k=1}^{n} m_k \Delta S_k$$

Clearly, $s_\Delta \leq S_\Delta$. We say that Δ' is a *refinement* of Δ if it is obtained from the latter by subdivision of the subregions of Δ.

Lemma 1.1. $S_\Delta \in \downarrow$, $s_\Delta \in \uparrow$ *under refinement of* Δ.

The proof follows as in §7.1.

Lemma 1.2. 1. Δ_1 *and* Δ_2 *are subdivisions of* R

$\Rightarrow$ $\qquad\qquad$ $s_{\Delta_1} \leq S_{\Delta_2}$, $s_{\Delta_2} \leq S_{\Delta_1}$

The proof follows as in §7.1, Chapter 5. It is at this stage that we need the assumption about the subdividing curves C_k made in §1.2. We now define s and S as the least upper bound of s_Δ and the greatest lower bound of S_Δ for all subdivisions Δ.

Lemma 1.3. 1. $f(x, y) \in C$ *in* R

$\Rightarrow$ $\qquad\qquad$ $s = S$

Let ϵ and δ be the numbers described in Definition 5. Then, if $\|\Delta\| < \delta$, we have by Theorem 8

$$0 \leq S_\Delta - s_\Delta = \sum_{k=1}^{n} (M_k - m_k) \Delta S_k \leq \epsilon A$$

Here A is the area of R.   As in §7.1, Chapter 5,

$$0 \le S - s \le \epsilon A$$

and the lemma is proved.

Lemma 1.4.   1. $f(x, y) \in C$ in R

$$\Rightarrow \qquad \lim_{\|\Delta\| \to 0} s_\Delta = \lim_{\|\Delta\| \to 0} S_\Delta = s = S$$

For, as in §7.1., Chapter 5, we have for $\|\Delta\| < \delta$

$$0 \le S_\Delta - S \le \epsilon A$$

$$0 \le s - s_\Delta \le \epsilon A$$

10.3 PROOF OF THEOREM I

Set σ_Δ equal to the sum appearing in equation (1) §1.2.   Then for any Δ

$$s_\Delta \le \sigma_\Delta \le S_\Delta$$

By Lemma 1.4, it is clear that

$$\lim_{\|\Delta\| \to 0} \sigma_\Delta = s = S$$

and the proof is complete.

10.4 AREA

Throughout this chapter we have assumed that the area of a plane region R is a known concept.   We conclude the chapter with a brief indication of the way in which it might be defined.   Assume the area of a square known. Cover R with a mesh of squares.   Denote the sum of the area of all squares consisting entirely of interior points of R by A_i and this sum plus the areas of all squares containing boundary points of R by A_e.   Clearly, the area A, which we seek to define, should lie between A_i and A_e.   If for all possible meshes of squares the least upper bound of A_i is equal to the greatest lower bound of A_e, the common value is defined as A.   We could now show that, if a subdivision of R is made into subregions, each of which has area, the sum of the areas of the subregion is equal to A.   This is the chief property of area which we have used in setting up the definition of a double integral.

It is interesting to observe that there are regions bounded by Jordan curves (§1, Chapter 7) which do not have area.* That is, $A_i \ne A_e$ for the region.   Of course, such regions are excluded from the discussions of the present chapter.   There is another definition of area due to H. Lebesgue under which every region bounded by a Jordan curve has area.   In fact,

---

* W. F. Osgood, "A Jordan curve of positive area." *Transactions of the American Mathematical Society*, Vol. 4 (1903), pp. 107–112.

every bounded closed point set has area (measure) under this more general definition.   If this definition of area is adopted, the definition of double integral in §1 is still valid and Theorem 1 still holds.   The resulting integral is then known as the *Lebesgue* rather than the *Riemann* integral.   However, for this new type of integral in its complete generality, the method of subdivision which we have employed is discarded in order to take care of integrands, which are very discontinuous.

E X E R C I S E S   (10)

1. Give an example of a function defined on a closed square that has no maximum value there.

2. Give an example of a continuous function $f(x, y)$ defined on the interior of a square and not bounded there.

3. State without proof the Heine–Borel theorem for two dimensions.   Use it to prove that a function $f(x, y)$ continuous in a region R (closed and bounded) is bounded there.

4. Give an example of a continuous function $f(x, y)$ that is not uniformly continuous.

5. Prove Theorem 8.

6. Prove Theorem 4 by use of Theorem 8.

7. Define uniform continuity for a function of three variables.   By use of a result corresponding to Theorem 8 prove the form of Duhamel's theorem required in §9.2.

7

Line and Surface
Integrals

§1. Introduction

In this chapter we generalize further the notion of integral. For the ordinary Riemann integral, the region of integration is an interval $a \leqq x \leqq b$. If the function to be integrated is defined along an arc of a curve in two or three dimensions, we can still define an integral over that region; the result is called a *line integral* or *curvilinear integral* over the arc. In like manner, the plane region of integration of a double integral can be replaced by a region on a curved surface, and the result is called a *surface integral*. In fact, these notions could be generalized to spaces of any number of dimensions.

1.1 CURVES

We shall be dealing with curves of various types. For easy reference let us introduce names for them. A *curve in the xy-plane* is a set of points (x, y) for which

$$(1) \qquad\qquad x = \varphi(t), \qquad y = \psi(t) \qquad\qquad a \leqq t \leqq b$$

217

where $\varphi(t) \in C$, $\psi(t) \in C$ in $a \leq t \leq b$. If $\varphi(a) = \varphi(b)$, $\psi(a) = \psi(b)$, the curve is *closed*. It is called a *Jordan curve* if it is closed and has no double points. That is, no two distinct values t in $a \leq t < b$ yield the same point (x, y). It can be shown that such a curve divides the plane into two parts, an exterior and an interior. See, for example, the *Cours d'Analyse* of de la Vallée Poussin, page 378 of the 1914 edition. This may seem obvious to the student, but he should recall that the curve is given by the pair of equations (1) and not by any drawing made on paper. There exist continuous curves (not Jordan curves) which pass through every point of a square. See, for example, *The Taylor Series* by P. Dienes, page 175. Of course, such a curve does not enclose an interior!

Definition 1. *The curve* (1) *is regular if it has no double points and if the interval* (a, b) *can be divided into a finite number of subintervals in each of which* $\varphi(t) \in C^1$, $\psi(t) \in C^1$ *and* $[\varphi'(t)]^2 + [\psi'(t)]^2 > 0$.

It is clear from elementary calculus that such a curve is sectionally smooth in the sense that it is composed of a finite number of arcs, each of which has a continuously turning tangent. Since $\varphi'(t)$ and $\psi'(t)$ do not vanish simultaneously, their quotient determines the direction of the tangent. Of course, the curve may have "corners" where the arcs are joined together. For example, the boundary of a rectangle is a regular curve. A Jordan curve can fail to be regular as, for example, when it contains a piece of the curve $y = x \sin (1/x)$ near the origin. It is evident that a regular curve has arc length.

Definition 2. *A region is regular if it is bounded and closed and if its boundary consists of a finite number of regular Jordan curves which have no points in common with each other. We shall denote such a region by the letter* S.

An example of a region S is the set of points (x, y) for which $1 \leq x^2 + y^2 \leq 4$. If from this region the points on the x-axis in the interval $1 < x < 2$ were removed, the region would no longer be closed and hence not regular. In particular, any region R of Chapter 6 is also a region S if its boundaries are regular curves. On the other hand, an R_x or R_y need not be regular ($R_x = R[-1, 1, -2, x \sin (1/x)]$, for example).

1.2 DEFINITION OF LINE INTEGRALS

Let a function $f(x, y)$ be defined at every point of the curve (1), which we shall denote by Γ. Make a subdivision Δ of the interval (a, b) by the points $t_0, t_1, \dots, t_n$. We define two types of line integrals indicated by the following notation:

(2) $$\int_\Gamma f(x, y)\, dx = \int_{x_0,y_0}^{x_1,y_1} f(x, y)\, dx$$

$$= \lim_{\|\Delta\| \to 0} \sum_{i=1}^n f(\varphi(t_i'),\, \psi(t_i'))[\varphi(t_i) - \varphi(t_{i-1})]$$

(3) $$\int_\Gamma f(x, y)\, dy = \int_{x_0,y_0}^{x_1,y_1} f(x, y)\, dy$$

$$= \lim_{\|\Delta\| \to 0} \sum_{i=1}^n f(\varphi(t_i'),\, \psi(t_i'))[\psi(t_i) - \psi(t_{i-1})]$$

Here $x_0 = \varphi(a)$, $y_0 = \psi(a)$, $x_1 = \varphi(b)$, $y_1 = \psi(b)$, $t_{i-1} \leq t_i' \leq t_i$. Both notations are incomplete. The first gives no indication of the direction of integration; the second does not show the dependence of the integral on the curve Γ. Usually no ambiguity results. Of course, for the line integrals to be defined the defining limits must exist. The line integrals (2) and (3) are, in fact,

$$\int_a^b f(\varphi(t),\, \psi(t))\, d\varphi(t), \qquad \int_a^b f(\varphi(t),\, \psi(t))\, d\psi(t)$$

respectively. But we shall give here an exposition that is independent of the theory of Stieltjes integrals.

Theorem 1. 1. Γ *is a regular curve*

2. $f(x, y) \in C$ *on* Γ

$\Rightarrow$ $$\int_\Gamma f(x, y)\, dx \text{ and } \int_\Gamma f(x, y)\, dy \text{ exist.}$$

It is no restriction to suppose that $\varphi(t)$, $\psi(t) \in C^1$ in $a \leq t \leq b$. Then by the law of the mean

$$\int_\Gamma f(x, y)\, dx = \lim_{\|\Delta\| \to 0} \sum_{i=1}^n f(\varphi(t_i),\, \psi(t_i))\varphi'(t_i')(t_i - t_{i-1})$$

where $t_{i-1} < t_i' < t_i$. By Duhamel's theorem the limit exists. In a similar way we see that the limit (3) exists. It is clear that by this result that both line integrals are equal to Riemann integrals (compare §2.3, Chapter 5),

$$\int_\Gamma f(x, y)\, dx = \int_a^b f(\varphi(t),\, \psi(t))\varphi'(t)\, dt$$

$$\int_\Gamma f(x, y)\, dy = \int_a^b f(\varphi(t),\, \psi(t))\psi'(t)\, dt$$

Hypothesis 1 may be altered in a variety of ways. For example, if the curve Γ is monotonic in the sense that $\varphi(t)$ and $\psi(t)$ are both monotonic in

(a, b), then the limits (2) and (3) both exist as Stieltjes integrals by Theorem 1 of Chapter 5,

$$\int_{\Gamma} f(x, y) \, dx = \int_a^b f(\varphi(t), \psi(t)) \, d\varphi(t)$$

$$\int_{\Gamma} f(x, y) \, dy = \int_a^b f(\varphi(t), \psi(t)) \, d\psi(t)$$

Furthermore, if $\varphi(t) = t$ and $\psi(t)$ belongs to C instead of to C^1, we see that

(4) $$\int_{\Gamma} f(x, y) \, dx = \int_a^b f(x, \psi(x)) \, dx$$

Thus it will be possible to extend the integral (2) over the boundary of a region R_x [see §1.1, Chapter 6], or the integral (3) over the boundary of a region R_y if $f(x, y) \in C$ there.

EXAMPLE A.   Compute $\int_{\Gamma} (x + y) \, dx$ if Γ is $x = \cos \theta$, $y = \sin \theta$, $0 \leq \theta \leq \pi/2$. Here the integration is intended to be from $(1, 0)$ to $(0, 1)$ along an arc of the unit circle.

$$\int_{\Gamma} (x + y) \, dx = - \int_0^{\pi/2} (\cos \theta + \sin \theta) \sin \theta \, d\theta = - \frac{1}{2} - \frac{\pi}{4}$$

We might also have used equation (4),

$$\int_{\Gamma} (x + y) \, dx = \int_1^0 (x + \sqrt{1 - x^2}) \, dx = - \frac{1}{2} - \frac{\pi}{4}$$

EXAMPLE B.   Compute $\int_{\Gamma} (x + y) \, dx$ if Γ is the two line segments $y = 0$, $0 \leq x \leq 1$; $x = 0$, $0 \leq y \leq 1$. The integration is again intended to be from $(1, 0)$ to $(0, 1)$ over the broken line.

$$\int_{\Gamma} (x + y) \, dx = \int_1^0 x \, dx = - \tfrac{1}{2}$$

These two examples show that a line integral may well depend upon the path and not merely on the end points of the path.

EXAMPLE C.   Extend the integral

$$\int_{\Gamma} (x + y) \, dx + (x - y) \, dy$$

over the two paths Γ of Examples A and B.   The sum of an integral (2) and an integral (3) is usually written in this way with a single integral sign Simple computations give for the circular arc

$$\int_0^{\pi/2} (\cos 2\theta - \sin 2\theta)\, d\theta = -1$$

and for the broken line

$$\int_1^0 x\, dx - \int_0^1 y\, dy = -1$$

We shall see later that in this case the integral is independent of the path.

1.3 WORK

One very natural application of the notion of a line integral is to the problem of defining the work done by a field of force on a particle moving along a curve through the field. Let the field be given by two functions $X(x, y)$ and $Y(x, y)$ which are to be the x- and y-components, respectively, of a force at the point (x, y). The magnitude of the force at the point is $\sqrt{X^2 + Y^2}$, and its direction is determined by the angle $\tan^{-1}(Y/X)$. Starting with the familiar definition of work as Fl if the particle moves in a straight line through a distance l under a constant force of magnitude F in the direction of motion, we can easily see how to make the definition in the general case.

Let the particle describe the regular curve (1) from $t = a$ to $t = b$. Make the subdivision Δ of §1.2 and let the arc length of the curve between the points $t = t_{i-1}$ and $t = t_i$ of the curve be Δs_i. Let θ_i be the angle between the direction of the force of the field at the point t_i and the direction of the tangent to the curve at t_i directed in the line of motion. It is natural to define the work done on the particle as it traverses the whole path as

$$(5) \qquad \lim_{\|\Delta\| \to 0} \sum_{i=1}^n \sqrt{X_i^2 + Y_i^2}\, \cos\theta_i\, \Delta s_i$$

$$X_i = X(\varphi(t_i), \psi(t_i)), \qquad Y_i = Y(\varphi(t_i), \psi(t_i))$$

The direction components of the tangent are $\varphi'(t_i)$, $\psi'(t_i)$ and of the direction of the force, X_i, Y_i, so that

$$\cos\theta_i = \frac{X_i\varphi'(t_i) + Y_i\psi'(t_i)}{\sqrt{X_i^2 + Y_i^2}\, \sqrt{[\varphi'(t_i)]^2 + [\psi'(t_i)]^2}}$$

$$\Delta s_i = \int_{t_{i-1}}^{t_i} \sqrt{[\varphi'(t)]^2 + [\psi'(t)]^2}\, dt = \Delta t_i \sqrt{[\varphi'(\xi_i)]^2 + [\psi'(\xi_i)]^2}$$

where $t_{i-1} < \xi_i < t_i$. By use of Duhamel's theorem we easily see that the limit (5) is the line integral

$$\int_\Gamma X(x, y)\, dx + Y(x, y)\, dy$$

For example, if $X = x + y$, $Y = x - y$, the work done by the field on a particle moving from $(1, 0)$ to $(0, 1)$ along *any* regular curve is -1. When the work is independent of the path, as in this example, the field is called *conservative*. The negative sign means that the particle has done work on the field. In other words, if the particle moved as a result of the forces of the field only, it would move in the opposite direction over most of the path.

EXERCISES   (1)

1. Compute the following integral first over the curve Γ of Example A and then over that of Example B,
$$\int_\Gamma xy \, dx + (x + y) \, dy$$

2. Solve the same problem for
$$\int_\Gamma y \, dx + x \, dy$$

3. Compute the integral of Exercise 1 where Γ is the boundary of the triangle with vertices $(0, 0)$, $(0, 2)$, $(1, 0)$, integration in the clockwise direction.

4. Compute
$$\int_\Gamma (x^2 + y) \, dx + (2x + y^2) \, dy$$
over the boundary of the square with vertices $(1, 1)$, $(1, 2)$, $(2, 2)$, $(2, 1)$ in the clockwise sense.

5. Compute
$$\int_{0,0}^{1,-1} (x + 2y) \, dx + yx \, dy$$
where the path is first the curve $y = -x^2$ and then the curve $x^3 = y^2$.

6. Compute
$$\int_\Gamma \frac{x \, dy - y \, dx}{x^2 + y^2}$$
where Γ is the entire curve $x = 1 + 2 \cos \theta$, $y = 2 \sin \theta$, integration counterclockwise.

7. Show that the integral of Example C, if Γ is the curve (1), has the value
$$\tfrac{1}{2}[\varphi^2(b) - \varphi^2(a) - \psi^2(b) + \psi^2(a)] + \varphi(b)\psi(b) - \varphi(a)\psi(a)$$
Check the results of Example C by this formula.

8. A field of force is set up by a particle situated at the origin (inverse square law). Find the work done on a particle moving over the path of Example A. Explain your answer.

9. Solve the problem if the particle moves along a straight line from the point $(1, 2)$ to the point $(2, 4)$.

10. Solve the same problem from the point $(1, 0)$ to the point $(0, 1)$.

11. Solve the same problem for the curve (1) assumed not to pass through the origin.

12. Show that the Stieltjes integrals of §1.2 are both equal to Riemann integrals when φ and ψ are strictly monotonic and $\in C$. *Hint:* Make a change of variable $x = \varphi(t)$ in the sum (2) and $y = \psi(t)$ in (3). What are you assuming about the inverse of a continuous, strictly monotonic function?

13. In the derivation of the limit (5), it was tacitly assumed that $\varphi'(t)$ and $\psi'(t)$ do not vanish simultaneously. How would you alter the discussion to take care of the curve $x = t^2$, $y = t^3$?

14. If $f(x, y) \in C$, $g(x, y) \in C^1$ on the curve (1) and if $x_i = \varphi(t_i)$, $y_i = \psi(t_i)$, evaluate the limit

$$\lim_{\|\Delta\| \to 0} \sum_{i=1}^{n} f(x_i, y_i)[g(x_i, y_i) - g(x_{i-1}, y_{i-1})]$$

§2. *Green's Theorem*

We shall prove here a result connecting a double integral over a region with a line integral over its boundary. It is sometimes referred to as *Gauss's theorem*. But it was brought to the attention of mathematicians by the work of G. Green and is more frequently known by his name.

2.1 A FIRST FORM

If a region is bounded by one or more curves the positive direction over the boundary is the one that leaves the region to the left. Thus for the region between two concentric circles the positive direction is counterclockwise for the outer boundary, clockwise for the inner one.

Theorem 2.   1. *R is a region* R_x *and also* R_y

           2. Γ *is the boundary of R*

           3. $P(x, y)$, $Q(x, y) \in C^1$ *in R*

(1) $\Rightarrow$ $$\int_{\Gamma} P\, dx + Q\, dy = \int\int_{R} [Q_1(x, y) - P_2(x, y)]\, dS$$

the line integral being taken in the positive sense.

Let $R_x = R[a, b, \varphi(x), \psi(x)]$. Then by Theorem 2, Chapter 6,

$$\int\int_{R} P_2(x, y)\, dS = \int_a^b dx \int_{\varphi(x)}^{\psi(x)} P_2(x, y)\, dy$$

$$= \int_a^b P(x, \psi(x))\, dx - \int_a^b P(x, \varphi(x))\, dx$$

By equation (4) of §1.2 we see that the right-hand side of this equation is equal to

$$- \int_\Gamma P(x, y)\, dx$$

the direction of integration being counterclockwise.   This proves the theorem in so far as it concerns $P(x, y)$.   The remainder is proved by using an iterated integral in the other order, and for this we need to know that $R = R_y$.

2.2 A SECOND FORM

Theorem 3. 1. *R is a region R_x and a regular region S*

        2. *Γ is the boundary of R*

        3. *$P(x, y), Q(x, y) \in C^1$ in R*

$$\Rightarrow \qquad \int_\Gamma P\, dx + Q\, dy = \int\int_R [Q_1(x, y) - P_2(x, y)]\, dS$$

the line integral being taken in the positive sense.

The region is now not known to be a region R_y, but it is known that Γ is a regular curve.   The previous proof applies in so far as it concerns $P(x, y)$.   The boundary Γ consists generally of four regular arcs.   Hence,

$$(2) \qquad \int_\Gamma Q(x, y)\, dy = \int_a^b Q(x, \varphi(x))\varphi'(x)\, dx - \int_a^b Q(x, \psi(x))\psi'(x)\, dx$$
$$+ \int_{\varphi(b)}^{\psi(b)} Q(b, y)\, dy - \int_{\varphi(a)}^{\psi(a)} Q(a, y)\, dy$$

and

$$(3) \qquad \int\int_R Q_1(x, y)\, dS = \int_a^b dx \int_{\varphi(x)}^{\psi(x)} Q_1(x, y)\, dy$$

Set

$$F(x) = \int_{\varphi(x)}^{\psi(x)} Q(x, y)\, dy$$

Then by Example B, §7.3, Chapter 10, we have

$$(4) \qquad F'(x) = \int_{\varphi(x)}^{\psi(x)} Q_1(x, y)\, dy + Q(x, \psi(x))\psi'(x) - Q(x, \varphi(x))\varphi'(x)$$

The first term on the right of equation (4) is the inner integral on the right of (3).   Hence,

$$\int\int_R Q_1(x, y)\, dS = F(b) - F(a) - \int_a^b Q(x, \psi(x))\psi'(x)\, dx$$
$$+ \int_a^b Q(x, \varphi(x))\varphi'(x)\, dx$$

We now complete the proof by comparing this equation with equation (2). Of course, R_x could be replaced by R_y in hypothesis 1.

2.3 REMARKS

If a regular region S is such that it can be divided into a finite number of regions R_x (or R_y) by cross cuts, equation (1) still holds where Γ is the total boundary, consisting of one or more regular closed curves. For example, the region between the circles $x^2 + y^2 = 1$ and $x^2 + y^2 = 4$ can be divided into four regions R_x by the lines $x = \pm 1$. If Theorem 2 is applied to each of these four regions, the line integral will be extended over each of the straight line segments twice, in opposite directions. Hence the line integrals over these segments add up to zero; the remaining line integrals add up to the left-hand side of equation (1). The sum of the four double integrals is equal to the right-hand side.

It can be shown that every regular region can be divided into a finite number of regular subregions R_x or R_y (W. F. Osgood, *Lehrbuch der Funktionentheorie*, 1923, p. 181). Hence, equation (1) is valid for every regular region. It should be noted that it is not always possible to subdivide a regular region into a finite number of regular subregions which are both R_x and R_y. Consider, for example, the region $R_x = R[0, 1, x^3 \sin(1/x), 1]$. It is for this reason that Theorem 3 is sometimes useful when Theorem 2 is not.

2.4 AREA

A useful application of Green's theorem is to the problem of finding the area of a region defined by the equations of its boundary curves. If R is a region to which Green's theorem applies and which is bounded by Γ, then the area of R is given by any of the three formulas

$$A = -\int_\Gamma y\,dx, \quad A = \int_\Gamma x\,dy, \quad A = \tfrac{1}{2}\int_\Gamma (-y)\,dx + x\,dy$$

the integration being in the positive sense. For, if formula (1) is applied to any one of these line integrals, we discover that it is equal to the double integral of unity over R.

EXAMPLE A.   Find the area of the ellipse $x = a \cos\theta$, $y = b \sin\theta$. Here

$$A = -\int_\Gamma y\,dx = ab \int_0^{2\pi} \sin^2\theta\,d\theta = \pi ab$$

EXERCISES (2)

1. Do Exercise 4 of §1 by use of Theorem 2.

2. Integrate by two methods

$$\int_{\Gamma} (x + y^2)\, dx + x^2 y\, dy$$

in the positive sense over the boundary of the region bounded by the curves $y^2 = x$ and $|y| = 2x - 1$.

3. Compute by two methods

$$\int\int_{R} 2xy\, dS$$

over the ellipse of Example A.

4. Prove Green's theorem in polar coordinates,

$$\int_{\Gamma} P(r, \theta)\, dr + Q(r, \theta)\, d\theta = \int\int_{R} r^{-1}[Q_1(r, \theta) - P_2(r, \theta)]\, dS$$

State carefully your hypotheses.

5. Find three line integrals for the area of a region bounded by a curve Γ whose equations are given in polar coordinates:

$$A = \frac{1}{2}\int_{\Gamma} r^2\, d\theta = -\int_{\Gamma} r\theta\, dr = \frac{1}{2}\int_{\Gamma} \frac{r^2}{2}\, d\theta - r\theta\, dr$$

State carefully your assumption about Γ. Why does only one of these formulas give the correct value for the area of the circle $r \leqq a$?

6. Find the area of the circle $r = a \cos \theta$.

7. Find the area of an ellipse by use of polar coordinates. Take a focus at the pole.

8. Find the area enclosed by the loop of the strophoid

$$x = a(1 - t^2)/(1 + t^2), \qquad y = xt$$

9. Solve the same problem for the folium $x^3 + y^3 = 3axy$. *Hint:* Use polar coordinates or the parametric equations

$$x = 3at(1 + t^3)^{-1}, \qquad y = xt$$

10. Find the area of a triangle by use of line integrals.

11. Prove Theorem 3 if R_x is replaced by R_y in hypothesis 1.

12. The boundary of a region R consists of the origin and the two arcs

$$y = x^3 \cos (2\pi/x) \qquad\qquad 0 < x \leqq 1$$
$$x = y^3 \cos (2\pi/y) \qquad\qquad 0 < y \leqq 1$$

Show how it can be divided into a finite number of subregions which are regions R_x or R_y.

13. State and prove sufficient conditions for the equation

$$\int_\Gamma PQ_1\, dx + PQ_2\, dy = \int\int_R \frac{\partial(P,\, Q)}{\partial(x,\, y)}\, dS$$

14. If S is a regular region bounded by a single regular closed curve $x = \varphi(s)$, $y = \psi(s)$, $0 \le s \le l$ and if $\Delta v = v_{11} + v_{22}$, show that

$$\int\int_S (u\, \Delta v + u_1 v_1 + u_2 v_2)\, dS = \int_0^l u\, \frac{\partial v}{\partial n}\, ds$$

where $\dfrac{\partial v}{\partial n}$ is a directional derivative in the direction of the exterior normal

(assuming that the curve is traced once in the positive sense as the arc length s varies from 0 to l).

15. Prove

$$\int\int_S (u\, \Delta v - v\, \Delta u)\, dS = \int_0^l \left(u\, \frac{\partial v}{\partial n} - v\, \frac{\partial u}{\partial n} \right) ds$$

16. If $\Delta u = 0$ in S, show that

$$\int_0^l \frac{\partial u}{\partial n}\, ds = 0$$

17. If $\Delta u = 0$ in S, show that

$$\int_0^l u\, \frac{\partial u}{\partial n}\, ds \ge 0$$

§3. *Application*

The line integral is a useful tool in the investigation of *exact differentials*. We wish to know when $P(x, y)\, dx + Q(x, y)\, dy$ is the differential of a function $F(x, y)$. Under what conditions will F exist such that $F_1 = P$, $F_2 = Q$, and how can one find F if it exists?

3.1 EXISTENCE OF EXACT DIFFERENTIALS

A domain D (see §1.1, Chapter 6) is *simply connected* if no Jordan curve in D contains in its interior a boundary point of D. Let us use the sign * as a superscript to the name of a domain to indicate that it is simply connected.

Theorem 4.   1. $P(x, y)$, $Q(x, y) \in C^1$ in D^*

           2. $Q_1(x, y) = P_2(x, y)$ in D^*

$\Leftrightarrow$ there exists $F(x, y) \in C^2$ in D^* such that

$$F_1 = P, \qquad F_2 = Q$$

To prove the necessity of the conditions assume the existence of F. Then the equality of F_{12} and F_{21} implies condition 2. Since $F \in C^2$, P and $Q \in C^1$. Conversely, if conditions 1 and 2 hold, we define $F(x, y)$ explicitly. Let (a, b) and (x_0, y_0) be points of D^*. Then

(1)
$$F(x_0, y_0) = \int_{a,b}^{x_0, y_0} P(x, y)\, dx + Q(x, y)\, dy$$

where the path of integration is a broken line. Such a line exists by the definition of a domain. In fact it is easy to see that the segments of the broken line may be taken parallel to the axes. We choose the path in this

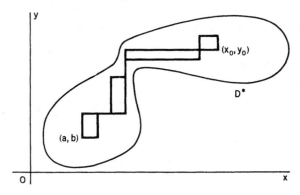

Fig. 22.

way in order to simplify the type of region that can be enclosed between two such broken lines (see Figure 22). We note first that F is a single-valued function, that the integral defining it does not depend upon the path. For consider two broken lines in D^* with segments parallel to the axes and joining (a, b) with (x_0, y_0). They will form the boundaries of a finite number of rectangles in which $Q_1 = P_2$. By Green's theorem the line integral (1) extended around the boundary of each rectangle will be zero. From this fact it is evident that the integral (1) is independent of the path.

Let us compute F_1 and F_2 at (x_0, y_0), a point of D^*. This point is the center of a circle K which lies entirely in D^*. Choose a point $(x_0 + \Delta x, y_0)$ inside K. Then

$$\frac{\Delta F}{\Delta x} = \frac{F(x_0 + \Delta x, y_0) - F(x_0, y_0)}{\Delta x} = \frac{1}{\Delta x} \int_{x_0, y_0}^{x_0 + \Delta x, y_0} P\, dx + Q\, dy$$

If the path of integration is taken to be a straight line, it is evident that the integral of Q is zero. Then by the law of the mean

$$\frac{\Delta F}{\Delta x} = P(x_0 + \theta \Delta x, y_0) \qquad\qquad 0 < \theta < 1$$

and $F_1(x_0, y_0) = P(x_0, y_0)$. Similarly, $F_2 = Q$.

3.2 EXACT DIFFERENTIAL EQUATIONS

It is now a simple matter to integrate the exact differential equation

$$P(x, y)\, dx + Q(x, y)\, dy = 0$$

where $Q_1 = P_2$ in D^*.   Clearly, the primitive is $F(x, y) = c$, where c is an arbitrary constant.

In the evaluation of the integral (1) it may be convenient to use regular paths which are not polygonal lines.   We must show that the value of the integral is not altered by the change in path.

Theorem 5.   1. $P(x, y), Q(x, y) \in C^1$ *in* D^*

2. $Q_1(x, y) = P_2(x, y)$ *in* D^*

3. Γ *is a regular curve in* D^* *joining* (a, b) *with* (x_0, y_0)

$\Rightarrow$ *the integral* (1) *extended over* Γ *is independent of* Γ.

For, let Γ have equations

$$x = \varphi(t), \qquad y = \psi(t) \qquad\qquad 0 \leq t \leq 1$$

Then by Theorem 4

$$\int_{a,b}^{x_0, y_0} P\, dx + Q\, dy = \int_0^1 \left[F_1(\varphi(t), \psi(t))\varphi'(t) + F_2(\varphi(t), \psi(t))\psi'(t) \right] dt$$

The integrand is $\dfrac{d}{dt} F(\varphi(t), \psi(t))$, so that

(2) $$\int_{a,b}^{x_0, y_0} P\, dx + Q\, dy = F(x_0, y_0) - F(a, b)$$

Since the final result does not depend on $\varphi(t)$ or $\psi(t)$, the proof is complete.

Observe that this result is analogous to the fundamental theorem of the integral calculus which enables one to evaluate a definite integral by use of an indefinite one.   If one can obtain $F(x, y)$ by inspection, equation (2) gives a simple way to evaluate the integral (1).

EXAMPLE A.   Do Example C of §1.2.   It is easy to see by regrouping terms that

$$(x + y)\, dx + (x - y)\, dy = d(x^2/2) - d(y^2/2) + d(xy)$$

so that the required integral is

$$\left[\frac{x^2}{2} - \frac{y^2}{2} + xy \right]_{(1,0)}^{(0,1)} = -1$$

3.3 A FURTHER RESULT

Theorem 6. *If $P(x, y)$, $Q(x, y) \in C^1$ in a domain D^*, then $Q_1 = P_2$ in D^**

$$\Leftrightarrow \qquad \int_\Gamma P\, dx + Q\, dy = 0 \text{ for every regular closed curve } \Gamma \text{ in } D^*.$$

The implication $\Rightarrow$ is an immediate result of Theorem 5. For, if (a, b) is any point of the curve Γ, then the value of the integral is $F(a, b) - F(a, b)$.

To prove the opposite implication, suppose that $Q_1 - P_2 > 0$ at a point (x_0, y_0) of D^*. By continuity this point is the center of a circle K of D^* with circumference C, throughout which $Q_1 - P_2 > 0$. By Green's theorem

$$\iint_K (Q_1 - P_2)\, dS = \int_C P\, dx + Q\, dy > 0$$

This contradicts the hypothesis. Similarly, if $Q_1 - P_2 < 0$ at (x_0, y_0), we obtain a contradiction. Hence, $Q_1(x_0, y_0) = P_2(x_0, y_0)$, and the proof is complete. Observe that the theorem remains true if the curve Γ is allowed to cut itself.

3.4 MULTIPLY CONNECTED REGIONS

In the previous theorems the simply connected character of the region was an essential part of the hypothesis. For, consider the integral

$$(3) \qquad \int_\Gamma \frac{x\, dy - y\, dx}{x^2 + y^2}$$

where Γ is the entire unit circle. Here P, $Q \in C^1$ and $P_2 = Q_1$ in the region $\frac{1}{4} \leq x^2 + y^2 \leq 4$, for example. The unit circle lies in the region. But the value of the integral is easily seen to be different from zero. Of course, the region considered is multiply connected. The results of the present section are easily applied to multiply connected regions by the introduction of cross cuts. For example, the integral (3) is zero if Γ is any regular closed curve in the region which does not cross the x-axis in the interval $\frac{1}{2} \leq x \leq 2$.

EXERCISES (3)

1. $\displaystyle\int_{0,0}^{1,\pi} e^x \cos y\, dx - e^x \sin y\, dy = ?$

2. $\displaystyle\int_{0,0}^{1,\pi} 2y \cos x\, dy - y^2 \sin x\, dx = ?$

3. $\displaystyle\int_\Gamma \frac{(y^2 - x^2)\, dx - 2xy\, dy}{(x^2 + y^2)} = ?$ Integration is in the positive sense, and Γ is the unit circle.

4. $\displaystyle\int_{a,b}^{x,y} \frac{x^2 - y^2}{x^2 y}\, dx + \frac{y^2 - x^2}{xy^2}\, dy = ?$ What restrictions do you impose on the

limits of integration and the path of integration?

5. If Γ is a regular closed curve lying entirely in the first quadrant ($x > 0$, $y > 0$), calculate

$$\int_\Gamma \frac{2(x^2 - y^2 - 1)\, dy - 4xy\, dx}{(x^2 + y^2 - 1)^2 + 4y^2}$$

6. $\displaystyle\int_\Gamma \frac{e^x(x \sin y - y \cos y)\, dx + e^x(x \cos y + y \sin y)\, dy}{x^2 + y^2} = ?$ The curve is

the same as in Exercise 3. *Hint:* Integrate over the circle $x^2 + y^2 = r^2$ and let $r \to 0$. You need not justify the process of taking the limit under the integral sign.

7. Evaluate the integral (3).

8. If $P, Q \in C^1$ and $Q_1 = P_2$ in the closed region between the concentric circles Γ_1 and Γ_2, show that

$$\int_{\Gamma_1} P\, dx + Q\, dy = \int_{\Gamma_2} P\, dx + Q\, dy$$

the integration being clockwise in both cases.

9. If $u(x, y) \in C^2$ and $\Delta u = 0$ in D^*, find a function $v(x, y)$ such that $u_1 = v_2$, $u_2 = -v_1$ in S^*. This function is said to be *conjugate* to u.

10. Find by line integration a conjugate to the following functions:

(a) $x^3 - 3xy^2$, (b) $e^y \cos x$, (c) $\dfrac{y}{x^2 + y^2}$

In case (c) specify the region D^*.

11. Give an example to show that the conjugate of a function need not be single-valued in a multiply connected region.

12. The equations defining *conjugate functions* in polar coordinates are

$$ru_1(r, \theta) = v_2(r, \theta), \qquad u_2(r, \theta) = -rv_1(r, \theta)$$

Find a sufficient condition on $u(r, \theta)$ in order that it should have a conjugate $v(r, \theta)$ and find $v(r, \theta)$ by line integration.

13. Illustrate Exercise 12 by $u = \log r$.

14. If $v(x, y)$ is conjugate to $u(x, y)$ show that the integrals

$$U(x, y) = \int_{a,b}^{x,y} u\, dx - v\, dy \qquad V(x, y) = \int_{a,b}^{x,y} v\, dx + u\, dy$$

are independent of the path and that $V(x, y)$ is conjugate to $U(x, y)$.

§4. *Surface Integrals*

Just as the Riemann integral generalizes to the line integral, so too the double integral over a plane area generalizes to a surface integral over an area of an arbitrary curved surface. We define the surface integral here and show how to compute it. We then generalize Green's theorem. This result will enable us to express a triple integral over a solid in terms of a surface integral over the surface bounding the solid.

4.1 DEFINITION OF SURFACE INTEGRALS

In defining a surface integral we follow the procedure of §1.2, Chapter 6, step by step, replacing the plane region R of that section by a region on an arbitrary surface (having area). Let Σ be a finite piece of such a surface. A *subdivision* Δ of Σ is a set of closed curves $\{C_k\}_1^n$ lying on Σ and dividing it into a set of n subregions of areas $\Delta\Sigma_k$, $k = 1, 2, \ldots, n$. The *diameter* of a region on Σ is the length of the largest straight line segment whose ends lie in the region. (Since Σ may be curved, the intermediate points of the segment need not lie on Σ). The *norm* of Δ, denoted by $||\Delta||$, is the largest of the n diameters of the subregions produced by the subdivision. Let $P(x, y, z)$ be a function defined at every point of Σ and let (ξ_k, η_k, ζ_k) be a point on Σ inside or on the boundary of the subregion bounded by C_k. Then the *surface integral of $P(x, y, z)$ over Σ* is

$$(1) \qquad \iint_\Sigma P(x, y, z)\, d\Sigma = \lim_{||\Delta|| \to 0} \sum_{k=1}^n P(\xi_k, \eta_k, \zeta_k) \Delta\Sigma_k$$

when this limit exists. If, in particular, Σ is a region R of the xy-plane, this limit reduces to the double integral of $P(x, y, 0)$ over R. A double integral is a special case of a surface integral.

One of the important uses of surface integrals is in the formulation of physical concepts. For example, if Σ is thought of as a smooth lamina of variable density given at a point (x, y, z) of Σ by the continuous function $\sigma(x, y, z)$, we could use the postulates of §5.2, Chapter 5, to show that the lamina has a uniquely determined moment of inertia about any axis. For the x-axis, say, it would be

$$I = \iint_\Sigma (y^2 + z^2)\sigma(x, y, z)\, d\Sigma$$

The actual calculation of a surface integral is frequently accomplished by reducing it to a double integral. For instance, let Σ be defined by the equation $z = f(x, y)$ when (x, y) lies in a region R of the xy-plane. That is, Σ is cut in a single point by a line parallel to the z-axis and has the projection R in the xy-plane. A subdivision Δ of Σ induces a corresponding subdivision (which we still call Δ) of R into subregions R_k. The point (ξ_k, η_k, ζ_k)

of Σ becomes $(\xi_k, \eta_k, f(\xi_k, \eta_k))$, where (ξ_k, η_k) is a point of R_k. If now $\Delta\Sigma_k$ can be expressed in terms of ΔS_k, the area of R_k, the sum (1) takes the form of the sum in Definition 3, Chapter 6. We are led in this way to the following result.

Theorem 7.   1. $P(x, y, z) \in C$ in V

       2. Σ is the surface $z = f(x, y)$ over the region R

       3. $f(x, y) \in C^1$ in R

       4. Σ lies in V

$\Rightarrow$ A. $\displaystyle\iint_\Sigma P(x, y, z)\, d\Sigma$ exists

       B. $\displaystyle\iint_\Sigma P(x, y, z)\, d\Sigma$

$$= \iint_R P(x, y, f(x, y))\sqrt{1 + f_1^2(x, y) + f_2^2(x, y)}\, dS$$

This theorem enables us to reduce a surface integral to an ordinary double integral. By equation (1), §7.1 of Chapter 6, and the law of the mean (Exercise 13, §2 of Chapter 6), we have

$$\Delta\Sigma_k = \iint_{R_k} \sqrt{1 + f_1^2(x, y) + f_2^2(x, y)}\, dS$$

$$= \sqrt{1 + f_1^2(a_k, b_k) + f_2^2(a_k, b_k)}\, \Delta S_k$$

where (a_k, b_k) is a point of R_k and ΔS_k is the area of that subregion. Substituting this value of $\Delta\Sigma_k$ in equation (1) and using Duhamel's theorem, we obtain the desired result.

There are obvious modifications of the theorem. The surface Σ might have the equation $x = f(y, z)$ or $y = f(x, z)$. In fact, the existence of the surface integral (1) is assured if Σ can be decomposed into a finite number of parts, each of which is cut only once by a line parallel to *some* axis and has a continuously turning tangent plane. The radical in equation B is equal to sec γ, where γ is the *acute* angle between the normal to Σ and the z-axis.

EXAMPLE A.   Compute $\displaystyle\iint_\Sigma xyz\, d\Sigma$, where Σ is the surface of the tetrahedron bounded by the planes

$$x = 0, \quad z = 0, \quad x + y = 2, \quad 2z = y$$

Since the integrand is zero on the coordinate faces, two of the integrals are zero. We evaluate the other two, Figure 23, by projection onto the region $R_x = R[0, 2, 0, (2 - x)/2]$ of the xz-plane.

$$\iint_{\Sigma_1} xyz \, d\Sigma = 2\sqrt{5} \iint_{R_x} xz^2 \, dS = \frac{2\sqrt{5}}{15}$$

$$\iint_{\Sigma_2} xyz \, d\Sigma = \sqrt{2} \iint_{R_x} z(2x - x^2) \, dS = \frac{\sqrt{2}}{5}$$

We have used the fact that $\sec \beta = \sqrt{5}$ on Σ_1 and $\sec \beta = \sqrt{2}$ on Σ_2.

EXAMPLE B. Compute $\iint_{\Sigma} x^2 z \cos \gamma \, d\Sigma$, where Σ is the unit sphere and γ is the angle between the exterior normal to the sphere and the positive z-axis. The two hemispheres have the equations $z = \pm\sqrt{1 - x^2 - y^2}$. Neither radical belongs to C^1 when $x^2 + y^2 = 1$, so that Theorem 7 is not directly applicable. However, the decomposition referred to above is easily performed here (compare §6.3 and Figure 26), so that the integral exists. In fact it may be evaluated by projection on a single plane if a limiting process is used. We first cut out the equator by the cylinder $x^2 + y^2 = (1 - \epsilon)^2$ and then let $\epsilon \to 0$. This is a familiar process in the evaluation of improper integrals (compare Chapter 10). On the whole sphere $\cos \gamma = z$. To reduce each surface integral (over the two hemispheres) to a double integral we must multiply by $|\sec \gamma| = 1/|z|$. Hence

$$\iint_{\Sigma} x^2 z \cos \gamma \, d\Sigma = \iint_R x^2 \sqrt{1 - x^2 - y^2} \, dS + \iint_R x^2 \sqrt{1 - x^2 - y^2} \, dS$$

where R is the disk $x^2 + y^2 \le 1$. Note that for the lower hemisphere $|\sec \gamma| \cos \gamma < 0$ and $z = -\sqrt{1 - x^2 - y^2} < 0$. Iterating in polar coordinates,

$$\iint_{\Sigma} x^2 z \cos \gamma \, d\Sigma = \lim_{\epsilon \to 0} 2 \int_0^{2\pi} \cos^2 \theta \, d\theta \int_0^{1 - \epsilon} r^3 \sqrt{1 - r^2} \, dr = \frac{4\pi}{15}$$

4.2 GREEN'S OR GAUSS'S THEOREM

We now prove a result analogous to Theorem 2. For the sake of simplicity of statement, let us introduce a further notation. A surface Σ will be denoted by Σ^* if it has the following properties. It is the boundary of a three dimensional region V, which is a region V_{xy}, V_{yz}, V_{zx} (§8.2, Chapter 6). In each case the defining functions are to belong to C^1. For example, if

(2) $V_{xy} = V(R, \varphi(x, y), \psi(x, y))$

then $\varphi, \psi \in C^1$ in R. Clearly Σ^* will have a continuously turning tangent plane over the parts of the surface corresponding to the defining functions.

Theorem 8.   1. $P(x, y, z)$, $Q(x, y, z)$, $R(x, y, z) \in C^1$ in V

2. V is bounded by Σ^*

3. α, β, γ are the direction angles of the exterior normal to Σ^*

$$\Rightarrow \quad \iiint_V [P_1(x, y, z) + Q_2(x, y, z) + R_3(x, y, z)]\, dV$$

$$= \iint_{\Sigma_*} [P(x, y, z) \cos \alpha + Q(x, y, z) \cos \beta + R(x, y, z) \cos \gamma]\, d\Sigma$$

If V is defined by equation (2), we have by Theorem 5, Chapter 6,

$$\iiint_V R_3\, dV = \iint_R dS \int_{\varphi(x,y)}^{\psi(x,y)} R_3\, dz$$

$$= \iint_R R(x, y, \psi(x, y))\, dS - \iint_R R(x, y, \varphi(x, y))\, dS$$

$$= \iint_{\Sigma_1} R(x, y, z)|\cos \gamma|\, d\Sigma - \iint_{\Sigma_2} R(x, y, z)|\cos \gamma|\, d\Sigma$$

Here Σ_1 and Σ_2 are the upper and lower nappes, respectively, of Σ^*.   Since $\cos \gamma > 0$ on Σ_1 and $\cos \gamma < 0$ on Σ_2, we have, if $\varphi = \psi$ on the boundary of R,

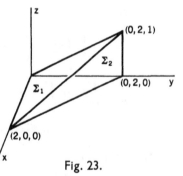

Fig. 23.

(3) $$\iiint_V R_3(x, y, z)\, dV$$

$$= \iint_{\Sigma_*} R(x, y, z) \cos \gamma\, d\Sigma$$

If $\varphi \neq \psi$ on the whole boundary of R, there is also a cylindrical surface bounding V_{xy} (compare Figure 23).   But such a surface can be projected on the other coordinate faces for evaluation of the surface integral.   However, the corresponding surface integral will be zero, since the factor $\cos \gamma$ of the integrand will be everywhere zero on the cylindrical surface.   Hence equation (3) holds in either case and the proof is complete insofar as it concerns $R(x, y, z)$.   The remainder of the proof is supplied by symmetry.

Corollary 8.   1. $R(x, y, z) \in C^1$ in $V_{xy} = V(R, \varphi(x, y), \psi(x, y))$

2. $\varphi, \psi \in C^1$ in R

3. γ is the angle between the positive z-axis and the exterior normal to Σ, the boundary of V_{xy}

$$\Rightarrow \quad \iiint_{V_{xy}} R_3\, dV = \iint_{\Sigma} R \cos \gamma\, d\Sigma$$

This result was proved above but is not included in the statement of the theorem. For, the present V_{xy} is not in general a V_{yz} nor a V_{xz}, so that Σ need not be a Σ^*. Of course, corresponding corollaries hold for regions V_{yz} or V_{xz}. These corollaries are useful in the subdivision of general regions (compare §4.3).

EXAMPLE C. Check Example B by Theorem 8. Take $P = Q = 0$, $R = x^2 z$. Then $R_3 = x^2$. Evaluating the triple integral by iteration in spherical coordinates we have $x = r \sin \varphi \cos \theta$ and

$$\iiint_V x^2 \, dV = \int_0^{2\pi} \cos^2 \theta \, d\theta \int_0^\pi \sin^3 \varphi \, d\varphi \int_0^1 r^4 \, dr = \frac{4\pi}{15}$$

EXAMPLE D. Evaluate by two methods

$$I = \iiint_V (xy + yz + zx) \, dV$$

where V is the region bounded by the planes $x = 0$, $y = 0$, $z = 0$, $z = 1$ and the cylinder $x^2 + y^2 = 1$.

By iteration

$$I = \iint_{R_x} \left(xy + \frac{y}{2} + \frac{x}{2} \right) dS \qquad R_x = R[0, 1, 0, \sqrt{1 - x^2}\,]$$

$$= \int_0^1 r^3 \, dr \int_0^{\pi/2} \cos \theta \sin \theta \, d\theta + \tfrac{1}{2} \int_0^1 r^2 \, dr \int_0^{\pi/2} \sin \theta \, d\theta$$

$$+ \tfrac{1}{2} \int_0^1 r^2 \, dr \int_0^{\pi/2} \cos \theta \, d\theta = \tfrac{11}{24}$$

By Theorem 8

$$(4) \qquad I = \iint_\Sigma \left(\frac{x^2 y}{2} \cos \alpha + \frac{y^2 z}{2} \cos \beta + \frac{z^2 x}{2} \cos \gamma \right) d\Sigma$$

Here Σ consists of four plane faces and a cylindrical surface. The only plane face that contributes a value not zero is $z = 1$. For it, $\alpha = \pi/2$, $\beta = \pi/2$, $\gamma = 0$. Hence, we obtain

$$\iint \frac{z^2 x}{2} \, d\Sigma = \iint_{R_x} \frac{x}{2} \, dS = \frac{1}{2} \int_0^1 r^2 \, dr \int_0^{\pi/2} \cos \theta \, d\theta = \frac{1}{6}$$

Finally, for the cylindrical surface, $\cos \alpha = x$, $\cos \beta = y$, $\cos \gamma = 0$. Here we have only to consider the first two terms of the integral (4) in this case.

The first can be expressed as a double integral over a unit square in the yz-plane, the second over a unit square in the xz-plane:

$$\iint \frac{x^2 y}{2} \cos \alpha \, d\Sigma = \frac{1}{2} \int_0^1 dz \int_0^1 (1 - y^2) y \, dy = \frac{1}{8}$$

$$\iint \frac{y^2 z}{2} \cos \beta \, d\Sigma = \frac{1}{2} \int_0^1 z \, dz \int_0^1 (1 - x^2) \, dx = \frac{1}{6}$$

$$I = \tfrac{1}{6} + \tfrac{1}{8} + \tfrac{1}{6} = \tfrac{11}{24}$$

4.3 EXTENSIONS

Surfaces Σ^* are evidently rather rare. A cube (if properly placed) has such a surface but a sphere does not. However, the applicability of Green's theorem may be extended to much more general regions by the same sort of decomposition as was suggested in §2.3. The method will be sufficiently illustrated by the region V:

$$x \geq 0, \qquad y \geq 0, \qquad z \geq 0, \qquad x^2 + y^2 + z^2 \leq 1$$

Theorem 8 is not directly applicable since each coordinate plane contains normals to the sphere. One simple way of decomposing the region into subregions is indicated in Figure 24, where V_1 is the part of V inside the cylinder $x^2 + y^2 = \rho^2, 0 < \rho < 1$. The rest of V is subdivided as indicated into three others by the planes $x = \sqrt{\rho}/2$ and $y = \sqrt{\rho}/2$. Then V_1 is a V_{xy}, V_2 a V_{yz}, V_3 a V_{xz} and V_4 either a V_{xz} or a V_{yz}. In each case the defining functions are of class C'. We can apply Green's theorem to each. The corresponding triple and surface integrals when added will combine to give a single triple integral over V and a single surface integral over its surface. That the surface integrals over the auxiliary surfaces which we introduced disappear in the end result should be evident from the fact that a normal to

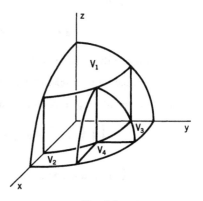

Fig. 24.

such a surface is at once an interior normal for one subregion and an exterior normal for the adjacent one. This example shows, of course, that Green's theorem may be made to apply to a complete sphere.

EXERCISES (4)

1. Check Green's theorem by computing both sides of the equation independently if $P = e^x$, $Q = R = 0$ and V is the tetrahedron bounded by the planes $x = 0, y = 0, z = 0, x + y + z = 1$.

2. Solve the same problem if $P = x^2$, $Q = R = 0$ and V is the unit sphere. Compute the triple integral by use of spherical coordinates.

3. Check Example A by projecting Σ_1 on the xy-plane, Σ_2 on the yz-plane.

4. Compute

$$\iint_\Sigma (x^2 y^2 + y^2 z^2 + z^2 x^2) \, d\Sigma$$

where Σ is that portion of the cone $x^2 + y^2 - z^2 = 0$ (2 nappes) cut out by the cylinder $x^2 + y^2 - 2x = 0$.

5. Show that the moment of inertia of a lamina in the form of a curved surface Σ about an axis is

$$I = \iint_\Sigma \rho r^2 \, d\Sigma$$

where r is the distance of a point of Σ from the axis and ρ is the density.

6. Find the moment of inertia of a spherical shell about a diameter.

7. Show that the volume of V in Theorem 8 is given by any of the integrals

$$\iint_{\Sigma^*} x \cos \alpha \, d\Sigma, \qquad \iint_{\Sigma^*} y \cos \beta \, d\Sigma, \qquad \iint_{\Sigma^*} z \cos \gamma \, d\Sigma$$

$$\tfrac{1}{3} \iint_{\Sigma^*} (x \cos \alpha + y \cos \beta + z \cos \gamma) \, d\Sigma$$

8. Compute the volume of the tetrahedron of Exercise 1 by use of Exercise 7.

9. Solve the same problem for the volume of a cone.

10. If $\Delta v = v_{11} + v_{22} + v_{33}$, show that

$$\iiint_V (u \, \Delta v + u_1 v_1 + u_2 v_2 + u_3 v_3) \, dV = \iint_{\Sigma^*} u \, \frac{\partial v}{\partial n} \, d\Sigma$$

where $\dfrac{\partial v}{\partial n}$ is a directional derivative in the direction of the exterior normal.

11. Prove

$$\iiint_V (u \, \Delta v - v \, \Delta u) \, dV = \iint_{\Sigma^*} \left(u \, \frac{\partial v}{\partial n} - v \, \frac{\partial u}{\partial n} \right) d\Sigma$$

12. If $\Delta u = 0$ in V, show that

$$\iint_{\Sigma^*} \frac{\partial u}{\partial n} \, d\Sigma = 0$$

13. If $\Delta u = 0$ in V, show that

$$\iint_{\Sigma^*} u \, \frac{\partial u}{\partial n} \, d\Sigma \geqq 0$$

14. Prove that the surface integral of Theorem 8 will be zero for every surface Σ^* if, and only if, $P_1 + Q_2 + R_3 = 0$. Make a precise statement of the result.

15. Let Σ be a surface $z = \varphi(x, y)$ bounded by a closed curve Γ. Show that if $P_1 + Q_2 + R_3 = 0$ the integral

$$\iint_{\Sigma} [P \cos \alpha + Q \cos \beta + R \cos \gamma] \, d\Sigma$$

has the same value for all Σ which have the same boundary Γ. For definiteness choose the direction of the normal so that γ is acute. [Compare Exercise 11, §6, for another method of proof.]

§5. *Change of Variable in Multiple Integrals*

For simple integrals we have, by the change of variable $x = \varphi(t)$,

$$(1) \qquad \int_{\varphi(a)}^{\varphi(b)} F(x) \, dx = \int_{a}^{b} F(\varphi(t)) \varphi'(t) \, dt$$

The interval (a, b) on the t-axis is transformed into the interval $(\varphi(a), \varphi(b))$ on the x-axis. We develop here a corresponding formula for a change of variable in multiple integrals.

5.1 TRANSFORMATIONS

Let the equations

$$(2) \qquad \begin{cases} x = g(u, v) \\ y = h(u, v) \end{cases}$$

define a one-to-one transformation of the region R_{uv} of the uv-plane into the region R_{xy} of the xy-plane. This means that to each point of either region corresponds just one point of the other by equations (2). Analytically, g and h are defined (single valued) in R_{uv}, and the equations (2) can be solved for u and v, the resulting functions being single valued in R_{xy}. For example, take $g(u, v) = v \cos u$, $h(u, v) = v \sin u$. The two regions would be as indicated in Figure 25. Let the boundary of R_{xy} be the curve Γ_{xy}:

$$(3) \qquad x = \varphi(t), \qquad y = \psi(t) \qquad\qquad 0 \leq t \leq \pi/2$$

Then the boundary curve Γ_{uv} of R_{uv} will be given by the equations

$$(4) \qquad \begin{aligned} \varphi(t) &= g(u, v) \\ \psi(t) &= h(u, v) \end{aligned}$$

These could be solved to obtain u and v as single valued functions of t.

Thus, in the above example, the curve $x^2 - 2x + y^2 = 0$ has the parametric equation $x = 1 + \cos t$, $y = \sin t$. Equations (4) become

$$1 + \cos t = v \cos u$$
$$\sin t = v \sin u$$

or

$$v = 2 \cos (t/2)$$
$$u = t/2$$

This is a piece of the curve $v = 2 \cos u$.

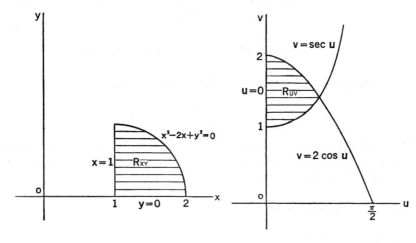

Fig. 25.

Let us investigate how a line integral is affected by the preceding transformation (2). We show that

(5) $$\int_{\Gamma_{xy}} Q(x, y)\, dy = \int_{\Gamma_{uv}} Q(g(u, v), h(u, v))[h_1(u, v)\, du + h_2(u, v)\, dv]$$

The direction of integration in one of these integrals is arbitrary; in the other it is determined by the transformation (2). In our example, the clockwise description of Γ_{xy} corresponds to the counterclockwise description of Γ_{uv}. The integral on the left of equation (5) is equal to

(6) $$\int_0^1 Q(\varphi(t), \psi(t))\psi'(t)\, dt$$

To evaluate the right-hand side we use the equations (4) of the curve Γ_{uv}. They give

$$\psi'(t) = h_1(u, v) \frac{du}{dt} + h_2(u, v) \frac{dv}{dt}$$

so that the line integral over Γ_{uv} is also equal to the ordinary integral (6).

5.2 DOUBLE INTEGRALS

Theorem 9.   1. $F(x, y) \in C$ in R_{xy}

2. $g(u, v), h(u, v) \in C^2$ in R_{uv}

3. $\dfrac{\partial(g, h)}{\partial(u, v)} \neq 0$ in R_{uv}

4. R_{xy} and R_{uv} correspond in a one-to-one fashion under transformation (2), and both are regular, simply connected region S

(7) $\Rightarrow$ $$\iint_{R_{xy}} F(x, y)\, dS_{xy} = \iint_{R_{uv}} F(g(u, v), h(u, v)) \left| \frac{\partial(g, h)}{\partial(u, v)} \right| dS_{uv}$$

Note the resemblance of equation (7) to equation (1). The region of integration is altered by the transformation in both cases. The factor $\varphi'(t)$ in the simple integral corresponds to the Jacobian in the double integral.

Let us first prove the theorem for the special case $F = 1$, when the left-hand side of (7) becomes the area A of R_{xy}. By Green's theorem with $Q = x$ we may express A as a line integral over Γ_{xy}, or by equation (5) as another line integral over Γ_{uv}:

$$A = \int_{\Gamma_{xy}} x\, dy = \int_{\Gamma_{uv}} g(u, v)[h_1(u, v)\, du + h_2(u, v)\, dv$$

Here the integration is counterclockwise. Apply Green's theorem to the latter integral to obtain

$$A = \pm \iint_{R_{uv}} \left[\frac{\partial}{\partial u}(gh_2) - \frac{\partial}{\partial v}(gh_1) \right] dS_{uv}$$

$$= \pm \iint_{R_{uv}} [g_1h_2 + gh_{21} - g_2h_1 - gh_{12}]\, dS_{uv} = \pm \iint_{R_{uv}} J(u, v)\, dS_{uv}$$

$$J(u, v) = \frac{\partial(g, h)}{\partial(u, v)}$$

The doubtful sign results from the ambiguity in the sense of description of Γ_{uv}, plus and minus corresponding, respectively, to counterclockwise and clockwise. By hypotheses 2 and 3 the Jacobian never changes sign. Since area is positive we must choose the plus sign when J is positive, the minus sign when J is negative:

$$A = \iint_{R_{uv}} |J(u, v)|\, dS_{uv}$$

This completes the proof in the special case. Incidentally, we have learned that when $J > 0$ clockwise description of Γ_{xy} corresponds to clockwise descriptions of Γ_{uv}.

Make a subdivision Δ of R_{xy} into subregions R_k of area ΔS_k, $k = 1, 2, \ldots, n$. By the transformation (2) there will correspond in the uv-plane a subdivision Δ' of R_{uv} into subregions R'_k of area $\Delta S'_k$. By the above proof the areas of these subregions are related as follows:

$$\Delta S_k = \int\int_{R_k'} |J(u, v)| \, dS_{uv} = |J(u_k, v_k)| \, \Delta S'_k$$

where the point (u_k, v_k) is in R'_k as required by the law of the mean (see Exercise 13, §2, Chapter 6). Let (x_k, y_k) be the point of R_k which corresponds to (u_k, v_k) under the transformation (2). By Theorem 1, Chapter 6,

$$\int\int_{R_{xy}} F(x, y) \, dS_{xy} = \lim_{\|\Delta\| \to 0} \sum_{k=1}^{n} F(x_k, y_k) \, \Delta S_k$$

$$= \lim_{\|\Delta'\| \to 0} \sum_{k=1}^{n} F(g(u_k, v_k), h(u_k, v_k))|J(u_k, v_k)| \, \Delta S'_k$$

Again by Theorem 1 the latter sum approaches the right-hand side of (7). That the norm of Δ approaches zero when the norm of Δ' does follows from the uniform continuity of $g(u, v)$ and $h(u, v)$ in R_{uv} (compare §6.4 of Chapter 5). Or one could establish the result by Theorem 3 of Chapter 1. This concludes the proof of the theorem. It should be observed that in equation (7) no derivative of g or h of higher order than the first appears. This suggests that C^2 may be replaced by C^1 in hypothesis 2. This is in fact the case, but the proof would need to be greatly altered to effect this generalization. Note also that the simply connected character of R_{xy} and R_{uv} is not an essential feature of the end result. For, if each can be subdivided into corresponding parts satisfying hypothesis 4, then equation (7) will hold for the original regions by virtue of property IV, §2.1, Chapter 6. Theorem 9 as stated is sufficiently general for most practical purposes.

EXAMPLE A.   Make the transformation $x = v \cos u$, $y = v \sin u$ to

$$\int\int_{R_{xy}} y \, dS_{xy}$$

where R_{xy} is the region shown in Figure 25. The Jacobian of the transformation is $-v$, so that the integral becomes

$$\int\int_{R_{uv}} v^2 \sin u \, dS_{uv}$$

Hence,

$$\int_1^2 dx \int_0^{\sqrt{2x - x^2}} y \, dy = \int_0^{\pi/4} \sin u \, du \int_{\sec u}^{2 \cos u} v^2 \, dv = \tfrac{1}{3}$$

5.3 AN APPLICATION

It is frequently required to evaluate a surface integral over a surface Σ which is given parametrically:

$$x = g(u, v), \qquad y = h(u, v), \qquad z = k(u, v)$$

Set

$$j_1 = \frac{\partial(h, k)}{\partial(u, v)}, \qquad j_2 = \frac{\partial(k, g)}{\partial(u, v)}, \qquad j_3 = \frac{\partial(g, h)}{\partial(u, v)}$$

$$D = \sqrt{j_1^2 + j_2^2 + j_3^2}$$

Let Σ correspond to the region R_{uv} of the uv-plane. Suppose that $D \neq 0$ in R_{uv}. Then j_1, j_2, j_3 do not vanish simultaneously. Suppose first that j_3 does not vanish. If γ is the acute angle between the normal to Σ and the z-axis, then $\sec \gamma = D/|j_3|$. If R_{xy} is the projection of Σ on the xy-plane, by Theorem 7

$$\iint_\Sigma P(x, y, z) \, d\Sigma = \iint_{R_{xy}} P(x, y, f(x, y)) \, \frac{D}{|j_3|} \, dS_{xy}$$

By Theorem 9 this is equal to

$$\iint_{R_{uv}} P(g(u, v), h(u, v), k(u, v)) D \, dS_{uv}$$

If it is j_1 or j_2 which does not vanish, we may project Σ on the yz- or xz-plane and obtain precisely the same formula. Finally, if no one of the Jacobians is different from zero throughout R_{uv}, we may divide this region into subregions in each of which *some* Jacobian does not vanish. Hence, we obtain in all cases

$$(8) \qquad \iint_\Sigma P(x, y, z) \, d\Sigma = \iint_{R_{uv}} P(g(u, v), h(u, v), k(u, v)) D \, dS_{uv}$$

The great advantage of this formula over that in Theorem 7 is that it no longer requires that the surface Σ be cut only once by a line parallel to the axis.

EXAMPLE B.   Find the area of the sphere

$$x = a \sin \varphi \cos \theta, \qquad y = a \sin \varphi \sin \theta, \qquad z = a \cos \varphi$$

Simple computation gives

$$D = a^2 \sin \varphi$$

Hence, the area is

$$A = \iint_{R_{\theta\varphi}} a^2 \sin \varphi \, dS_{\theta\varphi} = a^2 \int_0^{2\pi} d\theta \int_0^\pi \sin \varphi \, d\varphi = 4\pi a^2$$

5.4 REMARKS

The transformation (2) has another useful interpretation. It may be regarded as a change of coordinates. Thus (x, y) and (u, v), connected by equations (2), may be thought of as different coordinates of the same point. In our example, set $v = r$ and $u = \theta$. It then becomes the transformation to polar coordinates. There is then just one region of the plane under consideration. But its boundary has a different equation according as rectangular or polar coordinates are used. The Jacobian of the transformation is $-r$, and we obtain Theorem 3, Chapter 6, as a corollary of Theorem 9.

By use of Theorem 8, we could now extend Theorem 9 to three dimensions. The new factor introduced into the integral by the transformation would again be the absolute value of the Jacobian of the transformation. It is interesting to check that this factor is $r^2 \sin \varphi$ for spherical coordinates and r for cylindrical coordinates. This must follow from the results of Chapter 6.

5.5 AN AUXILIARY RESULT

In the application of Theorem 9, it is sometimes difficult to verify hypothesis 4. In view of Theorem 16, Chapter 1, it might be supposed that the nonvanishing of the Jacobian would be sufficient to guarantee the one-to-one nature of the transformation. But that result dealt with *local* properties, with *small* neighborhoods. Notice, for example, that the equations

$$(9) \qquad x = u^2 - v^2, \qquad y = 2uv$$

make the region $1 \leq u^2 + v^2 \leq 4$ correspond to the region $1 \leq x^2 + y^2 \leq 16$, that the Jacobian is not zero, and that the transformation is not one-to-one. The points $u = 1$, $v = 1$ and $u = -1$, $v = -1$ both correspond to the point $x = 0$, $y = 2$.

We state here without proof[*] a useful result that guarantees hypothesis 4. Let us suppose that the first three hypotheses of Theorem 9 hold. Suppose further that R_{uv} is bounded by a simple closed curve Γ_{uv} and that the transform of this curve under equations (2) is a simple closed curve Γ_{xy} traced once as Γ_{uv} is traced once. Let R_{xy} be the region inside Γ_{xy}. Then the correspondence between R_{xy} and R_{uv} is one-to-one. To apply this result, we have only to investigate the transform of a single closed curve.

As an example, consider the part of the region $1 \leq u^2 + v^2 \leq 4$ that lies in the first quadrant. By the transformation (9), its boundary becomes the boundary of the region $1 \leq x^2 + y^2 \leq 16$, $y \geq 0$. One sees this by transforming separately the two straight line segments and the two circular arcs of the boundary. By the result quoted, the two regions correspond in a one-to-one way.

---

[*] See, for example, the *Cours d'Analyse* of de la Vallée Poussin, 1923, Vol. 1, p. 355.

EXERCISES (5)

1. Compute the area of R_{xy} of Figure 25 first by use of the coordinates xy and then by use of the coordinates uv.

2. Solve the same problem for the area of R_{uv}.

3. From the region between the circles $x^2 + y^2 = 1$, $x^2 + y^2 = 4$ are removed the points for which $y^2 < 2x - x^2$ to form the region R_{xy}. Describe the region R_{uv}, corresponding to R_{xy} under the transformation $x = v \cos u$, $y = v \sin u$.

4. Find the area of R_{xy} in Exercise 3 by two methods.

5. Solve the same problem for R_{uv}.

6. Find the area of the ellipse

$$\frac{x^2}{a^2} + \frac{y^2}{b^2} = 1$$

by relating it to the area of a circle by the transformation $x = au$, $y = bv$.

7. Show analytically that areas are preserved under the rigid motion $x = a + u \cos \alpha - v \sin \alpha$, $y = b + u \sin \alpha + v \cos \alpha$.

8. Express the integral

$$\int_0^1 dx \int_{\sqrt{1-x^2}}^{\sqrt{4-x^2}} f(x, y)\, dy$$

as an iterated integral using uv-coordinates if $x = u^2 - v^2$, $y = 2uv$.

9. Evaluate the two integrals of Exercise 8 if $f = y$.

10. By use of equation (8) show that the area of the surface of revolution

$$x = u \cos v, \qquad y = u \sin v, \qquad z = f(u) \quad a \leq u \leq b, \quad 0 \leq v \leq 2\pi$$

is

$$2\pi \int_a^b u \sqrt{1 + [f'(u)]^2}\, du$$

11. Use the result of Exercise 10 to find the area of cylinder, cone, and sphere.

12. Find the area of a torus.

13. Prove the theorem of Pappus for the area of a surface of revolution by use of Exercise 10:

$$A = 2\pi h l$$

Here l is the length of the rotating curve and h is the distance of the center of gravity of the curve from the axis of rotation.

14. Compute the Jacobians for spherical and cylindrical coordinates mentioned in §5.4.

15. Show how a triple integral transforms under the transformation
$$x = g(u, v, w), \qquad y = h(u, v, w), \qquad z = k(u, v, w)$$
where
$$g_1 g_2 + h_1 h_2 + k_1 k_2 = 0$$
$$g_2 g_3 + h_2 h_3 + k_2 k_3 = 0$$
$$g_3 g_1 + h_3 h_1 + k_3 k_1 = 0$$
Show that the Jacobian of the transformation is $c_1 c_2 c_3$, where
$$c_i = \sqrt{g_i^2 + h_i^2 + k_i^2} \qquad\qquad i = 1, 2, 3$$

16. Illustrate Exercise 15 by the transformation of spherical coordinates.

17. Find the area of the region $1 \leqq x^2 + y^2 \leqq 16$, $y \geqq 0$ by integration in the uv-plane [transformation (9)].

18. Under the transformation (9) each point of the circle $u^2 + v^2 = 1$ is transformed into a point of the circle $x^2 + y^2 = 1$. Does this mean that the interiors of these circles correspond in a one-to-one way?

19. (*Generalization of a Theorem of Pappus.*) Find the area of the surface
$$X = x(s) + y'(s) \cos u, \qquad Y = y(s) - x'(s) \cos u, \qquad Z = \sin u$$
where $0 \leqq s \leqq l$, $0 \leqq u \leqq 2\pi$, $[x'(s)]^2 + [y'(s)]^2 = 1$. What is the curve?

§6. *Line Integrals in Space*

The line integral defined in §1 generalizes in an obvious way when the curve over which the integral is defined is no longer plane. In §4 we gave one generalization of Green's theorem to three dimensional space. There is another known as *Stokes's theorem*. This relates a line integral over a closed space curve to a surface integral over a surface spanning the curve. The relation reduces to Green's theorem for the plane when the curve lies in the xy-plane and the spanning surface is the plane itself. We prove Stokes's theorem here.

6.1 DEFINITION OF THE LINE INTEGRAL

Consider a curve Γ with parametric equations
$$(1) \qquad\qquad x = \varphi(t), \qquad y = \psi(t), \qquad z = \omega(t) \qquad\qquad a \leqq t \leqq b$$
It is *regular* if it has no double points and if the interval (a, b) can be divided into a finite number of subintervals in each of which $\varphi(t) \in C^1$, $\psi(t) \in C^1$, $\omega(t) \in C^1$. If $f(x, y, z)$ is defined on Γ, then with obvious notations we define the line integral
$$(2) \qquad \int_\Gamma f(x, y, z)\, dx = \lim_{\|\Delta\| \to 0} \sum_{i=1}^n f(\varphi(t_i'), \psi(t_i'), \omega(t_i'))[\varphi(t_i) - \varphi(t_{i-1})]$$
$$t_{i-1} \leqq t_i' \leqq t_i \qquad\qquad i = 1, 2, \ldots, n$$

whenever the limit exists.   Two other integrals, replacing dx by dy and dz, are defined in an analogous way.   As in the proof of Theorem 1, we show that when $f \in C$ on the regular curve Γ

$$\int_\Gamma f(x, y, z)\, dx = \int_a^b f(\varphi(t),\ \psi(t),\ \omega(t))\varphi'(t)\, dt$$

with similar equations for the other two integrals.   The direction of integration in (2) is that direction on Γ which corresponds to the motion of a point whose parametric value t moves from a to b.

EXAMPLE A.   Compute

$$\int_\Gamma x\, dx + xy\, dy + xyz\, dz$$

where Γ is the piece of the twisted cubic $x = t$, $y = t^2$, $z = t^3$ corresponding to the interval $0 \leq t \leq 1$.   The value is

$$\int_0^1 t\, dt + 2 \int_0^1 t^4\, dt + 3 \int_0^1 t^8\, dt = \tfrac{37}{30}$$

6.2 STOKES'S THEOREM

Theorem 10.   1. $f(x, y) \in C^2$

2. Σ *is the surface* $z = f(x, y)$ *bounded by the regular closed curve* Γ

3. $P(x, y, z)$, $Q(x, y, z)$, $R(x, y, z) \in C^1$ *on* Σ

4. α, β, γ *are direction angles of a directed normal to* Σ

$\Rightarrow$
$$\int_\Gamma P\, dx + Q\, dy + R\, dz$$

$$= \int\!\!\int_\Sigma [(R_2 - Q_3) \cos \alpha + (P_3 - R_1) \cos \beta + (Q_1 - P_2) \cos \gamma]\, d\Sigma$$

where the direction of integration is clockwise to an observer facing in the direction of the directed normal.

For definiteness choose the direction of the normal to Σ so as to make an acute angle with the positive direction on the z-axis.   Then

(3) $$f_1(x, y) = -\frac{\cos \alpha}{\cos \gamma}, \qquad f_2(x, y) = -\frac{\cos \beta}{\cos \gamma}$$

Let the projection of Σ and Γ on the xy-plane be R_{xy} and Γ_{xy}, respectively. The sense of description of Γ described in the theorem will give rise to a counterclockwise direction on Γ_{xy}. If a parametric representation of Γ_{xy} is $x = \varphi(t), \; y = \psi(t)$, then one for Γ is

$$x = \varphi(t), \qquad y = \psi(t), \qquad z = f(\varphi(t), \psi(t)) \qquad a \leq t \leq b$$

Then

$$\int_{\Gamma} P(x, y, z) \, dx = \int_a^b P(\varphi(t), \psi(t), f(\varphi(t), \psi(t))) \varphi'(t) \, dt$$

Also

$$\int_{\Gamma_{xy}} P(x, y, f(x, y)) \, dx = \int_a^b P(\varphi(t), \psi(t), f(\varphi(t), \psi(t))) \varphi'(t) \, dt$$

Hence,

$$\int_{\Gamma} P(x, y, z) \, dx = \int_{\Gamma_{xy}} P(x, y, f(x, y)) \, dx$$

where the sense of description over Γ_{xy} is counterclockwise. By Green's theorem for the plane

$$\int_{\Gamma_{xy}} P(x, y, f(x, y)) \, dx = -\iint_{R_{xy}} [P_2 + P_3 f_2] \, dS_{xy}$$

$$= -\iint_{\Sigma} [P_2(x, y, z) + P_3(x, y, z) f_2(x, y)] \cos \gamma \, d\Sigma$$

We have here made use of Theorem 7. By virtue of the second of equations (3) we see that

$$\int_{\Gamma} P(x, y, z) \, dx = \iint_{\Sigma} [P_3 \cos \beta - P_2 \cos \gamma] \, d\Sigma$$

This proves the theorem in so far as it concerns $P(x, y, z)$. A similar proof holds for $Q(x, y, z)$, using the first of equations (3). The projection is again made on the xy-plane. The proof for the function $R(x, y, z)$ is somewhat different. We give the equations used:

$$\int_{\Gamma} R(x, y, z) \, dz = \int_a^b R(\varphi(t), \psi(t), f(\varphi(t), \psi(t)))[f_1 \varphi' + f_2 \psi'] \, dt$$

$$= \int_{\Gamma_{xy}} R f_1 \, dx + R f_2 \, dy$$

$$= \iint_{R_{xy}} [R_1 f_2 + R_3 f_1 f_2 + R f_{12} - R_2 f_1 - R_3 f_1 f_2 - R f_{21}] \, dS_{xy}$$

$$= \iint_{\Sigma} [R_1 f_2 - R_2 f_1] \cos \gamma \, d\Sigma$$

$$= \iint_{\Sigma} [R_2 \cos \alpha - R_1 \cos \beta] \, d\Sigma$$

This completes the proof of the theorem.

Corollary 10.   1. Σ *is a surface bounded by the regular closed curve* Γ
2. Σ *has the three equations* $z = f(x, y)$, $x = g(y, z)$,
$y = h(z, x)$ *with* f, g, $h \in C^1$
3. *As in the theorem*
4. *As in the theorem*

$\Rightarrow$ *The conclusion of the theorem.*

For Theorem 10 we required that Σ should be cut only once by lines parallel to a *single* axis and that the single-valued defining function should belong to C^2. In contrast we are now demanding that Σ should be cut only once by lines parallel to *all three* axes but that the single-valued defining functions should belong only to C^1. The proof of the corollary is in effect already contained in the proof of the theorem. Insofar as P is concerned it is there word for word, for only class C^1 for f was used. Now permute symbols: $P \to Q \to R, f \to g \to h, x \to y \to z, 1 \to 2 \to 3, \alpha \to \beta \to \gamma$.

EXAMPLE B.   Compute in two ways the line integral

$$I = \int_\Gamma xyz \, dz$$

over the circle

$$x = \cos t, \qquad y = \frac{\sin t}{\sqrt{2}}, \qquad z = \frac{\sin t}{\sqrt{2}} \qquad 0 \leq t \leq 2\pi$$

in the direction of increasing t. Substitution gives

$$I = \frac{1}{2\sqrt{2}} \int_0^{2\pi} \sin^2 t \cos^2 t \, dt = \frac{\pi}{8\sqrt{2}}$$

The direction cosines of the directed normal to Σ, the plane of the circle, are $0, -\dfrac{\sqrt{2}}{2}, \dfrac{\sqrt{2}}{2}$. By Stokes's theorem

$$I = -\int\int_\Sigma yz \cos \beta \, d\Sigma$$

To evaluate this integral, project on the xz-plane. We have then to compute

$$I = \int\int_S z^2 \, dS$$

where S is the ellipse $x^2 + 2z^2 = 1$. Hence,

$$I = 4 \int_0^{1/\sqrt{2}} z^2 \, dz \int_0^{\sqrt{1-2z^2}} dx = \frac{\pi}{8\sqrt{2}}$$

6.3 REMARKS

Stokes's theorem clearly remains true if Σ is of a more complicated nature but still divisible into a finite number of parts, each of which satisfies

the conditions of the theorem or its corollary.   For example, suppose Σ is
the part of the unit sphere lying in the first octant, Figure 26.   Neither

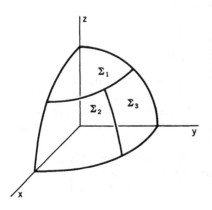

theorem nor corollary is directly applicable.   For example, if we take the
equation of Σ as $z = (1 - x^2 - y^2)^{1/2}$,
$f(x, y)$ is single-valued as required but
$f \notin C'$ along the unit circle of the
xy-plane.   The figure shows how Σ
can be decomposed into three parts.
We may project $\Sigma_1, \Sigma_2, \Sigma_3$ on the xy-,
yz-, zx-planes, respectively, to apply
Theorem 10.   Note that the line integrals over the auxiliary division lines
cancel each other, being executed in
opposite directions.   Thus Stokes's
equation is valid for the original

Fig. 26.   surface Σ bounded by the three circular
arcs.

But there is one type of surface that must be excluded even though it
permits of such subdivision.   This is the "one-sided" surface.   A sample of
such a surface can be made by joining together the opposite (far) edges of a

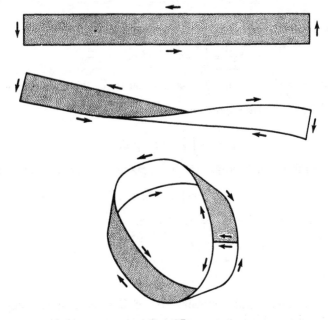

Fig. 27.

long strip of paper after a half twist has been made in the paper, Figure 27. In the figure the surface has been decomposed into two parts by introducing two cuts. To each, Stokes's theorem is applicable. But now the two line integrals over one of the cuts do not cancel each other. Moreover the single boundary of the surface is not described in the same sense over all its parts. Hence Stokes's theorem is not applicable to the original uncut surface.

6.4 EXACT DIFFERENTIALS

A solid region V is *simply connected* if any closed curve drawn in the region can be deformed continuously into a point of the region always lying entirely in the region. A similar definition might have been given for a plane region and shown to be equivalent to that of §3.1. As examples, the region between two concentric spherical surfaces is simply connected, whereas the region between two coaxial circular cylindrical surfaces is not. Denote a simply connected region by V^*. Precisely as in §3, we could prove the following results.

Theorem 11.   1. $P(x, y, z)$, $Q(x, y, z)$, $R(x, y, z) \in C^1$ in V^*

2. $Q_3 = R_2$, $R_1 = P_3$, $P_2 = Q_1$ in V^*

=►   *There exists* $F(x, y, z) \in C^2$ *in* V^* *such that*

$$F_1 = P, \qquad F_2 = Q, \qquad F_3 = R$$

Consider next the line integral

(4)
$$\int_{a,b,c}^{x_0,y_0,z_0} P \, dx + Q \, dy + R \, dz$$

Theorem 12.   1. $P(x, y, z)$, $Q(x, y, z)$, $R(x, y, z) \in C^1$ in V^*

2. $Q_3 = R_2$, $R_1 = P_3$, $P_2 = Q_1$ in V^*

3. Γ *is a regular curve in* V^* *joining* (a, b, c) *with* (x_0, y_0, z_0)

⇒   *The integral* (4) *extended over* Γ *is independent of* Γ

This result shows that the integral (4) defines a single-valued function of (x_0, y_0, z_0). Its differential is $P \, dx_0 + Q \, dy_0 + R \, dz_0$. That the simply connected character of the region is essential may be seen by consideration of the example

$$P = yz(x^2 + y^2)^{-1}, \quad Q = -xz(x^2 + y^2)^{-1}, \quad R = -\tan^{-1}(y/x)$$

When hypothesis 2 of these theorems is satisfied, then the equation

$$P \, dx + Q \, dy + R \, dz = 0$$

is called an *exact differential equation* and the left-hand side an *exact differential*. The solution of the equation is $F =$ constant, where F is the function of Theorem 11. If F is known, then the line integral (4) is equal to $F(x_0, y_0, z_0) - F(a, b, c)$, whatever the path of integration. Here F is usually most simply found by use of indefinite integrals, as illustrated in the following example.

EXAMPLE C.

$$\int_{1,1,1}^{e,0,1} \frac{(z^3 + zy^2)\, dx + xz^2\, dy + (xz^2 + xy^2 - xyz)\, dz}{xz^3 + xy^2z}$$

The differential is exact. Set

$$F_x = \frac{z^3 + zy^2}{xz^3 + xy^2z} = \frac{1}{x}, \qquad F = \log x + G(y, z)$$

$$F_y = G_y = \frac{z}{z^2 + y^2}, \qquad G = \tan^{-1}\frac{y}{z} + H(z)$$

$$F_z = G_z = \frac{-y}{y^2 + z^2} + H'(z) = \frac{1}{z} - \frac{y}{y^2 + z^2}, \qquad H = \log z$$

$$F = \tan^{-1}\frac{y}{z} + \log xz$$

The value of the integral is $1 - (\pi/4)$.

6.5 VECTOR CONSIDERATIONS

Both Green's theorem and Stokes's theorem take a particularly elegant form if vector notation is used. Besides being useful as a means of remembering the formulas, the vector form has the advantage of putting into evidence the invariant nature of the results. Both theorems were stated in such a way as to depend upon the particular choice of coordinate axes. The vector form will show that the results depend only on the curves, surfaces, and regions involved and upon the given functions defined there.

In order to express the theorems in vector form let us revert to the notation of Chapter 2. Let $\vec{y} = \vec{y}(x_1, x_2, x_3)$ be a vector function, defining a vector field. Let $\vec{\zeta}$ be the unit vector along the exterior normal to the surface Σ^* of Theorem 8. If the components of $\vec{y}$ take the place of the functions P, Q, R, then the conclusion of that theorem becomes

(5) $$\iiint \text{Div } \vec{y}\, dV = \iint_{\Sigma_*} \vec{y} \cdot \vec{\zeta}\, d\Sigma$$

Since the divergence appears as the integrand of the triple integral, Theorem 8 is often called the *divergence theorem*.

As an illustration of the physical meaning of equation (5) suppose that $\vec{y}$ defines a velocity field for a fluid. That is, the vector $\vec{y}$ at each point gives the velocity of the fluid there both in direction and magnitude (say, in feet per second). Thus if vectors $\vec{y}$ over a plane region R of area A are all perpendicular to R and of constant magnitude, then $A|\vec{y}|$ is the number of cubic feet per second of the fluid flowing through R. Recalling that

$$\vec{y} \cdot \vec{\zeta} = |\vec{y}| \cos \theta$$

where θ is the angle between $\vec{y}$ and $\vec{\zeta}$, we see that the integrand on the right of (5) is the component of the velocity vector $\vec{y}$ in the direction of the exterior normal. Since this is multiplied by the surface element $d\Sigma$, the surface integral (5) represents the number of cubic feet per second flowing out of the whole surface Σ^* (flowing into Σ^* if the number is negative). In particular if the fluid is incompressible the net rate of flow through Σ^* must be zero for every Σ^*, so that both integrals (5) must be zero. By a familiar argument div $\vec{y} = 0$, and we see that this equation is characteristic of an incompressible fluid. The divergence of $\vec{y}$ at a point is in some sense a measure of the failure of the fluid to be incompressible near the point. For example, if V is a sphere of radius r with center at a point $\vec{x}$, then the law of the mean for triple integrals (Exercise 11, §8, Chapter 6) gives

$$\iint_{\Sigma^*} \vec{y} \cdot \vec{\zeta} \, d\Sigma = \frac{4}{3} \pi r^3 \, \mathrm{Div} \, \vec{y} \Big|_{\vec{\xi}} \qquad (\xi_1, \xi_2, \xi_3) \in V$$

As $r \to 0$ the point $\vec{\xi}$ approaches $\vec{x}$ and

$$(6) \qquad \mathrm{Div} \, \vec{y} \Big|_{\vec{x}} = \lim_{r \to 0} \frac{3}{4\pi r^3} \iint_{\Sigma^*} \vec{y} \cdot \vec{\zeta} \, d\Sigma$$

It is in this sense that the divergence is a measure of the compressibility of the fluid, for the surface integral (6) is a measure of the imbalance between flow out of and flow into Σ^*. Equation (6) is often taken as the definition of divergence. It has the advantage of putting into evidence the invariance of the divergence under rigid motions.

For Theorem 10 we again let the components of $\vec{y}$ correspond to P, Q, R and let $\vec{\zeta}$ be the unit vector normal to Σ with the sense prescribed by the theorem. The conclusion becomes

$$(7) \qquad \int_\Gamma \vec{y} \cdot d\vec{x} = \iint_\Sigma \vec{\zeta} \cdot \mathrm{Curl} \, \vec{y} \, d\Sigma$$

Let $\vec{\alpha}$ be the unit tangent vector to Γ in the sense of the direction of integration. Since $\vec{\alpha} = d\vec{x}/ds$, we have

$$(8) \qquad \int_\Gamma \vec{y} \cdot d\vec{x} = \int_0^l \vec{y} \cdot \vec{\alpha} \, ds = \int_0^l |\vec{y}| \cos \theta \, ds$$

where l is the total length of Γ and θ is the angle between $\vec{y}$ and $\vec{\alpha}$. In particular if Σ is a circular disk of radius r and center at $\vec{x}$, the law of the mean applied as above gives

$$(9) \qquad \vec{\zeta} \cdot \operatorname{Curl} \vec{y}\Big|_{\vec{x}} = \lim_{r \to 0} \frac{1}{\pi r^2} \int_0^{2\pi r} |\vec{y}| \cos \theta \, ds$$

We thus have an invariant expression for the component of Curl $\vec{y}$ in an arbitrary direction $\vec{\zeta}$. The integration must be performed over a circle whose plane is perpendicular to $\vec{\zeta}$ and whose center is at $\vec{x}$.

If we again interpret $\vec{y}$ as defining a velocity field, equation (9) gives us a physical interpretation of Curl $\vec{y}$. In the special case in which $\vec{y}$ defines a constant rotation of w radians per second about $\vec{\zeta}$ we have $|\vec{y}| = wr$ and $\theta = 0$, so that (9) gives $\vec{\zeta} \cdot \operatorname{Curl} \vec{y} = 2w$ (compare Example E, §5, Ch. 2). Since the integrand (9) is the component of $\vec{y}$ in the direction of the tangent to Γ, we see that the component of Curl $\vec{y}$ in the direction $\vec{\zeta}$ may be interpreted as a measure of the tendency of the field to be a pure rotation about $\vec{\zeta}$ in the neighborhood of $\vec{x}$. In particular if Curl $\vec{y} \equiv 0$, the field is called *irrotational*.

As another interpretation we may think of $\vec{y}$ as defining a force field. Then the line integral (7) is the work performed by the field on a unit particle as it describes Γ in the sense of integration. If Curl $\vec{y} \equiv 0$ this work is zero for every closed curve and the field is called *conservative*. The integral over part of the path may be positive (when the field has done work on the particle) and negative over the rest (when the particle has done an equal amount of work on the field); thus total energy is *conserved*. For a conservative field Theorem 11 shows the existence of a function F for which Grad $F = \vec{y}$. This function F or its negative is called the *potential function* for the field.

EXERCISES (6)

1. Work Example A for the curve $x = \cos t$, $y = \cos t$, $z = \sin t$, $0 \leq t \leq 2\pi$.

2. Compute

$$\int_\Gamma x \, dx + z \, dy - y \, dz$$

over the curve (θ increasing from 0 to 2π)

$$\Gamma: \qquad x = 1 + \cos \theta, \qquad y = \sin \theta, \qquad z = 14 - 2 \cos \theta$$

Then by the theorem of Stokes express the integral as a surface integral over $\Sigma: x^2 + y^2 + z = 16$. Evaluate the surface integral by projection on the xy-plane. Indicate clearly which normal to Σ you are using.

3.

$$\int_{1,0,-1}^{1,2,3} yz \, dx + xz \, dy + xy \, dz = \, ?$$

Solve first by choosing a broken line path with segments parallel to the axes. Then solve by finding F as in Example C.

4. If $r = (x^2 + y^2 + z^2)^{1/2}$, compute

$$\int_{\pi, \, -\pi, \, \pi/2}^{2\pi/3, \, 2\pi/3, \, -\pi/3} \frac{\cos r}{r} (x \, dx + y \, dy + z \, dz)$$

5. Give details of the example following Theorem 12.

6. Prove Theorem 11.

7. Prove Theorem 12.

8. Extend the discussion of §1.3 to three dimensions.

9. Show that a three dimensional field of force due to the attraction of a particle (inverse square law) is conservative. Find the potential function.

10. Prove the converse of Theorem 11.

11. Show that the surface integral

$$\iint_{\Sigma} [P \cos \alpha + Q \cos \beta + R \cos \gamma] \, d\Sigma$$

is independent of Σ but depends only on Γ, the boundary curve of Σ, if $P_1 + Q_2 + R_3 = 0$. *Hint:* Solve

$$C_2 - B_3 = P, \qquad A_3 - C_1 = Q, \qquad B_1 - A_2 = R$$

for A, B, C. This may be done by choosing C arbitrarily.

12. Compute

$$\iint_{\Sigma} [x \cos \alpha + xy \cos \beta - z(1 + x) \cos \gamma] \, d\Sigma$$

over any surface Σ spanning the circle of Example B.

13. Solve

$$yz \, dx + zx \, dy - xy \, dz = 0$$

Hint: The equation becomes exact if multiplied by a suitable function of z.

14. Solve $yz \, dx + zx \, dy + dz = 0$.

15. Apply Stokes's theorem to two halves of a sphere to show that

$$\iint \vec{\zeta} \cdot \text{Curl } \vec{y} \, d\Sigma = 0$$

over the entire surface of any sphere. By Green's theorem conclude that div Curl $\vec{y} \equiv 0$. What continuity assumption are you making?

16. If $\vec{y} = \nabla F$ in Stokes's theorem, show that the line integral involved is zero over every closed curve Γ. Hence, show that Curl $\nabla F \equiv 0$. Discuss the continuity assumptions.

17. Show that, if $P\,dx + Q\,dy + R\,dz$ can be made exact by multiplication by a function $\lambda(x, y, z)$ of class C^1, then

$$P(Q_3 - R_2) + Q(R_1 - P_3) + R(P_2 - Q_1) = 0$$

Verify the equation for Exercises 13 and 14.

18. Compute directly and by use of Stokes's theorem

$$\int_\Gamma xy\,dx + x\,dy$$

where Γ is the unit circle. Use the spanning surface as a hemisphere and compute the double integral by the parametric method of §5.3.

8

Limits and
Indeterminate Forms

§1. *The Indeterminate Form* 0/0

The determination of the limit

(1) $$\lim_{x \to c} \frac{f(x)}{g(x)}$$

where $\lim f(x) = \lim g(x) = 0$, is traditionally referred to as the evaluation of the indeterminate form 0/0. This phraseology is misleading since division by zero is undefined. But the evaluation of the limit (1) is fundamental in the calculus. For example, the problem arises in the very definition of the derivative of a function

$$f'(x_0) = \lim_{\Delta x \to 0} \frac{f(x_0 + \Delta x) - f(x_0)}{\Delta x}$$

for, both numerator and denominator tend to zero with Δx. In computing the derivative of a given elementary function, some algebraic reduction or other device must always be employed to avoid the indeterminate character

257

of the limit. For example,

$$\lim_{\Delta x \to 0} \frac{\sqrt{1 + \Delta x} - 1}{\Delta x} = \lim_{\Delta x \to 0} \frac{\sqrt{1 + \Delta x} - 1}{\Delta x} \frac{\sqrt{1 + \Delta x} + 1}{\sqrt{1 + \Delta x} + 1}$$

$$= \lim_{\Delta x \to 0} \frac{1}{\sqrt{1 + \Delta x} + 1} = \frac{1}{2}$$

Other familiar examples from elementary calculus are

$$\lim_{x \to 0} \frac{\sin x}{x} = 1, \qquad \lim_{x \to 0} \frac{1 - \cos x}{x^2} = \frac{1}{2}$$

$$\lim_{x \to 0} \frac{\log (1 + x)}{x} = 1, \qquad \lim_{x \to 0} \frac{e^x - 1}{x} = 1$$

It is our purpose in this section to develop a general method for evaluating limits of the form (1).

1.1 THE LAW OF THE MEAN

The limit (1) may often be evaluated by a simple application of the law of the mean. Observe first that there is no *apriori* way of predicting the limit. The following examples show that it may be zero, different from zero, or indeed need not exist at all:

$$\lim_{x \to 0} \frac{\sin^2 x}{x} = 0, \qquad \lim_{x \to 0} \frac{\sin^2 x}{x^2} = 1$$

$$\lim_{x \to 0} \frac{\sin^2 x}{x^4} = +\infty, \qquad \lim_{x \to 0} \frac{x \sin (1/x)}{\sin x}$$

In the last two examples, the limit does not exist.

Theorem 1.   1. $f(x), g(x) \in C^1$ $a \le x \le b$

           2. $f(c) = g(c) = 0$ $a < c < b$

           3. $g'(c) \ne 0$

$\Rightarrow$ $\displaystyle \lim_{x \to c} \frac{f(x)}{g(x)} = \frac{f'(c)}{g'(c)}$

To prove this we use the law of the mean as follows:

$$(2) \quad \frac{f(c + h)}{g(c + h)} = \frac{f(c + h) - f(c)}{g(c + h) - g(c)} = \frac{f'(c + \theta h)h}{g'(c + \theta' h)h} \qquad 0 < \theta < 1, 0 < \theta' < 1$$

Here h is so chosen that $a \le c + h \le b$, $h \ne 0$, and is so small that $g'(c + \theta' h) \ne 0$. This is possible by virtue of hypotheses 1 and 3. Then no denominator in equation (2) is zero.   Now cancel h in the last quotient and allow h to approach zero.   We thus obtain the desired conclusion. If $g'(c) = 0$, $g'(x) \neq 0$ when $x \neq c$, $f'(c) \neq 0$, then

$$(3) \qquad\qquad \lim_{x \to c} \left| \frac{f(x)}{g(x)} \right| = +\infty$$

This is seen by applying the theorem to $g(x)/f(x)$.   Without the absolute value signs in equation (3) we could only be sure that the quotient becomes positively or negatively infinite as $x \to c+$ or $x \to c-$.

For example,

$$\lim_{x \to 0+} \frac{\sin x}{x^2} = +\infty, \quad \lim_{x \to 0-} \frac{\sin x}{x^2} = -\infty, \quad \lim_{x \to 0} \frac{|\sin x|}{x^2} = +\infty$$

If both $f'(c)$ and $g'(c)$ are zero, the theorem is not applicable.

EXAMPLE A. $\displaystyle \lim_{x \to 0} \frac{\log (1 + x)}{x} = \frac{1}{1 + x}\Big|_{x=0} = 1$

EXAMPLE B. $\displaystyle \lim_{x \to 0} \frac{\sin x}{x^3} = +\infty$

EXAMPLE C. $\displaystyle \lim_{h \to 0} \frac{f(c + 2h) - f(c - 2h)}{h}$

$$= [2f'(c + 2h) + 2f'(c - 2h)]_{h=0} = 4f'(c)$$

EXAMPLE D. $\displaystyle \lim_{x \to 1} \frac{x^3 + 3x + 2}{x^2 - x - 2} \neq \frac{3x^2 + 3}{2x - 1}\Big|_{x=1} = 6$

Here the form is not indeterminate, and the limit is -3.

EXAMPLE E. $\displaystyle \lim_{x \to 0} \frac{x^2 \sin (1/x)}{\sin x} \neq \lim_{x \to 0} \frac{2x \sin (1/x) - \cos (1/x)}{\cos x}$

$$\lim_{x \to 0} \frac{x^2 \sin (1/x)}{\sin x} = \lim_{x \to 0} \frac{x}{\sin x} \lim_{x \to 0} x \sin \left(\frac{1}{x} \right) = 0$$

Here Theorem 1 is not applicable, in view of the fact that $x^2 \sin (1/x) \notin C^1$. Yet the desired limit can be evaluated by inspection.   We thus see that the conditions of the theorem are sufficient but not necessary.

1.2 GENERALIZED LAW OF THE MEAN

In order to treat the case in which $f'(c) = g'(c) = 0$, we need a generalization of the law of the mean.

Theorem 2.   1. $f(x), g(x) \in C^1$ $a \leqq x \leqq b$

2. $a \leqq c \leqq b, a \leqq c + h \leqq b$

(4) $\Rightarrow$ $[f(c + h) - f(c)]g'(c + \theta h)$

$= [g(c + h) - g(c)]f'(c + \theta h)$ $0 < \theta < 1$

Notice that equation (2) would reduce to the above equation if $\theta = \theta'$. The very point of the generalization is that there is now but a single θ. We do not try to write the present equation as the equality of two quotients like those of equation (2), for there is nothing in our hypotheses to prevent the denominators from vanishing.

To prove the theorem, form the function

$$\varphi(x) = \begin{vmatrix} f(x) & g(x) & 1 \\ f(c) & g(c) & 1 \\ f(c + h) & g(c + h) & 1 \end{vmatrix}$$

Clearly $\varphi(c) = \varphi(c + h) = 0$. By Rolle's theorem,

$$\varphi'(c + \theta h) = \begin{vmatrix} f'(c + \theta h) & g'(c + \theta h) & 0 \\ f(c) & g(c) & 1 \\ f(c + h) & g(c + h) & 1 \end{vmatrix} = 0 \quad 0 < \theta < 1$$

The desired result is now obtained by expanding this determinant.

1.3 L'HOSPITAL'S RULE

We now treat the case, $f'(c) = g'(c) = 0$, which could not be handled by Theorem 1.

Theorem 3.   1. $f(x), g(x) \in C^1$ $a \leqq x \leqq b$

2. $f(c) = g(c) = 0$ $a < c < b$

3. $g'(x) \neq 0$ $x \neq c, \quad a \leqq x \leqq b$

4. $\lim\limits_{x \to c} \dfrac{f'(x)}{g'(x)} = A$ $[\pm\infty]$

$\Rightarrow$ $\lim\limits_{x \to c} \dfrac{f(x)}{g(x)} = A$ $[\pm\infty]$

From the law of the mean we have

$$g(c + h) = hg'(c + \theta_1 h) \qquad 0 < \theta_1 < 1$$

If $h \neq 0$, this shows by virtue of hypothesis 3 that $g(c + h) \neq 0$. Hence, from equation (4)

$$\frac{f(c + h)}{g(c + h)} = \frac{f'(c + \theta h)}{g'(c + \theta h)} \qquad 0 < \theta < 1$$

Clearly, the denominator on the right-hand side is not zero. Since

$$\lim_{h \to 0} \frac{f'(c + \theta h)}{g'(c + \theta h)} = \lim_{x \to c} \frac{f'(x)}{g'(x)}$$

we have the desired conclusion.

Observe why the above argument does not produce the conclusion that the existence of the limit $f(x)/g(x)$ implies that of $f'(x)/g'(x)$. Consider the example $(x^3 \sin x^{-1})/x^2$.

EXAMPLE F. $\lim_{x \to 0} \dfrac{1 - \cos x}{x^2} = \lim_{x \to 0} \dfrac{\sin x}{2x} = \lim_{x \to 0} \dfrac{\cos x}{2} = \dfrac{1}{2}$

EXAMPLE G. $\lim_{h \to 0} \dfrac{f(x + 2h) - 2f(x + h) + f(x)}{h^2}$

$$= \lim_{h \to 0} \frac{2f'(x + 2h) - 2f'(x + h)}{2h}$$

$$= \lim_{h \to 0} \frac{4f''(x + 2h) - 2f''(x + h)}{2} = f''(x)$$

We have thus far treated the case in which the variable approaches its limit from both sides. The case of one-sided limits could easily be included in the foregoing results. For example, if c is replaced by a or by b in Theorem 3, we should have to alter hypothesis 4 and the conclusion so as to have $x \to a+$ or $x \to b-$. Observe also that the case in which the independent variable $\to +\infty$ or $\to -\infty$ is also essentially included. For,

$$\lim_{x \to +\infty} \frac{f(x)}{g(x)} = \lim_{t \to 0+} \frac{f(1/t)}{g(1/t)} = \lim_{t \to 0+} \frac{f'(1/t)t^{-2}}{g'(1/t)t^{-2}}$$

$$= \lim_{x \to +\infty} \frac{f'(x)}{g'(x)}$$

EXAMPLE H. $\lim_{x \to +\infty} \dfrac{(\pi/2) - \tan^{-1} x}{x^{-1}} = \lim_{x \to +\infty} \dfrac{(1 + x^2)^{-1}}{x^{-2}}$

$$= \lim_{x \to +\infty} (1 + x^{-2})^{-1} = 1$$

Here successive differentiations would never attain the goal. An algebraic reduction of the quotient is the obvious procedure.

EXERCISES (1)

Determine the limits in Exercises 1–13.

1. $\lim\limits_{x \to 1/2} \dfrac{\log 2x}{2x - 1}$.

2. $\lim\limits_{x \to 3\pi} \dfrac{1 + \tan (x/4)}{\cos (x/2)}$.

3. $\lim\limits_{x \to 0} \dfrac{1 - \cos hx}{2^x - 3^x}$.

4. $\lim\limits_{x \to +\infty} \dfrac{\cot^{-1} x}{\tan^{-1} (x^{-1})}$.

5. $\lim\limits_{x \to +\infty} \dfrac{x^{10^{10}}}{e^{-x}}$.

6. $\lim\limits_{x \to -\infty} \dfrac{\log (1 + x^{-1})}{\sin (x^{-1})}$.

7. $\lim\limits_{x \to -\infty} \dfrac{\tan^{-1} x}{\cot^{-1} x}$.

8. $\lim\limits_{x \to 2+} \dfrac{x^2 - 4}{x^2 + 3x + 2}$.

9. $\lim\limits_{x \to 0} \dfrac{x^3 \cos (1/x)}{1 - \sec x}$.

10. $\lim\limits_{x \to 0+} \dfrac{x \log x}{\log (1 + ax)}$.

11. $\lim\limits_{x \to 0+} \dfrac{1 - \sec x}{x^3}$, $\lim\limits_{x \to 0-} \dfrac{1 - \sec x}{x^3}$.

12. $\lim\limits_{h \to 0} \dfrac{1}{h^4} \sum\limits_{k=0}^{4} (-1)^k \binom{4}{k} f(x + kh)$.

13. $\lim\limits_{h \to 0} \dfrac{1}{h^3} \begin{vmatrix} f(x) & g(x) & p(x) \\ f(x + h) & g(x + h) & p(x + h) \\ f(x + 2h) & g(x + 2h) & p(x + 2h) \end{vmatrix}$.

14. State and prove a result like Theorem 2 but involving three functions.

15. If $f(x) = x^3[2 + \sin (1/x)]$ when $x \neq 0$, $f(0) = 0$, $g(x) = x^2$, show that all hypotheses of Theorem 3 except the fourth are satisfied for $c = 0$ and that the conclusion of the theorem does hold for $A = 0$. Thus a converse of the theorem interchanging the fourth hypothesis and the conclusion is false.

16. Prove Theorem 2 if hypothesis 1 is replaced by

1a. $f(x), g(x) \in C$ $a \leq x \leq b$

1b. $f'(x), g'(x)$ exist $a < x < b$.

Compare Theorem 2, Chapter 1.

17. Prove:

1. $f(x), g(x) \in C$, $a \leq x \leq b$; $f'(x), g'(x)$ exist $a < x < b$

2. $f(a) = g(a) = 0$

3. $g'(x) \neq 0$ $a < x \leqq b$

4. $\displaystyle\lim_{x \to a+} \frac{f'(x)}{g'(x)} = A$ $[\pm \infty]$

$\Rightarrow \displaystyle\lim_{x \to a+} \frac{f(x)}{g(x)} = A$ $[\pm \infty]$

18. Use Exercise 17 to evaluate

$$\lim_{x \to 0+} \frac{\sqrt{x}}{\sin x}$$

and check directly without any differentiation, using $\lim (\sin x)/x = 1$.

19. Prove:

1. $f'(c), g'(c)$ exist, $g'(c) \neq 0$

2. $f(c) = g(c) = 0$

$\Rightarrow \displaystyle\lim_{x \to c} \frac{f(x)}{g(x)} = \frac{f'(c)}{g'(c)}$

Hint: Use the definition of derivative.

20. Discuss Example E by use of Exercise 19.

§2. *The Indeterminate Form* ∞ / ∞

We now turn to the limit

(1) $\displaystyle\lim_{x \to c} \frac{f(x)}{g(x)}$

where $f(x)$ and $g(x)$ both become infinite as x approaches c. This can, of course, be reduced to the form 0/0 by inverting:

(2) $\displaystyle\lim_{x \to c} \frac{1/g(x)}{1/f(x)} = \lim_{x \to c} \frac{g'(x)/g(x)^2}{f'(x)/f(x)^2}$

But it may be that this inversion is inconvenient. For example,

(3) $\displaystyle\lim_{x \to 0+} \frac{\log x}{\log 2x} = \lim_{x \to 0+} \frac{(\log 2x)^{-1}}{(\log x)^{-1}}$

Now differentiation of numerator and denominator of the latter quotient does not get rid of the logarithms but only makes each function *more* complicated. What we should like to know is that L'Hospital's rule applies equally well to both forms 0/0 and ∞/∞. Then we should have for the limit (1)

(4) $\displaystyle\lim_{x \to c} \frac{f(x)}{g(x)} = \lim_{x \to c} \frac{f'(x)}{g'(x)}$

when the limit on the right-hand side exists. For the limit (3) we should then have the value

$$\lim_{x \to 0+} \frac{x^{-1}}{x^{-1}} = 1$$

Observe that, if we know in advance that both limits (4) exist and are not zero, we can determine their equality by equation (2). For, set

$$B = \lim_{x \to c} \frac{f(x)}{g(x)}, \qquad A = \lim_{x \to c} \frac{f'(x)}{g'(x)}$$

Then equation (2) becomes

$$B = A^{-1}B^2$$

or $B = A$. But for a practical rule we must know that the existence of A *implies* the existence of B.

2.1 L'HOSPITAL'S RULE

We now prove a result analogous to Theorem 3. However, here we begin at once with the stronger theorem regarding *one-sided* limits.

Theorem 4. 1. $f(x), g(x) \in C^1$ $a < x \leqq b$

2. $\displaystyle\lim_{x \to a+} f(x) = \lim_{x \to a+} g(x) = +\infty$

3. $g'(x) \neq 0$ $a < x \leqq b$

4. $\displaystyle\lim_{x \to a+} \frac{f'(x)}{g'(x)} = A \qquad [\pm\infty]$

(5) $\Rightarrow$ $\displaystyle\lim_{x \to a+} \frac{f(x)}{g(x)} = A \qquad [\pm\infty]$

As in the proof of Theorem 3, we have for $a < x < y < b$

(6) $$\frac{f(y) - f(x)}{g(y) - g(x)} = \frac{f'(\xi)}{g'(\xi)} = \frac{f(x)}{g(x)} \frac{1 - f(y)f(x)^{-1}}{1 - g(y)g(x)^{-1}}$$

where $x < \xi < y$. Now let x and y both approach a, x, making the approach so much more rapidly than y that

$$\lim f(y)f(x)^{-1} = \lim g(y)g(x)^{-1} = 0$$

This is possible by virtue of hypothesis 2. As x and y approach a so must ξ. Hence,

$$\lim \frac{f(x)}{g(x)} = \left[\lim \frac{f'(\xi)}{g'(\xi)}\right]\left[\lim \frac{1 - g(y)g(x)^{-1}}{1 - f(y)f(x)^{-1}}\right] = A \qquad [\pm\infty]$$

This completes the proof. Since the proof has a novel feature, the use of the two related variables x and y, let us illustrate by an example. Take $a = 0$, and suppose that

$$g(x) = \frac{1}{x}, \quad f(x) = \log \frac{1}{x}$$

For $g(x)$ it is sufficient to choose $x = y^2$. Then

$$g(y)/g(x) = y, \quad \lim_{y \to 0+} g(y)/g(x) = 0$$

Here we could have chosen $x = h(y)$, where $h(y)$ approach zero more rapidly than y^2. If $x = y^2$, the quotient $f(y)/f(x)$ does not approach zero with y. We must choose a more rapid approach for x. Take $x = e^{-1/y}$. Then

$$f(y)/f(x) = y \log (1/y)$$
$$\lim_{y \to 0+} f(y)/f(x) = 0$$

The relation between x and y must depend on the functions $f(x)$ and $g(x)$. For another proof of Theorem 4 see §5.4.

At first sight it may seem that the theorem is illusory in view of the fact that the differentiation of a function which becomes infinite at a finite point can never produce a derivative which remains finite there. The theorem is none the less useful, for the quotient of the derived function may be subject to certain algebraic reductions to which the original quotient was not. The limit (3) is a case in point. By use of Theorem 4, we have

$$\lim_{x \to 0+} \frac{\log x}{\log 2x} = \lim_{x \to 0+} \frac{x^{-1}}{x^{-1}} = 1$$

Moreover, when the variable approaches $\pm \infty$, differentiation may decrease the "strength of an infinity." (See §4.3.)

Observe that hypothesis 3 is not a consequence of hypothesis 2. Consider

$$g(x) = \frac{1}{x} + \sin \frac{1}{x}$$

$$g'(x) = -\frac{1}{x^2}\left[1 + \cos \frac{1}{x}\right]$$

Here $g(0+) = +\infty$, but $g'(x)$ is zero infinitely often in every neighborhood of the origin.

EXAMPLE A. $\displaystyle \lim_{x \to \infty} \frac{x^2 + x + 1}{2x^2 - 1} = \lim_{x \to \infty} \frac{2x + 1}{4x} = \lim_{x \to \infty} \frac{2}{4} = \frac{1}{2}$

EXAMPLE B. $\displaystyle \lim_{x \to \infty} \frac{x^\alpha}{e^x} = 0$ for all α.

The method of proof is not the same for all α. If $\alpha \leq 0$, there is no indeterminancy, and an attempt to apply L'Hospital's rule would be incorrect. One sees by inspection that the limit is zero. If $\alpha > 0$, successive differentiations will always reduce the exponent of x to zero or to a number between -1 and 0. In either case, the limit is 0. In all problems involving a parameter, it is well to plot one's results. In the present example, we could indicate our results on an α-axis as follows:

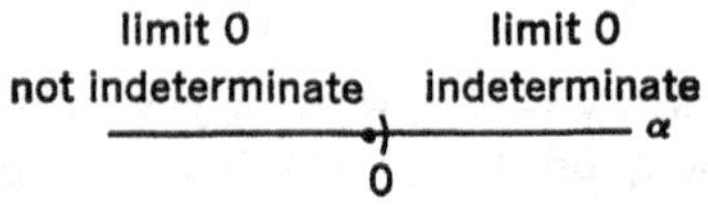

limit 0   limit 0
not indeterminate   indeterminate
$$\qquad \overset{)}{\underset{0}{\rule{5cm}{0.4pt}}} \alpha$$

Fig. 28.

The parenthesis, $)$, about the origin indicates that it should be included with the points to its left.

EXAMPLE C. $\displaystyle\lim_{x \to +\infty} \frac{(\log x)^\alpha}{x^\beta} = \lim_{x \to +\infty} \left(\frac{\log x}{x^{\beta/\alpha}} \right)^\alpha = 0$ $\alpha, \beta > 0$

The arrows of Figure 29 attached to the positive α-axis and to the negative β-axis, for example, indicate that these should be included in the fourth quadrant. The origin goes with none of the four quadrants, for,

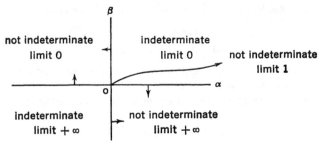

not indeterminate   indeterminate
limit 0   limit 0
                                                        not indeterminate
                                                        limit 1

indeterminate   not indeterminate
limit $+\infty$ limit $+\infty$

Fig. 29.

when $\alpha = \beta = 0$, the quotient reduces to 1 and, hence, has the limit 1, as indicated in the figure by the arrow coming from the origin.

EXAMPLE D. $\displaystyle\lim_{x \to \infty} \frac{x - \sin x}{x} = \lim_{x \to \infty} \left(1 - \frac{\sin x}{x} \right) = 1$

$$\neq \lim_{x \to \infty} \frac{1 - \cos x}{1}$$

Here one can evaluate the limit by inspection. Even though both numerator and denominator become infinite, Theorem 4 is not applicable. It is hypothesis 4 that fails. The conditions of the theorem are thus seen not to be necessary for the existence of the limit (5).

EXERCISES (2)

Evaluate the limits of Exercises 1–14.

1. $\lim\limits_{x \to 0+} \dfrac{\log x}{\cot \pi x}$.

2. $\lim\limits_{x \to 1-} \dfrac{\log (1 - x)}{\cot \pi x}$.

3. $\lim\limits_{x \to 0} \dfrac{\log |x|}{\log (3x^2)}$.

4. $\lim\limits_{x \to 0+} \dfrac{\sqrt{x} + \log x}{2 \tan x + \cot^{-1} x}$.   Is the answer $+\infty$ or $-\infty$?

5. $\lim\limits_{x \to 1+} \dfrac{\tan (\pi x/2) - \log (x - 1)}{\cot \pi x}$.

6. $\lim\limits_{x \to \pi/2+} \dfrac{\tan x}{\log (2x - \pi)}$.

7. $\lim\limits_{x \to +\infty} \dfrac{e^{e^x}}{e^x}, \ \lim\limits_{x \to -\infty} \dfrac{e^{e^x}}{e^x}$.

8. $\lim\limits_{x \to +\infty} \dfrac{e^{\alpha x}}{x^2}, \ \lim\limits_{x \to -\infty} \dfrac{e^{\alpha x}}{x^2}$.

9. $\lim\limits_{x \to +\infty} \dfrac{e^{\beta x}}{(\log x)^\beta}$.

10. $\lim\limits_{x \to +\infty} \dfrac{x (\log x)^\alpha}{e^{\beta x}}$.

11. $\lim\limits_{x \to +\infty} \dfrac{x^3}{x^2 - 3 \cos x}, \ \lim\limits_{x \to -\infty} \dfrac{x^3}{x^2 - 3 \cos x}$.

12. $\lim\limits_{x \to +\infty} \dfrac{\displaystyle\int_0^x e^{t^2} \, dt}{e^{x^2}}$.

13. $\lim\limits_{x \to +\infty} \dfrac{1}{x} \displaystyle\int_0^x \dfrac{|\sin t|}{t} \, dt$.

14. $\lim\limits_{x \to 0} \dfrac{1}{x} \displaystyle\int_0^x \dfrac{|\sin t|}{t} \, dt$.

15. Prove that when $\lim\limits_{x \to 0+} f(x) = +\infty$ then $f'(x)$ cannot remain finite as $x \to 0+$. *Hint:* Use the law of the mean.

§3. *Other Indeterminate Forms*

A variety of other indeterminate forms occur. Consider a function of the form

$$[f(x)]^{g(x)} \qquad\qquad f(x) \geqq 0$$

Let $f(x)$ and $g(x)$ tend to zero or to $+\infty$. We are thus led to the four possible forms, 0^0, 0^∞, ∞^0, ∞^∞. A little consideration will suffice to show that only two of these are indeterminate. Other indeterminate forms are $\infty - \infty$ and 1^∞. We can reduce all of these to the two cases already treated.

3.1 THE FORM $0 \cdot \infty$

Let

$$\lim_{x \to a} f(x) = 0, \qquad \lim_{x \to a} g(x) = +\infty$$

Then by writing

$$f(x)g(x) = f(x)/[g(x)]^{-1}$$

or

$$f(x)g(x) = g(x)/[f(x)]^{-1}$$

the indeterminate form $0 \cdot \infty$ is reduced to $0/0$ or to ∞/∞, respectively. Which of these to use will depend on the functions involved.

EXAMPLE A.
$$\lim_{x \to 0+} x^\alpha \log x = \lim_{x \to 0+} \frac{\log x}{x^{-\alpha}} \qquad\qquad (\alpha > 0)$$

$$= \lim_{x \to 0+} \frac{-1}{\alpha x^{-\alpha}} = 0$$

If we had reduced to $0/0$ instead of to ∞/∞,

$$x^\alpha \log x = \frac{x^\alpha}{[\log x]^{-1}}$$

L'Hospital's rule would have yielded no result. There is no guarantee that differentiation of the numerator and the denominator of a quotient will simplify it.

EXAMPLE B.
$$\lim_{x \to 0} x \cot x = \lim_{x \to 0} \frac{x}{\tan x} = \lim_{x \to 0} \frac{1}{\sec^2 x} = 1$$

Here we have reduced to $0/0$. Reduction to ∞/∞ would have led only to further complication.

3.2 THE FORM $\infty - \infty$

Here we consider

$$\lim_{x \to a} [f(x) - g(x)]$$

where

$$\lim_{x \to a} f(x) = \lim_{x \to a} g(x) = +\infty$$

By writing

(1)
$$f(x) - g(x) = \frac{g(x)^{-1} - f(x)^{-1}}{f(x)^{-1} g(x)^{-1}}$$

the form is reduced to 0/0. Actually, this reduction is not of great usefulness in practice, for it usually produces a quotient so complicated that the use of L'Hospital's rule is not feasible.

EXAMPLE C.
$$\lim_{x \to 0+} \left[\frac{1}{x} - \frac{1}{\sin x} \right] = \lim_{x \to 0+} \frac{\sin x - x}{x \sin x}$$

$$\lim_{x \to 0+} \frac{\cos x - 1}{\sin x + x \cos x} = \lim_{x \to 0+} \frac{-\sin x}{2 \cos x - x \sin x} = 0$$

EXAMPLE D.
$$\lim_{x \to \infty} [x\sqrt{x^2 + 1} - x^2] = \lim_{x \to \infty} \frac{x}{\sqrt{x^2 + 1} + x} = \frac{1}{2}$$

Here we have multiplied and divided by $\sqrt{x^2 + 1} + x$. It is evident that the general reduction (1) would have been useless.

3.3 THE FORMS 0^0, 0^∞, ∞^0, ∞^∞, 1^∞

Let

$$\lim_{x \to a} f(x) = \lim_{x \to a} g(x) = 0, \qquad \lim_{x \to a} h(x) = +\infty \qquad f(x) \geqq 0$$

Then
$$f(x)^{g(x)} = e^{g(x)\log f(x)}$$

$$\lim_{x \to a} f(x)^{g(x)} = e^c, \qquad c = \lim_{x \to a} g(x) \log f(x)$$

The form* 0^0 is reduced to the form $0 \cdot \infty$. In a similar way, we see that 0^∞ is not indeterminate:

$$\lim_{x \to a} f(x)^{h(x)} = \lim_{x \to a} e^{h(x)\log f(x)} = 0$$

The same logarithmic reduction reduces the form ∞^0 to the form $0 \cdot \infty$ and shows that ∞^∞ is not indeterminate.

---

* By convention the first term in a sum like $\sum_{k=0}^{n} x^k$ is *defined* as 1 whether x is zero or not. This is done to preserve continuity at $x = 0$; for x^0 approaches 1 as x approaches 0.

EXAMPLE E. $\displaystyle\lim_{x\to 0+} x^x = \lim_{x\to 0+} e^{x\log x} = 1$

EXAMPLE F. $\displaystyle\lim_{x\to 0+} x^{(1/x)} = \lim_{x\to 0+} e^{(\log x)/x} = 0$

EXAMPLE G. $\displaystyle\lim_{x\to 0+} (1/x)^x = \lim_{x\to 0+} e^{x\log(1/x)} = 1$

EXAMPLE H. $\displaystyle\lim_{x\to \infty} x^x = \lim_{x\to \infty} e^{x\log x} = \infty$

The form 1^∞ is also seen to be indeterminate. It is handled by the same logarithmic reduction.

EXAMPLE I. $\displaystyle\lim_{x\to \infty}\left(1 + \frac{a}{x}\right)^x = \lim_{x\to \infty} e^{x\log[1+(a/x)]} = e^a$ $a \neq 0$

If $a = 0$, the result is still accurate, but there is no indeterminate form. The function is constantly equal to unity and, hence, has unity for its limit.

EXERCISES (3)

Evaluate the limits in Exercises 1–22.

1. $\displaystyle\lim_{x\to 1} \log x \tan (\pi x/2)$.

2. $\displaystyle\lim_{x\to +\infty} x^{\log(1/x)}$.

3. $\displaystyle\lim_{x\to \pi/4} (\tan x)^{\tan 2x}$.

4. $\displaystyle\lim_{x\to +\infty} (\log x)^{\log(1-x^{-1})}$.

5. $\displaystyle\lim_{x\to 0}\left(\frac{\tan x}{x}\right)^{1/x^2}$.

6. $\displaystyle\lim_{x\to 0} [x^{-2} - \cot^2 x]$.

7. $\displaystyle\lim_{x\to 1+} (\log x)^{\sin(x-1)}$.

8. $\displaystyle\lim_{x\to 1+} (\log x)^{\tan(\pi x/2)}$.

9. $\displaystyle\lim_{x\to 1-} |\log x|^{\tan(\pi x/2)}$.

10. $\displaystyle\lim_{x\to \pi/2+} |\tan x|^{\tan x}$.

11. $\displaystyle\lim_{x\to \pi/2-} |\tan x|^{\tan x}$.

12. $\displaystyle\lim_{x\to 0+} x^{ax^b}$.

13. $\lim\limits_{x \to \infty} x^a e^{bx} (\log x)^c.$

14. $\lim\limits_{x \to -\infty} e^{ax} e^{be^x}.$

15. $\lim\limits_{x \to +\infty} [x^2 \sqrt{4x^4 + 5} - 2x^4].$

16. $\lim\limits_{x \to +\infty} [\sqrt[3]{x^9 - 7x^6} - x^3].$

17. $\lim\limits_{x \to a} (x - a)^{-1}[f(g(x, x), h(x, x)) - f(g(a, a), h(a, a))].$

18. $\lim\limits_{x \to 0} \left[\dfrac{\partial f(x \cos \alpha, x \sin \alpha)}{\partial \xi_\alpha} - \dfrac{\partial f(0, 0)}{\partial \xi_\alpha} \right] x^{-1}.$

19. $\lim\limits_{x \to 1-} \sqrt{1 - x} \log \log (1/x).$

20. $\lim\limits_{x \to 0+} \sqrt{x} \log \log (1/x).$

21. $\lim\limits_{x \to 0+} x \sqrt{\log (1/x)}\, e^{-\sqrt{\log(1/x)}}.$

22. $\lim\limits_{x \to 0} \dfrac{1}{x} \displaystyle\int_0^x (1 + \sin 2t)^{1/t}\, dt.$

23. Detect the fallacy:

$$\lim_{x \to 0} \frac{\log \dfrac{\sin x}{x}}{x^2} = \frac{\log \left(\lim \dfrac{\sin x}{x}\right)}{\lim x^2} = \frac{\log (\lim \cos x)}{\lim x^2}$$

$$= \lim \frac{\log \cos x}{x^2} = \lim \frac{-\tan x}{2x} = -\tfrac{1}{2}$$

§4. *Other Methods. Orders of Infinity*

In many cases the indeterminate form $0/0$ is not easily treated by use of L'Hospital's rule. The differentiation involved may be tedious, or indeed may serve to complicate the quotient in question. Certain other methods are available. We describe them below. By a study of the rapidity with which various functions become infinite, one may often evaluate the indeterminate form ∞/∞ without any differentiation at all.

4.1 THE METHOD OF SERIES

The following result may be regarded as a generalization of Theorem 1. As in Theorem 4, we shall deal here with one-sided limits.

Theorem 5.   1. $f(x), g(x) \in C^{n+1}$ $a \leqq x \leqq b$

           2. $f^{(k)}(a) = g^{(k)}(a) = 0$ $k = 0, 1, \ldots, n$

           3. $g^{(n+1)}(a) \neq 0$

$\Rightarrow$ $\displaystyle \lim_{x \to a+} \frac{f(x)}{g(x)} = \frac{f^{(n+1)}(a)}{g^{(n+1)}(a)}$

By Taylor's formula with remainder, we have

$$\frac{f(x)}{g(x)} = \frac{f^{(n+1)}(X)}{g^{(n+1)}(Y)} \qquad a < X < x, a < Y < x$$

Here we have chosen x so near to a that $g^{(n+1)}(Y) \neq 0$. This is possible by virtue of hypotheses 1 and 3. We now obtain the desired result by letting x approach a.

If $g^{(n+1)}(a) = 0$, $g^{(n+1)}(x) \neq 0$ $(a < x \leqq b)$, $f^{(n+1)}(a) \neq 0$, we obtain

$$\lim_{x \to a+} \left| \frac{f(x)}{g(x)} \right| = +\infty$$

When the Taylor expansions of the given functions are known, this theorem enables us to evaluate the form 0/0 without any differentiation.

EXAMPLE A. $\displaystyle \lim_{x \to 0} \frac{\sin x - x}{x^3} = -\frac{1}{6}$

Since we know the power series expansion of the numerator

$$\sin x - x = -\frac{x^3}{3!} + \frac{x^5}{5!} - \ldots$$

we know without computation that $f'''(0) = -1$. The technique suggested by Theorem 5 consists simply in replacing $f(x)$ and $g(x)$ by the first non-vanishing terms of their Taylor developments.

EXAMPLE B. $\displaystyle \lim_{x \to \infty} (x\sqrt{x^2 + 1} - x^2) = ?$

Here we must replace x by $1/y$ and let y approach zero in order to apply Theorem 5. Obviously, the same purpose will be served if we expand the original function in powers of $1/x$.

$$x\sqrt{x^2 + 1} - x^2 = x^2 \left[\left(1 + \frac{1}{x^2} \right)^{1/2} - 1 \right]$$

$$= x^2 \left[\frac{1}{2x^2} - \frac{1}{8x^4} + \ldots \right]$$

$$\lim_{x \to \infty} [x\sqrt{x^2 + 1} - x^2] = \frac{1}{2}$$

Care should be taken to expand the functions in question in power series which converge at the point that the variable is approaching. Thus, it would be incorrect to replace sin x by x, the first term in its MacLaurin development, in order to evaluate the limit

$$\lim_{x \to \infty} \frac{\sin x}{x}$$

EXAMPLE C. $\displaystyle \lim_{x \to 0} \frac{\csc x - \dfrac{1}{x} - \dfrac{x}{6}}{\sin^3 x} = \lim_{x \to 0} \frac{7x^3}{3.5!x^3} = \frac{7}{360}$

Here we have used formulas 846 and 851 from the Peirce-Foster tables.

4.2 CHANGE OF VARIABLE

A change of variable frequently simplifies the work of evaluating an indeterminate form.

EXAMPLE D. Show that, if $f(x) = e^{-1/x^2}$ $x \neq 0$

$$f(0) = 0$$

then $f'(x) \in C^1$. We have

$$f'(x) = \frac{2}{x^3} e^{-1/x^2} \qquad\qquad x \neq 0$$

$$f'(0) = \lim_{h \to 0} \frac{f(h) - f(0)}{h} = \lim_{h \to 0} \frac{e^{-1/h^2}}{h}$$

$$= \lim_{t \to +\infty} \frac{\sqrt{t}}{e^t} = 0$$

Here we have made the transformation $t = h^{-2}$ before using L'Hospital's rule. Direct application of the rule would have been useless. To show that $f'(x)$ is continuous, we must show that

$$\lim_{x \to 0} f'(x) = f'(0) = 0$$

But $\displaystyle \lim_{x \to 0} \frac{2e^{-1/x^2}}{x^3} = \lim_{t \to +\infty} \frac{2t^{3/2}}{e^t} = 0$

In like manner we could show that $f(x) \in C^\infty$ and that

$$f^{(k)}(0) = 0 \qquad\qquad k = 0, 1, 2, \ldots$$

4.3 ORDERS OF INFINITY

Let $f(x)$ and $g(x)$ be two functions which become positively infinite as the variable x approaches a finite limit or becomes infinite. Then we introduce the symbol $\prec$ by the following definition.

Definition 1. $f(x) \prec g(x) \Leftrightarrow \lim \dfrac{f(x)}{g(x)} = 0$

The relation may be read: "$f(x)$ is *weaker* than $g(x)$" or "$f(x)$ is a *lower order infinity* than $g(x)$." For example, if x is becoming infinite, then

$$(\log x)^{10} \prec \sqrt[10]{x}$$

We make a brief table of infinities arranged in the order of increasing strength:

$$\ldots \prec \log \log x \prec \log x \prec x \prec e^x \prec e^{e^x} \prec \ldots$$

The order of any infinity is increased by raising it to a power $p > 1$, is decreased if $0 < p < 1$. By use of this principle one could interpolate any number of functions between a given pair of the above table.

EXAMPLE E. $\log \log x \prec (\log x)^p$ $p > 0$

$$\lim_{x \to +\infty} \frac{\log \log x}{(\log x)^p} = \lim_{y \to +\infty} \frac{\log y}{y^p} = 0$$

EXAMPLE F.   Find an infinity stronger than all the functions x^p, where $0 < p < 1/2$, but weaker than $\sqrt{x}$. Such a function is $\sqrt{x}/\log x$. Obviously, for every $\epsilon > 0$ ($\epsilon < 1/2$), we have

$$\frac{\sqrt{x}}{x^\epsilon} \prec \frac{\sqrt{x}}{\log x} \prec \sqrt{x}$$

EXAMPLE G.   Which infinity, $e^{\sqrt{\log x}}$ or x, is of higher order? It can easily be shown that $f(x) \prec g(x) \Rightarrow e^{f(x)} \prec e^{g(x)}$. By use of this result, one sees that

$$e^{\sqrt{\log x}} \prec x$$

EXAMPLE H.   Arrange the infinities x^x, e^x, $x^{\log x}$ in the order of increasing strength. We have

$$(\log x)^2 \prec x \prec x \log x$$
$$x^{\log x} \prec e^x \prec x^x$$

EXAMPLE I.   Evaluate the limit

$$\lim_{x \to +\infty} \frac{\sqrt{\log x} \, \log (\log x)}{e^{\sqrt{x}}}$$

One easily recognizes the infinity in the denominator as the strongest of the three.   Since x^2 is weaker than this one but stronger than the other two, we have

$$\lim_{x \to \infty} \frac{(\sqrt{\log x}/x)[\log (\log x)/x]}{e^{\sqrt{x}}/x^2} = \frac{0 \cdot 0}{\infty} = 0$$

In evaluating limits of functions consisting of many factors, we should separate those which neither approach zero nor become infinite, since they have no effect on the indeterminate character of the product.

In conclusion, let us point out how the notion of strength of infinity enters into one of the famous problems of mathematics.   If $p_1 = 2$, $p_2 = 3, \ldots, p_n =$ the nth prime, Euler showed that the series

$$\sum_{n=1}^{\infty} \frac{1}{p_n}$$

diverges and that

$$p_n \prec n^{1+\epsilon}$$

for every positive ϵ.   He was able to conjecture from these two facts the strength of the infinity p_n.   The most obvious one satisfying the above conditions is $n \log n$:

$$\sum_{n=2}^{\infty} \frac{1}{n \log n} = \infty, \qquad n \log n \prec n^{1+\epsilon}$$

It was shown in 1898 that this conjecture is correct and further that

$$\lim_{n \to \infty} \frac{p_n}{n \log n} = 1$$

This latter result is known as the *prime-number theorem*.

It should be observed that the reciprocal of an infinity is an *infinitesimal* and that one could classify infinitesimals according to strength.   It is perhaps easier to transform all infinitesimals to infinities.

EXAMPLE J. $\lim_{x \to 0+} \sqrt[3]{x^4} \cot x \sqrt{\log (1/x)} = ?$

$$\lim_{x \to 0+} x^{1/3}(x \cot x)\sqrt{\log (1/x)} = 0$$

since $\sqrt{\log x} \prec x^{1/3}$ $x \to +\infty$

EXERCISES (4)

In the following exercises the familiar power series expansions, such as those given in the integral tables of B. O. Peirce and R. M. Foster, may be used.   Free use of the table of infinities given in §4.3 may also be made.

Evaluate the limits in Exercises 1–12.

1. $\lim\limits_{x \to 0} \dfrac{\sec x - 1}{x \sin 2x}$.

2. $\lim\limits_{x \to 0} \dfrac{x^2 + \log \cos^2 x}{(\log \cos x)^2}$.

3. $\lim\limits_{x \to 0} \dfrac{x^2 \cot x - x + (x^3/3)}{\tan^{-1} x - x + (x^3/3)}$.

4. $\lim\limits_{x \to +\infty} \dfrac{\tan^{-1} x - \pi/2 + x^{-1}}{\coth^{-1} x - x^{-1}}$.

5. $\lim\limits_{x \to 0} \dfrac{x^2 + 2 \log \cos x}{x^2 + 6 \log (x^{-1} \sin x)}$.

6. $\lim\limits_{x \to 0} \left[\dfrac{2}{x(e^x - 1)} - \dfrac{2}{x^2} + \dfrac{1}{x} \right]$.

7. $\lim\limits_{x \to 0} \left[\dfrac{1}{x^5} \int_0^x e^{-t^2}\, dt - \dfrac{1}{x^4} + \dfrac{1}{3x^2} \right]$.

8. $\lim\limits_{x \to 0} \dfrac{x - \displaystyle\int_0^x \cos t^2\, dt}{6 \sin^{-1} x - 6x - x^3}$.

9. $\lim\limits_{x \to +\infty} \dfrac{e^{\sqrt{\log x}}}{x^{\sqrt{\log\log x}}}$.

10. $\lim\limits_{x \to +\infty} \sin (1/x)(\log x)^{10} \sqrt{x}$.

11. $\lim\limits_{x \to +\infty} \dfrac{x^{\sqrt{\log x}}(\sqrt{\log x})^x}{(\sqrt{x})^{\log x}(\log x)^{\sqrt{x}}}$.

12. $\lim\limits_{x \to 0-} \dfrac{(\sin^2 x)e^{1-x} \tan^{-1} (1/x)}{(\sinh x)(e^{2x} - 1) \sec^{-1} (1/x)}$.

13. Prove that $f(x) \prec g(x) \Rightarrow e^{f(x)} \prec e^{g(x)}$.
Is the converse true?

14. Arrange in order of increasing strength the infinities:

$$x^{e^x},\ e^{x^x},\ (\log x)^{(\log x)^{\log x}}$$

15. Interpolate an infinity between e^x and every positive power of x.

16. Interpolate an infinity between $x(\log x)^p$ and $x(\log x)^{-q} \log \log x$, for all positive numbers p and q.

17. Show that $f(x) \in C^2$ in Example D.

18. Show that $f(x) \in C^{\infty}$ in Example D.

19. $\lim\limits_{\epsilon \to 0} \int_{\epsilon}^{\epsilon/(1+\epsilon)} e^{-u^2}/u^2 \, du = ?$

Hint: First find $\lim\limits_{u \to 0} (e^{-u^2} - 1)/u^2$.

20. Detect the fallacy:

$$\lim_{x \to \infty} \frac{x}{x} = \frac{\lim (x^2/x)}{\lim x} = \frac{\lim (2x/1)}{\lim x} = \lim \frac{2x}{x} = 2$$

§5. Superior and Inferior Limits

We introduce here certain notions concerning limit points of sets of points. We shall find these notions useful in establishing a fundamental criterion for the existence of a limit known as *Cauchy's criterion.*

5.1 LIMIT POINTS OF A SEQUENCE

We shall use the notation $\{S_n\}_1^{\infty}$ for the sequence of numbers

$$S_1, S_2, S_3 \ldots$$

Definition 2. *The sequence* $\{S_n\}_1^{\infty}$ *has a limit point* $A \Leftrightarrow$ *for every* $\epsilon > 0$ *there are infinitely many distinct integers* $n_1, n_2, n_3, \ldots$ *such that*

$$|S_{n_k} - A| < \epsilon \qquad\qquad k = 1, 2, \ldots$$

Note that the elements of the sequence $\{S_n\}_1^{\infty}$ need not be distinct. As a consequence, all the infinitely many elements S_{n_k} of Definition 2 may be the same number. For example, if

$$\{S_n\}_1^{\infty} = 1, -1, 1, -1, \ldots$$

then $A = 1$ is a limit point and the integers n_k may be taken, for example, as $1, 3, 5, \ldots$. Then

$$S_{n_k} = 1 \qquad\qquad k = 1, 2, 3, \ldots$$

In like manner, the number -1 is also a limit point of the above sequence.

Definition 3. *A sequence* $\{S_n\}_1^{\infty}$ *is bounded above (below)* $\Leftrightarrow$ *there exists a number M such that*

$$S_n < M \qquad (-M < S_n) \qquad\qquad n = 1, 2, \ldots$$

Theorem 6. *If* $\{S_n\}_1^{\infty}$ *is bounded above and below, it has at least one limit point.*

Let

$$|S_n| < M \qquad\qquad n = 1, 2, \ldots$$

There must be infinitely many elements of the sequence in at least one of the intervals $(-M, 0)$, $(0, M)$, say the latter. Then there must be infinitely many elements in at least one of the intervals $(0, M/2)$, $(M/2, M)$. By successive halving of intervals, we arrive thus at an infinite sequence of intervals, each being half of its predecessor and each containing infinitely many elements. The intervals of the sequence have one, and only one, common point A, which is a limit point of $\{S_n\}_1^\infty$.

Definition 4. *The limit superior (inferior) of the sequence $\{S_n\}_1^\infty$ is A,*

$$\overline{\lim_{n \to +\infty}} \, S_n = A \qquad (\underline{\lim_{n \to +\infty}} \, S_n = A)$$

⟺ *The sequence is bounded above (below) and has A as its largest (smallest) limit point.*

To justify this definition it must be observed that there actually exist a smallest and a largest limit point. For example, suppose the S_n are all positive numbers. Might not the points 1, 1/2, 1/3 ... be all the limit points? This sequence has no minimum, though it has the greatest lower bound zero. In this case zero must also be included among the limit points. For, in its every neighborhood is certainly one (infinitely many) of the above limit points, which in turn has infinitely many of the S_n clustering upon it. That is, zero is also a limit point and the greatest lower bound zero is also a minimum. Clearly this argument is always valid.

It is desirable to extend Definition 4 in such a way that every sequence will have both a limit superior and a limit inferior. This is done by making the convention that $+\infty$ is a limit point of a sequence which is unbounded above and that $-\infty$ is a limit point of a sequence which is unbounded below. If we now admit these two symbolic numbers into comparison with the finite numbers (with $+\infty$ greater than any finite number, $-\infty$ less than any finite number), we may still define the limit superior as the largest of the limit points, the limit inferior as the smallest of the limit points. With this convention every sequence will have both a limit superior and a limit inferior, as desired. For example, if a sequence like $\{-n\}_1^\infty$ is unbounded below and has no finite limit point, both its limit inferior and its limit superior (and hence its limit) are equal to $-\infty$. We formalize these conventions in the following definition.

Definition 5. *As $n \to \infty$*

$$\overline{\lim} \, S_n = +\infty \Leftrightarrow \{S_n\}_1^\infty \text{ is unbounded above}$$

$$\underline{\lim} \, S_n = -\infty \Leftrightarrow \{S_n\}_1^\infty \text{ is unbounded below}$$

$$\overline{\lim} \, S_n = -\infty \Leftrightarrow \lim S_n = -\infty$$

$$\underline{\lim} \, S_n = +\infty \Leftrightarrow \lim S_n = +\infty$$

EXAMPLE A. $\{S_n\}_1^\infty = 1, 0, -1, 2, 0, -2, \ldots$

This sequence has only one finite limit point, 0. But

$$\overline{\lim_{n \to +\infty}} S_n = +\infty, \qquad \underline{\lim_{n \to +\infty}} S_n = -\infty$$

EXAMPLE B. $$\overline{\lim_{n \to +\infty}} n \sin^2 \frac{n\pi}{2} = +\infty$$

$$\underline{\lim_{n \to +\infty}} n \sin^2 \frac{n\pi}{2} = 0$$

EXAMPLE C. $$\underline{\lim_{n \to \infty}} n = +\infty$$

$$\overline{\lim_{n \to \infty}} n \log (\cot^{-1} n) = -\infty$$

It should not be supposed that the elements of a sequence are always less (greater) than the limit superior (inferior) of the sequence.

EXAMPLE D. $$\overline{\lim_{n \to +\infty}} \left(1 + \frac{1}{n}\right) \cos n\pi = 1$$

$$\underline{\lim_{n \to +\infty}} \left(1 + \frac{1}{n}\right) \cos n\pi = -1$$

No element of this sequence lies in the interval $(-1, 1)$.

5.2 PROPERTIES OF SUPERIOR AND INFERIOR LIMITS

We list below some of the useful properties of the limits superior and inferior. They become immediately apparent if one represents the elements of the sequence as points on a line.

A. $\overline{\lim} S_n$ and $\underline{\lim} S_n$ *always exist or are* $\pm\infty$.

This is the great advantage that these operations enjoy over the limit operation. Note that $\lim \cos n\pi$ does not exist; nor does the limit equal $\pm\infty$.

B. $\underline{\lim} S_n \leq \overline{\lim} S_n$.

C. $\lim S_n = A \ (+\infty \ or \ -\infty) \Leftrightarrow \underline{\lim} S_n = \overline{\lim} S_n = A \ (+\infty \ or \ -\infty)$.

D. $S_n < T_n, n = 1, 2, \ldots \Rightarrow \underline{\lim} S_n \leq \underline{\lim} T_n, \overline{\lim} S_n \leq \overline{\lim} T_n$.

E. $\overline{\lim} S_n = A \Leftrightarrow for \ every \ \epsilon > 0$

(1) *there exists an integer m such that* $S_n < A + \epsilon, n > m$;

(2) *there exist infinitely many distinct integers* $n_1, n_2, \ldots$ *such that* $S_{n_k} > A - \epsilon, k = 1, 2, \ldots$

F. $A < S_n, n = 1, 2, \ldots ; \overline{\lim} S_n = A \Rightarrow \lim S_n = A$.

We emphasize the fact that $\lim S_n$ exists $\Leftrightarrow \{S_n\}_1^\infty$ is bounded above and below and has a single finite limit point. Also $\lim S_n = +\infty \Leftrightarrow \{S_n\}_1^\infty$ is unbounded above, is bounded below, and has no finite limit point. In either case there is a single limit point if the two ideal points $+\infty$ and $-\infty$ are admitted to consideration. But it must be understood that if $\lim S_n$ is $+\infty$ or $-\infty$, $\lim S_n$ *does not exist*.

Let us illustrate one way in which $\overline{\lim}$ is frequently used in analysis. Let us suppose that to an arbitrary $\epsilon > 0$ there corresponds an integer m such that for $n > m$

$$(1) \qquad\qquad |S_n| < \epsilon + \varphi(n) \qquad\qquad n > m$$

where $\varphi(n)$ is some function which tends to zero as n becomes infinite. Then by property D

$$(2) \qquad\qquad \overline{\lim_{n \to +\infty}} \; |S_n| \leqq \epsilon$$

We are at liberty to let n become infinite in inequality (1) since it holds for all large integers n. Since ϵ was arbitrary and since the left-hand side of (2) is a nonnegative number, we see that

$$\overline{\lim_{n \to +\infty}} \; |S_n| = 0$$

Then by property F

$$\lim_{n \to +\infty} |S_n| = \lim_{n \to +\infty} S_n = 0$$

5.3 CAUCHY'S CRITERION

Theorem 7. $\lim\limits_{n \to +\infty} S_n$ *exists* $\Leftrightarrow$ *to an arbitrary* $\epsilon > 0$ *corresponds an integer* m *such that when* $n, n' > m$

$$(3) \qquad\qquad |S_n - S_{n'}| < \epsilon$$

Let us first prove the implication $\Rightarrow$. We have given that

$$\lim_{n \to +\infty} S_n = A$$

This implies that, for an arbitrary $\epsilon > 0$, there is an integer m such that whenever $n > m$

$$|S_n - A| < \epsilon/2$$

If $n' > m$, we have by this same inequality

$$|S_n - S_{n'}| \leqq |S_n - A| + |S_{n'} - A| < \epsilon$$

This is what we were to prove.

For the opposite implication $\Leftarrow$ we begin with (3).   In particular, we may take $n' = m + 1$.   Then

$$(4) \qquad\qquad S_{m+1} - \epsilon < S_n \leqslant S_{m+1} + \epsilon \qquad\qquad n > m$$

By properties B and D above, we have

$$(5) \qquad\qquad S_{m+1} - \epsilon \leq \varliminf_{n \to +\infty} S_n \leq \varlimsup_{n \to +\infty} S_n \leq S_{m+1} + \epsilon$$

It is permitted to let n become infinite in (4) since the relation holds for all large n.   But (5) implies that $\varliminf S_n = \varlimsup S_n$, since ϵ was arbitrary.   The proof is now concluded by use of property C above.

We observe in conclusion that the notions of limit superior and limit inferior extend in an obvious way to functions.   For example,

$$\varlimsup_{x \to +\infty} \sin x = 1, \qquad \varliminf_{x \to +\infty} \sin x = -1$$

5.4 L'HOSPITAL'S RULE (CONCLUDED)

As another application of the present technique we give a new proof of Theorem 4.   For an arbitrary $\epsilon > 0$ there exists, by hypothesis 4, a neighborhood $a < x < a + \delta$ where

$$(6) \qquad\qquad A - \epsilon < \frac{f'(x)}{g'(x)} < A + \epsilon$$

If $a < x < y < a + \delta$, then equation (6) of §2.1 is

$$\frac{f(x)}{g(x)} = \frac{f'(\xi)}{g'(\xi)} h(x) \qquad\qquad x < \xi < y$$

$$h(x) = \frac{1 - g(y)g(x)^{-1}}{1 - f(y)f(x)^{-1}}$$

For fixed y, $h(x) \to 1$ as $x \to a+$, so that $h(x) > 0$ for x near a.   Since ξ is in the neighborhood where (6) holds, we have after multiplication by the *positive* function $h(x)$,

$$(A - \epsilon)h(x) < \frac{f'(\xi)}{g'(\xi)} h(x) = \frac{f(x)}{g(x)} < (A + \epsilon)h(x)$$

Now apply the operator $\varliminf$ to the first of these inequalities, $\varlimsup$ to the second:

$$(A - \epsilon) \leq \varliminf_{x \to a+} \frac{f(x)}{g(x)} \leq \varlimsup_{x \to a+} \frac{f(x)}{g(x)} \leq A + \epsilon$$

Since ϵ was arbitrary and since the numbers $\varlimsup f/g$ and $\varliminf f/g$ do not depend on ϵ, they must be equal to each other and to A.   That is,

$$\lim_{x \to a+} \frac{f(x)}{g(x)} = A$$

EXERCISES (5)

Obtain all the limit points, $\overline{\lim}$ and $\underline{\lim}$, for the sequences of Exercises 1–10, where $n = 1, 2, \ldots$.

1. $(-1)^n \left(1 + \dfrac{1}{n} \right)$.

2. $n \sin \dfrac{n\pi}{4}$.

3. $\dfrac{n + (-1)^n n^2}{n^2 + 1}$

4. $(1.5 + (-1)^n)^n$.

5. $[(-1)^n + 1] \sin \dfrac{n\pi}{4}$.

6. $(-n)^n (1 + n)^{-n}$.

7. $[1 - (-1)^n] \sin \dfrac{n\pi}{4}$.

8. $\sin \dfrac{n\pi}{4} \sin \dfrac{n\pi}{2}$.

9. $\sin \dfrac{n\pi}{4} + \sin \dfrac{n\pi}{2}$.

10. $e^n \sin (n\pi/4)$.

11. In the sequence of intervals described in the proof of Theorem 6, show that the sequences of left-hand and right-hand end points both approach the same limit, and thus establish the existence of the point A. Show that there are infinitely many elements of the sequence $\{S_n\}_1^\infty$ in every neighborhood of A.

12. Construct a sequence which has finite limit points $-2, 0, 1$ and for which

$$\underline{\lim} \, S_n = -\infty, \qquad \overline{\lim} \, S_n = 1$$

13. Prove property C.

14. Give an example illustrating D with the equality holding in the conclusion.

15. Prove property E.

16. State without proof a property analogous to E for $\underline{\lim}$. What does property E become for $\overline{\lim} \, S_n = +\infty$? $\underline{\lim} \, S_n = -\infty$? $\overline{\lim} \, S_n - \infty$? $\underline{\lim} \, S_n = +\infty$?

17. Prove that if $f(x) \in C^1$ and $|f'(x)| < 1$ in the interval $0 < x \leqq 1$ that $\lim\limits_{n \to +\infty} f(n^{-1})$ exists. *Hint:* Use Theorem 7 and the law of the mean.

18. Show that $\sum_{1}^{\infty} \frac{1}{n}$ diverges. *Hint:* Show that the sequence of partial sums $S_n = 1 + 1/2 + \ldots + 1/n$ does not approach a limit. In Cauchy's theorem choose $\epsilon = 1/2$ and show that

$$S_{2n} - S_n \geq \tfrac{1}{2} \qquad\qquad n = 1, 2, 3, \ldots$$

19. What does property E become if the sequence $\{S_n\}_1^{\infty}$ is replaced by a function $f(x)$?

20. If $u_n > 0$, $n = 1, 2, \ldots$, show that

$$\underline{\lim} \frac{u_{n+1}}{u_n} \leq \underline{\lim} \sqrt[n]{u_n} \leq \overline{\lim} \sqrt[n]{u_n} \leq \overline{\lim} \frac{u_{n+1}}{u_n}$$

21. If $\lim u_{n+1}/u_n = A$, show that $\lim \sqrt[n]{u_n} = A$ for any sequence of positive u_n.

22. Show that the converse of the previous result is false by use of the sequence $u_{2n} = u_{2n+1} = 2^{-n}$.

23. Check Exercise 21 for $u_n = n e^{2n} \log n$.

24. Same problem for $u_n = n^p$.

25. $\lim\limits_{n \to \infty} \frac{1}{n} \sqrt[n]{n!} = $?

26. $\lim\limits_{n \to \infty} \frac{1}{n} \sqrt[n]{(2n + 1)!/n!} = $?

9

Infinite Series

§1. *Convergence of Series. Comparison Tests*

The present chapter introduces briefly the theory of infinite series. Most students will have had an earlier acquaintance with the subject. The early part of the chapter may be regarded as a brief review preparatory to the study of improper integrals. In the study of such integrals, it is extremely useful to keep in mind the analogies between series and integrals. For this reason, it is desirable to have the fundamental facts about series in hand before studying improper integrals. The latter part of the chapter introduces the important notion of uniform convergence of series. We begin with definitions of convergence and the comparison tests for convergence.

1.1 CONVERGENCE AND DIVERGENCE

Consider the infinite series

$$(1) \qquad \sum_{k=1}^{\infty} u_k = u_1 + u_2 + u_3 + \dots$$

Denote the sum of the first n terms of this series by S_n,

$$(2) \qquad S_n = \sum_{k=1}^{n} u_k = u_1 + u_2 + \dots + u_n \qquad n = 1, 2, 3, \dots$$

284

Definition 1. *Series* (1) *converges* $\Leftrightarrow \lim_{n \to \infty} S_n = A$. *If* $\lim_{n \to \infty} S_n = A$, *the number* A *is the sum or value of the convergent series.*

Definition 2. *A series diverges if, and only if, it does not converge.*

Example A. $\sum_{k=1}^{\infty} \left(\dfrac{1}{k} - \dfrac{1}{k+1} \right)$ converges and has the value 1. For, $S_n = 1 - (n+1)^{-1}$, and this tends to 1 as n becomes infinite.

Example B. $\sum_{k=1}^{\infty} (-1)^k$ diverges. For, S_n is 0 when n is even, is -1 when n is odd. Hence, S_n approaches no limit.

Example C. $\sum_{k=1}^{\infty} 1 = 1 + 1 + 1 + \dots$. Here $S_n = n$; this increases without limit as n becomes infinite. Hence, the series diverges.

A series of particular interest is the geometric series. Since it can be used for comparison, let us give it a special designation.

Test series T_1: $\quad \sum_{k=0}^{\infty} r^k = \dfrac{1}{1-r}$

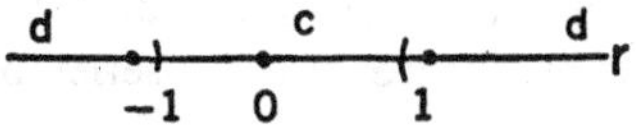

Fig. 30.

The diagram indicates that the series converges for $-1 < r < 1$ and diverges elsewhere. The parentheses indicate that the points $r = 1$ and $r = -1$ are included in the intervals of divergence. When a series contains a parameter, as here, convergence results should be indicated on a diagram.

Theorem 1. *Series* (1) *converges* $\Rightarrow \lim_{n \to \infty} u_n = 0$.

For, by hypothesis $\lim_{n \to \infty} S_n = A$. Hence,

$$\lim_{n \to \infty} u_n = \lim_{n \to \infty} (S_n - S_{n-1}) = A - A = 0$$

1.2 COMPARISON TESTS

Theorem 2. 1. $0 \leqq u_k \leqq v_k$ $\qquad\qquad\qquad\qquad k = 1, 2, \dots$

2. $\displaystyle\sum_{k=1}^{\infty} v_k < \infty$

$\Rightarrow \qquad\qquad \displaystyle\sum_{k=1}^{\infty} u_k < \infty$

We use the symbol $< \infty$ to indicate convergence of a series of *nonnegative* terms. It becomes meaningless for other series. Define S_n by equation (2) and set

$$T_n = \sum_{k=1}^{n} v_k \qquad\qquad n = 1, 2, 3, \ldots$$

$$\lim_{n \to \infty} T_n = B$$

Since the sequences $\{S_n\}_1^{\infty}$ and $\{T_n\}_1^{\infty}$ are both increasing, we have that

$$S_n \leqq T_n \leqq B$$

and that S_n approaches a limit $A \leqq B$.

Theorem 3.   1. $0 \leqq v_k \leqq u_k$ $k = 1, 2, \ldots$

        2. $\displaystyle\sum_{k=1}^{\infty} v_k = \infty$

$\Rightarrow$ $\displaystyle\sum_{k=1}^{\infty} u_k = \infty$

We use the symbols $= \infty$ for divergence of a series of *nonnegative* terms. The series (1) must diverge. For, if it converged, we could prove the convergence of the v-series by Theorem 2 by interchanging the roles of u_k and v_k.

Test series T_2: $\displaystyle\zeta(p) = \sum_{k=1}^{\infty} \frac{1}{k^p}$

Fig. 31.

The series converges for $p > 1$, and its value is denoted by $\zeta(p)$, a function which has been tabulated. The series diverges elsewhere. For $p = 1$ it is the divergent harmonic series. These facts will be proved later, but for the present the series may be used as a test series.

EXAMPLE D. $\displaystyle\sum_{k=3}^{\infty} \frac{1}{k^2 - 4}$ converges.

Take $v_k = 2k^{-2}$. Then

$$\frac{1}{k^2 - 4} < \frac{2}{k^2}$$

whenever

$$k^2 < 2k^2 - 8$$

that is, for all $k > 2$. But the v-series is T_2 with $p = 2$, except for the constant factor 2.

EXAMPLE E. $\displaystyle\sum_{k=1}^{\infty} \frac{1}{\sqrt{k + \pi}}$ diverges

Take $v_k = (2k)^{-1/2}$. Then

$$v_k = \frac{1}{\sqrt{2k}} < \frac{1}{\sqrt{k + \pi}} \qquad\qquad k = 4, 5, 6, \ldots$$

Hence, the original series, shorn of its first three terms, diverges by Theorem 3, using T_2 ($p = \frac{1}{2}$). Consequently, the complete series diverges.

EXERCISES (1)

Test the series of Exercises 1–10 for convergence.

1. $\displaystyle\sum_{k=0}^{\infty} \frac{e^{-k}}{\sqrt{k + \pi}}$

2. $\displaystyle\sum_{k=4}^{\infty} \frac{\sqrt{k - \pi}}{k + \pi}.$

3. $\displaystyle\sum_{k=1}^{\infty} \frac{k}{k^2 + 1}.$

4. $\displaystyle\sum_{k=1}^{\infty} \left(\frac{k^2 - 1}{k^2 + 1}\right)^{1/2}.$

5. $\displaystyle\sum_{k=4}^{\infty} \frac{k^2 + 1}{k^4 - 9}.$

6. $\displaystyle\sum_{k=0}^{\infty} \frac{\sqrt{k + 1}}{k^2 - 3k + 1}$

7. $\displaystyle\sum_{k=1}^{\infty} \frac{\log (k + 1) - \log k}{\tan^{-1} (2/k)}.$

8. $\displaystyle\sum_{k=1}^{\infty} \sin k^{-2}.$

9. $\displaystyle\sum_{k=1}^{\infty} (\sin k^{-k})^{-2}.$

10. $\displaystyle\sum_{k=1}^{\infty} \frac{k^3}{3^k}.$

Hint: Show $(\frac{2}{3})^x < x^{-3}$ for large x by consideration of $\lim_{x\to\infty} x^3(\frac{2}{3})^x$. Use T_1, $r = \frac{1}{2}$

11. Prove: $\displaystyle\sum_{k=1}^{\infty} u_k$ converges $\Rightarrow \displaystyle\sum_{k=1}^{\infty} cu_k$ converges (every c). State and prove a corresponding theorem for divergence.

12. Prove: $\displaystyle\sum_{k=1}^{\infty} u_k, \ \displaystyle\sum_{k=1}^{\infty} v_k$ converge $\Rightarrow \displaystyle\sum_{k=1}^{\infty} (u_k + v_k)$ converges. Does this imply that $u_1 + v_1 + u_2 + v_2 + \ldots$ converges?

13. Prove: $\displaystyle\sum_{k=1}^{\infty} u_k$ converges $\Leftrightarrow \displaystyle\sum_{k=m}^{\infty} u_k$ converges.

14. Prove:

1. $\displaystyle\sum_{k=1}^{\infty} (u_k + v_k)$ converges

2. $\displaystyle\lim_{k\to\infty} v_k = 0$

$\Rightarrow u_1 + v_1 + u_2 + v_2 + \ldots$ converges.

Give an example to show the result false if hypothesis 2 is omitted.

15. Prove:

    1. $u_k, v_k > 0$ $k = 1, 2, \ldots$

    2. $\lim_{k \to 8} u_k / v_k = A$

    3. $\sum_{k=1}^{\infty} v_k < \infty$

$\Rightarrow \sum_{k=1}^{\infty} u_k < \infty$

§2. Convergence Tests

We introduce here a number of the more useful tests for the convergence of series. In the present section, we are dealing only with series of positive terms.

2.1 D'ALEMBERT'S RATIO TEST

Theorem 4. 1. $u_k > 0$ $k = 1, 2, \ldots$

          2. $\lim_{k \to \infty} \dfrac{u_{k+1}}{u_k} = l < 1$ $(1 < l \leqq \infty)$

$\Rightarrow$ $\sum_{k=1}^{\infty} u_k < \infty$ $(= \infty)$

If $l = 1$, the test fails.

By hypothesis 2 with $l < 1$ we have for any number r between l and 1

$$\frac{u_{k+1}}{u_k} < r \qquad\qquad\qquad k = m, m + 1, \ldots$$

$$u_{m+p} < r^p u_m \qquad\qquad\qquad p = 1, 2, \ldots$$

Here m is some integer depending on r. By Theorem 2 and T_1 the series

$$\sum_{k=m+1}^{\infty} u_k < \infty$$

and hence the entire series, converges.

If $l > 1$, or if $l = +\infty$, the ratio u_{k+1}/u_k is greater than 1 when k is greater than some integer m. That is,

$$u_k > u_m > 0 \qquad\qquad\qquad k = m + 1, m + 2, \ldots.$$

Hence, $\lim_{k \to \infty} u_k \neq 0$

and the series diverges by Theorem 1.

To show that the test fails when $l = 1$, we observe that

$$\sum_{k=1}^{\infty} \frac{1}{k} = \infty, \qquad \sum_{k=1}^{\infty} \frac{1}{k^2} < \infty$$

and that in both cases $l = 1$.

2.2 CAUCHY'S TEST

Theorem 5.   1. $u_k > 0 \qquad k = 1, 2, \ldots$

2. $\lim_{k \to \infty} \sqrt[k]{u_k} = l < 1 \qquad (1 < l \leq \infty)$

$\Rightarrow \qquad \sum_{k=1}^{\infty} u_k < \infty \qquad (= \infty)$

The test fails if $l = 1$.

The proof is similar to that of Theorem 4, and is omitted.

2.3 MACLAURIN'S INTEGRAL TEST

Theorem 6.   1. $f(x) \geq 0, \in C, \downarrow \qquad\qquad\qquad 1 \leq x < \infty$

2. $\lim_{R \to \infty} \int_1^R f(x)\, dx = A \qquad (= \infty)$

$\Rightarrow \qquad \sum_{k=1}^{\infty} f(k) < \infty \qquad (= \infty)$

The symbol $\downarrow$ indicates that $f(x)$ is nonincreasing. If $f(x) \in C^1$, then $f(x) \in \downarrow \Leftrightarrow f'(x) \leq 0$. By hypothesis 1,

$$f(k + 1) \leq f(x) \leq f(k) \qquad\qquad k \leq x \leq k + 1$$

Integrating each term of these inequalities, we have

$$f(k + 1) \leq \int_k^{k+1} f(x)\, dx \leq f(k) \qquad\qquad k = 1, 2, \ldots, n$$

Adding these inequalities, we obtain

(1) $$\sum_{k=2}^{n+1} f(k) \leq \int_1^{n+1} f(x)\, dx \leq \sum_{k=1}^{n} f(k)$$

If we have hypothesis 2 with the finite number A, then by the positiveness of $f(x)$ and by inequalities (1)

$$\sum_{k=1}^{n+1} f(k) \leq A + f(1)$$

Hence,
$$\sum_{k=1}^{\infty} f(k) \leq A + f(1) < \infty$$

If we have the alternative hypothesis 2, then, letting n become infinite in (1), we obtain
$$\infty = \sum_{k=1}^{\infty} f(k)$$

Observe that there is no case here in which the test fails. The limit in hypothesis 2 must exist or else the integral must become positively infinite with R, since the integrand is nonnegative.

We can now establish the results stated about test series T_2. If $p > 1$ take $f(x) = x^{-p}$. Then $f'(x) = -px^{-p-1} < 0$. Also
$$\lim_{R \to \infty} \int_1^R x^{-p}\, dx = \frac{1}{p-1}$$

so that Theorem 6 assures convergence. If $0 \leq p \leq 1$, the test is still applicable, and
$$\lim_{R \to \infty} \int_1^R x^{-p}\, dx = \lim_{R \to \infty} \left[\frac{R^{1-p}}{1-p} - \frac{1}{1-p} \right] = \infty \qquad 0 \leq p < 1$$

$$= \lim_{R \to \infty} \log R = \infty \qquad\qquad p = 1$$

Divergence of the series is assured. For $p < 0$ the test is no longer applicable, for then $x^{-p} \in \uparrow$. But we see that the series diverges by Theorem 1.

Test series T_3: $\displaystyle\sum_{k=2}^{\infty} \frac{1}{k(\log k)^p}$

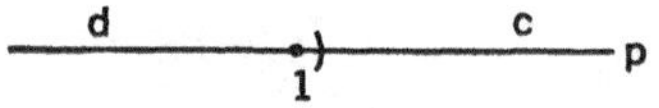

Fig. 32.

Here the discussion is much the same as for T_2. We have
$$\lim_{R \to \infty} \int_2^R \frac{1}{x(\log x)^p}\, dx = \frac{1}{(p-1)(\log 2)^{p-1}} \qquad p > 1$$

$$= \infty \qquad\qquad 0 \leq p \leq 1$$

Corollary 6.
$$A \leq \sum_{k=1}^{\infty} f(k) \leq A + f(1)$$

This follows from inequalities (1) by allowing n to become infinite. It frequently enables one to obtain estimates for the value of a series. For example,

(2)
$$\frac{1}{p-1} \leq \zeta(p) \leq \frac{p}{p-1} \qquad p > 1$$

In fact, since the terms of the series are all positive, the sum of the series is certainly greater than its first term. Hence,

$$1 \leq \zeta(p) \leq \frac{p}{p-1}$$

from which we conclude that $\zeta(p)$ tends to 1 as p becomes infinite. Inequalities (2) show that

$$\lim_{p \to 1+} \zeta(p) = +\infty$$

EXERCISES (2)

Test the series of Exercises 1–10 for convergence.

1. $\displaystyle\sum_{k=1}^{\infty} \frac{k+2}{k^2+k}$.

2. $\displaystyle\sum_{k=1}^{\infty} k^3 e^{-k}$.

3. $\displaystyle\sum_{k=1}^{\infty} \frac{\log k}{k^p}$.

4. $\displaystyle\sum_{k=3}^{\infty} \frac{1}{k(\log k)(\log \log k)^p}$

5. $\displaystyle\sum_{k=2}^{\infty} \frac{1}{(\log k)^k}$.

6. $\displaystyle\sum_{k=1}^{\infty} k^{-1} \log (1 + k^{-1})$.

7. $\displaystyle\sum_{k=1}^{\infty} \frac{(\log k)^2}{(\log 2)^k}$.

8. $\displaystyle\sum_{k=1}^{\infty} \cot^{-1} k$.

9. $\displaystyle\sum_{k=1}^{\infty} e^{-kp} \log k$.

10. $\displaystyle\sum_{k=1}^{\infty} (\log k)^p$.

11. Show that $\dfrac{\pi}{4} \leq \displaystyle\sum_{k=1}^{\infty} \dfrac{1}{k^2+1} \leq \dfrac{\pi}{4} + \dfrac{1}{4}$.

12. Test the series

$$\sum_{k=1}^{\infty} \frac{k}{2^k}$$

by use of Theorems 4, 5, and 6.

13. For what values of r is the integral test applicable to the geometric series? Apply it for these values.

14. By use of Corollary 6 prove that

$$\frac{-1}{\log r} \leq \frac{1}{1-r} \leq \frac{-1}{\log r} + 1 \qquad\qquad 0 < r < 1$$

Check geometrically or by the law of the mean.

15. Prove Theorem 5.

§3. *Absolute Convergence. Alternating Series*

We next consider series whose terms are not restricted to be positive, introducing the notion of absolute convergence. We then demonstrate a theorem of Leibniz useful for testing alternating series. By its use we exhibit series which converge but which fail to converge absolutely.

3.1 ABSOLUTE AND CONDITIONAL CONVERGENCE

Definition 3. *The series* $\sum_{k=1}^{\infty} u_k$ *converges absolutely* $\Leftrightarrow$ $\sum_{k=1}^{\infty} |u_k|$ *converges.*
For example, the series

$$\sum_{k=1}^{\infty} \frac{(-1)^k}{k^2}$$

converges absolutely. Of course, any series of positive terms which converges, converges absolutely.

Definition 4. *The series* $\sum_{k=1}^{\infty} u_k$ *converges conditionally* $\Leftrightarrow$ *it converges and*

$$\sum_{k=1}^{\infty} |u_k| = \infty$$

We shall show presently that the series

$$\sum_{k=1}^{\infty} \frac{(-1)^k}{k}$$

converges. Accordingly, it converges conditionally, since the harmonic series diverges.

Theorem 7. $\sum_{k=1}^{\infty} |u_k| < \infty \Rightarrow \sum_{k=1}^{\infty} u_k$ *converges.*

Since

$$- |u_k| \leqq u_k \leqq |u_k|$$

we have

$$0 \leqq |u_k| - u_k \leqq 2|u_k| \qquad\qquad k = 1, 2, \ldots$$

Hence, the series

(1)
$$\sum_{k=1}^{\infty} (|u_k| - u_k)$$

converges by Theorem 2, taking $v_k = 2|u_k|$. Subtract series (1) term by term from the convergent series $\sum_{k=1}^{\infty} |u_k|$. The resulting series $\sum_{k=1}^{\infty} u_k$ must also converge, and the proof is complete.

By use of this result we can at once extend the applicability of Theorems 4 and 5.

Theorem 4*.   1. $\lim\limits_{k \to \infty} \left| \dfrac{u_{k+1}}{u_k} \right| = l$

$\Rightarrow$ 2. $l < 1$ $(1 < l \leqq \infty)$

$$\sum_{k=1}^{\infty} u_k \text{ converges absolutely (diverges).}$$

Theorem 5*.   1. $\lim\limits_{k \to \infty} \sqrt[k]{|u_k|} = l < 1$ $(1 < l \leqq \infty)$

$\Rightarrow$

$$\sum_{k=1}^{\infty} u_k \text{ converges absolutely (diverges).}$$

These theorems are not restricted to series of positive terms. Their proofs are omitted.

3.2 LEIBNIZ'S THEOREM ON ALTERNATING SERIES

Theorem 8.   1. $v_k \in \ \downarrow$ $\hspace{5cm} k = 1, 2, \ldots$

2. $\lim\limits_{k \to \infty} v_k = 0$

(2) $\Rightarrow$

$$\sum_{k=1}^{\infty} (-1)^k v_k \text{ converges.}$$

Since the sequence $\{v_k\}_1^\infty$ is a decreasing sequence tending to zero, it is clear that

$$v_k \geqq 0 \hspace{4cm} k = 1, 2, \ldots$$

so that the series (2) is an *alternating series*. That is, its terms are alternately nonnegative and nonpositive. Since

(3)
$$S_{2n} = S_{2n-1} + v_{2n} = S_{2n-2} + v_{2n} - v_{2n-1}$$
$$S_{2n+1} = S_{2n} - v_{2n+1} = S_{2n-1} + v_{2n} - v_{2n+1}$$

we see that every S with even subscript is greater than every S with odd subscript. Moreover, since

$$v_{2n} - v_{2n-1} \leqq 0, \hspace{1cm} v_{2n} - v_{2n+1} \geqq 0 \hspace{2cm} n = 1, 2, \ldots$$

it follows that

$$S_{2n} \in \ \downarrow, \hspace{1cm} S_{2n+1} \in \ \uparrow$$

Hence, both sequences approach limits

$$\lim_{n \to \infty} S_{2n} = A, \hspace{1cm} \lim_{n \to \infty} S_{2n+1} = B$$

But, if we let n become infinite in equation (3), making use of hypothesis 2, we see that $A = B$, and the proof is complete.

EXAMPLE A. $\sum\limits_{k=1}^{\infty} \dfrac{(-1)^k}{k}$ converges.

Here $v_k = k^{-1}$ and it is clear that v_k tends monotonically to zero.

EXAMPLE B. $\sum\limits_{k=1}^{\infty} (-1)^k \dfrac{\log k}{\sqrt{k}}$ converges.

Here $v_k = f(k)$ where $f(x) = x^{-1/2} \log x$.   Since $f'(x) = x^{-3/2}(1 - \tfrac{1}{2}\log x)$, it is clear that $v_k \in \downarrow$ when $k > e^2$.   Hypothesis 2 is also clearly satisfied.

Corollary 8. $|R_n| = |A - S_n|$

$$= \left| \sum_{k=n+1}^{\infty} (-1)^k v_k \right| \leqq v_{n+1} \qquad\qquad n = 1, 2, \ldots$$

For, since A lies between any two consecutive elements of $\{S_n\}_1^{\infty}$,

$$|R_n| = |A - S_n| \leqq |S_{n+1} - S_n| = v_{n+1}$$

This result enables us to estimate the error when a partial sum is used for the correct value of the series (2).   It should be observed that, if the series is not alternating or otherwise fails to satisfy the hypotheses of Theorem 8, the present estimate of the remainder may not be used.   For example, in the series

$$\frac{2}{5} = 1 - \frac{1}{2} - \frac{1}{2^2} + \frac{1}{2^3} + \frac{1}{2^4} - \cdots$$

we have $|R_1| = |\tfrac{2}{5} - S_1| = \tfrac{3}{5}$, and this is not less than $\tfrac{1}{2}$, the absolute value of the first term of the series omitted.   If we introduce parentheses into the series

$$\frac{2}{5} = 1 - \left(\frac{1}{2} + \frac{1}{2^2} \right) + \left(\frac{1}{2^3} + \frac{1}{2^4} \right) - \cdots$$

the estimate again becomes applicable:

$$\left| \frac{2}{5} - 1 \right| < \frac{1}{2} + \frac{1}{2^2}$$

EXERCISES   (3)

Test the series of Exercises 1–8 for absolute and conditional convergence.

1. $\sum\limits_{k=1}^{\infty} (-1)^k k^p$.

2. $\sum\limits_{k=3}^{\infty} \dfrac{(-1)^k}{\sqrt{k} \log \log k}$.

3. $\sum\limits_{k=1}^{\infty} \dfrac{(1-k)^k}{k^{k+2}}$.

4. $\sum\limits_{k=2}^{\infty} (-1)^k \dfrac{\log \log k}{\sqrt{\log k}}$.

5. $\displaystyle\sum_{k=3}^{\infty} \frac{(-1)^k \log k}{k \log \log k}$.

6. $\displaystyle\sum_{k=3}^{\infty} \frac{(-1)^k}{k \log k (\log \log k)^p}$.

7. $\displaystyle\sum_{k=2}^{\infty} \frac{x^k}{\sqrt{\log k}}$

8. $\displaystyle\sum_{k=1}^{\infty} a^k k^a$.

9. Prove Theorems 4* and 5*. Note that to prove that a series diverges it is not enough to show that it fails to converge absolutely.

10. Give examples to show that neither hypothesis of Leibniz's theorem may be replaced by $v_k \geq 0$. *Hint:* Drop parentheses in the series $\displaystyle\sum_{k=1}^{\infty} \left(\frac{1}{k} - \frac{1}{k^2} \right)$.

11. Show that series (2) converges if

$$v_k = \int_{k\pi}^{(k+1)\pi} |\sin x|/x \, dx$$

Hence show that $\displaystyle\lim_{R\to\infty} \int_1^R (\sin x)/x \, dx$ exists.

12. For all values of p test for absolute and conditional convergence the series

$$\sum_{k=1}^{\infty} (-1)^k k^p \log k$$

13. Verify Corollary 8 for the geometric series $(-1 < r < 0)$ by actually computing the remainder R_n.

14. Add the first 10 terms of the following series and check that the sum differs from log 2 by less than the 11th term:

$$\log 2 = 1 - \tfrac{1}{2} + \tfrac{1}{3} - \tfrac{1}{4} + \cdots$$

15. Use Corollary 8 to prove

$$0 < x - \log (1 + x) < \frac{x^2}{2} \qquad\qquad 0 < x < 1$$

Verify the result by use of the Maclaurin development of $\log (1 + x)$ with the Lagrange remainder.

16. Test for conditional and absolute convergence

$$\sum_{k=2}^{\infty} \frac{(-1)^k}{k^p + (-1)^k} \qquad\qquad \text{(all } p > 0)$$

§4. *Limit Tests*

An exceedingly useful test, which we shall call the limit test, for the absolute convergence of infinite series is now developed. Although it is perhaps the easiest of all tests to apply, it has been somewhat neglected in textbooks. It is analogous to a very familiar test for the convergence of improper integrals.

4.1 LIMIT TEST FOR CONVERGENCE

Theorem 9.   1. $\lim_{k \to \infty} k^p u_k = A$ $\qquad\qquad\qquad\qquad\qquad$ $p > 1$

$\Rightarrow$ $\qquad$ $\sum_{k=1}^{\infty} u_k$ converges absolutely.

By hypothesis 1 we see that

$$\lim_{k \to \infty} k^p |u_k| = |A|$$

Hence, there exists an integer m such that

$$k^p |u_k| < |A| + 1 \qquad\qquad\qquad k = m, m+1, \ldots$$

$$|u_k| < (|A| + 1) k^{-p}$$

Hence, by Theorem 2, using test series T_2 we have

$$\sum_{k=m}^{\infty} |u_k| < \infty$$

from which the desired conclusion follows at once.

EXAMPLE A. $\sum_{k=1}^{\infty} \dfrac{(k+1)^{1/2}}{(k^5 + k^3 - 1)^{1/3}}$ converges.

For, taking $p = \frac{7}{6} > 1$, we have

$$\lim_{k \to \infty} u_k k^{7/6} = \lim_{k \to \infty} \frac{(1 + k^{-1})^{1/2}}{(1 + k^{-2} - k^{-5})^{1/3}} = 1$$

EXAMPLE B. $\sum_{k=1}^{\infty} (-1)^k \dfrac{\log k}{k^2}$ converges absolutely.

For,

$$\lim_{k \to \infty} u_k k^{3/2} = \lim_{k \to \infty} (-1)^k \frac{\log k}{\sqrt{k}} = 0$$

4.2 LIMIT TEST FOR DIVERGENCE

Theorem 10.   1. $\lim_{k \to \infty} k u_k = A \neq 0$ $\qquad$ (or $\pm\infty$)

$\Rightarrow$ $\qquad$ $\sum_{k=1}^{\infty} u_k$ diverges.

The test fails if $A = 0$.

CASE I. $A > 0$ (or $+\infty$). Then there exists an integer m such that

$$ku_k > \frac{A}{2} \text{ (or 1)} \qquad\qquad k = m, m+1, \ldots$$

Hence, by Theorem 3, comparing with the harmonic series, we obtain

$$\sum_{k=m}^{\infty} u_k = +\infty$$

from which the desired result follows.

CASE II. $A < 0$ (or $-\infty$). In this case the series

$$\sum_{k=1}^{\infty} (-u_k)$$

may be treated by Case I.

To see that the test fails when $A = 0$, consider the two series

$$\sum_{k=1}^{\infty} \frac{1}{k^2} < \infty, \qquad \sum_{k=2}^{\infty} \frac{1}{k \log k} = \infty$$

For each, $A = 0$.

EXAMPLE C. $\displaystyle\sum_{k=1}^{\infty} \frac{k \log k}{7 + 11k - k^2}$ diverges.

For,

$$\lim_{k \to \infty} ku_k = \lim_{k \to \infty} \frac{k^2 \log k}{7 + 11k - k^2} = -\infty$$

EXAMPLE D. Test for convergence the series

$$\sum_{k=1}^{\infty} (-1)^k k^\alpha e^{k\beta}$$

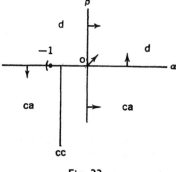

Fig. 33.

The results are contained in Figure 33. It indicates that the series diverges in quadrants I, II; converges absolutely in III, IV. The behavior on the axes is also shown. The series converges conditionally for $\beta = 0$, $-1 \leqq \alpha < 0$.

The limits involved when using the limit tests are easily evaluated by inspection of the orders of infinity of the various factors.

EXERCISES (4)

Test the series of Exercises 1–8 for convergence by use of the limit tests.

1. Exercises 3, 4, 5 of §1.

2. Exercises 8, 9, 10 of §1.

6. $\displaystyle\sum_{k=2}^{\infty} \frac{(-1)^k}{\sqrt{k^3 - 1}}$.

3. Exercises 1, 2, 3 of §2.

4. $\displaystyle\sum_{k=1}^{\infty} \frac{e^{-k\alpha}}{k}$.

7. $\displaystyle\sum_{k=2}^{\infty} k^\alpha (\log k)^\beta e^{\gamma k}$.

5. $\displaystyle\sum_{k=2}^{\infty} k^\alpha (\log k)^\beta$.

8. $\displaystyle\sum_{k=2}^{\infty} (\log k)^{-\log k}$.

9. Prove or disprove: $|u_k k| > 1,\ k = 1, 2, \ldots \Rightarrow \displaystyle\sum_{k=1}^{\infty} u_k$ diverges.

10. Prove or disprove: $f(x) \in C;\ |xf(x)| > 1,\ x \geq 1 \Rightarrow \displaystyle\sum_{k=1}^{\infty} f(k)$ diverges.

11. Prove: $\lim\limits_{k\to\infty} k (\log k)^p u_k = A,\ p > 1 \Rightarrow \displaystyle\sum_{k=1}^{\infty} |u_k| < \infty$.

12. Prove: $\lim\limits_{k\to\infty} k (\log k) u_k = A \neq 0 \Rightarrow \displaystyle\sum_{k=1}^{\infty} u_k$ diverges.

13. In Exercise 12 show that the test fails if $A = 0$.

14. Test for convergence the series

$$\sum_{k=3}^{\infty} \frac{\log (1 + k^{-1})}{|a|^{\log \log k}} \qquad\qquad a \neq 0$$

§5. *Uniform Convergence*

Consider a series of functions

(1)
$$f(x) = \sum_{k=1}^{\infty} u_k(x)$$

which we suppose convergent for every point x in the interval $a \leq x \leq b$. This property of convergence might be verified by applying some of the earlier tests for each value of x separately. There is a further type of convergence, known as *uniform convergence*, which has to do with the behavior of the series in the interval $a \leq x \leq b$. For series enjoying this type of convergence it is easier to infer properties, such as continuity, of the sum function $f(x)$ from the properties of the separate terms.

5.1 DEFINITION OF UNIFORM CONVERGENCE

Set

$$S_n(x) = \sum_{k=1}^{n} u_k(x)$$

Definition 5. *Series* (1) *converges uniformly to* $f(x)$ *in the interval* $a \leq x \leq b \Leftrightarrow$ *to an arbitrary* $\epsilon > 0$ *corresponds an integer* m *independent of* x *in* $a \leq x \leq b$ *such that when* $n > m$

(2) $|f(x) - S_n(x)| < \epsilon$ $a \leq x \leq b$

By the definition of limit, the series (1) converges at a point x_0 if to an arbitrary $\epsilon > 0$ corresponds an m such that (2) holds when x is replaced by x_0. If (1) converges for every x_0, we can determine an integer m for each x_0, but it will change in general as x_0 changes. If in particular it does not change, the series converges uniformly in the interval. In general, m is a function of ϵ and x,

$$m = m(\epsilon, x)$$

But, $m = m(\epsilon) \Leftrightarrow$ (1) converges uniformly.

EXAMPLE A. $\displaystyle\sum_{k=0}^{\infty} x^k$ converges uniformly to $\dfrac{1}{1-x}$ in $-a \leq x \leq a$, if $0 < a < 1$. Here

$$S_n(x) = \frac{1 - x^n}{1 - x}$$

$$|f(x) - S_n(x)| = \frac{|x|^n}{|1 - x|} \leq \frac{a^n}{1 - a} |x| \leq a$$

If $\epsilon > 0$ we have only to choose m so that

$$\frac{a^n}{1 - a} < \epsilon   n > m$$

In fact, we may choose for m any integer greater than

$$- \frac{\log\left[(1 - a)\epsilon\right]}{\log\left(1/a\right)}$$

Clearly it will depend in no way on x.

EXAMPLE B. $\displaystyle\sum_{k=1}^{\infty} [kxe^{-kx^2} - (k - 1)xe^{-(k-1)x^2}]$ converges on the interval $0 \leq x \leq 1$, but not uniformly. Here

$$\lim_{n \to \infty} S_n(x) = \lim_{n \to \infty} nxe^{-nx^2} = 0   0 \leq x \leq 1$$

Suppose the convergence were uniform on $0 \leq x \leq 1$. Then for any ϵ, say $\epsilon = 1$, there would exist an integer m such that

(3) $$|f(x) - S_n(x)| = nxe^{-nx^2} < 1 \qquad\qquad n > m$$

But

(4) $$\max_{0 \leq x \leq 1} nxe^{-nx^2} = \sqrt{\frac{n}{2e}}$$

so that we should have from inequality (3)

$$\sqrt{\frac{n}{2e}} < 1 \qquad\qquad n > m$$

Let n become infinite to obtain a contradiction.

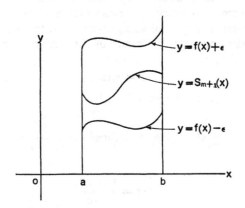

Fig. 34.

Graphically, inequality (2) means that the curves $y = S_n(x)$, $n > m$, lie between the curves $y = f(x) + \epsilon$ and $y = f(x) - \epsilon$ when $a \leq x \leq b$. See Figure 34. In Example B the curves $y = S_n(x)$ cannot be contained, for all large n, between the curves $y = f(x) \pm \epsilon$ in the interval $0 \leq x \leq 1$, even if ϵ is large, since the height to which these curves rise increases without limit as n becomes infinite.

EXAMPLE C. $\displaystyle\sum_{k=0}^{\infty} (1 - x)x^k = 1 \qquad\qquad 0 \leq x < 1$
$$= 0 \qquad\qquad x = 1$$

Since $f(x)$ is discontinuous, it is clear geometrically that for every n the continuous curve $y = S_n(x)$ must fail to lie between the curves $y = f(x) \pm \frac{1}{4}$ in the interval $0 \leq x \leq 1$. Analytically, we have

$$|f(x) - S_n(x)| = x^n \qquad\qquad 0 \leq x < 1$$
$$= 0 \qquad\qquad x = 0$$

If $\epsilon = \frac{1}{4}$, the inequality

$$x^n < \tfrac{1}{4} \qquad\qquad 0 \leq x < 1$$

is false for every fixed n. For, the left-hand side approaches 1 as x approaches 1.

5.2 WEIERSTRASS'S M-TEST

We now introduce one of the most useful methods of testing a series for uniform convergence.

Theorem 11.   1. $|u_k(x)| \leq M_k$ $a \leq x \leq b, \quad k = 1, 2, \ldots$

2. $\sum\limits_{k=1}^{\infty} M_k < \infty$

$\Rightarrow$ $\sum\limits_{k=1}^{\infty} u_k(x)$ converges uniformly in $a \leq x \leq b$.

Set

$$M = \sum_{k=1}^{\infty} M_k, \qquad T_n = \sum_{k=1}^{n} M_k, \qquad f(x) = \sum_{k=1}^{\infty} u_k(x)$$

This last series clearly converges for $a \leq x \leq b$ by Theorems 2 and 7. Then

$$|S_{n+p}(x) - S_n(x)| \leq \sum_{k=n+1}^{n+p} |u_k(x)| \leq \sum_{k=n+1}^{n+p} M_k$$
$$\leq T_{n+p} - T_n$$

Now let p become infinite:

$$|f(x) - S_n(x)| \leq M - T_n \qquad\qquad a \leq x \leq b$$

Given $\epsilon > 0$, we can determine m so that

(5) $M - T_n < \epsilon$ $n > m$

by hypothesis 2. Clearly, m does not depend on x since T_n does not. But inequality (5) implies the desired inequality (2).

EXAMPLE D. $\sum\limits_{k=1}^{\infty} \dfrac{\cos kx}{k^2}$ converges uniformly in $-R \leq x \leq R$, where R is any number. Take $M_k = k^{-2}$.

5.3 RELATION TO ABSOLUTE CONVERGENCE

If a series converges uniformly by virtue of Theorem 11, it clearly converges absolutely. One might be tempted to suppose that all uniformly convergent series are absolutely convergent. This is not the case. Example C is a series of positive terms. It converges absolutely but not uniformly in $0 \leq x \leq 1$.

EXAMPLE E. $\sum_{k=1}^{\infty} (-1)^k \dfrac{x^k}{k}$ converges uniformly but not absolutely in the
interval $0 \leq x \leq 1$. At $x = 1$ this series is the familiar alternating series
which converges conditionally to $\log \frac{1}{2}$. By Corollary 8,

$$|f(x) - S_n(x)| \leq \frac{x^{n+1}}{n+1} \leq \frac{1}{n+1} \qquad 0 \leq x \leq 1$$

If ϵ is an arbitrary positive number, we have only to choose m as the first
integer greater than ϵ^{-1}.

This example shows the limitations of Theorem 11. Even though the
present series is known to converge uniformly, it must be impossible to
find the sequence M_n required for the Weierstrass test.

EXERCISES (5)

Test the series of Exercises 1–10 for uniform convergence in the intervals indicated.

1. $\displaystyle\sum_{k=1}^{\infty} (2k + 1)^{-3/2} \sin 2kx$ $-\pi \leq x \leq \pi$

2. $\displaystyle\sum_{k=2}^{\infty} \frac{x^k}{k (\log k)^2}$ $-1 \leq x \leq 1$

3. $\displaystyle\sum_{k=2}^{\infty} \frac{(-1)^k e^{-kx}}{k(k-1)}$ $0 \leq x \leq 100$

4. $\displaystyle\sum_{k=1}^{\infty} \frac{x^k}{\sqrt{k}}$ $-1 \leq x \leq 0$

5. $\displaystyle\sum_{k=0}^{\infty} (-1)^k \frac{x^{2k+1}}{2k+1}$ $-1 \leq x \leq 1$

6. $\displaystyle\sum_{k=0}^{\infty} \frac{x^k}{k!}$ $-R \leq x \leq R$

7. $\displaystyle\sum_{k=1}^{\infty} \frac{1}{k} \left(\frac{x-1}{x} \right)^k$ $\dfrac{1}{2} \leq x \leq 1$

8. $\displaystyle\sum_{k=1}^{\infty} \left(\frac{1}{x+k} - \frac{1}{x+k+1} \right)$ $0 \leq x \leq 1$

9. $\displaystyle\sum_{k=0}^{\infty} \left(\frac{1}{kx+2} - \frac{1}{kx+x+2} \right)$ $0 \leq x \leq 1$

10. $\displaystyle\sum_{k=1}^{\infty} x(1+x)^{-k}$ $0 \leq x \leq 1$

11. In Example C show the convergence uniform in $0 \leq x \leq a$, where $a < 1$.

12. Establish equation (4).

13. Prove: $\max\limits_{a \leq x \leq 1} nxe^{-nx^2} = nae^{-na^2}$ if $(2n)^{-1/2} < a$.

14. In Example B show that the convergence is uniform in $a \leq x \leq 1$, where $a > 0$.

15. Define uniform convergence for $-\infty < x < \infty$ and show that the series of Example D has the property there but that the Maclaurin series for e^x does not.

§6. Applications

In Example C of §5 each term of the series was continuous. In fact, $u_k(x) \in C^\infty$, $-\infty < x < \infty$, $k = 1, 2, \ldots$. Yet the sum of the series was discontinuous in the interval $0 \leq x \leq 1$, though the series converged at each point of that interval. When can we infer the continuity of the sum of a convergent series from the continuity of the terms of the series? One answer is: when the series converges uniformly. This and similar applications of uniform convergence will be made in the present section.

6.1 CONTINUITY OF THE SUM OF A SERIES

Theorem 12.   1. $u_k(x) \in C$ $a \leq x \leq b, \quad k = 1, 2, \ldots$

         2. $f(x) = \sum\limits_{k=1}^{\infty} u_k(x)$ *uniformly in* $a \leq x \leq b$

$\Rightarrow$ $f(x) \in C$ $a \leq x \leq b$.

Let $a \leq x_0 \leq b, a \leq x_0 + \Delta x \leq b$. To an arbitrary $\epsilon > 0$ there corresponds, by hypothesis 2, an integer m such that

$$|f(x_0) - S_m(x_0)| < \epsilon/3$$

(1) $$|f(x_0 + \Delta x) - S_m(x_0 + \Delta x)| < \epsilon/3$$

In fact, uniform convergence implies more than this. These inequalities would hold if m were replaced by any larger integer; but we shall make no use of the fact. Since $S_m(x) \in C$ at x_0, there exists a number δ such that

$$|S_m(x_0) - S_m(x_0 + \Delta x)| < \epsilon/3 \qquad\qquad |\Delta x| < \delta$$

Combining these three inequalities, we obtain

(2) $$|f(x_0 + \Delta x) - f(x_0)| < \epsilon \qquad\qquad |\Delta x| < \delta$$

This implies $f(x) \in C$ at x_0. Observe where the uniform convergence entered the proof. We needed to know that inequality (1) was valid for *all* Δx such that $|\Delta x| < \delta$, $a \leq x_0 + \Delta x \leq b$, in order to be able to draw a like conclusion in inequality (2).

Example A of §5 illustrates the theorem. The sum, $1/(1 - x)$, of the series is continuous for $|x| \leqq a$, $a < 1$. That is, $1/(1 - x) \in C$ for $-1 < x < 1$. Example B of §5 shows that the conditions of Theorem 12 are not necessary. For, in that example the sum, 0, of the series is continuous even though the convergence is nonuniform.

6.2 INTEGRATION OF SERIES

Term by term integration of a convergent series of functions is not always valid, as we may show by use of Example B, §5. Here

$$\int_0^1 f(x)\, dx \neq \sum_{k=1}^{\infty} \int_0^1 u_k(x)\, dx$$

$$0 \neq \sum_{k=1}^{\infty} \tfrac{1}{2}(e^{-k+1} - e^{-k}) = \tfrac{1}{2}$$

It is clear geometrically why this happens. Remember that the curve $y = S_n(x)$ rises very high when n is large. It is this fact that enables the area under this curve to equal $\tfrac{1}{2} - e^{-n}$, a number, which approaches $\tfrac{1}{2}$ as n becomes infinite, even though each ordinate of the curve approaches zero.

Theorem 13.   1. $u_k(x) \in C$ $\qquad\qquad\qquad\qquad$ $a \leqq x \leqq b$, $\quad k = 1, 2, \ldots$

$\qquad\qquad$ 2. $f(x) = \sum_{k=1}^{\infty} u_k(x)$ $\qquad\qquad$ *uniformly in $a \leqq x \leqq b$*

$\Rightarrow \qquad\qquad \int_a^b f(x)\, dx = \sum_{k=1}^{\infty} \int_a^b u_k(x)\, dx$

To an arbitrary $\epsilon > 0$ corresponds an integer m, independent of x in $a \leqq x \leqq b$, such that when $n > m$

(3) $\qquad\qquad\qquad |f(x) - S_n(x)| < \epsilon/(b - a) \qquad\qquad a \leqq x \leqq b$

Hence, for $n > m$, we have

$$\left| \int_a^b f(x)\, dx - \int_a^b S_n(x)\, dx \right| \leqq \int_a^b |f(x) - S_n(x)|\, dx < \epsilon$$

That is,

$$\int_a^b f(x)\, dx = \lim_{n \to \infty} \int_a^b S_n(x)\, dx = \lim_{n \to \infty} \int_a^b \sum_{k=1}^{n} u_k(x)\, dx$$

$$= \lim_{n \to \infty} \sum_{k=1}^{n} \int_a^b u_k(x)\, dx = \sum_{k=1}^{\infty} \int_a^b u_k(x)\, dx$$

Observe that $f(x) \in C$ by Theorem 12 and is hence integrable. Note also that we needed uniform convergence to insure that inequality (3) should hold for *all* x in the interval when $n > m$.

EXAMPLE A. $\dfrac{1}{1+x} = \sum\limits_{k=0}^{\infty} (-x)^k$ uniformly in $0 \leqq x \leqq h, h < 1$.

Hence,

(4) $\log (1 + h) = \int_0^h \dfrac{dx}{1+x} = \sum\limits_{k=0}^{\infty} (-1)^k \dfrac{h^{k+1}}{k+1}$ $0 \leqq h < 1$

In Example E of §5, we showed that series (4) converges uniformly in the interval $0 \leqq h \leqq 1$. But we have not established equation (4) for $h = 1$. However, Theorem 12 assures us that the sum of the series must be continuous in $0 \leqq h \leqq 1$. But $\log (1 + h) \in C$ in $0 \leqq h \leqq 1$. Hence, equation (4) must also be valid at $h = 1$,

$$\log 2 = 1 - \tfrac{1}{2} + \tfrac{1}{3} \cdots$$

6.3 DIFFERENTIATION OF SERIES

Term by term differentiation of convergent series is not in general valid, even when all terms and the sum belong to C^1. Even uniform convergence of the given series does not validate the process.

EXAMPLE B. $x = \sum\limits_{k=1}^{\infty} \left(\dfrac{x^k}{k} - \dfrac{x^{k+1}}{k+1} \right)$ $0 \leqq x \leqq 1$

This series converges uniformly in $0 \leqq x \leqq 1$. Yet, when the series is differentiated,

$$\sum\limits_{k=1}^{\infty} (x^{k-1} - x^k)$$

we find that its sum is equal to the derivative of x in the interval $0 \leqq x < 1$ only. At $x = 1$ the sum of the series is 0.

Theorem 14.   1. $u_k(x) \in C^1$ $a \leqq x \leqq b, \quad k = 1, 2, \ldots$

2. $f(x) = \sum\limits_{k=1}^{\infty} u_k(x)$ $a \leqq x \leqq b$

3. $\sum\limits_{k=1}^{\infty} u_k'(x)$ *converges uniformly in* $a \leqq x \leqq b$

$\Rightarrow$ $f'(x) = \sum\limits_{k=1}^{\infty} u_k'(x)$ $a \leqq x \leqq b$

Observe that the conclusion includes the fact that $f(x) \in C^1$ in the interval $a \leqq x \leqq b$. It necessarily must refer only to the existence of right-hand and left-hand derivatives at the points a and b, respectively. Set

$$\varphi(x) = \sum\limits_{k=1}^{\infty} u_k'(x)$$

By Theorem 12 $\varphi(x) \in C$ in $a \leq x \leq b$. By Theorem 13

$$\int_a^h \varphi(x) \, dx = \sum_{k=1}^{\infty} [u_k(h) - u_k(a)] \qquad\qquad a \leq h \leq b$$

and by hypothesis 2 this series can be written as the difference of two convergent series,

$$\int_a^h \varphi(x) \, dx = f(h) - f(a)$$

Now differentiation with respect to h gives

$$f'(h) = \varphi(h) = \sum_{k=1}^{\infty} u_k'(h) \qquad\qquad a \leq h \leq b$$

This concludes the proof.

The conditions of the theorem are frequently abbreviated by the statement that the derived series must converge uniformly. This statement is not quite accurate, for it omits reference to the convergence of the given series contained in hypothesis 2. That hypothesis 3 does not imply hypothesis 2 is seen by the example $u_k(x) = 1 + x^k$, $0 \leq x \leq \frac{1}{2}$.

EXAMPLE C. $\dfrac{1}{1-x} = \displaystyle\sum_{k=0}^{\infty} x^k$ $\qquad\qquad -1 < x < 1$

The derived series $\qquad\qquad \displaystyle\sum_{k=0}^{\infty} kx^{k-1}$

converges uniformly in $-a \leq x \leq a$, $a < 1$, as we see by Theorem 11, $M_k = ka^{k-1}$. Hence,

$$(5) \qquad \frac{1}{(1-x)^2} = \sum_{k=0}^{\infty} kx^{k-1} \qquad\qquad -a \leq x \leq a$$

Since any given number x in the interval $-1 < x < 1$ can be included inside the closed interval $-a \leq x \leq a$ for some $a < 1$, equation (5) holds in $-1 < x < 1$. It can be checked by Taylor's expansion.

EXERCISES (6)

1. Prove: $\displaystyle\int_0^1 \sum_{k=0}^{\infty} \frac{x^k}{k!} \, dx = e - 1$

2. $\displaystyle\lim_{x \to 0} \sum_{k=1}^{\infty} \frac{\log k \sin kx}{\sqrt{k(2k-1)(k+3)}} = ?$

3. $\displaystyle\lim_{x \to 0} \sum_{k=2}^{\infty} \frac{\cos kx}{k(k+1)} = ?$

Which of the series in Exercises 4–8 can be differentiated term by term in the intervals indicated?

4. Exercise 3, §5.

5. Exercise 4, §5.

6. Exercise 8, §5.

7. $\displaystyle\sum_{k=1}^{\infty} (2k + 1)^{-5/2} \sin (2k + 1)x$ $\qquad\qquad -1 \leqq x \leqq 1$

8. $\displaystyle\sum_{k=1}^{\infty} \left(\frac{x}{x-1}\right)^k$ $\qquad\qquad\qquad -4 \leqq x \leqq -3$

9. Show that $f(x) \in C^{\infty}$ in the interval $1 \leqq x < \infty$ if

$$f(x) = \sum_{k=1}^{\infty} \sqrt{k}\, e^{-kx}$$

10. Show that the series

$$\sum_{k=0}^{\infty} (-1)^k x^{2k}$$

may be integrated term by term from 0 to h, $-1 < h < 1$, and thus prove the validity of the Maclaurin expansion of $\tan^{-1} h$ for $-1 < h < 1$.

11. Prove that

$$\frac{\pi}{4} = 1 - \frac{1}{3} + \frac{1}{5} - \cdots$$

12. In the proof of Theorem 14 it was tacitly assumed that, when a series converges uniformly in an interval, it does so in any smaller interval. Prove this fact.

13. Prove:

1. $f(x) = \displaystyle\sum_{k=1}^{\infty} u_k(x)$ $\qquad\qquad$ uniformly in $a \leqq x \leqq b$

2. $g(x) \in C$ $\qquad\qquad\qquad\qquad\quad$ $a \leqq x \leqq b$

$\Rightarrow f(x)g(x) = \displaystyle\sum_{k=1}^{\infty} g(x)u_k(x)$ $\qquad$ uniformly in $a \leqq x \leqq b$

14. The conclusion of Theorem 13 remains true if b is replaced by y, $a < y \leqq b$ Show that the resulting series is uniformly convergent in the interval $a \leqq y \leqq b$.

15. In the light of Exercise 14, why does not hypothesis 3 imply hypothesis 2 in Theorem 14?

§7. Divergent Series

If a series diverges, it may sometimes be used in computation. Even if it converges, its use in computation is an approximation process. Instead of the actual sum A of the series, one uses S_n, where n is so large that $|A - S_n|$ is within the limit of error allowed by the conditions of the problem. If the series diverges, it may be possible to use some other combination, not S_n, of the first n terms of the series as an approximation to the sum of the divergent series. In the present section we make a brief study of these summation processes.

7.1 PRECAUTION

Great care should be exercised in the use of divergent series. One must be careful not to carry over the obvious properties of convergent series to divergent ones. Let us illustrate.

Euler attached the value $\frac{1}{2}$ to the divergent series

$$(1) \qquad 1 - 1 + 1 - 1 + \dots$$

This value may be arrived at heuristically in many ways. For example,

$$(2) \qquad \frac{1}{1+x} = 1 - x + x^2 - \dots \qquad -1 < x < 1$$

For $x = 1$, the left-hand side has the value $\frac{1}{2}$ and the right-hand side becomes series (1). Note, however, that equation (2) is valid only for $-1 < x < 1$. Euler treated series (2) like a convergent series at $x = 1$.

Another way of guessing a sum for series (1) is to set it equal to the undetermined constant A,

$$A = 1 - 1 + 1 - 1 - \dots$$
$$A - 1 = -1 + 1 - 1 + 1 - \dots$$

Adding these series term by term, we get $2A - 1 = 0$ or $A = \frac{1}{2}$. But here again we have carried over to divergent series processes which are valid for convergent ones.

Observe that we could get very different results by processes which appear very similar. If we set $x = 1$ in the series

$$(3) \qquad \frac{1+x}{1+x+x^2} = 1 - x^2 + x^3 - x^5 + x^6 - \dots \qquad -1 < x < 1$$

we get

$$\tfrac{2}{3} = 1 - 1 + 1 - 1 + \dots$$

Also, if we insert parentheses in series (1), a process always valid for convergent series, we get

$$1 = 1 - (1 - 1) - (1 - 1) - \dots$$

or

$$0 = (1 - 1) + (1 - 1) + \dots$$

Thus, we have obtained the possible values $\frac{1}{2}$, $\frac{2}{3}$, 1, 0 for the series (1), according as we have chosen one or another of the valid properties of convergent series to apply to the divergent one. This should show clearly the need for caution. Obviously, we want only one sum for a series. We *must* proceed by definition and not by analogy.

7.2 CESÀRO SUMMABILITY

We now define a process of attaching a sum to a divergent series which is variously known as *the method of arithmetic means, Cesàro 1-summability, (C, 1)-summability*, etc. The meaning of the number 1 in this notation will appear later. Set

$$S_n = \sum_{k=1}^{n} u_k, \qquad \sigma_n = \frac{1}{n} \sum_{k=1}^{n} S_k \qquad n = 1, 2, \ldots$$

That is, σ_n is the average of the first n partial sums, and is, as a consequence, the following linear combination of the first n terms:

$$\sigma_n = \sum_{k=1}^{n} \left(1 - \frac{k-1}{n}\right) u_k$$

Definition 6. *The series* $\displaystyle\sum_{k=1}^{\infty} u_k$ *is summable* $(C, 1)$ *to* $A \Leftrightarrow$

$$\lim_{n \to \infty} \sigma_n = A \tag{4}$$

We also write equation (4) as

$$A = \sum_{k=1}^{\infty} u_k \tag{C, 1}$$

EXAMPLE A. $\frac{1}{2} = 1 - 1 + 1 - \ldots$ $\qquad\qquad\qquad\qquad$ $(C, 1)$

For,

$$S_n: \quad 1, 0, 1, 0, \ldots$$

$$t_n = \sum_{k=1}^{n} S_k: \quad 1, 1, 2, 2, \ldots$$

$$\sigma_n: \quad 1, \tfrac{1}{2}, \tfrac{2}{3}, \tfrac{1}{2}, \ldots$$

$$\sigma_{2n} = \frac{1}{2}, \qquad \sigma_{2n+1} = \frac{n+1}{2n+1} \qquad n = 1, 2, \ldots$$

$$\lim_{n \to \infty} \sigma_{2n} = \lim_{n \to \infty} \sigma_{2n+1} = \lim_{n \to \infty} \sigma_n = \tfrac{1}{2}$$

EXAMPLE B. $\frac{2}{3} = 1 + 0 - 1 + 1 + 0 - 1 + 1 + 0 - \ldots$ $\qquad$ (C, 1)

For,

$$S_{3n+1} = S_{3n+2} = 1, \qquad S_{3n+3} = 0 \qquad\qquad n = 0, 1, \ldots$$

$$\sigma_{3n+1} = \frac{2n + 1}{3n + 1}, \qquad\qquad \sigma_{3n+2} = \frac{2n + 2}{3n + 2}$$

$$\sigma_{3n+3} = \frac{2n + 2}{3n + 3}, \qquad\qquad \lim_{n \to \infty} \sigma_n = \tfrac{2}{3}$$

This example shows that the interpolation of zeros into a series may affect its Cesàro sum.

EXAMPLE C. $\displaystyle\sum_{k=1}^{\infty} (-1)^k k$ is not summable (C, 1).

For

$$t_{2n-1} = -n, \qquad\qquad t_{2n} = 0 \qquad\qquad n = 1, 2, \ldots$$

$$\lim_{n \to \infty} \sigma_{2n-1} = -\tfrac{1}{2}, \qquad \lim_{n \to \infty} \sigma_{2n} = 0$$

This shows the need for more powerful methods of summation. If

$$\lim_{n \to \infty} \frac{2}{n(n + 1)} \sum_{k=1}^{n} t_k = A$$

we say that

$$A = \sum_{k=1}^{\infty} u_k \qquad\qquad\qquad (C, 2)$$

In the case of Example C, we have

(5) $$-\frac{1}{4} = \sum_{k=1}^{\infty} (-1)^k k \qquad\qquad (C, 2)$$

Thus, a series may fail to be summable by one method and be summable by a stronger one.

7.3 REGULARITY

In Definition 6 there was no statement that the given series was divergent. What will be the Cesàro sum of a convergent series? We shall show that it is the same as the ordinary sum, $\lim_{n \to \infty} S_n$.

Definition 7. *A method of summability is regular $\Leftrightarrow$ it sums a convergent series to the ordinary sum.*

Theorem 15. *Cesàro summability is regular.*

Let
$$A = \sum_{k=1}^{\infty} u_k = \lim_{n \to \infty} S_n$$

We are to prove that σ_n also approaches A.

CASE I. $A = 0$. By hypothesis we know that to an arbitrary $\epsilon > 0$ corresponds an integer m such that

$$|S_{m+1}|, |S_{m+2}|, \ldots < \epsilon$$

Hence, if $n > m$

$$|\sigma_n| \leq \frac{1}{n} |S_1 + S_2 + \ldots + S_m| + \frac{1}{n} \{|S_{m+1}| + \ldots + |S_n|\}$$

(6) $$\leq \frac{1}{n} |S_1 + S_2 + \ldots + S_m| + \frac{n-m}{n} \epsilon$$

(7) $$\overline{\lim_{n \to \infty}} |\sigma_n| \leq \epsilon$$

We may let n become infinite in inequality (6) since it is valid for $n > m$. Inequality (7) shows that $\lim_{n \to \infty} \sigma_n = 0$.

CASE II. $A \neq 0$. Here the sequence $\{S_n - A\}_1^{\infty}$ tends to zero, and we may apply Case I to it. Hence,

$$\lim_{n \to \infty} \frac{1}{n} \sum_{k=1}^{n} (S_k - A) = \lim_{n \to \infty} (\sigma_n - A) = 0$$

7.4 OTHER METHODS OF SUMMABILITY

Many other methods of summing a divergent series have been devised. We mention only two in passing. The series $\sum_{k=0}^{\infty} u_k$ is summable to A by the *method of Abel* ⟺

$$\lim_{x \to 1-} \sum_{k=0}^{\infty} u_k x^k = A$$

Of course, for the method to be applicable, this power series must converge for $0 \leq x < 1$, and the above limit must exist. We have seen by use of equation (2) that series (1) is summable by Abel's method to the value $\frac{1}{2}$. Also, equation (3) shows that the series of Example B is summable by the method of Abel to the value $\frac{2}{3}$.

Finally, let us define the *method of Hölder* since it is so closely related to that of Cesàro. A series is summable $(H, 1)$ to the value $A \Leftrightarrow \lim_{n \to \infty} \sigma_n = A$.

In other words, $(C, 1)$ and $(H, 1)$ are the same process. A series is summable $(H, 2)$ to the value $A \Leftrightarrow$

$$\lim_{n \to \infty} \frac{1}{n} \sum_{k=1}^{n} \sigma_k = A$$

That is, we are dealing here with the average of the averages. It can be shown that $(C, 2)$ and $(H, 2)$ are equivalent in the sense that any series which is summable to A by one process is by the other also. Both $(C, 2)$ and $(H, 2)$ can be generalized in the obvious way.

EXERCISES (7)

Find the Cesàro sum of those of the series of Exercises 1–5 which are summable $(C, 1)$.

1. $1 - 2 + 2 - 2 + \ldots$ (all terms ± 2 after the first).

2. $1 - 1 + 0 + 1 - 1 + 0 + \ldots$

3. $1 - 1 + 2 - 2 + 3 - 3 + \ldots$

4. $1 + 0 - 1 + 0 + 1 + 0 - 1 + \ldots$

5. $1 + 0 + 0 - 1 + 1 + 0 + 0 - 1 + 1 + 0 + 0 - 1 + \ldots$.

6. Prove:

1. $A = \sum_{k=1}^{\infty} u_k$ $(C, 1)$

2. $B = \sum_{k=1}^{\infty} v_k$ $(C, 1)$

$\Rightarrow A + B = \sum_{k=1}^{\infty} (u_k + v_k)$ $(C, 1)$

7. Prove: $A = \sum_{k=0}^{\infty} u_k$ $(C, 1)$ $\Rightarrow$ $\sum_{k=1}^{\infty} u_k = A - u_0$ $(C, 1)$

$A = \sum_{k=1}^{\infty} u_k$ $(C, 1)$ $\Rightarrow$ $\sum_{k=0}^{\infty} u_k = A + u_0$ $(C, 1)$

8. Prove: $A = \sum_{k=1}^{\infty} u_k$ $(C, 1)$ $\Rightarrow$ $\sum_{k=1}^{\infty} B u_k = AB$ $(C, 1)$

9. Use Example A and Exercises 7 and 8 to find the $(C, 1)$ sum of the series in Exercise 1.

10. Prove: $A = \sum_{k=1}^{\infty} u_k$ $(C, 1)$ $\Rightarrow$ $\lim_{n \to \infty} \frac{S_n}{n} = 0$

Find $\lim_{k \to \infty} \frac{u_k}{k}$.

11. Use Exercise 10 to show that the series of Example C is not summable $(C, 1)$.

12. Prove: $\displaystyle\sum_{k=1}^{\infty} u_k^2 = A$ $(C, 1)$ $\Leftrightarrow$ $\displaystyle\sum_{k=1}^{\infty} u_k^2 = A$ (convergent)

13. Prove: $\displaystyle\sum_{k=1}^{\infty} u_k = A$ $(C, 2)$ $\Rightarrow$ $\displaystyle\lim_{n\to\infty} n^{-2} u_n = 0$

14. Prove: $\displaystyle\sum_{k=1}^{\infty} u_k = A$ $(C, 1)$ $\Rightarrow$ $\displaystyle\sum_{k=1}^{\infty} u_k = A$ $\qquad\qquad (C, 2)$

15. Prove that a finite number of zeros can be interpolated among the terms of an infinite series without altering its $(C, 1)$-sum.

16. Establish equation (5).

17. Show that a finite number of parentheses (enclosing two terms), but not an infinite number, may be inserted into a series without altering its $(C, 1)$-sum.

§8. Miscellaneous Methods

There are several tools that are indispensable in advanced studies of infinite series and that can be derived in an elementary fashion. They are introduced here.

8.1 CAUCHY'S INEQUALITY

Theorem 16.

(1) $$\left(\sum_{k=1}^{n} |a_k b_k|\right)^2 \leq \sum_{k=1}^{n} a_k^2 \sum_{k=1}^{n} b_k^2$$

This was proved by Cauchy and is known by his name. It is also sometimes called the *Schwarz inequality* because of its analogy with an integral inequality of that name. Observe that if (1) is established, then it will also be true after the absolute value signs are dropped. To prove it consider the positive quadratic form

$$\sum_{k=1}^{n} (|a_k| x + |b_k| y)^2 = x^2 \sum_{k=1}^{n} a_k^2 + 2xy \sum_{k=1}^{n} |a_k b_k| + y^2 \sum_{k=1}^{n} b_k^2$$

A quadratic form $Ax^2 + Bxy + Cy^2 \geq 0$ for all x, y if, and only if $B^2 - 4AC \leq 0$, $A \geq 0$, $C \geq 0$. The first of these inequalities is (1); the others hold trivially.

EXAMPLE A. $\displaystyle\frac{1}{\sqrt{n}} \sum_{k=1}^{n} \frac{1}{k^{3/4}} \leq \sqrt{\zeta(3/2)}$ $\qquad\qquad n = 1, 2, \ldots$

where $\zeta(p)$ is the function of §1.2. Take $a_k = k^{-3/4}$, $b_k = 1$ in (1). Then

$$\left(\sum_{k=1}^{n} \frac{1}{k^{3/4}}\right)^2 \leq \sum_{k=1}^{n} \frac{1}{k^{3/2}} \sum_{k=1}^{n} 1 < n\zeta(\tfrac{3}{2})$$

It is sometimes convenient to think of a sequence $\{a_k\}^n$ as an n-dimensional vector $\vec{\alpha}$. By analogy with §1.1, Chapter 2, we could define its *length* or *norm* as

$$\|\vec{\alpha}\| = \left(\sum_{k=1}^{n} a_k^2\right)^{1/2}$$

Again we would naturally define the inner product of $\vec{\alpha}$ and $\vec{\beta}$ as

$$\vec{\alpha} \cdot \vec{\beta} = \sum_{k=1}^{n} a_k b_k$$

Then from (1)

$$|\vec{\alpha} \cdot \vec{\beta}| \leq \sum_{k=1}^{n} |a_k b_k| \leq \|\vec{\alpha}\| \, \|\vec{\beta}\|$$

As an immediate consequence (Exercise 8) we could prove the *triangle inequality*

(2) $$\|\vec{\alpha} + \vec{\beta}\| \leq \|\vec{\alpha}\| + \|\vec{\beta}\|$$

8.2 HÖLDER AND MINKOWSKI INEQUALITIES

In the previous section the exponent 2 played a special role. It may be replaced by other exponents at the expense of some simplicity. For $\vec{\alpha}$ we define the *p-norm* as

$$\|\vec{\alpha}\|_p = \left(\sum_{k=1}^{n} |a_k|^p\right)^{1/p} \qquad 0 < p$$

Theorem 17.   1. $p > 1$

2. $\dfrac{1}{p} + \dfrac{1}{q} = 1$

(3) $\Rightarrow$ $$|\vec{\alpha} \cdot \vec{\beta}| \leq \sum_{k=1}^{n} |a_k b_k| \leq \|\vec{\alpha}\|_p \, \|\vec{\beta}\|_q$$

This is *Hölder's inequality*. Notice that (1) is the special case $p = q = 2$. If $a > 0$ and $p > 1$ are given numbers it is an elementary calculus problem to find the maximum of $ax - p^{-1}x^p$ for all $x \geq 0$. The maximum occurs when $x = a^{1/(p-1)}$, so that

$$ax \leq \frac{x^p}{p} + \frac{a^q}{q} \qquad 0 \leq x < \infty$$

In this inequality replace a by $|a_k|$ and x by $|b_k|$, $k = 1, 2, \ldots, n$, and add the resulting inequalities.

(4) $$\sum_{k=1}^{n} |a_k b_k| \leq \frac{1}{p} \sum_{k=1}^{n} |a_k|^p + \frac{1}{q} \sum_{k=1}^{n} |b_k|^q \leq \frac{X^p}{p} + \frac{Y^q}{q}$$

where $X = \|\vec{\alpha}\|_p$, $Y = \|\vec{\beta}\|_q$. If the elements of $\vec{\alpha}$ are multiplied by $\lambda > 0$ and those of $\vec{\beta}$ by $1/\lambda$ then $\vec{\alpha} \cdot \vec{\beta}$ is unchanged and (4) becomes

$$\sum_{k=1}^{n} |a_k b_k| \leq \lambda^p \frac{X^p}{p} + \frac{Y^q}{\lambda^q q}$$

It is again an elementary calculus problem to minimize the right-hand side for all $\lambda > 0$. The minimum occurs when $\lambda = Y^{1/p}/X^{1/q}$, so that

$$\sum_{k=1}^{n} |a_k b_k| \leq \frac{1}{p} \frac{YX^p}{X^{p/q}} + \frac{1}{q} \frac{XY^q}{Y^{q/p}} = XY$$

This is (3).

The triangle inequality (2) stated for the square norm is also valid for the more general norm and is then called *Minkowski's inequality*.

Theorem 18. 1. $p \geq 1$

$\Rightarrow$
$$\left(\sum_{k=1}^{n} |a_k + b_k|^p \right)^{1/p} \leq \left(\sum_{k=1}^{n} |a_k|^p \right)^{1/p} + \left(\sum_{k=1}^{n} |b_k|^p \right)^{1/p}$$

The result is trivial for $p = 1$ using $|a_k + b_k| \leq |a_k| + |b_k|$. Using the same inequality for $p > 1$ we have

$$\sum_{k=1}^{n} |a_k + b_k|^p \leq \sum_{k=1}^{n} |a_k| \, |a_k + b_k|^{p-1} + \sum_{k=1}^{n} |b_k| \, |a_k + b_k|^{p-1}$$

$$\leq \left(\sum_{k=1}^{n} |a_k|^p \right)^{1/p} \left(\sum_{k=1}^{n} |a_k + b_k|^{pq-q} \right)^{1/q} + \left(\sum_{k=1}^{n} |b_k|^p \right)^{1/p} \left(\sum_{k=1}^{n} |a_k + b_k|^{pq-q} \right)^{1/q}$$

Here we have applied Hölder's inequality twice. Dividing both sides by

$$\left(\sum_{k=1}^{n} |a_k + b_k|^{pq-q} \right)^{1/q}$$

we obtain the desired result since $qp - q = p$ and $1 - (1/q) = 1/p$.

EXAMPLE B. $\displaystyle\sum_{k=1}^{\infty} |a_k|^{3/2} < \infty \Rightarrow \sum_{k=1}^{\infty} |a_k| \, k^{-2/5} < \infty$

Take $p = 3/2$, $q = 3$, $b_k = k^{-2/5}$ in (3). Then

$$\sum_{k=1}^{n} |a_k| \, k^{-2/5} \leq \left(\sum_{k=1}^{n} |a_k|^{3/2} \right)^{2/3} \left(\sum_{k=1}^{n} k^{-6/5} \right)^{1/3}$$

The right-hand side remains bounded as $n \to \infty$.

8.3 PARTIAL SUMMATION

An operation for series analogous to integration by parts is known as *partial summation*. Compare §3.1 of Chapter 5.

Theorem 19.   1. $B_n = \sum_{k=1}^{n} b_k$ $k = 1, 2, \ldots$

           2. $1 < M < N$ $M, N = 2, 3, \ldots$

$\Rightarrow$ A. $\sum_{k=M}^{N} a_k b_k = a_N B_N - a_M B_{M-1} - \sum_{k=M}^{N-1} B_k(a_{k+1} - a_k)$

        B. $\sum_{k=1}^{N} a_k b_k = a_N B_N - \sum_{k=1}^{N-1} B_k(a_{k+1} - a_k)$

The similarity of these formulas to

$$\int_M^N a(x)b(x)\, dx = a(x)B(x)\,\Big|_M^N - \int_M^N B(x)a'(x)\, dx$$

$$B(x) = \int b(x)\, dx$$

is apparent. The proof follows by use of the equation $b_k = B_k - B_{k-1}$ as follows:

$$\sum_{k=M}^{N} a_k b_k = a_M(B_M - B_{M-1}) + a_{M+1}(B_{M+1} - B_M) + \ldots + a_N(B_N - B_{N-1})$$

$$= B_M(a_M - a_{M+1}) + B_{M+1}(a_{M+1} - a_{M+2}) + \ldots$$

$$+ B_{N-1}(a_{N-1} - a_N) + a_N B_N - a_M B_{M-1}$$

Here we have regrouped terms and obtained conclusion A. If we were to define $B_0 = 0$ and admit $M = 1$, conclusion B would be a special case of A.

Corollary 19.   1. $|B_n| \le M$ $n = 1, 2, \ldots$

           2. $a_k \ge 0$, $a_k \in \, \downarrow$ $k = 1, 2, \ldots$

$\Rightarrow$ $\left| \sum_{k=1}^{n} a_k b_k \right| \le a_1 M$

For, since $a_k - a_{k+1} \ge 0$, conclusion B gives

$$\left| \sum_{k=1}^{n} a_k b_k \right| \le a_n M + \sum_{k=1}^{n-1} M(a_k - a_{k+1}) = a_1 M$$

EXAMPLE C. If in addition to the hypotheses of Corollary 19 $a_k \to 0$ as $k \to \infty$, then $\sum_{k=1}^{\infty} a_k b_k$ converges. For,

$$\left| \sum_{k=m}^{m+p} a_k b_k \right| \le 2M a_m$$

The conclusion is now easily reached by use of Theorem 7, Chapter 8.

EXERCISES (8)

1. If $\vec{\alpha} = \{2, 1, 2\}$, $\vec{\beta} = \{2, 2, 1\}$, check inequalities (1) and (2).

2. If $\vec{\alpha} = \{2, 2, 1\}$, $\vec{\beta} = \{4, 4, 4\}$, $p = 3$, check Hölder's inequality.

3. Prove Theorem 16 for $n = 3$ by use of Lagrange's identity.

4. If $\vec{\alpha} = \{1, -1, 2, -1\}$, $\vec{\beta} = \{a, -a, 2a, -a\}$, show that equality holds in (1).

5. Same problem for $\vec{\beta} = a\vec{\alpha}$.

6. Prove:
$$\sum_{j=1}^{n} a_j^2 \sum_{k=1}^{n} b_k^2 - \left(\sum_{k=1}^{n} |a_k b_k| \right)^2 = \frac{1}{2} \sum_{j=1}^{n} \sum_{k=1}^{n} (|a_j b_k| - |a_k b_j|)^2$$
and thus obtain a new proof of Cauchy's inequality.

7. By use of Exercise 6 show that equality holds in (1) if, and only if the $|a_k|$ are proportional to the $|b_k|$. You may admit that the elements of one row of a two-row matrix are proportional to the elements of the other row if, and only if the matrix has rank <2.

8. Prove (2). *Hint:* square both sides.

9. Prove: $\displaystyle\sum_{k=1}^{\infty} a_k^4 < \infty \;\Rightarrow\; \sum_{k=1}^{\infty} |a_k|^3 k^{-1/3} < \infty$

10. Give the details of Example C.

11. Show that the conclusion of Example C is false if a_k does not approach zero.

12. Prove: $\displaystyle\sum_{k=1}^{\infty} |a_k| < \infty \;\Rightarrow\; \sum_{k=1}^{\infty} a_k^2 < \infty$
    Is the converse true?

13. Prove: $\displaystyle\sum_{k=1}^{\infty} a_k^2 < \infty \;\Rightarrow\; n^{-1} \sum_{k=1}^{n} \sqrt{k} a_k$ is bounded.

14. Prove: $\displaystyle\sum_{k=1}^{\infty} a_k^2 < \infty, \quad \sum_{k=1}^{\infty} b_k^2 < \infty \;\Rightarrow\; \sum_{k=1}^{\infty} |a_k - b_k|^p < \infty \qquad p \geqq 2.$

15. Prove that if $\displaystyle\sum_{k=1}^{\infty} a_k$ converges then $\displaystyle\sum_{k=1}^{\infty} a_k/k$ converges.

16. Use Example C to show that $\displaystyle\sum_{k=0}^{\infty} (-1)^k k^p$ converges for $p < 0$.

17. Use Example C to prove Theorem 8.

18. If $\displaystyle\sum_{k=1}^{\infty} a_k$ converges, $b_k > 0$, $b_k \in \downarrow$, show that $\displaystyle\sum_{k=1}^{\infty} a_k b_k$ converges.

19. Show that $\displaystyle\sum_{k=1}^{\infty} \frac{(-1)^k}{k} (\pi - \tan^{-1} k)$ converges by use of Exercise 18 and also by Theorem 8.

§9. *Power Series*

The importance of power series is well known through the familiar Taylor series. We introduce here several classical results about such series which facilitate operations with them.

9.1 REGION OF CONVERGENCE

Theorem 20. 1. $\overline{\lim_{k \to \infty}} \sqrt[k]{|a_k|} = \dfrac{1}{R} > 0$

$$(1) \quad \Rightarrow \quad \sum_{k=0}^{\infty} a_k x^k \qquad\qquad \textit{converges absolutely for } |x| < R$$

$$\textit{diverges for } |x| > R$$

There is no conclusion for $|x| = R$.

From the definition of $\overline{\lim}$ we know that for any $\epsilon > 0$ there exists an integer m such that

$$(2) \qquad\qquad \sqrt[k]{|a_k|} < \frac{(1 + \epsilon)}{R} \qquad\qquad k > m$$

$$|a_k x^k| < \frac{(1 + \epsilon)^k}{R^k} |x|^k \qquad\qquad k > m$$

By Theorem 2 and T_1 with $r = (1 + \epsilon)|x|/R$

$$\sum_{k=m+1}^{\infty} |a_k x^k|$$

converges if $(1 + \epsilon)|x| < R$. This proves the first conclusion.

If (1) converged for a fixed number x where $|x| = R(1 + \epsilon) > R$, then its general term would approach zero. Hence for some integer m,

$$|a_k| R^k (1 + \epsilon)^k < 1 \qquad\qquad k > m$$

$$|a_k|^{1/k} < \frac{1}{R(1 + \epsilon)} \qquad\qquad k > m$$

Taking $\overline{\lim}$ of both sides and using hypothesis 1 we have the contradiction $R(1 + \epsilon) \leq R$. This proves the second conclusion.

The series

$$\sum_{k=0}^{\infty} x^k, \qquad \sum_{k=1}^{\infty} \frac{x^k}{k}, \qquad \sum_{k=1}^{\infty} \frac{x^k}{k^2}$$

all have $R = 1$. The first diverges at $x = \pm 1$; the second converges conditionally at $x = -1$, diverges at $x = 1$; the third converges absolutely at $x = \pm 1$. Clearly hypothesis 1 cannot differentiate among these various possibilities. The number R is called the *radius of convergence* of (1).

Corollary 20. $\overline{\lim_{k \to \infty}} \sqrt[k]{|a_k|} = 0 \,(= \infty)$

$\Rightarrow$ (1) *converges absolutely for all x (diverges for $x \neq 0$).*

The proof is left for Exercise 9.

EXAMPLE A. $R = 1$ for $\sum_{k=1}^{\infty} k^2(x + 3)^k$

$$\overline{\lim} \; k^{2/k} = \lim k^{2/k} = 1$$

The series converges absolutely for $-4 < x < -2$ and diverges for $|x + 3| > 1$. This may also be checked in the present case by Theorems 4* and 5*· Contrast Exercises 8.

9.2 UNIFORM CONVERGENCE

Theorem 21.   1. $\overline{\lim_{k \to \infty}} \sqrt[k]{|a_k|} = \dfrac{1}{R} > 0$

2. $0 < S < R$

$\Rightarrow$ $\sum_{k=0}^{\infty} a_k x^k$ *converges uniformly in* $-S \leqq x \leqq S$.

This follows from the Weierstrass M-test taking $M_k = |a_k| S^k$. For then $\sum_{k=0}^{\infty} M_k < \infty$ by Theorem 20.

If $f(x)$ is the sum of series (1), then $f(x) \in C$ in $|x| \leqq S$ by Theorem 12 or in $|x| < R$ since S was arbitrary. Moreover, term by term integration of (1) between points $x = a$ and $x = b$ of $-R < x < R$ is also valid. Also

(3) $f'(x) = \sum_{k=1}^{\infty} k a_k x^{k-1}$ $-R < x < R$

To prove this we show that (3) converges uniformly in $|x| \leqq S$. This will follow from Theorem 21 if we can show that

(4) $\overline{\lim} \sqrt[k]{k|a_k|} = \dfrac{1}{R}$

From (2)

$$\sqrt[k]{k|a_k|} < \frac{(1 + \epsilon)}{R} k^{1/k}   k > m$$

$$\overline{\lim} \sqrt[k]{k|a_k|} \leqq \frac{1 + \epsilon}{R}$$

That is, the left-hand side of (4) $\leq 1/R$. Suppose it were $<1/R$. Then for some $\epsilon > 0$ and a corresponding m

$$\sqrt[k]{k|a_k|} < \frac{1-\epsilon}{R} \qquad\qquad k > m$$

$$\frac{1}{R} = \overline{\lim} \sqrt[k]{|a_k|} \leq \lim \frac{1-\epsilon}{Rk^{1/k}} = \frac{1-\epsilon}{R}$$

This contradiction completes the proof of (4).

It is thus clear that computations with power series are especially felicitous. Differentiations and integrations are always valid inside (strictly) the interval of convergence. In particular $f(x) \in C^\infty$ in $-R < x < R$.

EXAMPLE B. Find the sum of the series of Example A. By differentiating the geometric series we have

$$\frac{1}{(1-x)^2} = \sum_{k=0}^{\infty} kx^{k-1}, \qquad \frac{2}{(1-x)^3} = \sum_{k=2}^{\infty} k^2 x^{k-2} - \sum_{k=2}^{\infty} kx^{k-2}$$

$$\frac{2x^2}{(1-x)^3} = \sum_{k=0}^{\infty} k^2 x^k - \frac{x}{(1-x)^2} \qquad\qquad -1 < x < 1$$

$$\sum_{k=0}^{\infty} k^2 (x+3)^k = -\frac{x^2 + 7x + 12}{(x+2)^3} \qquad\qquad -4 < x < -2$$

9.3 ABEL'S THEOREM

We shall now prove a classical theorem of Abel to the effect that the sum of a power series is always a continuous function throughout the interval of convergence, even including an end point of the interval if the series converges there. Contrast this result with Example C of §5.1. The sum of that series is discontinuous at $x = 1$, an end point of the interval of convergence. But the series is not a power series.

Theorem 22.   1. $f(x) = \sum_{k=0}^{\infty} a_k x^k$ $\qquad\qquad -1 < x < 1$

$\qquad\qquad$ 2. $A = \sum_{k=0}^{\infty} a_k$

$\Rightarrow \qquad\qquad \lim_{x \to 1-} f(x) = A$

By Theorem 12 it will be sufficient to show that (1) converges uniformly in $0 \leq x \leq 1$. That is, for an arbitrary $\epsilon > 0$ we must find an integer m independent of x in $0 \leq x \leq 1$ such that

(5) $\qquad\qquad\qquad |f(x) - S_n(x)| \leq \epsilon \qquad\qquad n > m$

By hypothesis 2 and Cauchy's criterion for convergence we can choose m so that

$$|a_{n+1} + a_{n+2} + \dots + a_{n+p}| \leq \epsilon \qquad n > m, \quad p = 1, 2, \dots$$

Now apply Corollary 19, choosing $M = \epsilon$ and

$$B_p = a_{n+1} + a_{n+2} + \dots + a_{n+p}$$

Here it is the sequence x^k which is ≥ 0 and $\in\downarrow$. We conclude for $0 \leq x \leq 1$ and $n > m$ that

$$S_{n+p}(x) - S_n(x)| = |a_{n+1}x^{n+1} + a_{n+2}x^{n+2} + \dots + a_{n+p}x^{n+p}|$$
$$\leq \epsilon x^{n+1} \leq \epsilon.$$

Allowing p to become infinite, $S_{n+p}(x) \to f(x)$ and we obtain (5). Since m was determined by Cauchy's criterion applied to a convergent sequence of *constants* it is certainly independent of x.

EXAMPLE C. $\log 2 = 1 - \frac{1}{2} + \frac{1}{3} - \frac{1}{4} + \dots$
Taylor's series gives

$$\log (1 + x) = x - \frac{x^2}{2} + \frac{x^3}{3} - \dots \qquad -1 < x < 1$$

and hypothesis 2 holds by Theorem 8.

We have stated Abel's result for a series with unit radius of convergence. It is true for any power series (Exercise 16).

EXERCISES (9)

Find the region of convergence for the series in Exercises 1–6.

1. $\displaystyle\sum_{k=1}^{\infty} k^3(3x)^k$.

2. $\displaystyle\sum_{k=1}^{\infty} \left(1 + \frac{1}{k}\right)^{k^2} x^k$.

3. $\displaystyle\sum_{k=1}^{\infty} k^k(x - 4)^k$.

4. $\displaystyle\sum_{k=1}^{\infty} (kx)^{-k}$.

5. $\displaystyle\sum_{k=1}^{\infty} \frac{(ex)^k}{2^{\sqrt{k}}(1 - x)^k}$.

6. $\displaystyle\sum_{k=1}^{\infty} \frac{2^{\sqrt{k}}x^k}{(e - ex)^k}$.

7. If $a_{2k} = 2$, $a_{2k+1} = 1$, find R. Show that Theorem 4* is not applicable.

8. If $a_{2k} = 2^k$, $a_{2k+1} = 1$, find R. Show that Theorem 5* is not applicable.

9. Prove Corollary 20.

10. If $a_k \to 1$ as $k \to \infty$, show that $R = 1$ for series (1).

11. If $b \neq 0$, $a_k b^k \to c$ as $k \to \infty$, show that $R = |b|$ for series (1).

12. Prove $\overline{\lim} \sqrt[k]{|a_k|} = 1/R \Rightarrow \overline{\lim} \sqrt[k]{|a_k|/(k+1)} = 1/R$.   That is, the radius of convergence of (1) is unchanged by integration.

13. Find the sum of $\sum\limits_{k=0}^{\infty} (k + 3)x^k$.   *Hint:* multiply by x^2 and integrate.

14. Find the sum of $\sum\limits_{k=0}^{\infty} \dfrac{x^{2k+1}}{2k + 1}$.   *Hint:* differentiate.

15. Find the sum of $\sum\limits_{k=1}^{\infty} (k^2 + 2k)x^k$.

16. Restate Abel's theorem for a power series with general R.   Prove.

17. $\lim\limits_{x \to +\infty} \sum\limits_{k=1}^{\infty} \dfrac{1}{k}\left(\dfrac{1 - x}{x}\right)^k = \,?$

18. $\lim\limits_{x \to 4+} \sum\limits_{k=0}^{\infty} \dfrac{1}{2k + 1}\left(\dfrac{x - 5}{x - 3}\right)^k = \,?$

10

Convergence of
Improper Integrals

§1. *Introduction*

In this chapter, we shall discuss definite integrals that are *improper* either by virtue of an infinite limit of integration or on account of a discontinuity of the integrand between the limits of integration. To show why such integrals need special attention consider the integral

$$\int_{-1}^{1} \frac{dx}{x^2}$$

If we try to evaluate this by use of an indefinite integral, as we could do if the integrand were continuous, we obtain

$$-\frac{1}{x}\Big|_{-1}^{1} = -2$$

This is clearly an absurd result, since the integrand is positive. In this first section we shall begin with integrals in which one of the limits of integration is infinite.

323

1.1 CLASSIFICATION OF IMPROPER INTEGRALS

For convenience, let us classify all improper integrals into four types as follows:

TYPE I. $\displaystyle\int_a^\infty f(x)\,dx$; $f(x) \in C$, $a \leqq x < \infty$.

TYPE II. $\displaystyle\int_{-\infty}^b f(x)\,dx$; $f(x) \in C$, $-\infty < x \leqq b$.

TYPE III. $\displaystyle\int_{a+}^b f(x)\,dx$; $f(x) \in C$, $a < x \leqq b$, $\displaystyle\lim_{x \to a+} f(x)$ does not exist.

TYPE IV. $\displaystyle\int_a^{b-} f(x)\,dx$; $f(x) \in C$, $a \leqq x < b$, $\displaystyle\lim_{x \to b-} f(x)$ does not exist.

If a limit of integration $a+$ or $b-$ appears, it will be apparent that the integral is improper. However, the signs $+$, $-$ are not always used, so that the integral must sometimes be recognized as improper by the discontinuities. Besides, the discontinuity of the integrand may occur at an interior point of the interval of integration. For example, the integral

$$\int_{-\infty}^\infty \frac{dx}{x(x-1)}$$

can be considered as the sum of six other integrals corresponding to the intervals $(-\infty, -1)$, $(-1, 0)$, $(0, \frac{1}{2})$, $(\frac{1}{2}, 1)$, $(1, 2)$, $(2, \infty)$. The types are II, IV, III, IV, III, I, respectively.

Note that a "finite jump" in the integrand causes no difficulty (see §7.3 of Chapter 5). Thus if $f(x) \in C$ on $a \leqq x \leqq b$ except that $f(c+) \neq f(c-)$, $a < c < b$, the integral of $f(x)$ from a to b is considered proper.

1.2 TYPE I. CONVERGENCE

The improper integral of Type I resembles in many respects an infinite series. It is interesting to consider the analogies between the results of Chapter 9 and those about to be obtained. The very notation used in the two cases emphasizes the similarities:

$$\int(\ )\,dx \text{ corresponds to } \sum$$

$$\int_a^\infty(\ )\,dx \text{ corresponds to } \sum_{k=m}^\infty$$

$$x \text{ corresponds to } k$$

$$f(x) \text{ corresponds to } u_k$$

$$\int_a^R f(x)\,dx \text{ corresponds to } S_n = \sum_{k=1}^n u_k$$

Since the variables x and R vary continuously, whereas the variables k and n vary through the integers only, some important differences in the two cases may be expected. It is just on this account that the natural analogue of Theorem 1, Chapter 9, is false, as we shall see later.

Let $f(x) \in C$ in the interval $a \leq x < \infty$ and let us define the convergence of the integral

$$(1) \qquad \int_a^\infty f(x)\, dx$$

Definition 1. *The integral* (1) *converges* $\Leftrightarrow \lim\limits_{R\to\infty} \int_a^R f(x)\, dx = A$. *If* $\lim\limits_{R\to\infty} \int_a^R f(x)\, dx = A$, *then A is the value of the integral* (1).

Definition 2. *The integral* (1) *diverges* $\Leftrightarrow$ *it does not converge*.

EXAMPLE A. $\int_1^\infty \dfrac{dx}{x^2}$ converges and has the value 1.   For

$$\lim_{R\to\infty} \int_1^R \frac{dx}{x^2} = \lim_{R\to\infty} \left(1 - \frac{1}{R}\right) = 1$$

EXAMPLE B. $\int_0^\infty \sin x\, dx$ diverges.   For $\lim\limits_{R\to\infty}(1 - \cos R)$ does not exist.

Test integral T_1: $\displaystyle\int_0^\infty e^{-rx}\, dx = \frac{1}{r}$

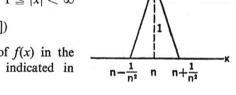

Fig. 35.

Test integral T_2: $\displaystyle\int_1^\infty \frac{1}{x^p}\, dx = \frac{1}{p - 1}$

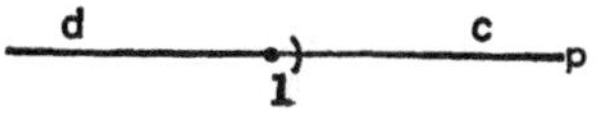

Fig. 36.

The analogue of Theorem 1, Chapter 9 would be that when the integral (1) converges, $\lim\limits_{x\to\infty} f(x) = 0$. The following example will show this false. Set

$$g(x) = 1 - |x| \qquad 0 \leq |x| \leq 1$$
$$= 0 \qquad\quad 1 \leq |x| < \infty$$

$$f(x) = \sum_{k=2}^\infty g(k^2[x - k])$$

It is easy to see that the graph of $f(x)$ in the neighborhood of $x = n$ is as indicated in Figure 37. It is clear that

$$\int_0^\infty f(x)\, dx = \sum_{k=2}^\infty \frac{1}{k^2} = \zeta(2) - 1$$

Fig. 37.

Yet $f(x)$ does not tend to zero since $f(n) = 1$ for $n = 2, 3, \ldots$.

1.3 COMPARISON TESTS

Theorems 1, 2 of this paragraph correspond, respectively, to Theorems 2, 3 of Chapter 9.

Theorem 1.   1. $f(x), g(x) \in C$ $a \leqq x < \infty$

             2. $0 \leqq f(x) \leqq g(x)$ $a \leqq x < \infty$

             3. $\displaystyle\int_a^\infty g(x)\, dx < \infty$

$\Rightarrow$ $\displaystyle\int_a^\infty f(x)\, dx < \infty$

As in the case of series, the symbols $< \infty$ and $= \infty$ may be used for *converges* and *diverges*, respectively, only when the integrand is nonnegative. If B is the value of the integral in hypothesis 3, it is clear that

$$F(R) = \int_a^R f(x)\, dx \leqq \int_a^R g(x)\, dx \leqq B$$

Since $F(R) \in \uparrow$, we see that

$$\lim_{R \to \infty} F(R) = A \leqq B$$

and the result is proved.

Theorem 2.   1. $f(x), g(x) \in C$ $a \leqq x < \infty$

             2. $0 \leqq g(x) \leqq f(x)$ $a \leqq x < \infty$

             3. $\displaystyle\int_a^\infty g(x)\, dx = \infty$

$\Rightarrow$ $\displaystyle\int_a^\infty f(x)\, dx = \infty$

For, if the latter integral converged, we could use Theorem 1 to show that the integral in hypothesis 3 converged.

EXAMPLE C. $\displaystyle\int_2^\infty \frac{x^2\, dx}{\sqrt{x^7 + 1}} < \infty$

For,

$$0 < \frac{x^2}{\sqrt{x^7 + 1}} < \frac{1}{x^{3/2}} \qquad\qquad 2 \leqq x < \infty$$

By T_2 with $p = \frac{3}{2} > 1$, we have our result.

EXAMPLE D. $\displaystyle\int_{2}^{\infty} \frac{x^3\,dx}{\sqrt{x^7 + 1}} = \infty$

For

$$\frac{x^3}{\sqrt{x^7 + 1}} = \frac{1}{\sqrt{x}\sqrt{1 + x^{-7}}}$$

$$\geq \frac{1}{\sqrt{x}\,\sqrt{1 + 2^{-7}}} \qquad 2 \leq x < \infty$$

By T_2' with $p = \frac{1}{2} < 1$, we have our result.

1.4 ABSOLUTE CONVERGENCE

Theorem 3 of this paragraph corresponds to Theorem 7 of Chapter 9.

Definition 3.   *Integral* (1) *converges absolutely* $\Leftrightarrow \displaystyle\int_{a}^{\infty} |f(x)|\,dx$ *converges.*

Definition 4.   *Integral* (1) *converges conditionally* $\Leftrightarrow$ *it converges, but not absolutely.*

EXAMPLE E. $\displaystyle\int_{1}^{\infty} \frac{\sin x}{x^2}\,dx$ converges absolutely.

For,

$$\frac{|\sin x|}{x^2} \leq \frac{1}{x^2}$$

Hence, by Theorem 1 and test-integral T_2 with $p = 2$

$$\int_{1}^{\infty} \frac{|\sin x|}{x^2}\,dx < \infty$$

EXAMPLE F. $\displaystyle\int_{0}^{\infty} \frac{\sin x}{x}\,dx$ converges conditionally.

We defer the proof of convergence.   We show here that

(2) $$\int_{0}^{\infty} \frac{|\sin x|}{x}\,dx = \infty$$

In the interval $k\pi \leq x \leq (k + 1)\pi$, $k = 0, 1, 2, \ldots$, we have

$$\frac{|\sin x|}{x} \geq \frac{|\sin x|}{(k + 1)\pi}$$

Hence,

$$\int_{k\pi}^{(k+1)\pi} \frac{|\sin x|}{x}\,dx \geq \frac{1}{(k + 1)\pi} \int_{k\pi}^{(k+1)\pi} |\sin x|\,dx$$

$$= \frac{2}{(k + 1)\pi}$$

If $n\pi \leq R < (n+1)\pi$,

(3)
$$\int_0^R \frac{|\sin x|}{x}\, dx \geq \frac{2}{\pi} \sum_{k=0}^{n-1} \frac{1}{k+1}$$

As R becomes infinite, so does n; and so does the right-hand side of inequality (3). This proves (2).

Theorem 3.   1. $f(x) \in C$ $a \leq x < \infty$

2. $\displaystyle\int_a^\infty |f(x)|\, dx < \infty$

$\Rightarrow$ $\displaystyle\int_a^\infty f(x)\, dx$ converges.

Since

$$0 \leq |f(x)| - f(x) \leq 2|f(x)| \qquad a \leq x < \infty$$

we have by Theorem 1 that the integral

$$\int_a^\infty \{|f(x)| - f(x)\}\, dx$$

converges. If we subtract it from the convergent integral of hypothesis 2, we get the convergent integral (1), thus completing the proof.

EXERCISES (1)

Test the integrals of Exercises 1–6 for convergence.

1. $\displaystyle\int_0^\infty \frac{x}{x^2 + 1}\, dx.$

2. $\displaystyle\int_{-7}^\infty \frac{x^2 - 1}{x^2 + 1}\, dx.$

3. $\displaystyle\int_2^\infty \frac{x^2 + 1}{x^4 - 9}\, dx.$

4. $\displaystyle\int_1^\infty \sin x^{-2}\, dx.$

5. $\displaystyle\int_2^\infty \frac{\cos x}{x(\log x)^2}\, dx.$

6. $\displaystyle\int_1^\infty (\log x)e^{-x}\, dx.$

Which of the integrals in Exercises 7–9 converges absolutely?

7. $\displaystyle\int_0^\infty \frac{2\cos^2 x - 3\sin x + 1}{2x^3 + x + 1}\, dx.$

8. $\displaystyle\int_1^\infty \left(\frac{\cos \pi x}{x}\right)^{1/3} dx.$

9. $\displaystyle\int_0^\infty x^2 2^{-x} \sin (2x)\, dx.$

10. Find the area under the curve $y = 1/x$ from 1 to ∞, and the volume of revolution obtained by rotating this area about the x-axis.

11. Prove:

    1. $f(x), g(x) \in C$ $a \leqq x < \infty$

    2. $\lim\limits_{x \to \infty} \dfrac{f(x)}{g(x)} = A$

    3. $\displaystyle\int_a^\infty |g(x)| \, dx < \infty$

    $\Rightarrow$ $\displaystyle\int_a^\infty f(x) \, dx$ converges absolutely.

12. If $g(x) = 1$ $-1 \leqq x \leqq 1$

             $= 0$ $|x| > 1$

$$f(x) = \sum_{k=2}^\infty g(k^2[x - k])$$

    $f(x)$ is discontinuous. Find the value of $\displaystyle\int_0^\infty f(x) \, dx$.

13. Solve the same problem for $\displaystyle\int_0^\infty f^2(x) \, dx$.

14. If $g(x)$ is defined as in Exercise 12 and if

$$f(x) = \sum_{k=1}^\infty g([2k + 1][x - 2k - 1])$$

$$F(x) = \sum_{k=1}^\infty g(2k[x - 2k])$$

    show

$$\int_0^\infty f(x) \, dx = \int_0^\infty F(x) \, dx = \infty$$

    Does $\displaystyle\int_0^\infty [f(x) - F(x)] \, dx$

    converge?

15. Prove:

    1. $f(x) \in C, \downarrow$ $a \leqq x < \infty$

    2. $\displaystyle\int_a^\infty f(x) \, dx$ converges

    $\Rightarrow$ $\lim\limits_{x \to \infty} f(x) = 0$

16. Can an improper integral ever be transformed into a proper integral by a change of variable?

§2. *Type I.   Limit Tests*

In this section we prove two useful limit tests analogous to those of Theorems 9 and 10, Chapter 9, for series.

2.1 LIMIT TEST FOR CONVERGENCE

Theorem 4.   1. $f(x) \in C$ $\qquad\qquad\qquad\qquad a \leqq x < \infty$

2. $\lim\limits_{x \to \infty} x^p f(x) = A$ $\qquad\qquad\qquad\qquad p > 1$

(1) $\Rightarrow$ $\qquad \displaystyle\int_a^\infty |f(x)|\, dx < \infty$

For, hypothesis 2 implies
$$\lim_{x \to \infty} x^p |f(x)| = |A|$$

Hence, there exists a number b such that
$$x^p |f(x)| \leqq |A| + 1 \qquad\qquad\qquad b \leqq x < \infty$$

Now Theorem 1 and T_2 with $p > 1$ give
$$\int_b^\infty |f(x)|\, dx < \infty$$

whence the result (1) follows.

2.2 LIMIT TEST FOR DIVERGENCE

Theorem 5.   1. $f(x) \in C$ $\qquad\qquad\qquad\qquad a \leqq x < \infty$

2. $\lim\limits_{x \to \infty} x f(x) = A \neq 0$ $\qquad$ (or $= \pm\infty$)

$\Rightarrow$ $\qquad \displaystyle\int_a^\infty f(x)\, dx$ *diverges.*

The test fails if $A = 0$.

CASE I. $A > 0$ (or $A = +\infty$).   Then a number b exists such that

(2) $\qquad\qquad\qquad\qquad\qquad x f(x) > \dfrac{A}{2} \qquad\qquad\qquad b \leqq x < \infty$

(If $A = \pm\infty$, the right-hand side of (2) may be taken as any number, in particular 1.)   Now by use of Theorem 2 and T_2, $p = 1$, we obtain
$$\int_b^\infty f(x)\, dx = +\infty$$

whence the desired conclusion follows.

CASE II. $A < 0$ (or $A = -\infty$). In this case the integral

$$\int_a^\infty [-f(x)]\, dx$$

may be treated by Case I.

To show that the test fails if $A = 0$, we exhibit two integrals:

$$\int_1^\infty \frac{dx}{x^2} < \infty, \qquad \int_2^\infty \frac{dx}{x \log x} = \infty$$

In each case $A = 0$.

EXAMPLE A. $\displaystyle\int_0^\infty e^{-x^2}\, dx < \infty$

For, taking $p = 2 > 1$,

$$\lim_{x \to \infty} x^2 f(x) = \lim_{x \to \infty} x^2 e^{-x^2} = 0$$

EXAMPLE B. $\displaystyle\int_0^\infty \frac{\cos x}{\sqrt{1 + x^3}}\, dx$ converges absolutely.

Here we cannot take $p = \frac{3}{2}$. Any smaller value of $p > 1$ will suffice:

$$\lim_{x \to \infty} x^{5/4} f(x) = \lim_{x \to \infty} \frac{\cos x}{x^{1/4}(1 + x^{-3})^{1/2}} = 0$$

EXAMPLE C. $\displaystyle\int_0^\infty \frac{dx}{\sqrt{1 + 2x^2}} = +\infty$

For,

$$\lim_{x \to \infty} x f(x) = \lim_{x \to \infty} \frac{x}{\sqrt{1 + 2x^2}} = \frac{1}{\sqrt{2}} \neq 0$$

EXAMPLE D. $\displaystyle\int_0^\infty \frac{7e^{-x} - 1}{\sqrt[3]{1 + 2x^2}}\, dx = -\infty$

For,

$$\lim_{x \to \infty} x f(x) = \lim_{x \to \infty} \frac{(7e^{-x} - 1)x^{1/3}}{\sqrt[3]{2 + x^{-2}}} = -\infty$$

EXAMPLE E. $\displaystyle\int_{1/2}^\infty \frac{\log x}{\sqrt[p]{1 + x^3}}\, dx$

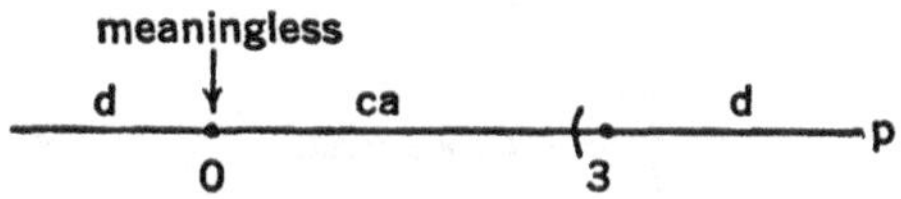

Fig. 38.

The diagram means that the integral converges absolutely for $0 < p < 3$, diverges elsewhere except at $p = 0$, where the integral has no meaning. In case $0 < p < 3$, choose q so that $p < q < 3$. Then

$$\lim_{x \to \infty} x^{3/q} f(x) = \lim_{x \to \infty} \frac{\log x}{x^{3/p - 3/q} \sqrt[q]{1 + x^{-3}}} = 0$$

Since $3/q > 1$, we may use Theorem 4 to establish absolute convergence. If $p < 0$ or if $p \geq 3$,

$$\lim_{x \to \infty} x f(x) = +\infty$$

so that the given integral diverges.

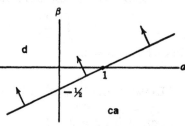

Fig. 39.

EXAMPLE F. $\displaystyle\int_{1}^{\infty} \frac{(e^{1/x} - 1)^{\alpha}}{[\log (1 + x^{-1})]^{2\beta}} \, dx$

The diagram means that the integral diverges for $\alpha - 2\beta \leq 1$, converges absolutely elsewhere. To see this, write the integrand as follows:

$$f(x) = \left(\frac{e^{1/x} - 1}{1/x}\right)^{\alpha} \left(\frac{\log (1 + x^{-1})}{1/x}\right)^{-2\beta} x^{2\beta - \alpha}$$

Since the first two factors tend to unity as x becomes infinite, it is easy to evaluate

$$\lim_{x \to \infty} x^{p} f(x)$$

for $p \geq 1$.

EXERCISES (2)

Test the integrals of Exercises 1–10 for convergence by use of the limit tests.

1. Exercises 1, 2, 3 of §1.

2. Exercises 4, 7, 9 of §1.

3. $\displaystyle\int_{1}^{\infty} t^{x-1} e^{-t} \, dt.$

4. $\displaystyle\int_{1}^{\infty} t^{x} e^{-t} (\log t)^{k} \, dt.$ $k = 1, 2, \ldots$

5. $\displaystyle\int_{2}^{\infty} x(\log x)^{\beta} \, dx.$

6. $\displaystyle\int_{2}^{\infty} x(\log x)^{\beta} e^{-x} \, dx.$

7. $\displaystyle\int_{1}^{\infty} x\left(1 - \cos \frac{1}{x}\right)^{\beta} \, dx.$

8. $\displaystyle\int_{0}^{\infty} e^{-\nu t} \log (1 + e^{-tx^2}) \, dt.$

9. $\displaystyle\int_{0}^{\infty} e^{-\nu t} \log (1 + e^{xt}) \, dt.$

10. $\displaystyle\int_{1}^{\infty} t^{-\nu}(1 + t) e^{xt} \, dt.$

11. Prove or disprove:

   1. $f(x) \in C$

   2. $|x f(x)| > 1, \quad x > 1$

$$\Rightarrow \int_1^\infty f(x) \, dx \text{ diverges.}$$

12. Prove:

   1. $f(x) \in C, \quad a \leqq x < \infty$

   2. $\lim_{x \to \infty} x \, (\log x)^p f(x) = A, \quad p > 1$

$$\Rightarrow \int_a^\infty |f(x)| \, dx < \infty$$

13. Prove:

   1. $f(x) \in C, \quad a \leqq x < \infty$

   2. $\lim_{x \to \infty} x(\log x) f(x) = A \neq 0 \text{ (or } \pm\infty)$

$$\Rightarrow \int_a^\infty f(x) \, dx \text{ diverges.}$$

14. Devise two examples to show that the test of Exercise 13 fails if $A = 0$.

15. Devise two examples which can be tested by use of Exercises 12 and 13 but not by Theorems 4 and 5.

16. In Example E find the limit of the integral as $p \to 0+$. *Hint:* If $0 < p < \dfrac{1}{2}$,

$$\frac{1}{(1 + x^3)^{1/p}} \leqq \frac{1}{(1 + 2^{-3})^{(1/p) -1}} \frac{1}{(1 + x^3)} \text{ in the interval } \tfrac{1}{2} \leqq x < \infty.$$

§3. *Type I. Conditional Convergence*

In this section we develop a result analogous to Leibniz's theorem concerning the convergence of alternating series. In the present case a trigonometric factor, such as sin x or cos x, in the integrand takes the place of the factor $(-1)^k$ in the general term of the alternating series.

3.1 INTEGRAND WITH OSCILLATING SIGN

   Theorem 6.   1. $g(x) \in C$ $a \leqq x < \infty$

                     2. $g(x) \in \downarrow$ $a \leqq x < \infty$

                     3. $\lim_{x \to \infty} g(x) = 0$

(1) $\Rightarrow$ $\displaystyle\int_a^\infty g(x) \sin x \, dx \ \text{converges}$

We observe first that since $g(x)$ is nonincreasing and approaches zero it is necessarily ≥ 0. Let $a < m\pi < n\pi < R \leq (n + 1)\pi$, where m and n are integers. Then

$$(2) \qquad \int_a^R g(x) \sin x \, dx = \int_a^{m\pi} g(x) \sin x \, dx + \sum_{k=m}^{n-1} \int_{k\pi}^{(k+1)\pi} g(x) \sin x \, dx$$

$$+ \int_{n\pi}^R g(x) \sin x \, dx$$

Keeping m fixed, we let R, and hence n, become infinite. Since

$$\left| \int_{n\pi}^R g(x) \sin x \, dx \right| \leq g(n\pi) \int_{n\pi}^{(n+1)\pi} |\sin x| \, dx = 2g(n\pi)$$

it is clear by hypothesis 3 that the last term on the right-hand side of the equation (2) approaches zero. To prove our result, it will consequently be sufficient to prove that the second term on the right-hand side approaches a limit, or that the series

$$\sum_{k=m}^{\infty} \int_{k\pi}^{(k+1)\pi} g(x) \sin x \, dx$$

converges. But we shall show that this series satisfies all conditions of Theorem 8, Chapter 9. Since $\sin x$ does not change sign for $k\pi \leq x \leq (k + 1)\pi$,

$$v_k = \left| \int_{k\pi}^{(k+1)\pi} g(x) \sin x \, dx \right| = \int_{k\pi}^{(k+1)\pi} g(x) |\sin x| \, dx$$

Because $g(x) \in \downarrow$ we have

$$g(k\pi + \pi) \int_{k\pi}^{(k+1)\pi} |\sin x| \, dx \leq v_k \leq g(k\pi) \int_{k\pi}^{(k+1)\pi} |\sin x| \, dx$$

$$(3) \qquad\qquad 2g(k\pi) \leq v_{k-1} \leq 2g(k\pi - \pi)$$

Combining these inequalities, we see that

$$0 \leq v_k \leq v_{k-1} \leq 2g(k\pi - \pi)$$

Hence, $v_k \in \downarrow$ and $\lim_{k \to \infty} v_k = 0$. This completes the proof.

This theorem enables us to exhibit integrals which are conditionally convergent. Example F of §1.4 is a case in point. We have already shown that that integral does not converge absolutely. Since $1/x \notin C$ in $0 \leq x < \infty$, we break the integral into two parts:

$$\int_0^\infty \frac{\sin x}{x} \, dx = \int_0^1 \frac{\sin x}{x} \, dx + \int_1^\infty \frac{\sin x}{x} \, dx$$

In spite of the discontinuity in $1/x$ the first integral on the right is not improper since the integrand approaches 1 as $x \to 0+$. The second integral on the right may be tested by Theorem 6 with $a = 1$, $g(x) = x^{-1}$.

3.2 SUFFICIENT CONDITIONS FOR CONDITIONAL CONVERGENCE

By the addition of a further hypothesis to those of Theorem 6 we can be sure that the integral (1) converges conditionally.

Corollary 6.1.   *Hypotheses* 1, 2, 3 *of Theorem* 6

$$4. \quad \int_a^\infty g(x)\, dx = \infty \qquad (<\infty)$$

$$\Rightarrow \qquad \int_a^\infty g(x)\, |\sin x|\, dx = \infty \qquad (<\infty)$$

Replace $\sin x$ by $|\sin x|$ in equation (2). The last term on the right still approaches zero with R^{-1}. Since $g(x) \in \downarrow$, inequality (3) gives

$$v_{k-1} \geq 2g(k\pi) \geq \frac{2}{\pi} \int_{k\pi}^{(k+1)\pi} g(x)\, dx$$

$$\sum_{k=m+1}^{n} v_{k-1} \geq \frac{2}{\pi} \int_{(m+1)\pi}^{(n+1)\pi} g(x)\, dx \to \infty \qquad (R \to \infty)$$

This proves the result with the assumption $=\infty$ in hypothesis 4. With the alternative assumption the absolute convergence of (1) follows, without the intervention of series, by Theorem 1 since $|\sin x| \leq 1$.

The conditional convergence of Example F, §1.4, is an immediate result of Corollary 6.1.

Corollary 6.2.   *Hypotheses* 1, 2, 3

$$\Rightarrow \qquad \int_a^\infty g(x) \sin (\alpha x + \beta)\, dx$$

$$\int_a^\infty g(x) \cos (\alpha x + \beta)\, dx \qquad\qquad \alpha \neq 0$$

converge.

The proof is made by changing the variable.

We may also obtain a result analogous to Corollary 8, Chapter 9.

Corollary 6.3.   *Hypotheses* 1, 2, 3

4. *n an integer* $> a/\pi$

$$\Rightarrow \qquad \left| \int_{n\pi}^\infty g(x) \sin x\, dx \right| \leq 2g(n\pi)$$

For,
$$\int_{n\pi}^{\infty} g(x) \sin x \, dx = \pm \sum_{k=n}^{\infty} (-1)^k v_k$$
Hence, we may apply Corollary 8, Chapter 9, to the series on the right and make use of inequality (3).

EXAMPLE A. $\int_0^{\infty} \sin x^2 \, dx$ converges conditionally.   Set $x^2 = t$,
$$\frac{1}{2} \int_0^{\infty} \frac{\sin t}{\sqrt{t}} \, dt$$
Apply Theorem 6 and Corollary 6.1 with $a = 1$, $g(t) = t^{-1/2}$.   Then $g(t) \in \downarrow$, $g(t) \to 0$, and
$$\int_1^{\infty} g(t) \, dt = \infty$$

EXAMPLE B. $\left| \frac{\pi}{2} - \int_0^{n\pi} \frac{\sin x}{x} \, dx \right| = \left| \int_{n\pi}^{\infty} \frac{\sin x}{x} \, dx \right| \leqq \frac{2}{n\pi}.$
We shall show later that
$$\frac{\pi}{2} = \int_0^{\infty} \frac{\sin x}{x} \, dx$$
This admitted, the result follows from Corollary 6.3.

EXERCISES (3)

Test the integrals of Exercises 1–9 for absolute and conditional convergence.

1. $\int_0^{\infty} \frac{\sin x}{\sqrt[3]{x^2 + x + 1}} \, dx.$

2. $\int_1^{\infty} \frac{\cos(1 - 2x)}{\sqrt{x} \sqrt[3]{x^2 + 1}} \, dx.$

3. $\int_1^{\infty} \frac{\sin 2x}{\sqrt{x} \log(x + 1)} \, dx.$

4. $\int_2^{\infty} \frac{\sin x}{x \log x} \, dx.$

5. $\int_0^{\infty} \frac{e^{-x} - 1}{e^{-x} + 1} \sin x \, dx.$

6. $\int_0^{\infty} \frac{\sin x - \cos x}{1 + e^{-x}} \, dx.$

7. $\int_3^{\infty} \frac{\sin x}{x \log x (\log \log x)^p} \, dx.$

8. $\int_2^{\infty} \frac{\log \log x}{\log x} \cos 2x \, dx.$

9. $\int_0^{\infty} \cos x^2 \, dx.$

10. Discuss the integral
$$\int_a^{\infty} g(x) \cos x \, dx$$
directly without reducing to the integral (1). State and prove results analogous to Theorem 6, Corollaries 6.1, 6.3.

11. Illustrate Exercise 10 by taking $g(x) = (x + 1)^{-1/2}$, $a = 0$.

12. Show

$$\left| \int_0^R \frac{\sin x}{x} \, dx \right| \leqq \pi \qquad\qquad 0 \leqq R < \infty$$

Hint: use the second mean-value theorem on $(\pi/2, R)$ if $R > \pi/2$.

13. Find a finite bound for the integral

$$\left| \int_R^S \frac{\sin x}{x} \, dx \right|$$

that will hold for all positive numbers R and S.

14. Show that neither hypothesis 2 nor hypothesis 3 may be omitted in Theorem 6. *Hint:* Take $g(x) = (\sin x)/x$.

15. Test for conditional convergence, absolute convergence, or divergence

$$\int_1^\infty \frac{\sin x}{\sqrt{x} + \sin x} \, dx, \qquad \int_1^\infty \frac{\sin x}{x + \sin x} \, dx$$

16. Use the second mean-value theorem to prove Theorem 6. Compare Example A, §4 of Chapter 5.

17. Prove Theorem 6 if $\sin x$ is replaced by $s(x)$, where

$$\left| \int_a^b s(x) \, dx \right| < M \qquad\qquad \text{all } a \text{ and } b$$

§4. *Type III*

In integrals of Type III the integrand is continuous at every point of the interval of integration (a, b) except at the left-hand end point. If the limit of the integrand exists as the variable approaches a, we call the integral *proper* even though there is a discontinuity at a. For example,

$$\int_0^1 \frac{\sin x}{x} \, dx$$

is proper even though the integrand is not defined at $x = 0$. It could be defined as 1 at $x = 0$ so as to make the integrand continuous. Such discontinuities are called *removable*. They have no effect on the behavior of an integral. Integrals of Type III could be treated by reducing them to Type I by a change of variable. We prefer to treat them directly.

4.1 CONVERGENCE

Let $f(x) \in C$ in the interval $a < x \leq b$ and let the limit

$$(1) \qquad\qquad \lim_{x \to a+} f(x)$$

fail to exist. We consider the improper integral

$$(2) \qquad\qquad \int_{a+}^{b} f(x) \, dx$$

Definition 1*. *The integral* (2) *converges* $\Leftrightarrow \lim_{\epsilon \to 0+} \int_{a+\epsilon}^{b} f(x) \, dx = A.$
If $\lim_{\epsilon \to 0+} \int_{a+\epsilon}^{b} f(x) \, dx = A,$ *then* A *is the value of the integral* (2).

Definition 2*. *The integral* (2) *diverges* $\Leftrightarrow$ *it does not converge.*

Test integral T_2^*: $\displaystyle\int_{a+}^{b} \frac{dx}{(x-a)^p}$

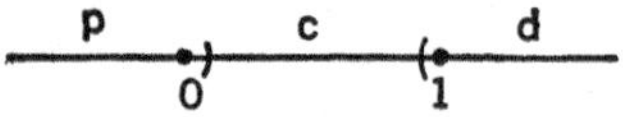

Fig. 40.

The diagram indicates that the integral converges in $0 < p < 1$, diverges in $1 \leq p < \infty$, and is proper in $-\infty < p \leq 0$. The value of the integral for $-\infty < p < 1$ is

$$\int_{a+}^{b} \frac{dx}{(x-a)^p} = \lim_{\epsilon \to 0} \left[\frac{(b-a)^{-p+1}}{1-p} - \frac{\epsilon^{-p+1}}{1-p} \right] = \frac{(b-a)^{1-p}}{1-p}$$

4.2 COMPARISON TESTS

The tests for convergence of integrals of Type III are very similar to those of Type I. We number the theorems so as to emphasize the analogy. We assume throughout that the limit (1) does not exist. The theorems would be true without this assumption, but the integrals would not be improper.

Theorem 1*.   1. $f(x), g(x) \in C$ $\qquad\qquad\qquad\qquad a < x \leq b$

2. $0 \leq f(x) \leq g(x)$ $\qquad\qquad\qquad\qquad a < x \leq b$

3. $\displaystyle\int_{a+}^{b} g(x) \, dx < \infty$

$\Rightarrow \qquad \displaystyle\int_{a+}^{b} f(x) \, dx < \infty$

For, if $\epsilon > 0$,

$$\int_{a+\epsilon}^{b} f(x) \, dx \leq \int_{a+\epsilon}^{b} g(x) \, dx \leq \int_{a+}^{b} g(x) \, dx$$

As $\epsilon \to 0+$ the integral on the left increases, but remains bounded. Consequently, it approaches a limit.

Let us give an alternative proof by use of Theorem 1. We make the change of variable

$$x = a + t^{-1}$$

It is then clear that the integral (2) converges $\Leftrightarrow$ the integral

$$\int_{(b-a)^{-1}}^{\infty} f(a + t^{-1})t^{-2}\, dt$$

converges. We now use Theorem 1, noting that

$$0 \leq f(a + t^{-1})t^{-2} \leq g(a + t^{-1})t^{-2} \qquad 0 < (b - a)^{-1} \leq t < \infty$$

Theorem 2*.   1. $f(x), g(x) \in C$ $\qquad\qquad\qquad\qquad a < x \leq b$

2. $0 \leq g(x) \leq f(x)$ $\qquad\qquad\qquad\qquad\qquad a < x \leq b$

3. $\displaystyle\int_{a+}^{b} g(x)\, dx = \infty$

$\Rightarrow$ $\qquad\displaystyle\int_{a+}^{b} f(x)\, dx = \infty$

The proof is similar to that of Theorem 2 and is omitted.

4.3 ABSOLUTE CONVERGENCE

The definitions for absolute and conditional convergence are obtained from Definitions 3 and 4 by changing the limits of integration.

Theorem 3*.   1. $f(x) \in C$ $\qquad\qquad\qquad\qquad\qquad\qquad a < x \leq b$

2. $\displaystyle\int_{a+}^{b} |f(x)|\, dx < \infty$

$\Rightarrow$ $\qquad\displaystyle\int_{a+}^{b} f(x)\, dx$ converges.

The proof is the same as that of Theorem 3 except for a change in the limits of integration.

4.4 LIMIT TESTS

Theorem 4*.   1. $f(x) \in C$ $\qquad\qquad\qquad\qquad\qquad\qquad a < x \leq b$

2. $\displaystyle\lim_{x \to a+} (x - a)^p f(x) = A$ $\qquad\qquad 0 < p < 1$

$\Rightarrow$ $\qquad\displaystyle\int_{a+}^{b} |f(x)|\, dx < \infty$

For, hypothesis 2 implies the existence of a number c such that

$$(x - a)^p |f(x)| \leqq |A| + 1 \qquad a < x \leqq c < b$$

Then by Theorem 1* and test integral T_2^* with $0 < p < 1$, we see that

$$\int_{a+}^{c} |f(x)|\, dx < \infty$$

It follows that the integral (2) converges absolutely.

Theorem 5*.   1. $f(x) \in C$ $a < x \leqq b$

                2. $\lim\limits_{x \to a+} (x - a)f(x) = A \neq 0 \quad (or \pm \infty)$

$\Rightarrow \qquad \int_{a+}^{b} f(x)\, dx$ *diverges.*

The test fails if $A = 0$.

The proof is like that of Theorem 5 and is omitted. To show that the test fails when $A = 0$, we may exhibit the two integrals

$$\int_{0+}^{1} \frac{dx}{\sqrt{x}} < \infty, \qquad \int_{0+}^{1/2} \frac{dx}{x \log (1/x)} = \infty$$

EXAMPLE A. $\int_{0+}^{1/2} \left(\log \frac{1}{x} \right)^{\alpha} dx$

Fig. 41.

Apply Theorem 4* with $p = \frac{1}{2}$ when $\alpha > 0$:

$$\lim_{x \to 0+} \sqrt{x}\, f(x) = 0$$

If $\alpha \leqq 0$, the integrand approaches a limit when $x \to 0+$, so that the integral is not improper.

EXAMPLE B. $\int_{0+}^{1} t^{x-1} e^{-t}\, dt$

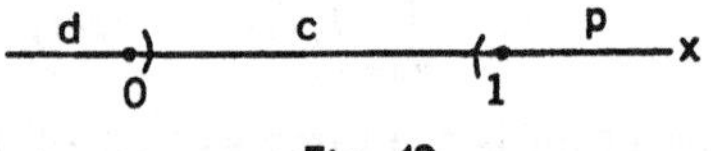

Fig. 42.

We have

$$\lim_{t \to 0+} f(t) = 0 \qquad\qquad x > 1$$

$$= 1 \qquad\qquad x = 1$$

$$\lim_{t \to 0+} t^{1-x} f(t) = 1 \qquad\qquad 0 < x < 1$$

$$\lim_{t \to 0+} t f(t) = 1 \neq 0 \qquad\qquad x = 0$$

$$= +\infty \qquad\qquad x < 0$$

4.5 OSCILLATING INTEGRANDS

Theorem 6*. 1. $g(x) \in C$ $\hspace{3cm} a < x \leqq b$

2. $g(x)(x - a)^2 \in \uparrow$ $\hspace{2cm} a < x \leqq b$

3. $\lim_{x \to a+} g(x)(x - a)^2 = 0$

(3) $\Rightarrow$ $\displaystyle\int_{a+}^{b} g(x) \sin \frac{1}{x - a}\, dx$ converges.

This could be proved directly, but we shall reduce the integral to one of Type I and apply Theorem 6. As we saw in §4.2, the integral (3) converges $\Leftrightarrow$

(4) $\displaystyle\int_{(b-a)^{-1}}^{\infty} g(a + t^{-1})t^{-2} \sin t\, dt$

converges. But

$$\lim_{t \to +\infty} g(a + t^{-1})t^{-2} = \lim_{x \to a+} g(x)(x - a)^2 = 0$$

Also, under the transformation $x = a + t^{-1}$ the variable t decreases when x increases. Hence, hypothesis 2 is equivalent to

$$g(a + t^{-1})t^{-2} \in \downarrow \hspace{2cm} b - a \leqq t < \infty$$

Consequently, we are in a position to apply Theorem 6 to show that the integral (4) converges.

EXAMPLE C. $\displaystyle\int_{0+}^{1} \frac{\sin 1/x}{x^{3/2}}\, dx$ converges conditionally.

Take $g(x) = x^{-3/2}$ in Theorem 6*. Since

$$g(x)x^2 = \sqrt{x} \in \uparrow$$
$$\lim_{x \to 0+} g(x)x^2 = 0$$

the convergence of the integral follows. To see that the convergence is conditional, make the change of variable $x = t^{-1}$ and apply Corollary 6.1 to the integral

$$\int_{1}^{\infty} \frac{\sin t}{\sqrt{t}}\, dt$$

E X E R C I S E S (4)

Test for convergence the integrals of Exercises 1–8.

1. $\displaystyle\int_{0+}^{1} \frac{\log x}{\sqrt{x}}\, dx$. $\hspace{3cm}$ **2.** $\displaystyle\int_{1+}^{2} \frac{\sqrt{x}}{\log x}\, dx$.

3. $\displaystyle\int_{1+}^{2} x\,[\log(1+x)]^{\beta}\,dx.$

4. $\displaystyle\int_{0+}^{1} x^{2}e^{1/x}\,dx.$

5. $\displaystyle\int_{0+}^{1/3} \left(\log\log\frac{1}{x}\right)^{\alpha}\,dx.$

6. $\displaystyle\int_{-1}^{-1/2} |\log x^{2}|^{\alpha}(1+x)^{\beta}\,dx.$

7. $\displaystyle\int_{0+}^{1} \frac{\sin(1/x)}{x^{3/2}\log(1+x^{-1})}\,dx.$

8. $\displaystyle\int_{0+}^{1/3} \frac{\sin(1/x)}{x\log(1/x)\log\log(1/x)}\,dx.$

9. Discuss absolute convergence of the integrals of Exercises 7 and 8.

10. Prove Theorem 5* directly and by change of variable.

11. State and prove for integrals of Type III a theorem analogous to Exercise 11, §4, Chapter 9.

12. Solve the same problem for Exercise 12, §4, Chapter 9.

13. Solve the same problem for Corollary 6.1.

14. Show that Theorem 6* remains true if hypothesis 3 is replaced by

$$3'.\ \int_{a+}^{b} g(x)\,dx \text{ converges.}$$

Hint: Use Exercise 15, §1.

15. Test for conditional or absolute convergence

$$\int_{0+}^{1} \frac{\cos(1/x)}{x^{5/4}+x^{2}\sin(1/x)}\,dx.$$

§5. *Combination of Types*

In this section we shall discuss briefly improper integrals of Types II and IV and integrals which are made up of combinations of various types. Types II and IV could be treated directly by a group of theorems analogous to the first six theorems of the present chapter, but it is usually as convenient to reduce these types to I and III, respectively, by a change of variable. In using the limit tests, however, it is perhaps a little quicker to make the appropriate changes in Theorems 4, 5, 4*, 5*.

5.1 TYPE II

Here $f(x) \in C$ in $-\infty < x \leq b$. The integral

$$\int_{-\infty}^{b} f(x)\,dx$$

becomes

(1) $$\int_{-b}^{\infty} f(-t)\,dt$$

when we set $x = -t$. If

$$\lim_{t \to +\infty} f(-t)t^p = \lim_{x \to -\infty} f(x)(-x)^p = A \qquad\qquad p > 1$$

the integral (1) converges absolutely. If

$$\lim_{t \to +\infty} f(-t)t = \lim_{x \to -\infty} -f(x)x = A \neq 0 \text{ (or } \pm\infty)$$

the integral (1) diverges.

5.2 TYPE IV

Here $f(x) \in C$ in $a \leq x < b$ and $\lim_{x \to b-} f(x)$ does not exist. The integral

(2)
$$\int_a^{b-} f(x)\, dx$$

becomes
$$\int_{0+}^{b-a} f(b - t)\, dt$$

when we set $x = b - t$. The transformation $x = -t$ would have been equally good. The integral (2) converges absolutely if

$$\lim_{t \to 0+} f(b - t)t^p = \lim_{x \to b-} (b - x)^p f(x) = A \qquad 0 < p < 1$$

diverges if
$$\lim_{t \to 0+} f(b - t)t = \lim_{x \to b-} (b - x)f(x) = A \neq 0 \text{ (or } \pm\infty)$$

5.3 SUMMARY OF LIMIT TESTS

For convenience of reference we summarize the limit tests for the four types.

Absolute convergence.

TYPE I.	$\lim_{x \to +\infty} x^p f(x) = A$	$p > 1$
TYPE II.	$\lim_{x \to -\infty} (-x)^p f(x) = A$	$p > 1$
TYPE III.	$\lim_{x \to a+} (x - a)^p f(x) = A$	$0 < p < 1$
TYPE IV.	$\lim_{x \to b-} (b - x)^p f(x) = A$	$0 < p < 1$

Divergence

Type I.	$\lim\limits_{x \to +\infty} xf(x) = A \neq 0 \text{ (or } \pm\infty)$
Type II.	$\lim\limits_{x \to -\infty} xf(x) = A \neq 0 \text{ (or } \pm\infty)$
Type III.	$\lim\limits_{x \to a+} (x - a)f(x) = A \neq 0 \text{ (or } \pm\infty)$
Type IV.	$\lim\limits_{x \to b-} (b - x)f(x) = A \neq 0 \text{ (or } \pm\infty)$

Observe that in the limit tests for convergence the factor preceding $f(x)$ is always a *positive* quantity raised to power p. This fact provides a convenient memory rule, for, if the factor were altered to make the quantity negative, imaginary numbers might be introduced ($p = \frac{1}{2}$, for example).

5.4 COMBINATIONS OF INTEGRALS

We gave an example in §1 to show that an improper integral may be a combination of integrals of various types. It is clear that every improper integral for which the integrand has at most a finite number of discontinuities can be decomposed into a finite number of integrals of the four types.

Definition 5. *An integral which is the sum of a finite number of improper integrals of Types I, II, III, IV converges* $\Leftrightarrow$ *each of these integrals converges.*

Definition 6. *An integral which is the sum of a finite number of improper integrals of Types I, II, III, IV diverges* $\Leftrightarrow$ *one or more of these integrals diverges.*

In the example of §1 the integral diverges since

$$\int_{0+}^{1/2} \frac{dx}{x(x - 1)} = -\infty$$

It is clear that in testing such composite improper integrals one should look first for a divergent part.

At first sight it may seem that these definitions of convergence and divergence are not the most practical ones. It is conceivable that it would be convenient to describe an integral as convergent if it is the sum of two parts, the first of which diverges to $+\infty$, the second to $-\infty$. A case in point would be the integral

$$(3) \qquad \int_{-1}^{1} \frac{1}{x}\, dx$$

According to Definition 6 this integral diverges, even though

$$(4) \qquad \lim_{\epsilon \to 0}\left[\int_{-1}^{-\epsilon} \frac{dx}{x} + \int_{\epsilon}^{1} \frac{dx}{x}\right] = 0$$

From a certain point of view it might be convenient to say that the area under the curve $y = 1/x$ from -1 to 1 is zero. On the other hand, integral (3) does not enjoy all the properties of convergent or proper integrals. For example, if we deduct from the interval of integration the interval $(0, \delta)$, the value of the integral becomes $-\infty$, however small δ may be. This is at odds with our feeling that the value of an integral should change continuously as the length of the interval of integration does. The limit (4) is sometimes described as the *Cauchy-value* of the divergent integral (3).

Let us arrange our integrals in a sort of descending "social scale"; that is, in the order of decreasing number of desirable properties:

p proper

ac absolutely convergent

cc conditionally convergent

d divergent.

A little consideration will make it clear that if an integral is the sum of several others from various levels of this scale, that integral belongs to the lowest level of any of its parts. For example,

$$\int_0^\infty \frac{\cos x}{\sqrt{x}}\, dx = \int_0^{2\pi} + \int_{2\pi}^{4\pi} + \int_{4\pi}^\infty$$
$$= ac + p + cc = cc$$

The second equation is meant to be symbolic, but is easily interpreted.

EXAMPLE A. $\displaystyle\int_{0+}^{1-} \left(\log \frac{1}{x}\right)^\alpha dx = \int_{0+}^{1/2} + \int_{1/2}^{1-}$

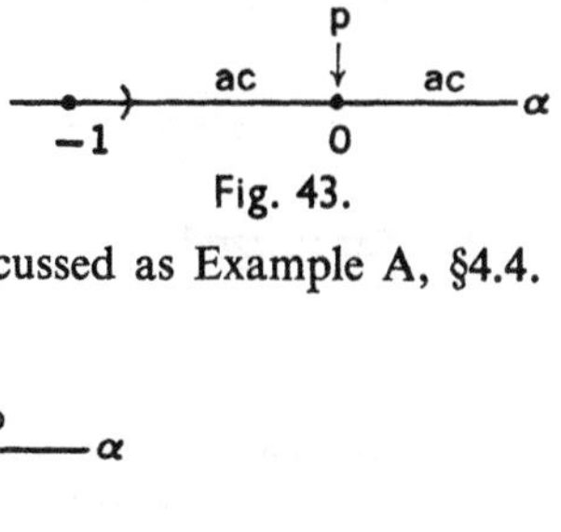

Fig. 43.

The first of the integrals on the right we discussed as Example A, §4.4. For the second we have the diagram

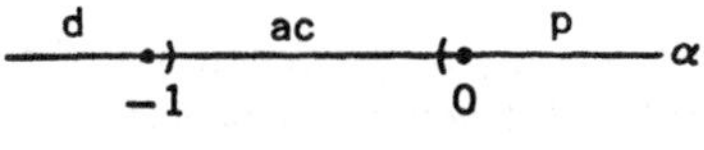

Fig. 44.

For,

$$\lim_{x \to 1-} (1 - x)^{-\alpha} f(x) = 1 \qquad\qquad -1 < \alpha < 0$$

$$\lim_{x \to 1-} (1 - x) f(x) = 1 \neq 0 \qquad\qquad \alpha = -1$$

$$= +\infty \qquad\qquad \alpha < -1$$

Combining the two results as explained above, we have the final result indicated in Figure 43.

EXAMPLE B. $\displaystyle\int_{0+}^{\infty} t^{x-1}e^{-t}\,dt$

$$\xrightarrow{\overset{d}{}\overset{)}{0}\overset{ca}{}}x$$

Fig. 45.

The integral is the sum of two others corresponding to the intervals $(0, 1)$ and $(1, \infty)$. The first of these was Example B, §4.4. The second converges absolutely for all x, since

$$\lim_{t \to +\infty} t^2 f(t) = 0$$

EXERCISES (5)

Test the integrals of Exercises 1–10 using the symbols p, ac, cc, d, as indicated in the text.

1. $\displaystyle\int_{-\infty}^{\infty} |x|^\alpha\,dx$.

2. $\displaystyle\int_{-\infty}^{\infty} \frac{\sin x}{\sqrt[3]{x}}\,dx$.

3. $\displaystyle\int_{-\infty}^{\infty} \frac{\cos x}{\sqrt[3]{x}}\,dx$.

4. $\displaystyle\int_{-\infty}^{\infty} \frac{\sin(1/x)}{x^{4/3}}\,dx$.

5. $\displaystyle\int_{-1}^{1} |\sin x|^\alpha \sqrt[3]{x}\,dx$.

6. $\displaystyle\int_{0}^{2} \log|\log x|\,dx$.

7. $\displaystyle\int_{1}^{3} \frac{dx}{\log\log x}$.

8. $\displaystyle\int_{-1}^{1} e^{-1/x}\,dx$.

9. $\displaystyle\int_{0}^{1} \left(\log\frac{2}{x}\right)^{-\alpha^2}\,dx$.

10. $\displaystyle\int_{-1}^{2} |\sin x|^\alpha |\cos x|^\beta\,dx$.

11. The integral

$$\int_{0}^{\infty} x^{-2} \sqrt[3]{\tan x}\,dx$$

does not come within any of the definitions of convergence or divergence thus far given. Why? Introduce reasonable definitions that will be applicable to this integral.

12. Does the divergent integral

$$\int_{-1}^{1} x^{-2}\,dx$$

have a Cauchy-value?

13. Find the Cauchy-value of the divergent integral

$$\int_{0}^{3} \frac{dx}{1-x}$$

14. Define the Cauchy-value for a divergent integral

$$\int_{-\infty}^{\infty} f(x)\, dx \qquad\qquad f(x) \in C, \quad -\infty < x < \infty$$

Illustrate by the integral

$$\int_{-\infty}^{\infty} \sin x\, dx$$

15. Test for convergence (all p)

$$\int_0^{\infty} \frac{x\, |\log x|^p\, dx}{x^2 + 1}$$

§6. Uniform Convergence

The notion of uniform convergence of improper integrals can be introduced by the analogy with infinite series which we set up in §1. Let us consider first integrals of Type I,

$$(1) \qquad\qquad \int_a^{\infty} f(x, t)\, dt$$

Let us suppose that this integral converges for each fixed x in the interval $A \leqq x \leqq B$ and has the value $F(x)$. Set

$$S_R(x) = \int_a^R f(x, t)\, dt$$

Definition 7. *The integral* (1) *converges uniformly to* $F(x)$ *in the interval* $A \leqq x \leqq B \Leftrightarrow to\ an\ arbitrary\ \epsilon > 0\ corresponds\ a\ number\ Q\ independent\ of\ x\ in\ A \leqq x \leqq B\ such\ that\ when\ R > Q,$

$$|F(x) - S_R(x)| < \epsilon \qquad\qquad A \leqq x \leqq B$$

For integrals of Type III,

$$(2) \qquad\qquad F(x) = \int_{a+}^{b} f(x, t)\, dt \qquad\qquad A \leqq x \leqq B$$

set

$$S_r(x) = \int_r^b f(x, t)\, dt \qquad\qquad a < r \leqq b$$

Definition 7*. *The integral* (2) *converges uniformly to* $F(x)$ *in the interval* $A \leqq x \leqq B \Leftrightarrow to\ an\ arbitrary\ \epsilon > 0\ corresponds\ a\ number\ q\ independent\ of\ x\ in\ A \leqq x \leqq B\ such\ that\ when\ a < r < q$

$$|F(x) - S_r(x)| < \epsilon \qquad\qquad A \leqq x \leqq B$$

EXAMPLE A. $\int_0^{\infty} e^{-xt}\, dt$ converges uniformly to $1/x$ in the interval $1 \leqq x \leqq 2$. For,

$$\left| \frac{1}{x} - S_R(x) \right| = \frac{e^{-xR}}{x} \leqq e^{-R} < \epsilon \qquad\qquad R > \log \frac{1}{\epsilon}$$

EXAMPLE B. $\int_0^\infty xe^{-xt}\, dt$ does not converge uniformly in the interval $0 \leqq x \leqq 1$, though it converges at each point of the interval.   Here

$$F(x) = 1 \qquad\qquad\qquad x > 0$$
$$= 0 \qquad\qquad\qquad x = 0$$

Then

$$|F(x) - S_R(x)| = e^{-xR} \qquad\qquad 0 < x \leqq 1$$
$$= 0 \qquad\qquad\qquad x = 0$$

Choose $\epsilon = \frac{1}{2}$. If the number Q of Definition 7 existed, we should have for $R > Q$

$$|F(x) - S_R(x)| = e^{-xR} < \tfrac{1}{2} \qquad\qquad 0 < x \leqq 1$$

This is false, since for every $R > 0$

$$\lim_{x \to 0+} e^{-xR} = 1$$

6.1 THE WEIERSTRASS M-TEST

Theorem 7.   1. $f(x, t) \in C$ \qquad\qquad $a \leqq t < \infty,\quad A \leqq x \leqq B$

2. $M(t) \in C$ \qquad\qquad\qquad\qquad $a \leqq t < \infty$

3. $|f(x, t)| \leqq M(t)$ \qquad\qquad $a \leqq t < \infty,\quad A \leqq x \leqq B$

4. $\int_a^\infty M(t)\, dt < \infty$

$\Rightarrow \qquad \int_a^\infty f(x, t)\, dt$ *converges uniformly in* $A \leqq x \leqq B$.

For,

$$|F(x) - S_R(x)| \leqq \int_R^\infty |f(x, t)|\, dt \leqq \int_R^\infty M(t)\, dt$$

Since the last integral is independent of x in $A \leqq x \leqq B$ and tends to zero with $1/R$, the result is immediate.

Theorem 7*.   1. $f(x, t) \in C$ \qquad\qquad $a < t \leqq b,\quad A \leqq x \leqq B$

2. $M(t) \in C$ \qquad\qquad\qquad\qquad $a < t \leqq b$

3. $|f(x, t)| \leqq M(t)$ \qquad\qquad $a < t \leqq b,\quad A \leqq x \leqq B$

4. $\int_{a+}^b M(t)\, dt < \infty$

$\Rightarrow \qquad \int_{a+}^b f(x, t)\, dt$ *converges uniformly in* $A \leqq x \leqq B$.

The proof is omitted. In Example A above we may choose the function $M(t)$ as e^{-t}. In Example B we have for a fixed $t > 0$

$$\max_{0 \leq x \leq 1} f(x, t) = \frac{1}{te}$$

Since

$$\int_1^\infty \frac{1}{te}\, dt = \infty$$

the M-test fails. This does not prove nonuniform convergence.

EXAMPLE C. $\displaystyle\int_{0+}^1 \frac{\sin (t/x)}{\sqrt{t}}\, dt$ converges uniformly in any interval $0 <$ $A \leq x \leq B$. For, we may take the function $M(t)$ of Theorem 7* equal to $t^{-1/2}$.

EXERCISES (6)

Show the integrals of Exercises 1–6 uniformly convergent in the intervals indicated.

1. $\displaystyle\int_1^\infty \frac{x\, dt}{x^2 + t^2}$ $\qquad\qquad\qquad\qquad$ $1 \leq x \leq 2$

   Do this first integral in two ways: first by Definition 7, then by Theorem 7.

2. $\displaystyle\int_1^\infty \frac{\sin (xt)}{t^2}\, dt$ $\qquad\qquad\qquad\quad$ $-10 \leq x \leq 10$

3. $\displaystyle\int_0^\infty \frac{\cos xt}{1 + t^2}\, dt$ $\qquad\qquad\qquad\quad$ $A \leq x \leq B$

4. $\displaystyle\int_0^\infty e^{-x^2 t^2}\, dt$ $\qquad\qquad\qquad\quad$ $0.1 \leq x \leq 100$

5. $\displaystyle\int_0^1 e^{-t} t^{x-1}\, dt$ $\qquad\qquad\qquad\quad$ $0.1 \leq x \leq 1$

6. $\displaystyle\int_{0+}^1 (\log xt)^{1/3}\, dt$ $\qquad\qquad\qquad\quad$ $1 \leq x \leq 3$

7. Prove Theorem 7*.

8. Give an example of a convergent integral of Type III which does not converge uniformly.

9. Does the integral of Exercise 1 converge uniformly in $-\infty < x < \infty$?

10. If all conditions of Theorem 7 are satisfied, and if in addition $M(t) \in\, \downarrow$, does the series

$$\sum_{k=m}^\infty f(x, k) \qquad\qquad\qquad a \leq m$$

converge uniformly in $A \leq x \leq B$?

11. Show that

$$\int_2^\infty \frac{\sin t}{t^x} dt$$

converges uniformly in $\frac{1}{2} \leqq x \leqq 1$. *Hint:* Use Corollary 6.3.

12. Prove: 1. $g(t) \in C$ $\qquad\qquad\qquad\qquad\qquad\qquad\qquad$ $0 \leqq t < \infty$

$\qquad$ 2. $\displaystyle\int_0^\infty g(t)\, dt$ converges

$\Rightarrow$ $\displaystyle\int_0^\infty e^{-xt} g(t)\, dt$ converges uniformly in $0 \leqq x \leqq r$ (any r).

Hint: Set $h(t) = \displaystyle\int_t^\infty g(y)\, dy$. Then for $x \geqq 0$

$$\int_R^\infty e^{-xt} g(t)\, dt = e^{-xR} h(R) - x \int_R^\infty e^{-xt} h(t)\, dt$$

Choose Q of Definition 7 so that $|h(t)| < \epsilon$ when $t > Q$.

§7. *Properties of Proper Integrals*

In order to make the applications of uniform convergence analogous to those given for series in §6, Chapter 9, we discuss here first certain properties of proper definite integrals. In particular, when the integrand contains a parameter, we shall study the continuity and differentiable properties of the integral considered as a function of the parameter.

7.1 INTEGRAL AS A FUNCTION OF ITS LIMITS OF INTEGRATION

Theorem 8. 1. $f(x) \in C$ $\qquad\qquad\qquad\qquad\qquad\qquad\qquad$ $a \leqq x \leqq b$

$\qquad$ 2. $F(x) = \displaystyle\int_c^x f(t)\, dt$ $\qquad\qquad\qquad\qquad$ $a \leqq c, \quad x \leqq b$

$\Rightarrow$ $\qquad$ A. $F(x) \in C^1$ $\qquad\qquad\qquad\qquad\qquad\qquad$ $a \leqq x \leqq b$

$\qquad$ B. $F'(x) = f(x)$ $\qquad\qquad\qquad\qquad\qquad$ $a \leqq x \leqq b$

It is understood, of course, that $F'(a)$ and $F'(b)$ are right-hand and left-hand derivatives, respectively. To prove this, form the difference quotient

$$\frac{F(x_0 + \Delta x) - F(x_0)}{\Delta x} = \frac{1}{\Delta x} \int_{x_0}^{x_0 + \Delta x} f(t)\, dt \qquad a \leqq x_0, \quad x_0 + \Delta x \leqq b$$

Now apply the mean-value theorem for integrals and let $\Delta x \to 0$:

$$\frac{F(x_0 + \Delta x) - F(x_0)}{\Delta x} = f(x_0 + \theta \Delta x) \qquad 0 < \theta < 1$$

$$F'(x_0) = \lim_{\Delta x \to 0} \frac{F(x_0 + \Delta x) - F(x_0)}{\Delta x} = f(x_0)$$

If $x_0 = a$, for example, we must have $\Delta x \to 0+$. Since $f(x) \in C$, we have $F'(x) \in C$ or $(Fx) \in C^1$.

Corollary 8. 1. $f(x) \in C$ $\qquad\qquad\qquad\qquad\qquad\qquad a \leqq x \leqq b$

$\qquad\qquad$ 2. $F(x) = \displaystyle\int_x^c f(t)\, dt$ $\qquad\qquad\qquad\qquad a \leqq c, \quad x \leqq b$

$\Rightarrow\cdot$ $\qquad\qquad F'(x) = -f(x)$

7.2 INTEGRAL AS A FUNCTION OF A PARAMETER

Theorem 9. 1. $f(x, t) \in C$ $\qquad\qquad\qquad\qquad a \leqq t \leqq b, \quad A \leqq x \leqq B$

$\qquad\qquad$ 2. $F(x) = \displaystyle\int_a^b f(x, t)\, dt$ $\qquad\qquad\qquad\qquad A \leqq x \leqq B$

$\Rightarrow\cdot$ $\qquad\qquad F(x) \in C$ $\qquad\qquad\qquad\qquad\qquad\qquad A \leqq x \leqq B$

Since $f(x, t) \in C$ in the closed rectangle $a \leqq t \leqq b$, $A \leqq x \leqq B$, it is uniformly continuous there.[*] To an arbitrary $\epsilon > 0$ corresponds a number δ such that if

$$A \leqq x_0 \leqq B, \qquad A \leqq x_0 + \Delta x \leqq B, \qquad |\Delta x| < \delta$$

then for $a \leqq t \leqq b$

$$|f(x_0 + \Delta x, t) - f(x_0, t)| < \epsilon/(b - a)$$

Hence, for $|\Delta x| < \delta$

$$|F(x_0 + \Delta x) - F(x_0)| \leqq \int_a^b |f(x_0 + \Delta x, t) - f(x_0, t)|\, dt < \epsilon$$

This completes the proof.

Corollary 9. *Under the conditions of Theorem 9*

$$\lim_{x \to x_0} \int_a^b f(x, t)\, dt = \int_a^b \lim_{x \to x_0} f(x, t)\, dt = \int_a^b f(x_0, t)\, dt \qquad A \leqq x_0 \leqq B$$

That is, it is permissible to take the limit operation under the sign of integration.

---

[*] Compare Theorem 8, Chapter 5. The result given there is for functions of a single variable, but an analogous theorem can be proved for functions of any number of variables.

EXAMPLE A. $\lim\limits_{x\to 0+} \int_0^1 \dfrac{x}{(x+t)^2}\, dt = ?$

The integral is proper for each $x > 0$. Moreover,

$$\lim_{x\to 0+} f(x, t) = \lim_{x\to 0+} \frac{x}{(x+t)^2} = 0 \qquad\qquad 0 < t$$

But it is not permissible to take the limit under the integral sign. For,

$$\lim_{x\to 0+} \int_0^1 \frac{x}{(x+t)^2}\, dt = \lim_{x\to 0+} \frac{1}{x+1} = 1$$

But
$$\int_0^1 \lim_{x\to 0+} f(x, t)\, dt = 0$$

Of course, $f(x, t) \notin C$ in the square $0 \leq x \leq 1, 0 \leq t \leq 1$.

Theorem 10.   1. $f(x, t), f_1(x, t) \in C$ $a \leq t \leq b, \quad A \leq x \leq B$

         2. $F(x) = \displaystyle\int_a^b f(x, t)\, dt$ $A \leq x \leq B$

$\Rightarrow$ A. $F(x) \in C^1$ $A \leq x \leq B$

         B. $F'(x) = \displaystyle\int_a^b f_1(x, t)\, dt$ $A \leq x \leq B$

For,

$$\frac{F(x_0 + \Delta x) - F(x_0)}{\Delta x} = \frac{1}{\Delta x} \int_a^b [f(x_0 + \Delta x, t) - f(x_0, t)]\, dt$$

By the law of the mean we obtain

$$\frac{\Delta F}{\Delta x} = \frac{F(x_0 + \Delta x) - F(x_0)}{\Delta x} = \int_a^b f_1(x_0 + \theta\, \Delta x, t)\, dt \qquad 0 < \theta < 1$$

In general, a different value of θ will be needed for each t; that is, θ is a function of $t, x_0, \Delta x$. Subtracting from $\Delta F/\Delta x$ its alleged limit, we obtain

(1) $\left| \dfrac{\Delta F}{\Delta x} - \displaystyle\int_a^b f_1(x_0, t)\, dt \right| \leq \displaystyle\int_a^b |f_1(x_0 + \theta\, \Delta x, t) - f_1(x_0, t)|\, dt$

Since $f_1(x, t) \in C$ in the closed rectangle $a \leq t \leq b, \ A \leq x \leq B$, it is uniformly continuous there. Hence to an arbitrary number $\epsilon > 0$ there corresponds another, δ, such that when

$$A \leq x_0 \leq B, \qquad A \leq x_0 + \theta\, \Delta x \leq B, \qquad |\Delta x| < \delta$$

then for $a \leq t \leq b$,

$$|f_1(x_0 + \theta\, \Delta x, t) - f_1(x_0, t)| < \epsilon/(b - a)$$

Hence for $|\Delta x| < \delta$ the right-hand side of inequality (1) is $< \epsilon$, and the proof is complete.

7.3 INTEGRALS AS COMPOSITE FUNCTIONS

Theorem 11.   1. $f(x, t) \in C^1$ $a \leq t \leq b, \quad A \leq x \leq B$

2. $F(x, y, z) = \int_y^z f(x, t)\, dt$ $a \leq y, z \leq b, A \leq x \leq B$

$\Rightarrow$ $F_1(x, y, z) = \int_y^z f_1(x, t)\, dt$

$F_2(x, y, z) = -f(x, y)$

$F_3(x, y, z) = f(x, z)$ $a \leq y, \quad z \leq b, \quad A \leq x \leq B$

These results are direct consequences of Theorems 8 and 10 and Corollary 8.

As a consequence of Theorem 11, we may now compute derivatives and differentials of a variety of functions defined by integrals.

EXAMPLE B.   Find $G'(x)$ if $G(x) = \int_{g(x)}^{h(x)} f(x, t)\, dt$

If $F(x, y, z)$ is defined as in Theorem 11, we have

$$G(x) = F(x, g(x), h(x))$$

so that

$$G'(x) = F_1 + F_2 g' + F_3 h'$$

$$= \int_{g(x)}^{h(x)} f_1(x, t)\, dt - f(x, g(x))g'(x) + f(x, h(x))h'(x)$$

EXAMPLE C. $\dfrac{d}{dx} \displaystyle\int_{x^3}^{x^2} \dfrac{dt}{x + t} = - \int_{x^3}^{x^2} \dfrac{dt}{(x + t)^2} - \dfrac{3x^2}{x + x^3} + \dfrac{2x}{x + x^2}$

$$= \frac{2x + 1}{x + x^2} - \frac{3x^2 + 1}{x + x^3}$$

Here we can check the result directly by performing the integration indicated before the differentiation,

$$\frac{d}{dx} \int_{x^3}^{x^2} \frac{dt}{x + t} = \frac{d}{dx} \left[\log (x + x^2) - \log (x + x^3) \right]$$

$$= \frac{2x + 1}{x + x^2} - \frac{3x^2 + 1}{x + x^3}$$

Of course, the chief usefulness of the present method occurs when the given integral cannot be evaluated in terms of the elementary functions, as in the following example.

EXAMPLE D. $d\displaystyle\int_{\sin x}^{\log y} \frac{\sin xt}{ty}\, dt = ?$

Here, the given integral is a function of two variables, $F(x, y)$, and

$$F_1(x, y) = -\frac{\cos x \sin (x \sin x)}{y \sin x} + \int_{\sin x}^{\log y} \frac{\cos xt}{y}\, dt$$

$$F_2(x, y) = \frac{\sin (x \log y)}{y^2 \log y} - \int_{\sin x}^{\log y} \frac{\sin xt}{ty^2}\, dt$$

$$dF(x, y) = F_1(x, y)\, dx + F_2(x, y)\, dy$$

It is interesting to observe that $F_1(x, y)$ can be evaluated in terms of the elementary functions even though the given integral cannot.

7.4 APPLICATION TO TAYLOR'S FORMULA

An interesting use of the above theory is the establishment of Taylor's formula with exact remainder.*   Let $f(x) \in C^{n+1}$, and set

$$R(x) = \int_0^x \frac{(x - t)^n}{n!} f^{(n+1)}(t)\, dt$$

Then, by the foregoing theory

$$R^{(k)}(x) = \int_0^x \frac{(x - t)^{n-k}}{(n - k)!} f^{(n+1)}(t)\, dt \qquad k = 0, 1, \ldots, n$$

(2) $$\qquad\qquad R^{(k)}(0) = 0 \qquad\qquad\qquad k = 0, 1, \ldots, n$$

(3) $$\qquad\qquad R^{(n+1)}(x) = f^{(n+1)}(x)$$

Now if we integrate both sides of equation (3) successively, using equation (2) to determine the constants of integration, we have

$$R(x) = f(x) - f(0) - f'(0)x - f''(0)\frac{x^2}{2!} - \cdots - f^{(n)}(0)\frac{x^n}{n!}$$

(4) $$\qquad f(x) = \sum_{k=0}^{n} f^{(k)}(0)\frac{x^k}{k!} + \int_0^x \frac{(x - t)^n}{n!} f^{(n+1)}(t)\, dt$$

EXERCISES (7)

Compute the derivatives and differentials in Exercises 1–7.

1. $\displaystyle\frac{d}{dx}\int_{-x}^{x} \frac{dt}{x^2 + t + 1}$ (2 ways)

---

* See Theorem 11, Chapter 1.

2. $\dfrac{d}{dx}\displaystyle\int_{x^2}^{-\sin x} e^{xt}\, dt$ (2 ways)

3. $\dfrac{\partial}{\partial x}\displaystyle\int_0^x \sqrt{1 + y^3 t^3}\, dt, \quad \dfrac{\partial}{\partial y}\displaystyle\int_0^x \sqrt{1 + y^3 t^3}\, dt.$

4. $d\displaystyle\int_{t^2+y^2}^{t^2-y^2} (x^2 + y^2)\, dx$ (2 ways)

5. $d\displaystyle\int_{xy}^{x^2+y} \dfrac{\sin (x + y)}{x^2 + y^2}\, dx$

Suggestion: To avoid confusion, use some new letter for the integration variable.

6. $\dfrac{d}{dx}\log\left(\displaystyle\int_0^{x^2} \dfrac{\sin xt}{t}\, dt\right)$

7. $dF(x, y)$, if

$$x = \int_{r^2}^{-s} \dfrac{\log (r^2 + s^2 + 3t^2)}{t}\, dt$$

$$y = \int_{-s}^{2} \dfrac{\sin rst}{t}\, dt$$

8. Find $\dfrac{dy}{dx}$ if

$$\int_{\sqrt{x}}^{2y} e^{-xt^2}\, dt = -2$$

9. Find $\dfrac{\partial z}{\partial x}, \dfrac{\partial z}{\partial y}$ if

$$\int_{g(x,y,z)}^{h(x,y)} f(y, z, t)\, dt = 0$$

10. Find $\dfrac{dz}{dx}, \dfrac{dy}{dx}$ if

$$\int_0^x f(x, y, z, t)\, dt = 0$$

$$\int_{xy}^{z} g(z, t)\, dt = 0$$

11. Find:

$$\lim_{x \to 0} \dfrac{1}{x^3}\int_0^x (x - t)^2 f(t)\, dt$$

12. Find:

$$\lim_{x \to 0+} \int_0^1 \dfrac{x}{x + t}\, dt$$

Can Corollary 9 be applied to this example?

13. Find:

$$\lim_{x \to 0+} x \int_x^{1/x} \frac{1 + e^{-t}}{t^\alpha} \, dt$$

14. Find:

$$\lim_{x \to 0} \sin^{-3} x \int_0^{x^3} (e^{t^2} + 2) \, dt$$

15. Find $\displaystyle \lim_{x \to +\infty} \frac{\dfrac{1}{\log x} \displaystyle\int_2^x \dfrac{e^{-t} + 1}{t} \, dt}{\displaystyle\int_2^x \dfrac{e^{-t} + 2}{t \log t} \, dt}$

16. Use equation (4) to show that

$$\sum_{k=0}^n \frac{x^k}{k!} = e^x - \int_0^x \frac{(x - t)^n}{n!} e^t \, dt$$

17. Use equation (4) to show that

$$\sum_{k=0}^n x^k = \frac{1}{1 - x} - (n + 1) \int_0^x \frac{(x - t)^n}{(1 - t)^{n+2}} \, dt$$

Make the change of variable $y = (1 - t)^{-1}$ to reduce this to the usual form for the sum of a finite geometric progression.

18. Prove that $f'(0) = 0$ if

$$f(x) = \int_0^x \cos t^{-1} \, dt$$

Hint: use the definition of $f'(0)$ and integration by parts, integrating $t^{-2} \cos t^{-1}$.

§8. *Application of Uniform Convergence*

In this section we shall discuss the continuity and differentiability properties of a function defined by an improper integral, obtaining results quite analogous to those obtained for series in §6, Chapter 9. As in that section, uniform convergence will be the useful tool. We shall discuss only integrals of Type I. The corresponding results for integrals of the other types will be evident to the reader.

8.1 CONTINUITY

Theorem 12. 1. $f(x, t) \in C$ $\qquad\qquad a \leq t < \infty, \quad A \leq x \leq B$

2. $\displaystyle\int_a^\infty f(x, t) \, dt$ *converges uniformly to* $F(x)$ *in* $A \leq x \leq B$

$\Rightarrow \qquad F(x) \in C$ $\qquad\qquad\qquad\qquad\qquad\qquad A \leq x \leq B$

Let ϵ be an arbitrary positive number. Then by hypothesis 2 there exists a number R *independent* of x in $A \leq x \leq B$ such that

$$(1) \qquad \left| \int_R^\infty f(x, t) \, dt \right| < \epsilon \qquad\qquad A \leq x \leq B$$

Hence, if $A \leq x_0 \leq B$,

$$|F(x) - F(x_0)| \leq \int_a^R |f(x, t) - f(x_0, t)| \, dt + \left| \int_R^\infty f(x, t) \, dt \right| + \left| \int_R^\infty f(x_0, t) \, dt \right|$$

$$\leq \int_a^R |f(x, t) - f(x_0, t)| \, dt + 2\epsilon$$

Since this inequality holds for all x in the interval $A \leq x \leq B$, we may let $x \to x_0$. The integral on the right-hand side will approach zero by Theorem 9. Hence,

$$\overline{\lim_{x \to x_0}} \, |F(x) - F(x_0)| \leq 2\epsilon$$

and, since ϵ was arbitrary, this becomes

$$\lim_{x \to x_0} F(x) = F(x_0)$$

and the proof is complete.

EXAMPLE A. $\dfrac{1}{x} = \displaystyle\int_0^\infty e^{-xt} \, dt$ is continuous for $0 < x < \infty$. In this case, we have an explicit expression for the value of the integral, so that the continuity is easily checked. But we could obtain the result without evaluating the integral. Let x_0 be an arbitrary positive number. It can be included in a closed interval $0 < A \leq x \leq B$. But the given integral converges uniformly there, as we saw in Example A, §6.

EXAMPLE B. $F(x) = \displaystyle\int_0^\infty x^2 t e^{-xt} \, dt$

The integrand is continuous in any finite rectangle. Yet $F(x) \notin C$. Direct integration gives

$$F(x) = 1 \qquad\qquad x > 0$$
$$F(0) = 0$$

The convergence is not uniform in any interval containing the origin.

8.2 INTEGRATION

Theorem 13. 1. $f(x, t) \in C$ $\qquad\qquad a \leq t < \infty, \quad A \leq x \leq B$

2. $\displaystyle\int_a^\infty f(x, t) \, dt$ *converges uniformly to* $F(x)$ *in* $A \leq x \leq B$

$\Rightarrow \qquad \displaystyle\int_A^B F(x) \, dx = \int_a^\infty dt \int_A^B f(x, t) \, dx$

We have in this theorem a criterion for interchanging the order of integration in iterated integrals. Since inequality (1) holds for all R greater than some number Q, we have

$$\left| \int_A^B dx \int_R^\infty f(x, t)\, dt \right| < \epsilon(B - A) \qquad\qquad R > Q$$

That is,

$$\lim_{R \to \infty} \int_A^B \left[F(x) - \int_a^R f(x, t)\, dt \right] dx = 0$$

$$\int_A^B F(x)\, dx = \lim_{R \to \infty} \int_A^B dx \int_a^R f(x, t)\, dt$$

$$= \lim_{R \to \infty} \int_a^R dt \int_A^B f(x, t)\, dx$$

$$= \int_a^\infty dt \int_A^B f(x, t)\, dx$$

This completes the proof.

EXAMPLE C. $\displaystyle \int_0^\infty \frac{e^{-pt} - e^{-qt}}{t}\, dt = \log \frac{q}{p}$ $0 < p < q$

For, the integral

$$\frac{1}{x} = \int_0^\infty e^{-xt}\, dt$$

converges uniformly for $p \leq x \leq q$. Hence,

$$\log \left(\frac{q}{p} \right) = \int_0^\infty dt \int_p^q e^{-xt}\, dx$$

$$= \int_0^\infty \frac{e^{-pt} - e^{-qt}}{t}\, dt$$

8.3 DIFFERENTIATION

Theorem 14. 1. $f(x, t), f_1(x, t) \in C$ $a \leq t < \infty,\ A \leq x \leq B$

2. $\displaystyle \int_a^\infty f(x, t)\, dt$ converges to $F(x)$ $A \leq x \leq B$

3. $\displaystyle \int_a^\infty f_1(x, t)\, dt$ converges uniformly in $A \leq x \leq B$

$\Rightarrow$ $\displaystyle F'(x) = \int_a^\infty f_1(x, t)\, dt$

Set $\displaystyle \varphi(x) = \int_a^\infty f_1(x, t)\, dt$ $A \leq x \leq B$

By Theorem 12 $\varphi(x) \in C$ in $A \leq x \leq B$, and by Theorem 13

$$\int_A^h \varphi(x)\, dx = \int_a^\infty [f(h, t) - f(A, t)]\, dt \qquad A \leq h \leq B$$
$$= F(h) - F(A)$$

Consequently, we have by Theorem 8 that

$$F'(h) = \varphi(h)$$

and this is the result we wished to prove.

EXAMPLE D. $\dfrac{d}{dx} \displaystyle\int_0^\infty e^{-xt}\, dt = -\int_0^\infty e^{-xt} t\, dt \qquad 0 < x < \infty$

For, if $x_0 > 0$, choose constants A and B such that $0 < A < x_0 < B$. Then the integral

$$\int_0^\infty e^{-xt}\, dt$$

converges in $A \leq x \leq B$, and

$$\int_0^\infty e^{-xt} t\, dt$$

converges uniformly in $A \leq x \leq B$. (Take $M(t) = e^{-At} t$ in Theorem 7.) Since both integrals may be evaluated in terms of the elementary functions, the result may be checked directly.

EXERCISES (8)

1. If

$$F(x) = \int_0^\infty e^{-xt^2}\, dt$$

show by Theorem 12 that $F(x) \in C$ in $0 < x < \infty$.   Check by Peirce-Foster No. 507.

2. Find $F'(x)$ in Exercise 1 by Theorem 14.   Show $F(x) \in C^1$ in $0 < x < \infty$. Check by No. 509.

3. Find by Theorem 13 $\displaystyle\int_1^2 F(x)\, dx$ in Exercise 1.

4. Prove:

1. $g(x) \in C$ $\qquad\qquad\qquad\qquad\qquad\qquad\qquad 0 \leq x < \infty$
2. $\lim\limits_{x \to \infty} x^p g(x) = A$ $\qquad\qquad\qquad\qquad\qquad$ *for some $p \leq 0$*
3. $F(x) = \displaystyle\int_0^\infty e^{-xt} g(t)\, dt$

$\Rightarrow$ $F(x) \in C^\infty$ $\qquad\qquad\qquad\qquad\qquad\qquad\qquad 0 < x < \infty$

5. In Exercise 4 find

$$\int_1^x F(t)\, dt \qquad\qquad 0 < x < \infty$$

6. Show that the integral

$$F(x) = \int_0^\infty \frac{\sin xt}{t}\, dt$$

converges for all x.   Show that $F(x) \notin C$ at $x = 0$.   *Hint:* Set $xt = u$ and use Example B, §3.

7. From the equation

$$\frac{1}{1 + x^2} = \int_0^\infty e^{-t} \cos xt\, dt$$

prove

$$\frac{2x}{(1 + x^2)^2} = \int_0^\infty e^{-t} t \sin xt\, dt$$

For what values of x are these two equations valid?

8. Prove Theorem 12 in a way analogous to the proof of Theorem 12, Chapter 9.

9. Prove a theorem for integrals of Type III analogous to Theorem 12.

10. Solve the same problem for Theorem 13.

11. Solve the same problem for Theorem 14.

12. From the equation

$$\frac{1}{y + 1} = \int_0^1 x^y\, dx \qquad\qquad y > -1$$

show

$$\frac{1}{(y + 1)^2} = -\int_0^1 x^y \log x\, dx \qquad\qquad y > -1$$

13. In Example B, show without the use of Theorem 12 that the convergence is not uniform.

14. If $f(x) \in C, |f(x)| < M$ for $-\infty < x < \infty$, show that the integral

$$\int_{-\infty}^\infty e^{-(x-y)^2} f(x)\, dx$$

converges uniformly in any finite interval.

15. Under the assumptions of the previous exercise prove that

$$\frac{d}{dy} \int_{-\infty}^\infty e^{-x^2/y} f(x)\, dx = \int_{-\infty}^\infty \frac{\partial}{\partial y} e^{-x/y^2} f(x)\, dx \qquad 0 < y < \infty$$

16. The value of the integral of Exercise 9, §6, is discontinuous at the origin. How do you reconcile this fact with Theorem 12?

§9. Divergent Integrals

Just as in the case of infinite series we may study divergent improper integrals, defining a process of summability. We have already done this for integrals of a very special type when we introduced the Cauchy-value. We wish now to introduce the Cesàro method, or the method of arithmetic means. We shall treat integrals of Type I only.

9.1 CESÀRO SUMMABILITY

Let $f(x) \in C$ for $a \leq x < \infty$. Consider the integral

(1)
$$\int_a^\infty f(x)\, dx$$

Set
$$S(R) = \int_a^R f(x)\, dx, \qquad \sigma(R) = \frac{1}{R - a} \int_a^R S(t)\, dt$$

Definition 8. *The integral* (1) *is summable* $(C, 1)$ *to* $A \Leftrightarrow$

(2)
$$\lim_{R \to \infty} \sigma(R) = A$$

We also write equation (2) as

$$A = \int_a^\infty f(x)\, dx \qquad\qquad (C, 1)$$

Note that an inversion in the order of iterated integrals gives

$$\sigma(R) = \frac{1}{R - a} \int_a^R S(t)\, dt = \int_a^R \frac{R - x}{R - a} f(x)\, dx$$

EXAMPLE A. $\displaystyle\int_0^\infty \sin x\, dx = 1$ $\qquad\qquad (C, 1)$

For,
$$\sigma(R) = \int_0^R \left(1 - \frac{x}{R}\right) \sin x\, dx = 1 - \frac{\sin R}{R}$$

9.2 REGULARITY

We shall show here that the convergence of an integral implies its $(C, 1)$-summability.

Theorem 15. $\displaystyle A = \int_a^\infty f(x)\, dx \;\Rightarrow\; A = \int_a^\infty f(x)\, dx$ $\qquad (C, 1)$

We have given that
$$A = \lim_{R \to \infty} S(R)$$

and we wish to prove that

$$A = \lim_{R \to \infty} \frac{1}{R - a} \int_a^R S(x) \, dx$$

CASE I. $A = 0$.   Given $\epsilon > 0$; there exists Q such that

$$|S(x)| < \epsilon \qquad\qquad x > Q$$

Let $R > Q$.   Then

$$\left| \frac{1}{R - a} \int_a^R S(x) \, dx \right| \leq \left| \frac{1}{R - a} \int_a^Q S(x) \, dx \right| + \frac{1}{R - a} \int_Q^R |S(x)| \, dx$$

$$\leq \left| \frac{1}{R - a} \int_a^Q S(x) \, dx \right| + \epsilon \frac{R - Q}{R - a}$$

$$\varlimsup_{R \to \infty} \left| \frac{1}{R - a} \int_a^R S(x) \, dx \right| \leq \epsilon$$

$$\lim_{R \to \infty} \frac{1}{R - a} \int_a^R S(x) \, dx = 0$$

CASE II. $A \neq 0$.   Apply Case I to the functions $S(x) - A$.

9.3 OTHER METHODS OF SUMMABILITY

A method analogous to that of Abel for series is the following.   If the integral

$$\int_0^\infty e^{-xt} f(t) \, dt$$

converges for $x > 0$, and if

$$\lim_{x \to 0+} \int_0^\infty e^{-xt} f(t) \, dt = A$$

then the integral (1) is summable to the value A.   Let us apply this method to the integral of Example A.   We have by use of the indefinite integral

$$\frac{1}{1 + x^2} = \int_0^\infty e^{-xt} \sin t \, dt \qquad\qquad x > 0$$

Since the left-hand side tends to 1 as $x \to 0$, we get the same value for the divergent integral as before.

EXERCISES   (9)

1. $\displaystyle \int_0^\infty \sin ax \, dx = \, ?$ $\qquad\qquad\qquad\qquad\qquad$ $(C, 1)$

2. $\displaystyle \int_a^\infty \sin x \, dx = \, ?$ $\qquad\qquad\qquad\qquad\qquad$ $(C, 1)$

3. $\displaystyle\int_0^\infty \sin(a + x)\, dx = ?$ $\hspace{3cm}$ (C, 1)

4. $\displaystyle\int_0^\infty \cos x\, dx = ?$ $\hspace{3.5cm}$ (C, 1)

5. Find the (C, 1)-sum of the series

$$\sum_{k=0}^\infty \int_{k\pi}^{(k+1)\pi} \sin x\, dx$$

6. Same problem for

$$\int_0^{\pi/2} \cos x\, dx + \sum_{k=0}^\infty \int_{(2k+1)\pi/2}^{(2k+3)\pi/2} \cos x\, dx$$

7. Is the integral

$$\int_0^\infty x \sin x\, dx$$

summable (C, 1)?

8. Definition: $\displaystyle\int_0^\infty f(x)\, dx = A$ (C, 2)

$\langle\Rightarrow\rangle$ $\hspace{2cm}$ $\displaystyle\lim_{R\to\infty} \frac{2}{R^2} \int_0^R (R - t)S(t)\, dt = A$

Show that the integral of Exercise 4 is summable (C, 2). To what value?

9. Prove: $\displaystyle\int_0^\infty f(x)\, dx = A$ (C, 1) $\Rightarrow$ $\displaystyle\int_0^\infty f(x)\, dx = A$ $\hspace{1cm}$ (C, 2)

10. Prove: $\displaystyle\int_0^\infty [f(x)]^2\, dx = A$ (C, 1) $\Rightarrow$ $\displaystyle\int_0^\infty [f(x)]^2\, dx = A$

11. Prove: $\displaystyle\int_0^\infty f(x)\, dx = A$ (C, 1) $\Rightarrow$ $\displaystyle\int_a^\infty f(x)\, dx = A - \int_0^a f(x)\, dx$ (C, 1)

12. Prove: $\displaystyle\int_0^\infty f(x)\, dx = A$ $\Rightarrow$ $\displaystyle\lim_{x\to 0+} \int_0^\infty e^{-xt}f(t)\, dt = A$

Hint: Integrate by parts to obtain

$$\int_0^\infty e^{-xt}f(t)\, dt = x\int_0^\infty e^{-xt}S(t)\, dt \hspace{2cm} x > 0$$

First take $A = 0$; break the integral into two parts, the second being integrated over the interval where $|S(t)| < \epsilon$.

13. If $f(t) \in C$, $0 \leqq t < \infty$ and $f(\infty) = A$, show that

$$\lim_{x\to\infty} \frac{1}{x} \int_{0+}^x f(t) \log\frac{x}{t}\, dt = A$$

§10. *Integral Inequalities*

The inequalities of Cauchy, Hölder, and Minkowski, which were discussed in §8 of Chapter 9, have their counterparts in integral theory and are discussed here. The integral analogue of Cauchy's inequality is usually called the *Schwarz inequality*.

10.1 THE SCHWARZ INEQUALITY

Theorem 16.   1. $f(t), g(t) \in C$ $\qquad\qquad\qquad\qquad a \leqq x \leqq b$

(1) $\Rightarrow$ $\qquad \left(\int_a^b |f(t)g(t)|\, dt \right)^2 \leqq \int_a^b [f(t)]^2\, dt \int_a^b [g(t)]^2\, dt$

The proof follows by observing that the discriminant of the positive quadratic form in x and y

$$\int_a^b (x|f(t)| + y|g(t)|)^2\, dt$$

must be $\leqq 0$, as in the proof of Cauchy's inequality.

EXAMPLE A. $\quad \displaystyle\int_1^\infty [f(t)]^2 t^{-2}\, dt < \infty \quad \Rightarrow \quad \int_1^\infty |f(t)| t^{-2}\, dt < \infty$

For, take $a = 1$, $b = R$, $g(t) = 1/t$ in (1):

$$\left(\int_1^R \frac{|f(t)|}{t} \frac{1}{t}\, dt \right)^2 \leqq \int_1^R \frac{[f(t)]^2}{t^2}\, dt \int_1^R \frac{dt}{t^2}$$

$$\int_1^\infty \frac{|f(t)|}{t^2}\, dt \leqq \left[\int_1^\infty \frac{[f(t)]^2}{t^2}\, dt \right]^{1/2}$$

We may think of a function $f(t)$ as the continuous analogue of a vector. It is natural to define the *p-norm* of $f(t)$ as

$$\|f(t)\|_p = \left\{ \int_a^b [f(t)]^p\, dt \right\}^{1/p}$$

and the *inner product* of $f(t)$ and $g(t)$ as

$$\langle f(t), g(t) \rangle = \int_a^b f(t)g(t)\, dt$$

With this notation (1) becomes

$$\langle |f(t)|, |g(t)| \rangle \leqq \|f(t)\|_2 \|g(t)\|_2$$

The *triangle inequality* is

(2) $\qquad\qquad\qquad \|f(t) + g(t)\|_2 \leqq \|f(t)\|_2 + \|g(t)\|_2$

10.2 THE HÖLDER INEQUALITY

Theorem 17.   1. $f(t), g(t) \in C$ $\qquad\qquad\qquad\qquad a \leqq x \leqq b$

$\qquad\qquad$ 2. $\dfrac{1}{p} + \dfrac{1}{q} = 1$ $\qquad\qquad\qquad\qquad p > 1, \quad q > 1$

(3) $\Rightarrow \qquad \displaystyle\int_a^b |f(t)g(t)|\, dt \leqq \left(\int_a^b |f(t)|^p\, dt \right)^{1/p} \left(\int_a^b |g(t)|^q\, dt \right)^{1/q}$

In the proof of Hölder's inequality for sums we proved that

$$xy \leqq \frac{x^p}{p} + \frac{y^q}{q} \qquad\qquad x > 0, \quad y > 0$$

Replace x by $|f(t)|$, y by $|g(t)|$ and integrate:

$$\int_a^b |f(t)g(t)|\, dt \leqq \frac{X^p}{p} + \frac{Y^q}{q}$$

where $X = \|f(t)\|_p$, $Y = \|g(t)\|_q$. Replacing $f(t)$ by $\lambda f(t)$ and $g(t)$ by $g(t)/\lambda$, $\lambda > 0$, gives

$$\int_a^b |f(t)g(t)|\, dt \leqq \frac{\lambda^p X^p}{p} + \frac{1}{\lambda^q} \frac{Y^q}{q}$$

We proved in §8.2 of Chapter 9 that for proper choice of λ the right-hand side becomes XY. This proves the theorem.

EXAMPLE B. $\displaystyle\int_{0+}^1 |f(t)|^b\, dt < \infty \quad \Rightarrow \quad \int_{0+}^1 |f(t)|^a\, dt < \infty \qquad 0 < a < b$

From (3) with $g(t) = 1$ we have for $0 < \epsilon < 1$

$$\int_\epsilon^1 |f(t)|^a\, dt \leqq \left(\int_\epsilon^1 |f(t)|^{ap}\, dt \right)^{1/p} \left(\int_\epsilon^1 1^q\, dt \right)^{1/q}$$

Choose p so that $ap = b$. Then $p > 1$ as required in Theorem 17. The desired result follows when $\epsilon \to 0+$.

10.3 THE MINKOWSKI INEQUALITY

Theorem 18.   1. $f(t), g(t) \in C$ $\qquad\qquad\qquad\qquad a \leqq x \leqq b$

$\qquad\qquad$ 2. $p \geqq 1$

$\Rightarrow \quad \left(\displaystyle\int_a^b |f(t) + g(t)|^p\, dt \right)^{1/p} \leqq \left(\int_a^b |f(t)|^p\, dt \right)^{1/p} + \left(\int_a^b |g(t)|^p\, dt \right)^{1/p}$

The proof follows as in §8.2, Chapter 9, from

$$\int_a^b |f(t) + g(t)|^p\, dt \leqq \int_a^b |f(t)|\, |f(t) + g(t)|^{p-1}\, dt + \int_a^b |g(t)|\, |f(t) + g(t)|^{p-1}\, dt$$

by application of Hölder's inequality to each integral on the right.

EXERCISES (10)

1. Check inequality (1) if $f(t) = \sin t$, $g(t) = \cos t$, $a = 0$, $b = \pi$.

2. Same problem for inequality (2).

3. Check Hölder's inequality for $f(t) = t^2$, $g(t) = t^3$, $p = 3$, $a = 0$, $b = 1$.

4. Check Minkowski's inequality for $f(t) = t^{7/3}$, $g(t) = 1$, $p = 3$, $a = 0$, $b = 1$.

5. Prove for the square norm ($p = 2$) that
$$\|f + g\|^2 + \|f - g\|^2 = 2\|f\|^2 + 2\|g\|^2$$

6. Prove for the square norm
$$\|f + g\|^2 = \|f\|^2 + \|g\|^2 \quad \Leftrightarrow \quad \langle f, g \rangle = 0$$
What does this mean geometrically if f and g are replaced by vectors?

7. Prove that
$$\int_a^b f^2 \, dx \int_a^b g^2 \, dx - \left(\int_a^b |fg| \, dx \right)^2 = \frac{1}{2} \int_a^b dx \int_a^b (|f(x)g(y)| - |f(y)g(x)|)^2 \, dy$$
and thus give a new proof of the Schwarz inequality.

8. Prove:
$$\int_1^\infty \frac{[f(t)]^2}{t(\log t)^2} \, dt < \infty \quad \Rightarrow \quad \int_1^\infty \frac{|f(t)|}{t(\log t)^2} \, dt < \infty$$
If $f(t) = (\log t)^p$, what values of p will give illustrations of the result?

9. Prove:
$$\int_{0+}^1 [f(t)]^2 \sqrt{t} \, dt < \infty \quad \Rightarrow \quad \int_{0+}^1 |f(t)| \, dt < \infty$$

10. Prove:
$$\int_{0+}^1 [f(t)]^2 t^p \, dt < \infty, \quad p < 1 \quad \Rightarrow \quad \int_{0+}^1 |f(t)| \, dt < \infty$$

11. Prove inequality (2).

12. Give details in the proof of Theorem 18.

13. Find the minimum value of $F(x) = \|f(t) + xg(t)\|_2$ for $-\infty < x < \infty$.

14. If in Exercise 13, $F(x) \geq F(x_0)$, $-\infty < x < \infty$, show that
$$\langle f(t) + x_0 g(t), \, x_0 g(t) \rangle = 0$$

15. In Exercise 14 show that $\|f + x_0 g\|^2 + \|x_0 g\|^2 = \|f\|^2$.

16. Solve a vector problem analogous to Exercises 13, 14, 15. Solve it either by calculus or by elementary geometry.

17. If $F(t)$ and $G(t)$ are positive and continuous for $1 \leq t < \infty$, prove:
$$\int_1^\infty F(t) \, dt = \infty \quad \Rightarrow \quad \text{either} \int_1^\infty F(t)G(t) \, dt = \infty \quad \text{or} \int_1^\infty F(t)/G(t) \, dt = \infty$$
Hint: Take $f = (FG)^{1/2}$ and $g = (F/G)^{1/2}$ in (1). *American Mathematical Monthly*, 1958, p. 126.

11

The Gamma Function.

Evaluation of Definite

Integrals

§1. *Introduction*

In this chapter we shall define a function known as the *gamma function*, $\Gamma(x)$, which has the property that $\Gamma(n) = (n-1)!$ for every positive integer n. It may be regarded then as a generalization of factorial n to apply to values of the variable which are not integers. The function is defined in terms of an improper integral. This integral cannot be evaluated in terms of the elementary functions. It has great importance in analysis and in the applications. As a consequence, it has been tabulated and very carefully studied.

We shall also discuss methods of finding the value of improper definite integrals when it is impossible to find an indefinite integral in terms of the elementary functions. Certain of these integrals are related to $\Gamma(x)$ and can be expressed in terms of that function.

1.1 THE GAMMA FUNCTION

Definition 1.

(1)
$$\Gamma(x) = \int_{0+}^{\infty} e^{-t} t^{x-1}\, dt \qquad\qquad 0 < x < \infty$$

If $0 < x < 1$, the integrand becomes infinite as $t \to 0+$. The integral corresponding to the interval $(0, 1)$ is convergent or proper for $0 < x$, while that corresponding to $(1, \infty)$ converges for all x. Hence, $\Gamma(x)$ is well defined by the integral (1) for $x > 0$.

Theorem 1.

(2)
$$\Gamma(x + 1) = x\Gamma(x) \qquad\qquad 0 < x < \infty$$

For, integration by parts gives

$$\int_{\epsilon}^{R} e^{-t} t^{x-1}\, dt = \frac{t^x}{x} e^{-t}\Big|_{\epsilon}^{R} + \frac{1}{x}\int_{\epsilon}^{R} e^{-t} t^{x}\, dt$$

Now allowing R to become infinite and ϵ to approach $0+$ we obtain

$$x\int_{0+}^{\infty} e^{-t} t^{x-1}\, dt = \int_{0+}^{\infty} e^{-t} t^{x}\, dt \qquad\qquad x > 0$$

We shall usually abbreviate this sort of calculation as follows:

$$\int_{0+}^{\infty} e^{-t} t^{x-1}\, dt = \frac{t^x}{x} e^{-t}\Big|_{0+}^{\infty} + \frac{1}{x}\int_{0+}^{\infty} e^{-t} t^{x}\, dt$$

This equation will have a meaning if at least five of the six limits involved are known to exist. The sixth will then automatically exist.

Corollary 1. $\Gamma(x + p) = (x + p - 1)(x + p - 2) \dots x\Gamma(x)$
$$x > 0, \quad p = 1, 2, \dots$$

Theorem 2. $\Gamma(n + 1) = n!$ $n = 0, 1, 2, \dots$

Factorial zero is defined as 1. From Corollary 1 we have

$$\Gamma(n + 1) = n!\,\Gamma(1)$$

But
$$\Gamma(1) = \int_{0}^{\infty} e^{-t}\, dt = 1$$

Theorem 3. $\Gamma(0+) = +\infty$

Since the integrand of the integral (1) is positive, we have

$$\Gamma(x) > \int_{0+}^{1} t^{x-1} e^{-t}\, dt > e^{-1}\int_{0+}^{1} t^{x-1}\, dt = (ex)^{-1} \quad 0 < x < \infty$$

This inequality establishes the result.

Theorem 4. $\Gamma(x) \in C$ $0 < x < \infty$

For, let x_0 be an arbitrary positive number.   Determine A, B so that $0 < A < x_0 < B$.   Then the integral

$$\int_1^\infty e^{-t} t^{x-1}\, dt$$

converges uniformly in $A \leq x \leq B$ (take $M(t) = e^{-t} t^{B-1}$ in Theorem 7, Chapter 10).   The integral

$$\int_{0+}^1 e^{-t} t^{x-1}\, dt$$

is either proper ($A \geq 1$) or converges uniformly in $A \leq x \leq B$ (take $M(t) = t^{A-1}$ in Theorem 7*, Chapter 10).   The continuity of $\Gamma(x)$ at x_0 now follows from Theorem 12, Chapter 10, and its analogue for integrals of Type III.   Since x_0 was arbitrary, the proof is complete.

Theorem 5. $\lim\limits_{x \to 0+} x\Gamma(x) = 1$

This follows in an obvious way from Theorems 1, 2 ($n = 1$), and 4. Note that Theorem 3 is included in Theorem 5.

1.2 EXTENSION OF DEFINITION

Definition 2.   *For* $n = 1, 2, \ldots$,

(3) $\Gamma(x) = \dfrac{\Gamma(x + n)}{x(x + 1) \ldots (x + n - 1)}$ $-n < x < -n + 1$

Thus we have defined $\Gamma(x)$ for all x except $x = 0, -1, -2, \ldots$.   Observe that when $n = 1$ the right-hand side of (3) depends on the values of $\Gamma(x)$ in the interval $0 < x < 1$.   It is clear that $\Gamma(x)$ has been defined for negative x in such a way that equation (2) will hold for all x.

Theorem 6. $\Gamma(x + 1) = x\Gamma(x)$ $x \neq 0, -1, -2, \ldots$

From this result it is evident that it is necessary to tabulate the function only in an interval of length 1.   This is done on p. 136 of Peirce's Tables,† for example.   It is easy to plot the curve in character for $x > 0$ by use of Theorems 2 and 3 and from the fact that the curve is convex there.   The latter fact follows from the equation

(4) $\Gamma''(x) = \displaystyle\int_{0+}^\infty e^{-t} t^{x-1} (\log t)^2\, dt > 0$ $0 < x < \infty$

The curve may then be plotted for $x < 0$ by use of Theorem 6.   The graph of the function $y = \Gamma(x)$ is given accurately in Figure 46.

---

* See footnote in §7.1, Chapter 3.

$$y = \Gamma(x + 1)$$

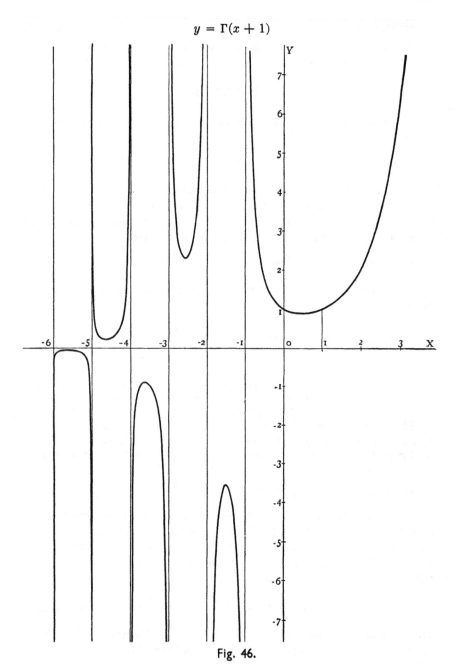

Fig. 46.

1.3 CERTAIN CONSTANTS RELATED TO $\Gamma(x)$

We shall show that $\Gamma(\tfrac{1}{2}) = \sqrt{\pi}$. In order to do this, we compute first the so-called *probability integral*.

Theorem 7. $\displaystyle\int_0^\infty e^{-x^2}\,dx = \tfrac{1}{2}\sqrt{\pi}$

To prove this, consider the double integral of $e^{-x^2-v^2}$ over the two circular sectors D_1 and D_2 and the square S indicated in Figure 47. Since the integrand is positive, we have

(5)
$$\iint_{D_1} < \iint_S < \iint_{D_2}$$

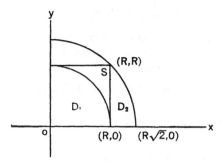

Fig. 47.

Now evaluate these integrals by iteration, the center one in rectangular coordinates, the other two in polar coordinates:

$$\int_0^R e^{-r^2} r\,dr \int_0^{\pi/2} d\theta < \int_0^R e^{-x^2}\,dx \int_0^R e^{-v^2}\,dy < \int_0^{R\sqrt{2}} e^{-r^2} r\,dr \int_0^{\pi/2} d\theta$$

$$\frac{\pi}{4}(1 - e^{-R^2}) < \left(\int_0^R e^{-x^2}\,dx\right)^2 < \frac{\pi}{4}(1 - e^{-2R^2})$$

Now let R become infinite and obtain

$$\left(\int_0^\infty e^{-x^2}\,dx\right)^2 = \pi/4$$

whence the desired result follows.

Theorem 8. $\Gamma(\tfrac{1}{2}) = \sqrt{\pi}$

For,
$$\Gamma(\tfrac{1}{2}) = \int_0^\infty e^{-t} t^{-1/2}\,dt$$
$$= 2\int_0^\infty e^{-v^2}\,dy = \sqrt{\pi} \qquad\qquad t = y^2$$

It is clear from the graph that the curve $y = \Gamma(x)$ has a minimum in the interval $(1, 2)$. The position of the minimum was computed by Gauss and found to be

$$x_0 = 1.461632145 \ldots$$

The minimum value of $\Gamma(x)$ in the interval $(0, \infty)$ is

$$\Gamma(x_0) = \min_{0 < x < \infty} \Gamma(x) = 0.885603 \ldots$$

A further fact of interest is the slope of the curve at $x = 1$. It can be shown (Exercise 23 of §4) that

$$\Gamma'(1) = -\gamma$$

where γ is *Euler's constant*, defined as follows:

Definition 3. $\quad \gamma = \lim_{n \to \infty} \left(\sum_{k=1}^{n} \frac{1}{k} - \log n \right)$

The limit is in the indeterminate form $\infty - \infty$. Its existence will be established later. The value of the number has been computed by J. C. Adams to 263 places of decimals:

$$= .57721566490153286061 \ldots$$

1.4 OTHER EXPRESSIONS FOR $\Gamma(x)$

Theorem 9. $\quad \Gamma(x) = r^x \int_{0+}^{\infty} e^{-rt} t^{x-1} \, dt \qquad\qquad 0 < r, \quad 0 < x$

This follows from Definition 1 by the change of variable $rt = y$. It is formula No. 508 in Peirce's Tables.

Theorem 10. $\quad \Gamma(x) = 2 \int_{0+}^{\infty} e^{-t^2} t^{2x-1} \, dt \qquad\qquad 0 < x < \infty$

Set $t^2 = y$. This is essentially No. 509.

EXERCISES (1)

In the following problems, numerical results should be obtained by use of p. 136 of Peirce's Tables.

1. Compute: $\Gamma(-\tfrac{1}{2})$, $\Gamma(\tfrac{7}{2})$, $\Gamma(2.135)$, $\Gamma(-3.728)$.

2. Compute: $\displaystyle\int_0^{\infty} e^{-t} \sqrt{t} \, dt$
   3. Compute: $\displaystyle\int_0^{\infty} e^{-st} \sqrt[3]{t^2} \, dt$

4. Compute: $\displaystyle\int_{0+}^{\infty} e^{-st} t^{-0.428} \, dt$
   5. Compute: $\displaystyle\int_{0+}^{\infty} e^{-t^2} \sqrt{t} \, dt$

6. $\displaystyle\lim_{x \to 0+} (1 - \cos x)^{1/2} \Gamma(x) = ?$
   7. $\displaystyle\lim_{x \to (-n)+} (x + n) \Gamma(x) = ?$

8. $\lim\limits_{x \to 0+} \Gamma(x) \int_0^x \dfrac{|\sin 2t|}{t} \, dt = ?$ **9.** $\lim\limits_{x \to +\infty} \dfrac{1}{\Gamma(1/x)} \int_0^x \dfrac{|\sin 2t|}{t} \, dt = ?$

10. $\lim\limits_{x \to 0} x^{n+1} \Gamma(x) \Gamma(x-1) \ldots \Gamma(x-n) = ?$

11. Prove: $\Gamma(x) = \displaystyle\int_{0+}^{1-} \left(\log \dfrac{1}{t} \right)^{x-1} dt$ $0 < x < \infty$

12. Prove: $\displaystyle\int_0^1 (\log x)^n \, dx = (-1)^n n!$

13. Prove: $\displaystyle\int_0^1 \left(\log \dfrac{1}{x} \right)^{1/2} dx = \dfrac{\sqrt{\pi}}{2}$

14. Prove: $\displaystyle\int_0^1 \left(\log \dfrac{1}{x} \right)^{-1/2} dx = \sqrt{\pi}$

15. Prove: $\displaystyle\int_0^1 t^{x-1} \left(\log \dfrac{1}{t} \right)^{y-1} dt = \Gamma(y) x^{-y}$ $x, y > 0$

16. Compute: $\displaystyle\int_{0+}^{1-} \dfrac{dx}{\sqrt{x \log (1/x)}}$

17. Prove equation (4).

18. Compute: $\displaystyle\int_0^{1-} (x/\log x)^{1/3} \, dx$

19. Prove: $\Gamma\left(n + \dfrac{1}{2} \right) = \dfrac{(2n)! \sqrt{\pi}}{4^n n!}$ $n = 0, 1, 2, \ldots$

20. Prove: $\Gamma(x) \in C^\infty$ $0 < x < \infty$

§2. The Beta Function

In this section we shall introduce a useful function of two variables known as the *beta function*. Its usefulness is considerably overshadowed by that of $\Gamma(x)$. In fact, we shall show that it can be evaluated in terms of the latter function. As a consequence, it would be unnecessary to introduce it as a new function. Since it occurs so frequently in analysis, a special designation for it is accepted.

2.1 DEFINITION AND CONVERGENCE

(1) **Definition 4.** $B(x, y) = \displaystyle\int_{0+}^{1-} t^{x-1} (1 - t)^{y-1} \, dt$ $0 < x, \; 0 < y$

To show that the integral converges for $0 < x, 0 < y$, we break it into two parts:

(2) $B(x, y) = \displaystyle\int_{0+}^{1/2} t^{x-1} (1 - t)^{y-1} \, dt + \int_{1/2}^{1-} t^{x-1} (1 - t)^{y-1} \, dt$

The first integral on the right clearly diverges for $x \le 0$, converges for $0 < x < 1$, and is proper for $1 \le x < \infty$, no matter what the value of y may be. If we set $u = 1 - t$, we have

$$\int_{1/2}^{1-} t^{x-1}(1-t)^{y-1}\,dt = \int_{0+}^{1/2} u^{y-1}(1-u)^{x-1}\,du$$

so that the discussion of the second integral on the right of equation (2) is reduced to that of the first. The results for $B(x, y)$ are indicated in Figure 48.

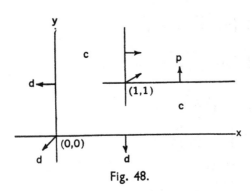

Fig. 48.

2.2 OTHER INTEGRAL EXPRESSIONS

Theorem 11. $B(x, y) = B(y, x)$ $\qquad\qquad 0 < x < \infty, 0 < y < \infty$

This follows by the change of variable $1 - t = u$ in equation (1).

Theorem 12. *For* $0 < x < \infty$, $0 < y < \infty$

$$B(x, y) = 2\int_{0+}^{\pi/2-} (\sin t)^{2x-1}(\cos t)^{2y-1}\,dt$$

To prove this, set $t = \sin^2 u$ in the integral (1).

Theorem 13. *For* $0 < x < \infty$, $0 < y < \infty$

$$B(x, y) = \int_{0+}^{\infty} \frac{t^{x-1}}{(1+t)^{x+y}}\,dt$$

Here the change of variable $t = u(1 + u)^{-1}$ suffices. This result is No. 497 of Peirce's Tables.

EXAMPLE A. $\displaystyle\int_0^\infty \frac{t^3}{(1+t)^7}\,dt = \frac{1}{60}$

By Theorem 13 the value of this integral is $B(4, 3)$. But, when x and y are positive integers, $B(x, y)$ can be evaluated by use of the binomial expansion.

$$B(4, 3) = \int_0^1 t^3(1 - t)^2\,dt = \int_0^1 (t^3 - 2t^4 + t^5)\,dt = \frac{1}{4} - \frac{2}{5} + \frac{1}{6} = \frac{1}{60}$$

2.3 RELATION TO $\Gamma(x)$

Theorem 14. $B(x, y) = \dfrac{\Gamma(x)\Gamma(y)}{\Gamma(x + y)}$ $\qquad 0 < x < \infty, \quad 0 < y < \infty$

We give a proof first when x and y are positive integers. As was evident in Example A above, the computation of $B(x, y)$ is particularly simple in this case. If $f(m)$ is a function of the integer m, we define a difference operator Δ upon it:

$$\Delta f(m) = f(m + 1) - f(m)$$

For example, if $m = 1, 2, 3, \ldots$

$$\Delta \frac{1}{m} = \frac{1}{m + 1} - \frac{1}{m} = \frac{-1}{m(m + 1)}$$

$$\Delta^2 \frac{1}{m} = \Delta \frac{-1}{m(m + 1)} = \frac{-1}{(m + 1)(m + 2)} + \frac{1}{m(m + 1)}$$

$$= \frac{2!}{m(m + 1)(m + 2)}$$

(3) $\qquad \Delta^n \dfrac{1}{m} = \Delta\left(\Delta^{n-1} \dfrac{1}{m}\right) = \dfrac{(-1)^n n!}{m(m + 1) \ldots (m + n)} \qquad n = 1, 2, \ldots$

Observe the analogy between $\Delta f(m)$ and $\dfrac{d}{dx} f(x)$, and specifically between $\Delta^n\left(\dfrac{1}{m}\right)$ and $\dfrac{d^n}{dx^n}\left(\dfrac{1}{x}\right)$.

Now it is clear by direct integration that

$$\frac{1}{m} = \int_0^1 t^{m-1} \, dt \qquad\qquad m = 1, 2, \ldots$$

and that

$$\Delta \frac{1}{m} = \int_0^1 (t^m - t^{m-1}) \, dt = -\int_0^1 t^{m-1}(1 - t) \, dt$$

$$\Delta^{n-1} \frac{1}{m} = (-1)^{n-1} \int_0^1 t^{m-1}(1 - t)^{n-1} \, dt$$

Hence, by equation (3)

$$\int_0^1 t^{m-1}(1 - t)^{n-1} \, dt = \frac{(n - 1)!}{m(m + 1) \ldots (m + n - 1)}$$

$$= \frac{(n - 1)!(m - 1)!}{(m + n - 1)!} = \frac{\Gamma(m)\Gamma(n)}{\Gamma(m + n)}$$

$$m, n = 1, 2, \ldots$$

This completes the proof when $x = m$, $y = n$.

When x and y are arbitrary positive numbers, the proof proceeds as follows. Form the double integral of the nonnegative function $t^{2x-1}u^{2y-1}e^{-t^2-u^2}$ over the three regions D_1, D_2, and S of Figure 47. Now, however, t and u are the variables; x and y, positive constants. We have relation (5) of §1.3 as before. Again we evaluate the central double integral by iteration in rectangular coordinates; the other two, in polar coordinates:

$$\int_0^{\pi/2} \cos^{2x-1}\theta \sin^{2y-1}\theta\, d\theta \int_0^R e^{-r^2} r^{2x+2y-1}\, dr < \int_0^R t^{2x-1}e^{-t^2}dt \int_0^R u^{2y-1}e^{-u^2}\, du$$

$$< \int_0^{\pi/2} \cos^{2x-1}\theta \sin^{2y-1}\theta\, d\theta \int_0^{R\sqrt{2}} e^{-r^2} r^{2x+2y-1}\, dr$$

Now, if we let R become infinite and use Theorems 10 and 12, we obtain

$$\frac{1}{2}B(y,x)\frac{1}{2}\Gamma(x+y) = \frac{\Gamma(x)}{2}\frac{\Gamma(y)}{2} \qquad 0 < x, \quad 0 < y$$

This completes the proof of the theorem. Note how Theorem 14 reveals the symmetry between x and y which was proved in Theorem 11.

2.4 WALLIS'S PRODUCT

As an application of the above results, let us establish an infinite product for $\pi/2$ known as *Wallis's product*.

Theorem 15. $\dfrac{\pi}{2} = \dfrac{2}{1}\dfrac{2}{3}\dfrac{4}{3}\dfrac{4}{5}\dfrac{6}{5}\dfrac{6}{7} \cdots \dfrac{2k}{2k-1}\dfrac{2k}{2k+1} \cdots$

By this is meant that if P_n is the product of the first n factors on the right-hand side,

$$\lim_{n\to\infty} P_n = \frac{\pi}{2}$$

By Theorems 12 and 14

$$\int_0^{\pi/2} \sin^{2n}x\, dx = \frac{\sqrt{\pi}\,\Gamma(n+\tfrac{1}{2})}{2(n!)} \qquad n = 0, 1, \ldots$$

(4)
$$\int_0^{\pi/2} \sin^{2n+1}x\, dx = \frac{\sqrt{\pi}\,n!}{2\Gamma(n+\tfrac{3}{2})} \qquad n = 0, 1, \ldots$$

Hence, the quotient of these two integrals is

(5)
$$\frac{\displaystyle\int_0^{\pi/2}\sin^{2n}x\,dx}{\displaystyle\int_0^{\pi/2}\sin^{2n+1}x\,dx} = \frac{\Gamma(n+\tfrac{1}{2})}{n!}\frac{\Gamma(n+\tfrac{3}{2})}{n!}$$

$$= \frac{2n+1}{2n}\frac{2n-1}{2n}\frac{2n-1}{2n-2} \cdots \frac{3}{4}\frac{3}{2}\frac{1}{2}\frac{\pi}{2}$$

$$= \frac{1}{P_{2n}}\frac{\pi}{2}$$

We shall now show that the left-hand side of equation (5) approaches 1 as $n \to \infty$. By equation (4) formed for n and for $n - 1$ we have

(6)
$$\int_0^{\pi/2} \sin^{2n+1} x \, dx = \frac{2n}{2n+1} \int_0^{\pi/2} \sin^{2n-1} x \, dx$$

Since $0 \leq \sin x \leq 1$ in the interval $(0, \pi/2)$, we have

$$0 < \int_0^{\pi/2} \sin^{2n+1} x \, dx < \int_0^{\pi/2} \sin^{2n} x \, dx < \int_0^{\pi/2} \sin^{2n-1} x \, dx$$

Dividing this inequality by the first of its integrals and allowing n to become infinite, we have by equation (6) that the left-hand side of equation (5) approaches 1. Hence,

$$\lim_{n \to \infty} P_{2n} = \frac{\pi}{2}$$

Also
$$\lim_{n \to \infty} P_{2n+1} = \lim_{n \to \infty} \frac{2n+2}{2n+1} P_{2n} = \frac{\pi}{2}$$

and the proof is complete.

Corollary 15. $\lim\limits_{n \to \infty} \dfrac{(n!)^2 2^{2n}}{(2n)! \sqrt{n}} = \sqrt{\pi}$

To prove this, multiply and divide the right-hand side of the equation

$$P_{2n} = \frac{2}{1} \frac{2}{3} \cdots \frac{2n}{2n-1} \frac{2n}{2n+1}$$

by $2 \cdot 2 \ldots 2n \cdot 2n$, thus introducing factorials in the denominator. If then factors 2 are segregated in the numerator, the result becomes apparent.

EXERCISES   (2)

1. Compute: $\displaystyle\int_0^1 t^3 (1 - t)^3 \, dt$ (2 ways)

2. Compute: $\displaystyle\int_0^1 \sqrt[3]{t(1 - t)} \, dt$

3. Compute: $\displaystyle\int_{0+}^1 \left(1 - \frac{1}{t}\right)^{1/3} dt$

4. Compute: $\displaystyle\int_0^{\pi/2-} \sqrt{\tan x} \, dx$

5. Compute: $\displaystyle\int_0^{\pi/2} (\sin 2x)^{1/4} \, dx$

6. Compute: $\displaystyle\int_{0+}^\infty \frac{1}{\sqrt{t}(1 + t)} \, dt$

7. Compute: $\displaystyle\int_0^\infty \frac{t\,dt}{(1+t)^3}$ (2 ways)

8. Compute: $\displaystyle\int_0^\infty \frac{dt}{(1+t)^2\sqrt{1+(1/t)}}$

9. Compute: $\displaystyle\int_{0+}^{\pi/2^-} (\sin 2x)^{2t-1}\,dx$ $0 < t < \infty$

10. Compute: $\displaystyle\int_0^1 t^{x-1}(\log t)(1-t)^{y-1}\,dt$ $0 < x,\ \ 0 < y$

Details involving uniform convergence may be omitted.

11. Prove: $B(x,x) = 2^{1-2x}B(x,\tfrac{1}{2})$ $0 < x < \infty$

   Hint: $B(x,x) = 2\displaystyle\int_0^{1/2} (t-t^2)^{x-1}\,dt.$ Set $t - t^2 = u$.

12. Prove: $\sqrt{\pi}\,\Gamma(2x) = 2^{2x-1}\Gamma(x)\Gamma(x+\tfrac{1}{2})$ $0 < x < \infty$

13. Show by direct computation that $\displaystyle\sum_{k=0}^3 (-1)^k \binom{3}{k}\frac{1}{k+6} = \frac{\Gamma(6)\Gamma(4)}{\Gamma(10)}$.
Check by use of $B(6,4)$.

14. Prove: $\displaystyle\sum_{k=0}^n (-1)^k \binom{n}{k}\frac{1}{m+k+1} = \frac{m!\,n!}{(m+n+1)!}$ $m, n = 1, 2, \ldots$

15. Complete the proof of Corollary 15.

16. Try Wallis's product on a slide rule.

17. Find the area inside the curve

$$x^{2/3} + y^{2/3} = 1$$

18. Find the numerical value of

$$\int_0^1 \frac{dx}{\sqrt[3]{1+x^6}}$$

   Hint: Set $x^3 + x^{-3} = 2/\sqrt{t}$.

§3. *Evaluation of Definite Integrals*

The values of many definite integrals can be obtained even when there exists no corresponding indefinite integral in terms of the elementary functions. Great ingenuity is frequently required, each integral demanding some special device. Certain general methods can be described, however, and we illustrate them here by examples.

3.1 DIFFERENTIATION WITH RESPECT TO A PARAMETER

EXAMPLE A. $f(x) = \int_0^\infty e^{-t^2} \cos xt \, dt$ $-\infty < x < \infty$

The integral converges absolutely for all x.   Then

(1) $f'(x) = -\int_0^\infty e^{-t^2} t \sin xt \, dt$

by Theorem 14, Chapter 10.   Integration by parts gives

$$f'(x) = -\frac{x}{2} \int_0^\infty e^{-t^2} \cos xt \, dt = -\frac{x}{2} f(x)$$

Integrating this differential equation, we obtain

$$f(x) = Ce^{-x^2/4}$$

To determine the constant of integration, set $x = 0$ and use Theorem 7:

$$\int_0^\infty e^{-t^2} \cos xt \, dt = \frac{\sqrt{\pi}}{2} e^{-x^2/4}$$

This is essentially No. 523, Peirce's Tables.

EXAMPLE B. $f(x) = \int_0^\infty e^{-t^2 - x^2 t^{-2}} \, dt$ $-\infty < x < \infty$

(2) $f'(x) = -2x \int_0^\infty e^{-t^2 - x^2 t^{-2}} t^{-2} \, dt$

Assume first that $x > 0$, and make the substitution $x = tu$:

$$f'(x) = -2 \int_0^\infty e^{-u^2 - x^2 u^{-2}} \, du = -2f(x)   0 < x < \infty$$

Integrating this differential equation, we have

$$f(x) = Ce^{-2x}   0 < x < \infty$$

To determine the constant of integration C let $x \to 0+$.   We know that $f(x)$ is continuous at $x = 0$, since the given integral is obviously uniformly convergent in any finite interval.   Clearly $C = \sqrt{\pi}/2$.   Finally, observing that $f(-x) = f(x)$, we obtain

$$\int_0^\infty e^{-t^2 - x^2 t^{-2}} \, dt = \frac{\sqrt{\pi}}{2} e^{-2|x|}   -\infty < x < \infty$$

This is No. 510, Peirce's Tables.

3.2 USE OF SPECIAL LAPLACE TRANSFORMS

A *Laplace transform* is an integral of the form

$$(3) \qquad f(x) = \int_0^\infty e^{-xt}\varphi(t)\, dt$$

It may be regarded as an operation which *transforms* one function, $\varphi(t)$, into another, $f(x)$. For example, if $\varphi(t) = t^n/n!$, we see by Theorem 9 that $f(x) = x^{-n-1}$. As another example, let us obtain the Laplace transform of $\varphi(t) = \sin at$. By use of the indefinite integral or by two integrations by parts, we obtain

$$\int_0^\infty e^{-xt} \sin at\, dt = \frac{a}{a^2 + x^2}$$

$$-\infty < a < \infty, \quad 0 < x < \infty$$

In like manner,

$$\int_0^\infty e^{-xt} \cos at\, dt = \frac{x}{a^2 + x^2}$$

$$-\infty < a < \infty, \quad 0 < x < \infty$$

These are No. 522 and No. 521, respectively.

If a definite integral includes as a factor of the integrand a power t^{-n} or a quotient $a/(a^2 + t^2)$ or $t/(a^2 + t^2)$, the value of the integral can sometimes be obtained by expressing that factor itself as the integral (3) and then interchanging the order of integration.

EXAMPLE C. $\displaystyle \int_0^\infty \frac{\sin x}{x}\, dx = \frac{\pi}{2}$

For, we have

$$\int_0^\infty \frac{\sin x}{x}\, dx = \int_0^\infty \sin x\, dx \int_0^\infty e^{-xt}\, dt = \int_0^\infty dt \int_0^\infty e^{-xt} \sin x\, dx$$

By No. 522

$$\int_0^\infty \frac{\sin x}{x}\, dx = \int_0^\infty \frac{1}{1 + t^2}\, dt = \frac{\pi}{2}$$

To justify this procedure show first (Exercise 19) that the integral

$$(1+X^2)^{-1} = \int_0^\infty e^{-xt} \sin t\, dt$$

converges uniformly on $\epsilon \leqq X \leqq R$, $\epsilon > 0$. Then by Theorem 13, Chapter 10,

$$\tan^{-1} R - \tan^{-1} \epsilon = \int_0^\infty e^{-\epsilon t}\, \frac{\sin t}{t}\, dt - \int_0^\infty e^{-Rt}\, \frac{\sin t}{t}\, dt$$

$$= I_1\,(\epsilon) - I_2\,(R).$$

Since $(\sin t)/t \leqq 1$ we have

$$|I_2(R)| \leqq 1/R \to 0 \qquad R \to \infty$$

Since the integral $I_1(0)$ converges by Theorem 6, Chapter 10, we may apply Exercise 6, p. 350, to show that the integral $I_1(\epsilon)$ converges uniformly for $0 \leqq \epsilon \leqq 1$. Then by Theorem 12, Chapter 10,

$$\lim_{\epsilon \to 0}\left(\frac{\pi}{2} - \tan^{-1}\epsilon\right) = \lim_{\epsilon \to 0} I_1(\epsilon) = \int_0^\infty \frac{\sin t}{t}\, dt.$$

3.3 THE METHOD OF INFINITE SERIES

In some cases it is useful to expand the integral in infinite series and to integrate the series term by term. The following series will be found useful; the sums given will be verified in §4.3, Chapter 12.

$$\frac{\pi^2}{6} = 1 + \frac{1}{2^2} + \frac{1}{3^2} + \cdots$$

$$\frac{\pi^2}{24} = \frac{1}{2^2} + \frac{1}{4^2} + \frac{1}{6^2} + \cdots$$

$$\frac{\pi^2}{8} = 1 + \frac{1}{3^2} + \frac{1}{5^2} + \cdots$$

EXAMPLE D. $\displaystyle\int_{0+}^1 \frac{\log x}{1-x}\, dx = -\frac{\pi^2}{6}$

For,

$$\frac{\log x}{1-x} = \sum_{k=0}^\infty x^k \log x$$

$$\int_{0+}^1 \frac{\log x}{1-x}\, dx = \sum_{k=0}^\infty \int_{0+}^1 x^k \log x\, dx = -\sum_{k=1}^\infty \frac{1}{k^2}$$

by No. 536. To justify the term-by-term integration, it will be sufficient to show that

(4) $$\lim_{n \to \infty} \int_{0+}^1 \frac{x^{n+1} \log x}{1-x}\, dx = 0$$

as we see by use of the remainder of a geometric series. Since

(5) $$\max_{0 \leqq x \leqq 1}\left|\frac{x \log x}{1-x}\right| = 1$$

we have

$$\left|\int_{0+}^1 \frac{x^{n+1} \log x}{1-x}\, dx\right| \leqq \int_0^1 x^n\, dx = \frac{1}{n+1}$$

whence equation (4) follows immediately.

EXERCISES (3)

*In Exercises 1–14, details involving uniform convergence may be omitted.
Prove the following results,*

1. $\displaystyle\int_0^\infty \frac{\sin xt}{t}\, dt = \frac{\pi}{2}$, if $x > 0$; 0, if $x = 0$; $-\frac{\pi}{2}$, if $x < 0$.

2. $\displaystyle\int_0^\infty \frac{\sin t \cos xt}{t}\, dt = 0$, if $|x| > 1$; $\frac{\pi}{2}$, if $|x| < 1$; $\frac{\pi}{4}$, if $x = \pm 1$.

3. $\displaystyle\int_0^\infty \frac{\sin^2 t}{t^2}\, dt = \frac{\pi}{2}$ (2 ways). *Hint:* (a) Integrate by parts; (b) use the method of §3.2.

4. $\displaystyle\int_0^\infty \frac{\cos t}{\sqrt{t}}\, dt = \int_0^\infty \frac{\sin t}{\sqrt{t}}\, dt = \sqrt{\frac{\pi}{2}}$. *Hint:* The method of §3.2 leads to the integral $\displaystyle\int_{0+}^\infty (1 + t^2)^{-1} t^{-1/2}\, dt$. This may be evaluated by partial fractions after the substitution $t = u^2$. Or set $t^2 = u$ and then use Theorem 13 and Exercise 12, §2.

5. $\displaystyle\int_0^\infty \cos{(t^2)}\, dt = \int_0^\infty \sin{(t^2)}\, dt = \frac{\sqrt{\pi}}{2\sqrt{2}}$. *Hint:* Use the previous exercise.

6. $\displaystyle\int_0^\infty \frac{\cos xt}{1 + t^2}\, dt = \frac{\pi}{2} e^{-|x|}$. *Hint:* Use the method of §3.2. The resulting integral is the derivative of original integral except for sign.

7. $\displaystyle\int_0^\infty \frac{dt}{\cosh xt} = \frac{\pi}{2|x|}$ $x \neq 0$

Hint: Set $e^{xt} = u$.

8. $\displaystyle\int_0^1 \frac{\log t}{1 + t}\, dt = -\frac{\pi^2}{12}$.

9. $\displaystyle\int_0^1 \frac{\log t}{1 - t^2}\, dt = -\frac{\pi^2}{8}$.

10. $\displaystyle\int_0^\infty \frac{t\, dt}{\sinh xt} = \frac{\pi^2}{4x^2}$ $x \neq 0$

Hint: Use the previous exercise.

11. $\displaystyle\int_0^1 \log\left(\frac{1 + t}{1 - t}\right)\frac{dt}{t} = \frac{\pi^2}{4}$.

12. $\displaystyle\int_0^{\pi/2} \log{(\sin t)}\, dt = \int_0^{\pi/2} \log{(\cos t)}\, dt = -\frac{\pi}{2}\log 2$. *Hint:* Add the two integrals; then set $2t = y$.

13. $\int_0^1 \dfrac{\log t}{\sqrt{1 - t^2}} \, dt = -\dfrac{\pi}{2} \log 2.$

14. $\int_0^\pi t \log (\sin t) \, dt = -\dfrac{\pi^2}{2} \log 2.$ *Hint:* Write $\int_0^\pi = \int_0^{\pi/2} + \int_{\pi/2}^\pi$; set $\pi - t = y$ in the last integral.

15. By the method of §3.1 show that

$$F(x) = \int_0^\infty e^{-xt} \frac{\sin t}{t} \, dt = \cot^{-1} x$$

If the constant of integration is determined by $F(\infty) = 0$, a new proof of Example C is obtained by letting $x \to 0+$ (Exercise 12, §6, Chapter 10).

16. Give details in the proof of equation (1).

17. Solve the same problem for equation (2).

18. Solve the same problem for equation (5).

19. Show that the integral

$$\int_0^\infty e^{-xt} \sin t \, dt$$

converges uniformly on $\epsilon \leqq \times \leqq R, \epsilon > 0$.

20. Show that the answer to Exercise 3, §8, Chapter 10, is $(\sqrt{2} - 1)\sqrt{\pi}$.

§4. *Stirling's Formula*

In this section we shall obtain an estimate of the rate at which $n!$ becomes infinite with n.   Observe that when n is large it is extremely difficult to compute $n!$, even with the help of logarithms.   For example, if one wished to determine the number of possible shuffles of an ordinary deck of cards, $52!$, one's task would be time consuming.   We shall show that in a certain precise sense $(n/e)^n\sqrt{2\pi n}$ is a good approximation for $n!$ when n is large. The value of this function is very easily computed for any n if logarithm tables are available.   The equation

(1) $$\lim_{n \to \infty} \frac{(n/e)^n\sqrt{2\pi n}}{n!} = 1$$

is known as *Stirling's formula*.

21. Show that series $\Sigma \times^k \log \times$ of Example D does not converge uniformly on $0 \leqq \times \leqq 1$.

4.1 PRELIMINARY RESULTS

For greater clarity in the proof of equation (1), we introduce several simple lemmas.

Lemma 16.1. $\quad \log \left(1 + \dfrac{1}{n}\right) > \dfrac{2}{2n + 1}$ $\qquad n = 1, 2, \ldots$

This is clear from Figure 49.

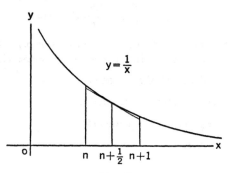

Fig. 49.

Since the curve $y = 1/x$ is convex, the area under the curve from $x = n$ to $x = n + 1$ is greater than the area of the trapezoid bounded by these two ordinates, the x-axis, and the tangent to the curve at the point $\left(n + \dfrac{1}{2}, \dfrac{2}{2n + 1}\right)$:

$$\int_n^{n+1} \frac{dx}{x} = \log \left(1 + \frac{1}{n}\right) > \frac{2}{2n + 1}$$

The area of a trapezoid is equal to the product of the length of the median by the length of the base.

Lemma 16.2. $\quad$ 1. $a_n = \dfrac{n!}{(n/e)^n \sqrt{n}}$ $\qquad n = 1, 2, \ldots$

$\Rightarrow \qquad \lim\limits_{n \to \infty} a_n$ *exists.*

Note first that the sequence $\{a_n\}_1^\infty \in \downarrow$. For,

$$\frac{a_n}{a_{n+1}} = \frac{\left(1 + \dfrac{1}{n}\right)^{n+1/2}}{e} > 1$$

since by Lemma 16.1

$$\left(n + \frac{1}{2}\right) \log \left(1 + \frac{1}{n}\right) > 1$$

Since $a_n > 0$ for all n, the proof is complete.

Lemma 16.3. $\lim\limits_{n \to \infty} a_n > 0$

To prove this, observe that the sum of the areas of the circumscribed trapezoids and of the two rectangles at the ends is greater than the area under the curve in Figure 50.   The altitudes of the two rectangles at the ends of the figure are 2 and $\log n$; (note that $2 > \log 1.5$).   The tops of the trapezoids are segments of tangents to the curve at points with integral abscissas and are terminated by the lines $x = k + \frac{1}{2}$, $k = 1, 2, \ldots, n - 1$.   It is unessential to the argument that these segments do not form a continuous broken line.

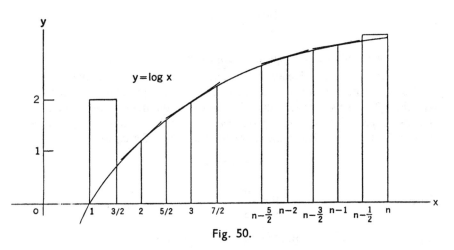

Fig. 50.

The area of the trapezoids and the two rectangles is

$$1 + \log 2 + \log 3 + \ldots + \log(n - 1) + \tfrac{1}{2} \log n = 1 + \log n! - \log \sqrt{n}$$

The area under the curve is

$$\int_1^n \log x \, dx = n \log n - n + 1 = \log(n/e)^n + 1$$

Hence,

$$\log \left(\frac{n}{e}\right)^n < \log \frac{n!}{\sqrt{n}}$$

$$\frac{(n/e)^n \sqrt{n}}{n!} < 1 \qquad\qquad n = 1, 2, \ldots$$

Consequently,

$$a_n > 1, \qquad \lim_{n \to \infty} a_n \geqq 1$$

We have proved more than stated.   It is only the nonvanishing of the limit which is needed.

4.2 PROOF OF STIRLING'S FORMULA

Theorem 16. $\displaystyle\lim_{n\to\infty} \frac{(n/e)^n\sqrt{2\pi n}}{n!} = 1$

We need only show that

$$\lim_{n\to\infty} a_n = r = \sqrt{2\pi}$$

We use Corollary 15 to evaluate r. The function of n appearing in Corollary 15 can be rewritten in terms of a_n as follows:

(2)
$$\frac{(n!)^2 2^{2n}}{(2n)!\sqrt{n}} = \frac{a_n^2}{a_{2n}}\frac{1}{\sqrt{2}}$$

As n becomes infinite, this quotient approaches $\sqrt{\pi}$ on the one hand and $r^2/(r\sqrt{2})$ on the other. Hence, $r = \sqrt{2\pi}$, and the proof is complete.

Observe where Lemma 16.3 enters the proof. In taking the limit in equation (2), we use the fact that the limit of a quotient is the quotient of the limits, *provided the limit of the denominator is not zero.* Suppose we neglect the latter proviso in the following example. Set $a_n = e^{-n}$ and

$$r = \lim_{n\to\infty} a_n$$

Then
$$\lim_{n\to\infty} \frac{a_n^2}{a_{2n}} = \frac{r^2}{r} = r$$

But
$$\lim_{n\to\infty} \frac{a_n^2}{a_{2n}} = \lim_{n\to\infty} \frac{(e^{-n})^2}{e^{-2n}} = 1$$

and
$$r = \lim_{n\to\infty} e^{-n} = 0$$

so that we have "proved" that $1 = 0$.

EXAMPLE A. $\displaystyle\lim_{n\to\infty} \frac{(2n)!e^{2n}}{(2n)^{2n}} = +\infty$

For,
$$\lim_{n\to\infty} \frac{(2n)!e^{2n}}{(2n)^{2n}} = \lim_{n\to\infty} \left(\frac{(2n)!}{(2n)^{2n}e^{-2n}\sqrt{4\pi n}}\right)\sqrt{4\pi n} = +\infty$$

In calculating limits involving factorials, one should not indiscriminately replace $n!$ by $(n/e)^n\sqrt{2\pi n}$; rather the quotient a_n should be introduced. See Exercise 10.

EXAMPLE B. $\displaystyle\frac{(n+p)!}{n!} \sim n^p$ $\qquad\qquad n\to\infty, \quad p = 1, 2, \ldots$

The symbol $\sim$ is here read "is asymptotic to."   We say that $a_n \sim b_n$, $n \to \infty$, $\Leftrightarrow \lim\limits_{n \to \infty} (a_n/b_n) = 1$. We can prove this result in two ways.   By Stirling's formula

$$\lim_{n \to \infty} \frac{(n - p)!}{n! n^p}$$

$$= \lim_{n \to \infty} \frac{(n + p)!}{(n + p)^{n+p} e^{-n-p} \sqrt{2\pi(n + p)}} \frac{n^n e^{-n} \sqrt{2\pi n}}{n!} \frac{(1 + (p/n))^{n+p+1/2}}{e^p} = 1$$

Each of the three quotients on the right clearly approaches 1.   On the other hand,

$$\frac{(n + p)!}{n! n^p} = \left(1 + \frac{p}{n}\right)\left(1 + \frac{p - 1}{n}\right) \cdots \left(1 + \frac{1}{n}\right)$$

and each of the p factors on the right approaches 1 as $n \to \infty$.

EXAMPLE C. $\lim\limits_{n \to \infty} \dfrac{1}{n} \sqrt[n]{n!} = \dfrac{1}{e}$

By Stirling's formula,

$$\lim_{n \to \infty} \frac{\sqrt[n]{n!}}{n} = \lim_{n \to \infty} \left(\frac{n!}{(n/e)^n \sqrt{2\pi n}}\right)^{1/n} \frac{(2\pi n)^{1/(2n)}}{e} = \frac{1}{e}$$

Assuming that the limit exists, we can check its value by use of the series

$$\sum_{n=1}^{\infty} \frac{n!}{n^n} x^n$$

By the ratio test, it converges for $|x| < e$ and diverges for $|x| > e$:

$$\lim_{n \to \infty} \frac{(n + 1)! n^n}{(n + 1)^{n+1} n!} |x| = \lim_{n \to \infty} \left(1 + \frac{1}{n}\right)^{-n} |x| = e^{-1}|x|$$

But by the root test we have, if the required limit is r, that

$$\lim_{n \to \infty} \sqrt[n]{\frac{n!}{n^n} |x|^n} = |x| r$$

Hence, r must be e^{-1} to conform with the known convergence facts.

4.3 EXISTENCE OF EULER'S CONSTANT

A result related to the foregoing consideration is the following.

Theorem 17.   1. $g(x) \in C, \downarrow$ $1 \leqq x < \infty$

2. $g(x) \geqq 0$ $1 \leqq x < \infty$

$\Rightarrow$ $\lim\limits_{n \to \infty} \left[\sum\limits_{k=1}^{n} g(k) - \int_{1}^{n} g(x)\, dx\right]$ exists.

For, by hypothesis 1,

(3) $$g(k) \leq \int_{k-1}^{k} g(x)\, dx \leq g(k-1) \qquad k = 2, 3, \ldots, n$$

(4) $$\sum_{k=2}^{n} g(k) \leq \int_{1}^{n} g(x)\, dx \leq \sum_{k=1}^{n-1} g(k)$$

Set $$C_n = \sum_{k=1}^{n} g(k) - \int_{1}^{n} g(x)\, dx \qquad n = 1, 2, \ldots$$

Then by inequalities (4)

$$0 \leq g(n) \leq C_n \leq g(1)$$

Moreover, $\{C_n\}_1^{\infty} \in \downarrow$, since by inequalities (3)

$$C_n - C_{n-1} = g(n) - \int_{n-1}^{n} g(x)\, dx \leq 0$$

Since the sequence $\{C_n\}_1^{\infty}$ is nonnegative and nonincreasing, $\lim_{n \to \infty} C_n$ exists.

EXAMPLE D. $\lim_{n \to \infty} \left(\sum_{k=1}^{n} \dfrac{1}{k^2} + \dfrac{1}{n} \right)$ exists.

The result is obvious here since the form is not indeterminate. In fact, we know from other considerations that the limit is $\pi^2/6$, but the existence of the limit follows from Theorem 17 if $g(x) = x^{-2}$. This example shows that the theorem is of interest only if

$$\int_{1}^{\infty} g(x)\, dx = \infty$$

EXAMPLE E.   Euler's constant exists.   Take $g(x) = x^{-1}$.   Then

$$\lim_{n \to \infty} \left(\sum_{k=1}^{n} \frac{1}{k} - \log n \right)$$

exists.

4.4 INFINITE PRODUCTS

In §2.4 appeared an example of an infinite product. The value of that product was there defined as a specific limit, but before proceeding to an infinite product for $\Gamma(x)$ let us give a general definition for the convergence of such a product. We admit into consideration only those infinite products

(5) $$\prod_{k=1}^{\infty} p_k = p_1 p_2 p_3 \cdots$$

for which at most a finite number of factors p_k vanish. That is, we assume that there exists a constant N such that $p_k \neq 0$ when $k \geq N$.

Definition 5. *The infinite product* (5) *converges*

$$\Leftrightarrow \qquad \lim_{n \to \infty} \prod_{k=N}^{N+n} p_k = A \neq 0$$

If (5) *converges its value is*

$$A \prod_{k=1}^{N-1} p_k$$

The product (5) *diverges* $\Leftrightarrow$ *it does not converge.*

The requirement that $A \neq 0$ is made so that certain basic properties of finite products will carry over to infinite products. For example, under Definition 5 the value of a convergent product cannot be zero unless some factor vanishes. Also, since $A \neq 0$ the reciprocal of a convergent product no factor of which is zero is convergent.

EXAMPLE F. $\displaystyle\prod_{k=2}^{\infty} \frac{k^2}{k^2 - 1}$ converges and has the value 2.

For,

$$\prod_{k=2}^{n} \frac{k^2}{k^2 - 1} = \frac{2n}{n + 1} \to 2 \qquad\qquad (n \to \infty)$$

EXAMPLE G. $\displaystyle\prod_{k=1}^{\infty} \frac{k}{k + 1}$ diverges.

For,

$$\prod_{k=1}^{n} \frac{k}{k + 1} = \frac{1}{n + 1} \to 0 \qquad\qquad (n \to \infty)$$

EXAMPLE H. $\displaystyle\prod_{k=1}^{\infty} \sin \frac{k\pi}{2}$

Here convergence and divergence are undefined, since the product has infinitely many vanishing factors.

4.5 AN INFINITE PRODUCT FOR $\Gamma(x)$

Lemma 18.1. $\displaystyle 0 \leqq x \leqq 1 \Rightarrow \lim_{n \to \infty} \frac{\Gamma(x + n)}{\Gamma(n) \, n^x} = 1$

First observe, by integration by parts, that

(6) $$\int_n^\infty e^{-t} t^n \, dt = e^{-n} n^n + n \int_n^\infty e^{-t} t^{n-1} \, dt$$

Since $x \geqq 0$ and $x - 1 \leqq 0$

$$t^x \leqq n^x, \qquad n^{x-1} \leqq t^{x-1} \qquad\qquad 0 \leqq t \leqq n$$

Multiply the first inequality by $e^{-t}t^{n-1}$, the second by $e^{-t}t^n$ and integrate:

$$(7) \qquad n^{x-1}\int_0^n e^{-t}t^n\,dt \leqq \int_0^n e^{-t}t^{n+x-1}\,dt \leqq n^x\int_0^n e^{-t}t^{n-1}\,dt$$

In a similar way

$$n^x \leqq t^x, \qquad t^{x-1} \leqq n^{x-1} \qquad\qquad n \leqq t < \infty$$

$$(8) \qquad n^x\int_n^\infty e^{-t}t^{n-1}\,dt \leqq \int_n^\infty e^{-t}t^{n+x-1}\,dt \leqq n^{x-1}\int_n^\infty e^{-t}t^n\,dt$$

By use of (6), inequalities (8) become

$$-e^{-n}n^{n+x-1} + n^{x-1}\int_n^\infty e^{-t}t^n\,dt \leqq \int_n^\infty e^{-t}t^{n+x-1}\,dt$$

$$\leqq n^x\int_n^\infty e^{-t}t^{n-1}\,dt + e^{-n}n^{n+x-1}$$

Adding these inequalities to (7) and dividing by $n^x\Gamma(n)$, we obtain

$$-\frac{e^{-n}n^n}{n!} + 1 \leqq \frac{\Gamma(x+n)}{\Gamma(n)\,n^x} \leqq 1 + \frac{e^{-n}n^n}{n!}$$

Both extremes approach 1 as $n \to \infty$, by Stirling's formula, so that the proof is complete.

Lemma 18.2. $\quad 0 \leqq x < \infty \Rightarrow \lim_{n\to\infty} \dfrac{\Gamma(x+n)}{\Gamma(n)\,n^x} = 1$

The result follows from a proof by induction, Exercise 22.

Theorem 18. $\quad x > 0 \Rightarrow \dfrac{1}{\Gamma(x)} = e^{\gamma x}x\prod_{k=1}^\infty \left(1 + \dfrac{x}{k}\right)e^{-x/k}$

Set

$$P_n = x\prod_{k=1}^{n-1}\left(1 + \frac{x}{k}\right)e^{-x/k} = \frac{e^{-c_n x}}{\Gamma(n)}\prod_{k=0}^{n-1}(x+k)$$

$$c_n = 1 + \frac{1}{2} + \dots + \frac{1}{n-1}$$

By Theorem 1

$$\frac{1}{\Gamma(x)} = \frac{\displaystyle\prod_{k=0}^{n-1}(x+k)}{\Gamma(x+n)} = \frac{P_n\Gamma(n)\,e^{c_n x}}{\Gamma(x+n)}$$

$$= P_n e^{(c_n - \log n)x}\left[\frac{\Gamma(n)\,n^x}{\Gamma(x+n)}\right]$$

Here we have multiplied and divided by $n^x = e^{x\log n}$ in order to introduce the quotient of Lemma 18.2. By Example E, $c_n - \log n \to \gamma$. Hence $P_n \to e^{-\gamma x}/\Gamma(x)$ as desired.*

---

*A. Pringsheim, *Zur Theorie der Gamma-Functionen*, Mathematische Annalen, Vol. 31, 1888, p. 458.

EXERCISES (4)

1. Compute a_n, §4.1, for $n = 2$ and for $n = 10$.

2. Compute $(n/e)^n \sqrt{2\pi n}$ for $n = 52$.

3. Prove equation (2).

4. $\lim\limits_{n \to \infty} \sqrt[n]{n!} = ?$

5. $\lim\limits_{n \to \infty} (n!)^{1/(n \log n)} = ?$

6. $\lim\limits_{n \to \infty} \left(\dfrac{1}{n} \log n! - \log n \right) = ?$

7. $\dbinom{n}{p} \sim ?$ $\qquad\qquad\qquad\qquad\qquad n \to \infty, \; p = 1, 2, \ldots$

8. $\dbinom{2n}{n} \sim ?$ $\qquad\qquad\qquad\qquad\qquad\qquad n \to \infty$

9. $\dbinom{3n}{n} \sim ?$ $\qquad\qquad\qquad\qquad\qquad\qquad n \to \infty$

10. If a_n is defined as in Lemma 16.2, it can be proved that $(a_n/\sqrt{2\pi}) - 1 \sim \dfrac{1}{12n}$, $n \to \infty$. Assuming this, prove

$$\lim_{n \to \infty} (a_n/\sqrt{2\pi})^n = \sqrt[12]{e}$$

This example shows that it is not always legitimate to replace $n!$ by $(n/e)^n \sqrt{2\pi n}$ in the calculation of limits.

11. Prove that the following limit exists:

$$\lim_{n \to \infty} \left(\sum_{k=2}^{n} \frac{1}{k \log k} - \log \log n \right)$$

12. Obtain a result similar to that of Exercise 11 involving the function $\log \log \log n$.

13. Prove by use of inequalities (4) that

$$\frac{1}{\log 2} \leq \sum_{k=2}^{\infty} \frac{1}{k(\log k)^2} \leq \frac{1}{\log 2} + \frac{1}{2(\log 2)^2}$$

14. Prove: $\lim\limits_{n \to \infty} \left| \dfrac{u_{n+1}}{u_n} \right| = A \Rightarrow \lim\limits_{n \to \infty} \sqrt[n]{|u_n|} = A$. *Hint:* Take the $(n + p)$th root of each term of the inequalities

$$(A - \epsilon)^p |u_n| < |u_{n+p}| < |u_n|(A + \epsilon)^p$$

and then apply $\overline{\lim}\limits_{p \to \infty}$ and $\lim\limits_{p \to \infty}$. Compare Exercise 20, §5, Chapter 8.

15. Prove that the converse of the result of Exercise 14 is not true. *Hint:* Take $u_{2n} = u_{2n+1} = 2^n$.

16. Use Exercise 14 to evaluate $\lim_{n \to \infty} (2n)^{1/n}$.

17. Same problem for

$$\lim_{n \to \infty} \frac{1}{(2n+1)^2} \left[\frac{(4n+1)!}{(2n+3)!} \right]^{1/n}$$

Check by Stirling's formula.

18. $\lim_{n \to \infty} \dfrac{1}{\log an} \sum_{k=1}^{n} \dfrac{1}{k} = ?$ $\qquad (0 < a)$

19. Prove:

$$\lim_{n \to \infty} \sum_{k=pn}^{qn} \frac{1}{k} = \log \frac{q}{p} \qquad p < q, \ p, q = 1, 2, 3, \ldots$$

20. Prove:

$$\sum_{k=n+1}^{2n} \frac{1}{k} = \sum_{k=1}^{2n} \frac{(-1)^{k+1}}{k} \to \log 2 \qquad\qquad n \to \infty$$

21. $\lim_{n \to \infty} \left[\displaystyle\int_0^1 \frac{1 - t^n}{1 - t} \, dt - \int_1^n \frac{(1+t)^2}{t+t^3} \, dt \right] = ?$

22. Prove Lemma 18.2.

23. Prove $\Gamma'(1) = -\gamma$. *Hint:* Take the logarithm of each side of the equation of Theorem 18, differentiate, and set $x = 1$. Details of rigor may be omitted.

12
Fourier Series

§1. Introduction

In this chapter we shall be discussing series of the form

(1) $$\frac{A_0}{2} + \sum_{k=1}^{\infty}(A_k \cos kx + B_k \sin kx)$$

We shall be interested particularly in discussing what functions $f(x)$ can be expressed as the sum of such a series. Series of this type occur very frequently in the problems of mathematical physics. They were applied by Fourier to the study of heat conduction, and, as a consequence, certain of the series (1) are known as *Fourier series*. We shall study in some detail one physical application of Fourier series, the problem of a vibrating string. Finally, we shall give a brief discussion of the Fourier integral, an integral representation of a function, analogous to the series above.

1.1 DEFINITIONS

Definition 1. *The series* (1) *is a trigonometric series.*

Definition 2. *The series*

(2) $$\frac{a_0}{2} + \sum_{k=1}^{\infty}(a_k \cos kx + b_k \sin kx)$$

393

is the Fourier series of the function f(x)

$\Leftrightarrow$

$$a_k = \frac{1}{\pi} \int_{-\pi}^{\pi} f(x) \cos kx \, dx \qquad k = 0, 1, 2, \ldots$$

(3)

$$b_k = \frac{1}{\pi} \int_{-\pi}^{\pi} f(x) \sin kx \, dx \qquad k = 1, 2, 3, \ldots$$

EXAMPLE A. Let $f(x) = \dfrac{\pi}{4}$ when $0 < x \le \pi$ and $f(x) = -\dfrac{\pi}{4}$ when $-\pi \le x \le 0$. Then

$$a_k = 0 \qquad\qquad\qquad\qquad\qquad\qquad k = 0, 1, 2, \ldots$$

$$b_k = \frac{1}{2} \int_0^{\pi} \sin kx \, dx = \frac{1}{2k} [1 - \cos k\pi] \qquad k = 1, 2, 3, \ldots$$

Hence, the Fourier series for this function $f(x)$ is

$$\sin x + \frac{\sin 3x}{3} + \frac{\sin 5x}{5} + \cdots$$

Observe that in the definition of a Fourier series no mention of convergence, much less of the sum of the series, is made. A Fourier series is a trigonometric series whose coefficients bear a definite relation (3) to some function $f(x)$.

EXAMPLE B. $\displaystyle\sum_{k=2}^{\infty} \frac{\sin kx}{\log k}$ is a trigonometric series which is not a Fourier series. It can be shown* that there is no function $f(x)$ related to the coefficients by equations (3). The series converges for all x.

1.2 ORTHOGONALITY RELATION

We recall that two vectors $\vec{f}$ and $\vec{g}$ with components f_k and g_k, $k = 1, 2, 3$, are orthogonal if, and only if,

$$\vec{f} \cdot \vec{g} = \sum_{k=1}^{3} f_k g_k = 0$$

that they are of unit length or *normed* if, and only if,

$$\vec{f} \cdot \vec{f} = \sum_{k=1}^{3} f_k^2 = 1, \qquad \vec{g} \cdot \vec{g} = \sum_{k=1}^{3} g_k^2 = 1$$

These notions could be extended to a space of n dimensions by extending the above sums over n rather than three terms. It is possible to conceive of a function $f(x)$ as a vector with infinitely many components corresponding

---

* Titchmarsh, E. C., *The Theory of Functions* 1939, p. 420.

to the infinitely many points of a line segment (a, b).   It is such notions
that lead to the terminology in the following definitions.

Definition 3.   *The functions $f(x)$ and $g(x)$ are orthogonal on the interval*
$a \leqq x \leqq b$

$$\Leftrightarrow \qquad \int_a^b f(x)g(x)\, dx = 0$$

Definition 4.   *The function $f(x)$ is normed on the interval $a \leqq x \leqq b$*

$$\Leftrightarrow \qquad \int_a^b f^2(x)\, dx = 1$$

The terms of series (1) form good examples of orthogonal functions.
Each term is orthogonal to each other term on the interval $(-\pi, \pi)$.   A
given term will be normed only if the corresponding coefficient A_k or B_k is
suitable.   We have, in fact, the following orthogonality and normality
relations:

$$\int_{-\pi}^{\pi} \cos mx \cos nx\, dx = 0 \qquad m \neq n; \quad m, n = 0, \pm 1, \pm 2, \ldots$$

$$\int_{-\pi}^{\pi} \sin mx \sin nx\, dx = 0 \qquad m \neq n; \quad m, n = 0, \pm 1, \pm 2, \ldots$$

$$\int_{-\pi}^{\pi} \cos mx \sin nx\, dx = 0 \qquad \qquad m, n = 0, \pm 1, \pm 2, \ldots$$

$$\frac{1}{\pi} \int_{-\pi}^{\pi} \sin^2 nx\, dx = 1 \qquad \qquad n = 1, 2, \ldots$$

$$\frac{1}{\pi} \int_{-\pi}^{\pi} \cos^2 nx\, dx = 1 \qquad \qquad n = 1, 2, \ldots$$

Let us prove the first and last of these equations only.   From elementary
trigonometry we have

$$\cos mx \cos nx = \tfrac{1}{2} \cos (m - n)x + \tfrac{1}{2} \cos (m + n)x$$

Integrating over $(-\pi, \pi)$, we obtain the desired result.

By use of these facts, we can obtain a useful relation between trigonometric
series and Fourier series.

Theorem 1.   1. *Series* (1) *converges uniformly to $f(x)$ in $-\pi \leqq x \leqq \pi$*
$\Rightarrow$ *It is the Fourier series of $f(x)$.*

For, if we multiply the series by $\cos nx$, it remains uniformly convergent
in $-\pi \leqq x \leqq \pi$ and can be integrated term by term:

$$\int_{-\pi}^{\pi} f(x) \cos nx\, dx = \frac{A_0}{2} \int_{-\pi}^{\pi} \cos nx\, dx + \sum_{k=1}^{\infty} A_k \int_{-\pi}^{\pi} \cos kx \cos nx\, dx$$

$$+ \sum_{k=1}^{\infty} B_k \int_{-\pi}^{\pi} \sin kx \cos nx\, dx$$

By the orthogonality and normality relation, we have

$$\frac{1}{\pi} \int_{-\pi}^{\pi} f(x) \cos nx \, dx = A_n \qquad n = 0, 1, 2, \ldots$$

The constants B_n are determined in a similar way.

This theorem shows a relation between the defining function of a Fourier series and its sum. We shall be able to show that for a very large class of functions the Fourier series converges to its defining function.

1.3 FURTHER EXAMPLES OF FOURIER SERIES

When we compute the coefficients of a Fourier series from its defining function, it is useful to recall the following facts:

$$f(x) \text{ is } even \Leftrightarrow f(-x) = f(x)$$
$$f(x) \text{ is } odd \Leftrightarrow f(-x) = -f(x)$$
$$f(x) \text{ is even} \Rightarrow \int_{-a}^{a} f(x) \, dx = 2 \int_{0}^{a} f(x) \, dx$$
$$f(x) \text{ is odd} \Rightarrow \int_{-a}^{a} f(x) \, dx = 0$$

The numbers a_k, b_k of equations (3) are known as the *Fourier coefficients* of $f(x)$.

EXAMPLE C. $f(x) = x$ in the interval $-\pi \leq x \leq \pi$. The Fourier coefficients of this function are

$$a_k = 0, \qquad \text{since } f(x) \text{ is odd}$$
$$b_k = \frac{2}{\pi} \int_{0}^{\pi} x \sin kx \, dx = (-1)^{k+1} 2/k \qquad k = 1, 2, \ldots$$

The Fourier series for x is

$$2 \sin x - \frac{2 \sin 2x}{2} + \frac{2 \sin 3x}{3} - \frac{2 \sin 4x}{4} + \cdots$$

EXAMPLE D. $f(x) = |x|$ $\qquad\qquad\qquad -\pi \leq x \leq \pi$

$$a_k = \frac{2}{\pi} \int_{0}^{\pi} x \cos kx \, dx = \frac{2}{\pi k^2} [\cos k\pi - 1] \qquad k = 1, 2, \ldots$$
$$a_0 = \pi$$
$$b_k = 0, \qquad \text{since } f(x) \text{ is even.}$$

The Fourier series for $|x|$ is

$$\frac{\pi}{2} - \frac{4}{\pi}\left(\cos x + \frac{\cos 3x}{3^2} + \frac{\cos 5x}{5^2} + \cdots \right)$$

EXERCISES (1)

Find the Fourier series corresponding to the functions of Exercises 1–6.

1. $f(x) = x^2$ $\hspace{6cm} -\pi \leqq x \leqq \pi$

2. $f(x) = x^2$ $\hspace{6.3cm} 0 \leqq x \leqq \pi$

$\hspace{1cm} = -x^2$ $\hspace{6cm} -\pi \leqq x \leqq 0$

3. $f(x) = \dfrac{\pi}{4}$ $\hspace{6cm} -\pi \leqq x \leqq \pi$

4. $f(x) = \sin^2 x$ $\hspace{5.5cm} -\pi \leqq x \leqq \pi$

5. $f(x) = x$ $\hspace{6.3cm} 0 \leqq x \leqq \pi/2$

$\hspace{1cm} = \pi - x$ $\hspace{5.7cm} \pi/2 \leqq x \leqq \pi$

$\hspace{1cm} = -f(-x)$ $\hspace{5.7cm} -\pi \leqq x \leqq 0$

6. $f(x) = \cos cx$ $\hspace{5.7cm} -\pi \leqq x \leqq \pi$

Here c may or may not be an integer.

Answer: If c is not an integer, the Fourier series for $\cos cx$ is

$$\frac{2c}{\pi} \sin c\pi \left(\frac{1}{2c^2} + \sum_{k=1}^{\infty} \frac{(-1)^k \cos kx}{c^2 - k^2} \right)$$

7. Plot carefully the two functions

$$f(x) = \frac{\pi}{4} x \hspace{4cm} 0 \leqq x \leqq \pi/2$$

$$= \frac{\pi}{4} (\pi - x) \hspace{3.2cm} \pi/2 \leqq x \leqq \pi$$

$$g(x) = \sin x - \frac{\sin 3x}{3^2} \hspace{2.3cm} 0 \leqq x \leqq \pi$$

Note that $g(x)$ is the sum of the first two terms in the Fourier series for $f(x)$.

8. Prove the rest of the orthogonality and normality relations.

9. Prove that, if a function is multiplied by a constant, each of its Fourier coefficients is multiplied by that constant. What happens to the Fourier coefficients when the constant is added to the function?

10. Is the following series a Fourier series?

$$\sum_{k=2}^{\infty} \frac{\cos (2k^2 - k + 7)x}{k(\log k)^2}$$

11. Prove:

1. $\lim_{k \to \infty} k^p A_k = A$ $\hspace{4.5cm} p > 1$

2. $\lim_{k \to \infty} k^p B_k = B$ $\hspace{4.5cm} p > 1$

$\Rightarrow$ Series (1) is a Fourier series.

12. Express the Fourier coefficients of $f'(x)$ and $f''(x)$ in terms of those of $f(x)$, assuming $f \in C^2$ and $f(x + 2\pi) = f(x)$.

13. Prove: The sequence of Fourier coefficients of a continuous function is a bounded sequence.

14. Prove:

1. $f(x) \in C^p$ $-\pi \leqq x \leqq \pi$

2. $f^{(k)}(-\pi) = f^{(k)}(\pi)$ $k = 0, 1, \ldots, p - 1$

3. $|f^{(p)}(x)| \leqq M$ $-\pi \leqq x \leqq \pi$

$\Rightarrow |a_k| \leqq 2Mk^{-p}$, $|b_k| \leqq 2Mk^{-p}$ $k = 1, 2, \ldots$

Hint: Integrate the integrals (3) by parts.

15. Prove: $f(x) \in C^2$ and $f(x + 2\pi) = f(x) \Rightarrow$ the Fourier series for $f(x)$ converges uniformly in every finite interval. Does this prove, by the aid of Theorem 1, that the sum of the series is $f(x)$?

16. If $f(x)$ is defined in $(0, \pi)$, its definition in $(-\pi, 0)$ can be given so as to make $f(x)$ either even or odd. What do the Fourier series and the Fourier coefficients of $f(x)$ become in the two cases?

§2. Several Classes of Functions

Examples A, C, and D of §1 illustrate classes of functions which frequently appear in the theory of Fourier series. We shall be able to show that the series of those three examples actually converge to their defining function, at least at points of continuity of the interval $(-\pi, \pi)$. But each term of the Fourier series has period 2π, so that the sum of the series must also have that period. Hence, if the series of Example D, say, is to converge to $f(x)$ from $-\infty$ to $+\infty$, $f(x)$ must be defined outside the interval $(-\pi, \pi)$ so that $f(x + 2\pi) = f(x)$ for all x. If this is done, the graph of the function in Example D will have a saw-toothed appearance. The continuous graph is really composed of infinitely many line segments joined together. Before the time of Fourier such a graph was not thought to define a function at all, but many different functions. It may well be imagined that the mathematicians of Fourier's time experienced a severe shock with the knowledge that such a saw-toothed combination could be represented from $-\infty$ to $+\infty$ by a Fourier series, each term of which belongs to C^∞ in $(-\infty, \infty)$. Example A must have been even more surprising, for there the sum of the series is discontinuous.

Let us point out the properties of the functions of Examples A, C, D that are essential for the convergence of their Fourier series to these defining functions. The functions are continuous except for a finite number of points in every finite interval. They have period 2π. At all but at most a finite number of points of each finite interval, the graphs of the functions have definite slopes. Indeed at all points, even at points of discontinuity, the graphs have right-hand and left-hand slopes. In order to avoid repetition of these various properties, we shall define several new classes of functions.

2.1 THE CLASSES P, D, D¹

In the rest of this chapter we shall suppose, unless otherwise stated, that all functions are defined from $-\infty$ to $+\infty$. The classes we are about to introduce include such functions only.

Definition 5. $f(x) \in P \iff f(x + 2\pi) = f(x)$ $\qquad -\infty < x < \infty$

EXAMPLE A. $\quad f(x) = \pi/4$ $\qquad\qquad 2k\pi < x < (2k + 1)\pi$
$$k = 0, \pm1, \pm2, \ldots$$
$$= -\pi/4 \qquad\qquad (2k - 1)\pi < x < 2k\pi$$
$$k = 0, \pm1, \pm2, \ldots$$
$$= 0 \qquad\qquad x = k\pi; \quad k = 0, \pm1, \pm2, \ldots$$

We shall show that the Fourier series for this function (see Example A, §1) converges to the function for *all* x. Observe that the definition of $f(x)$ can be changed at any finite number of points of $-\pi \le x \le \pi$ without affecting the Fourier coefficients of the function. We have altered the definition from that given in §1 (at $x = \pi$ and $x = -\pi$) so as to get convergence at *all* points. It is obvious by inspection that the sum of the Fourier series is 0 at $x = 0, \pm\pi, \pm2\pi, \ldots$. Clearly $f(x) \in P$.

Definition 6. $f(x)$ *has a finite jump at* $x = c$

$\iff$ $\qquad$ A. $f(c+)$ *exists*

$\qquad$ B. $f(c-)$ *exists*

$\qquad$ C. $f(c+) \ne f(c-)$

In Example A $f(x)$ has a finite jump at $x = 0, \pm\pi, \pm2\pi, \ldots$. For example, $f(0+) = \pi/4$, $f(0-) = -\pi/4$. The function $1/x$ has an *infinite jump* at $x = 0$.

Definition 7. $f(x) \in D \iff f(x) \in C$ *except for at most a finite number of finite jumps in every finite interval.*

To show that $f(x) \in D$, it will be sufficient to show that $f(c+)$ and $f(c-)$ exist for every c and that the two values are equal to $f(c)$ with but few exceptions. If $f(x) \in P$, the exceptions must be finite in number in the interval $-\pi \le x \le \pi$. Of course, a single discontinuity in that interval produces infinitely many in $(-\infty, \infty)$. The functions of Examples A, C, D all belong to D. Obviously, $f \in C \Rightarrow f \in D$.

Definition 8. $f(x) \in D^1$

$\Leftrightarrow$ A. $f(x) \in D$

       B. *The graph of $f(x)$ has a right-hand and left-hand slope at every point.*

The geometric language needs analytic elucidation, especially when $f(x) \notin C$. If $f(x) \in C$ at $x = c$, clearly the right-hand slope and the left-hand slope at c are, respectively,

$$\lim_{\Delta x \to 0+} \frac{f(c + \Delta x) - f(c)}{\Delta x}, \qquad \lim_{\Delta x \to 0-} \frac{f(c + \Delta x) - f(c)}{\Delta x}$$

But for the function of Example A these limits do not exist at $c = 0$ (they are $+\infty$ and $-\infty$). Yet we wish to agree that the graph in that case does have right-hand and left-hand slopes ($= 0$) at every point. Clearly, what is needed is the following:

(1)   right-hand slope of $f(x)$ at $c = \lim\limits_{\Delta x \to 0+} \dfrac{f(c + \Delta x) - f(c+)}{\Delta x}$

(2)   left-hand slope of $f(x)$ at $c = \lim\limits_{\Delta x \to 0-} \dfrac{f(c + \Delta x) - f(c-)}{\Delta x}$

To show that $f \in D^1$ we must show that $f \in D$ and that the limits (1) and (2) exist for all c. At most points these limits will be computed by the ordinary rules of differentiation.

EXAMPLE B. $f(x) = x \sin (1/x)$ $x \neq 0$

             $f(0) = 0$

Here $f(x) \in C, f(x) \in D, f(x) \notin D^1$. Clearly,

$$\lim_{\Delta x \to 0+} \frac{f(0 + \Delta x) - f(0+)}{\Delta x} = \lim_{\Delta x \to 0+} \sin (1/\Delta x)$$

does not exist. Note that the functions of Examples A, C, D, §1, all belong to D^1. The function of Example A does not belong to C.

EXAMPLE C. $f(x) = x \sin (1/x)$ $0 < x < \infty$

           $= 1$ $-\infty < x \leq 0$

Here $f(x) \in D, f(x) \notin C, f(x) \notin D^1$. Clearly, $f(x) \in C^\infty$ except at $x = 0$. But $f(0+) = 0 \neq f(0-) = 1$. Finally, the limit (1) does not exist, though the limit (2) is zero.

2.2 RELATION AMONG THE CLASSES

The interrelations among the various classes of functions which we have considered are best kept in mind by use of Figure 51.   Each point inside a given circle is thought of as corresponding to a function of the class that the whole circle represents.   A point common to several circles indicates the existence of a function belonging to all of the corresponding classes.   To show that the classes have the relation indicated in the figure, one must show the existence of at least one function corresponding to the various regions into which the plane is divided by the circles. These examples are inserted in the figure. The class corresponding to a given circle is marked on the circumference of that circle.   Obviously, the choice of a circle for the region is unimportant.   In logic Figure 51 would be called a Venn diagram   We shall show that $f \in P$, $f \in D^1 \Rightarrow$ the Fourier series for $f(x)$ converges to $f(x)$ at all points of continuity.

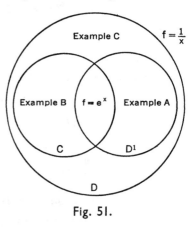

Fig. 51.

2.3 ABBREVIATIONS

To shorten the writing in subsequent work, we shall introduce the following abbreviations:

$$C_0(x) = a_0/2$$

$$C_k(x) = a_k \cos kx + b_k \sin kx \qquad\qquad k = 1, 2, \dots$$

(3) $$S_n(x) = \sum_{k=0}^{n} C_k(x) \qquad\qquad n = 0, 1, 2, \dots$$

Here a_k and b_k are defined by equations (3), of §1.1, so that the notation applies only to Fourier series of a given function $f(x)$.   This Fourier series can now be written as

$$\sum_{k=0}^{\infty} C_k(x)$$

and we shall want to prove that, if $f(x) \in P$, $f(x) \in D^1$, then at points of continuity of $f(x)$

$$\lim_{n \to \infty} S_n(x) = f(x)$$

EXERCISES (2)

To which, if any, of the classes C, C^1, D, D^1, P do the functions of Exercises 1–10 belong?

1. $f(x) = \sin 17x$; $f(x) = \sin 3\pi x$; $f(x) = \cos x/2$.

2. $f(x) = \sqrt{|x|}$.

3. $f(x) = \sqrt{|x|^3}$.

4. $f(x) = e^{-1/x}$ (define $f(x)$ at $x = 0$).

5. $f(x) = xe^{-1/x}$ (define $f(x)$ at $x = 0$).

6. $f(x) = \sum\limits_{k=1}^{\infty} \dfrac{\cos kx}{k^3}$.

7. $f(x) = \int_0^x \dfrac{e^t}{\sqrt{|t|}}\, dt$.

8. $f(x) = [\pi x]$; $[a]$ means the largest integer $\leq a$.

9. $f(x) = x^{-1}\sin(1/x)$, $x \neq 0$; $f(0) = 0$.

10. $f(x) = 1$, x rational; $f(x) = 0$, x irrational.

11. Prove:

    1. $f(x) \in D$
    2. $g(x) \in C$
    $\Rightarrow$ $f(x)g(x) \in D$.

12. Give an example where $f(x) \in D$, $f(x) \notin C$, $g(x) \in C$, $f(x)g(x) \in C$.

13. Prove:

    1. $f(x) \in D^1$
    2. $g(x) \in C^1$
    $\Rightarrow$ $f(x)g(x) \in D^1$.

14. Insert a circle for the class C^1 in Figure 51 and insert the examples necessary to show the correctness of your drawing.

15. Solve the same problem for P. The new region need not be a circle.

16. Prove:

    1. $f(x) \in D$ $-\infty < x < \infty$
    2. $f(x) \in C^1$ $x \neq 0$
    3. $\lim\limits_{x \to 0} f'(x) = A$
    $\Rightarrow$ $f(x) \in D^1$

17. In Exercise 16, give an example to show $f(x) \notin C$. Give another to show that hypothesis 3 cannot be omitted.

§3. *Convergence of a Fourier Series to Its Defining Function*

In this section we shall prove that, if $f(x) \in P, f(x) \in D^1$, then the Fourier series for $f(x)$ converges to $f(x)$ at points of continuity. To prove this result, we shall need certain preliminary results, which are of interest in themselves.

3.1 BESSEL'S INEQUALITY

Theorem 2. 1. $f(x) \in D$

$$\Rightarrow \qquad \frac{a_0^2}{2} + \sum_{k=1}^{n} (a_k^2 + b_k^2) \leqq \frac{1}{\pi} \int_{-\pi}^{\pi} f^2(x)\, dx \qquad\qquad n = 1, 2, \ldots$$

Of course, the a_k and b_k are the Fourier coefficients of $f(x)$ defined by equations (3), §1. By these equations and by the orthogonality relations, §1.2, we have for any positive integer n

(1) $\quad \dfrac{a_0^2}{2} = \dfrac{1}{\pi} \int_{-\pi}^{\pi} \dfrac{a_0}{2} f(t)\, dt = \dfrac{1}{\pi} \int_{-\pi}^{\pi} \dfrac{a_0}{2} S_n(t)\, dt$

(2) $\quad a_k^2 = \dfrac{1}{\pi} \int_{-\pi}^{\pi} a_k f(t) \cos kt\, dt = \dfrac{1}{\pi} \int_{-\pi}^{\pi} a_k S_n(t) \cos kt\, dt \qquad k = 1, 2, \ldots, n$

(3) $\quad b_k^2 = \dfrac{1}{\pi} \int_{-\pi}^{\pi} b_k f(t) \sin kt\, dt = \dfrac{1}{\pi} \int_{-\pi}^{\pi} b_k S_n(t) \sin kt\, dt \qquad k = 1, 2, \ldots, n$

Here $S_n(t)$ is defined by equation (3), §2.3. Adding all the equations (1), (2), (3) ($k = 1, 2, \ldots, n$), we obtain

(4) $\qquad \dfrac{a_0^2}{2} + \sum_{k=1}^{n} (a_k^2 + b_k^2) = \dfrac{1}{\pi} \int_{-\pi}^{\pi} f(t) S_n(t)\, dt = \dfrac{1}{\pi} \int_{-\pi}^{\pi} S_n^2(t)\, dt$

Since

(5) $\quad \displaystyle\int_{-\pi}^{\pi} (f(t) - S_n(t))^2\, dt = \int_{-\pi}^{\pi} f^2(t)\, dt - 2\int_{-\pi}^{\pi} f(t) S_n(t)\, dt + \int_{-\pi}^{\pi} S_n^2(t)\, dt$

and since the left-hand side of equation (5) is nonnegative we have, by equation (4), that

$$\frac{1}{\pi} \int_{-\pi}^{\pi} f^2(t)\, dt - \frac{a_0^2}{2} - \sum_{k=1}^{n} (a_k^2 + b_k^2) \geqq 0$$

and the proof is complete. The hypothesis $f(x) \in D$ insures that the integrals involved in the proof all exist.

EXAMPLE A. Take $f(x) = \pm\pi/4$ as in Example A, §1. Then Bessel's inequality becomes

$$1 + \frac{1}{3^2} + \frac{1}{5^2} + \cdots + \frac{1}{(2n+1)^2} \leqq \frac{1}{\pi} \int_{-\pi}^{\pi} \frac{\pi^2}{16}\, dt = \frac{\pi^2}{8}$$

This is evident directly since

$$\frac{\pi^2}{8} = \sum_{k=0}^{\infty} \frac{1}{(2k+1)^2}$$

Corollary 2. $f(x) \in D \Rightarrow \dfrac{a_0^2}{2} + \sum_{k=1}^{\infty} (a_k^2 + b_k^2) < \infty$

3.2 THE RIEMANN-LEBESGUE THEOREM

A result proved first by Riemann for continuous functions and later extended by Lebesgue to more general functions is now called the *Riemann-Lebesgue theorem*.   The result that we shall prove here is a special case of the general theorem, but entirely adequate for the convergence theorem in the proof of which it is needed.   A more general result will be proved later.

Theorem 3. 1. $f(x) \in D$

$$\Rightarrow \qquad \lim_{k \to \infty} \int_{-\pi}^{\pi} f(t) \cos kt \, dt = \lim_{k \to \infty} \int_{-\pi}^{\pi} f(t) \sin kt \, dt = 0$$

By Corollary 2 it is clear that

$$\lim_{k \to \infty} a_k = \lim_{k \to \infty} b_k = 0$$

since the general term of a convergent series approaches zero.   This proves the theorem.

Corollary 3. $\displaystyle\lim_{k \to \infty} \int_{-\pi}^{\pi} f(t) \sin (k + \tfrac{1}{2})t \, dt = 0$

This is proved by expanding $\sin (k + \tfrac{1}{2})t$,

$$\sin (k + \tfrac{1}{2})t = \sin kt \cos \tfrac{1}{2}t + \cos kt \sin \tfrac{1}{2}t$$

and applying the theorem to the functions $f(t) \cos (t/2)$ and $f(t) \sin (t/2)$.

EXAMPLE B.   In Corollary 3 take $f(t) = 1 \ (t > 0), f(t) = -1 \ (t < 0)$:

$$\int_{-\pi}^{\pi} f(t) \sin \left(k + \frac{1}{2}\right) t \, dt = \frac{4}{2k+1}$$

It is evident that this tends to zero when k becomes infinite.

EXAMPLE C. $f(x) = 0 \ (x \leq 0), \qquad f(x) = x^{-1} \ (x > 0)$
Here $f(x) \notin D$.   We have

$$\int_{-\pi}^{\pi} f(t) \sin kt \, dt = \int_{0}^{\pi} \frac{\sin kt}{t} \, dt = \int_{0}^{k\pi} \frac{\sin t}{t} \, dt$$

Hence,

$$\lim_{k \to \infty} \int_{-\pi}^{\pi} f(t) \sin kt \, dt = \int_{0}^{\infty} \frac{\sin t}{t} \, dt = \frac{\pi}{2}$$

This example shows that hypothesis 1 cannot be omitted.

3.3 THE REMAINDER OF A FOURIER SERIES

A compact integral form of the remainder of a Fourier series can now be obtained. It depends on the following trigonometric identity.

Theorem 4. *For* $-\infty < x < \infty$

(6) $\frac{1}{2} + \cos x + \cos 2x + \ldots + \cos nx = D_n(x) = \dfrac{\sin (n + \frac{1}{2})x}{2 \sin (x/2)}$

At points where $\sin (x/2) = 0$, it is understood that the indeterminate form on the right is to be replaced by its limiting value. To prove the identity, note that

$$2 \sin \tfrac{1}{2}x \cos kx = \sin (k + \tfrac{1}{2})x - \sin (k - \tfrac{1}{2})x$$

Hence,

$$2 \sin \frac{x}{2} \left[\frac{1}{2} + \sum_{k=1}^{n} \cos kx \right] = \sin \frac{x}{2} + \left[\sin \frac{3x}{2} - \sin \frac{x}{2} \right]$$

$$+ \ldots + \left[\sin \left(n + \frac{1}{2} \right)x - \sin \left(n - \frac{1}{2} \right)x \right]$$

$$= \sin (n + \tfrac{1}{2})x$$

The result is now evident when $\sin (x/2) \neq 0$. If $x = 0, \pm 2\pi, \pm 4\pi, \ldots$, it is sufficient to apply a limiting process:

$$n + \frac{1}{2} = \lim_{x \to 0} \frac{\sin (n + \frac{1}{2})x}{2 \sin (x/2)}$$

The function $D_n(x)$ is called the *Dirichlet kernel*.

Corollary 4.1. $D_n(x) \in P$ $n = 0, 1, 2, \ldots$

The functions in numerator and denominator have the period 4π, but the quotient has period 2π.

Corollary 4.2. $\dfrac{1}{\pi} \displaystyle\int_{-\pi}^{\pi} D_n(x) \, dx = 1$

This is obtained by integrating both sides of equation (6).

Theorem 5.   1. $f(x) \in D$

2. $f(x) \in P$

$\Rightarrow \qquad f(x) - S_n(x) = \dfrac{1}{\pi} \displaystyle\int_{-\pi}^{\pi} [f(x) - f(x+t)] D_n(t) \, dt$

$$n = 0, 1, 2, \ldots$$

By Corollary 4.2 it is sufficient to prove that

$$S_n(x) = \frac{1}{\pi} \int_{-\pi}^{\pi} f(x+t) D_n(t) \, dt$$

By the definition of $S_n(x)$ and by Theorem 4 we have

$$S_n(x) = \sum_{k=0}^{n} C_k(x)$$

$$= \frac{1}{\pi} \int_{-\pi}^{\pi} f(t) \left[\frac{1}{2} + \cos(x-t) + \ldots + \cos n(x-t) \right] dt$$

$$= \frac{1}{\pi} \int_{-\pi}^{\pi} f(t) D_n(x-t) \, dt$$

Now set $t - x = u$:

(7) $\qquad S_n(x) = \dfrac{1}{\pi} \displaystyle\int_{-\pi-x}^{\pi-x} f(x+u) D_n(u) \, du$

But by Corollary 4.1 the integrand, considered as a function of u, has period 2π. Hence, the limits of integration on the integral (7) may be replaced by $-\pi$ and π.

3.4 THE CONVERGENCE THEOREM

Theorem 6.   1. $f(x) \in P$

2. $f(x) \in D^1$

3. $f(x) \in C \qquad$ *at* $x = x_0$

$\Rightarrow \qquad f(x_0) = \displaystyle\sum_{k=0}^{\infty} C_k(x_0)$

By Theorem 5 we need show only that

$$\lim_{n \to \infty} \int_{-\pi}^{\pi} [f(x_0 + t) - f(x_0)] \frac{\sin\left(n + \frac{1}{2}\right)t}{2 \sin(t/2)} \, dt = 0$$

This will follow by Corollary 3 if

$$g(t) = \frac{f(x_0 + t) - f(x_0)}{2 \sin(t/2)} \in D$$

Since $g(t + 2\pi) = -g(t)$, it will be sufficient to show that $g(t)$ has at most a finite number of finite jumps in the interval $-\pi \leq x \leq \pi$. But in that interval $g(t)$ has the same discontinuities as $f(x_0 + t)$ with a possible additional one at $t = 0$. But

$$g(0+) = \lim_{t \to 0+} \frac{f(x_0 + t) - f(x_0)}{t} \lim_{t \to 0+} \frac{t}{2 \sin (t/2)}$$

$$= \lim_{t \to 0+} \frac{f(x_0 + t) - f(x_0)}{t}$$

For $g(0-)$ replace $t \to 0+$ by $t \to 0-$. Since $f(x) \in D^1$, these limits exist and are, in fact, the right-hand and left-hand slopes of $f(x)$ at x_0. This completes the proof.*

Observe where hypothesis 3 enters the proof. If, for example, $f(x_0+) > f(x_0)$, then $g(0+) = +\infty$ even though $f(x) \in D^1$.

EXAMPLE D. If $f(x)$ is the function of Example A, it is clear that $f(x) \in P, f(x) \in D^1$. Hence, for $n = 0, \pm 1, \pm 2, \ldots$

$$\sum_{k=0}^{\infty} \frac{\sin (2k + 1)x}{2k + 1} = \frac{\pi}{4} \qquad 2n\pi < x < (2n + 1)\pi$$

$$= -\frac{\pi}{4} \qquad (2n - 1)\pi < x < 2n\pi$$

$$= 0 \qquad x = n\pi$$

The value of the sum of the series at the points of discontinuity of $f(x)$ cannot be found by Theorem 6. For the present simple example the value can be determined by inspection.

EXERCISES (3)

1. What does Bessel's inequality become for Example C, §1? Verify the result.

2. What does Bessel's inequality become for Example D, §1?

3. $\lim\limits_{k \to \infty} \int_0^\pi \dfrac{\sin x \cos kx}{\sqrt[3]{x^2}} \, dx = ?$

4. $\lim\limits_{k \to \infty} \int_{-\pi}^\pi \sin^2 kx \, dx = ?$

5. $\lim\limits_{k \to \infty} \int_0^\pi \sqrt{\sin x} \sin^2 kx \, dx = ?$

---

* Note that hypothesis 2 is stronger than needed. For the convergence of the Fourier series to $f(x_0)$ at x_0, it is sufficient to know the behaviour of $f(x)$ in a *neighborhood* of x_0.

6. $\displaystyle\lim_{k\to\infty}\int_0^1 x\sqrt[3]{\log x}\,\cos^2 kx\,dx = ?$

7. $\displaystyle\lim_{k\to\infty}\frac{1}{k}\int_0^\pi \frac{\sin^2 kt}{t^2}\,dt = ?$

8. Write out the remainder in integral form for the series of Example C and of Example D, §1.

9. Prove: $\displaystyle\sum_{k=1}^{n}\sin kx = \frac{\cos(x/2) - \cos(n + \frac{1}{2})x}{2\sin(x/2)}$

$$= \frac{\sin[(nx)/2]\sin[(n + 1)x/2]}{\sin(x/2)}$$

10. Prove: $\displaystyle\sum_{k=1}^{n}\cos(2k - 1)x = \frac{\sin 2nx}{2\sin x}$.   Are there exceptions?

11. Prove analytically:

1. $f(x) \in C$
2. $f(x) \in P$

$$\Rightarrow \quad \int_0^{2\pi} f(x)\,dx = \int_{a-\pi}^{a+\pi} f(x)\,dx \qquad\qquad -\infty < a < \infty$$

12. Apply Theorem 6 to Examples C and D, §1.

13. Prove Theorem 6 if hypothesis 2 is replaced by 2′: $f(x) \in D$ and $f'_+(x_0)$, $f'_-(x_0)$ both exist.

14. Apply the previous exercise to Example B of §2.   For what values of x_0 will the Fourier series of $f(x)$ converge to $f(x_0)$, as guaranteed by that exercise?

15. Same problem for Example C of §2.

§4. *Extensions and Applications*

In this section, we shall make several applications of Theorem 6 and extend it to include points of discontinuity.   In addition, we shall extend Riemann's theorem to include the case in which the interval of integration is arbitrary instead of $(-\pi, \pi)$ and in which the variable becomes infinite continuously instead of through the integers.

4.1 POINTS OF DISCONTINUITY

Theorem 7.   1. $f(x) \in P$

2. $f(x) \in D^1$

(1) $\Rightarrow$ $\qquad\qquad \dfrac{f(x+) + f(x-)}{2} = \displaystyle\sum_{k=0}^{\infty} C_k(x) \qquad\qquad -\infty < x < \infty$

At points of continuity of $f(x)$, the left-hand side of equation (1) is equal to $f(x)$ and the result is given by Theorem 6.   Equation (1) is clearly true for Example D, §3.   Denote the sum of the Fourier series of that example by $g(x)$.   Then $g(0+) = \pi/4$, $g(0-) = -\pi/4$, $g(0) = (\pi - \pi)/8 = 0$. Theorem 7 is obviously valid for $g(x)$.

Let $x = c$ be an arbitrary point of discontinuity of $f(x)$.   Consider the function

(2) $$h(x) = f(x) - \frac{2J}{\pi} g(x - c) \qquad J = f(c+) - f(c-)$$

Then

$$h(c) = f(c), \qquad h(c+) = f(c+) - \frac{2J}{\pi}\frac{\pi}{4} = \frac{f(c+) + f(c-)}{2}$$

$$h(c-) = \frac{f(c+) + f(c-)}{2}$$

If we alter* the definition of $h(x)$, if necessary, so that

$$h(c) = h(c+) = h(c-)$$

then $h(x) \in C$ at $x = c$.   By Theorem 6 the sum of the Fourier series for $h(x)$ at $x = c$ is equal to $h(c)$.   By equation (2) the Fourier series for $h(x)$ is the sum of that for $f(x)$ and the one for $-2J\, g(x - c)/\pi$.   But the sum of the latter series at $x = c$ is zero.

Hence,

$$\frac{f(c+) + f(c-)}{2} = \sum_{k=1}^{\infty} C_k(c)$$

Since c was an arbitrary point of discontinuity, the proof is complete.

4.2 RIEMANN'S THEOREM

Theorem 8. 1. $f(x) \in D$

$$\Rightarrow \qquad \lim_{x \to +\infty} \int_a^b f(t) \sin xt \, dt = \lim_{x \to +\infty} \int_a^b f(t) \cos xt \, dt = 0$$

Let us treat one of the integrals only.   Set

$$I(x) = \int_a^b f(t) \sin xt \, dt$$

Set $xt = xu + \pi$, so that

$$I(x) = -\int_{a-\frac{\pi}{x}}^{b-\frac{\pi}{x}} f\left(u + \frac{\pi}{x}\right) \sin xu \, du$$

$$2I(x) = -\int_{a-\frac{\pi}{x}}^{a} f\left(t + \frac{\pi}{x}\right) \sin xt \, dt + \int_{b-\frac{\pi}{x}}^{b} f(t) \sin xt \, dt$$

$$+ \int_a^{b-\frac{\pi}{x}} \left[f(t) - f\left(t + \frac{\pi}{x}\right) \right] \sin xt \, dt$$

---

* Alteration of a function at isolated points cannot alter the Fourier coefficients of the function

Since $f(x) \in D$, we can decompose (a, b) into a finite number of intervals in each of which $f(x) \in C$. The integral will be the sum of integrals corresponding to these intervals and we need only show that each of these integrals approaches zero. Hence, there is no restriction in supposing $f(x) \in C$ in $a \leq x \leq b$. Then there exists a constant M such that $|f(x)| < M$ in $a \leq x \leq b$. By uniform continuity there corresponds to an arbitrary $\epsilon > 0$ a number δ such that the relations

$$a \leq x' \leq b, \qquad a \leq x'' \leq b, \qquad |x' - x''| < \delta$$

imply
$$|f(x') - f(x'')| < \epsilon$$

Choose
$$x' = t, \qquad x'' = t + \frac{\pi}{x}, \qquad |x' - x''| = \frac{\pi}{|x|} < \delta$$

This can clearly be done by choosing x sufficiently large. Then

$$2|I(x)| < 2\frac{M\pi}{x} + \epsilon(b - a)$$

$$\varlimsup_{x \to +\infty} 2|I(x)| \leq \epsilon(b - a)$$

$$\lim_{x \to +\infty} I(x) = 0$$

This completes the proof.

4.3 APPLICATIONS

EXAMPLE A.   From Example A of §1 we have at $x = \pi/2$, by Theorem 6,

$$\frac{\pi}{4} = 1 - \frac{1}{3} + \frac{1}{5} - \dots$$

This can be checked by Maclaurin's series for $\tan^{-1} x$.

EXAMPLE B.   In §3.3 of Chapter 11 we gave without proof the values of certain series. We can now supply the proofs. In Example D, §1, $f(x) \in P$, $f(x) \in D^1$, $f(x) \in C$. Hence, we may apply Theorem 6 at $x = 0$ to obtain

$$0 = \frac{\pi}{2} - \frac{4}{\pi}\left(1 + \frac{1}{3^2} + \frac{1}{5^2} + \dots\right)$$

$$\frac{\pi^2}{8} = 1 + \frac{1}{3^2} + \frac{1}{5^2} + \dots$$

Set
$$A = 1 + \frac{1}{2^2} + \frac{1}{3^2} + \dots$$

Then

$$\frac{A}{4} = \frac{1}{2^2} + \frac{1}{4^2} + \frac{1}{6^2} + \cdots$$

and by addition

$$\frac{A}{4} + \frac{\pi^2}{8} = 1 + \frac{1}{2^2} + \frac{1}{3^2} + \cdots = A$$

Hence, $A = \pi^2/6$, and $A/4 = \pi^2/24$, so that the three series have the values attributed to them in the previous chapter.

EXAMPLE C. If $f(x) = \cos cx$, $-\pi \le x \le \pi$, and $f(x + 2\pi) = f(x)$, $-\infty < x < \infty$, then $f(x)$ satisfies the hypotheses of Theorem 6. Setting $x = 0$ in the Fourier series for $f(x)$, Exercise 6, §1, we have

$$\frac{\pi}{\sin \pi c} = 2c\left(\frac{1}{2c^2} - \frac{1}{c^2 - 1^2} + \frac{1}{c^2 - 2^2} - \cdots\right)$$

By Theorems 13 and 14 of Chapter 11,

$$\Gamma(c)\Gamma(1 - c) = B(c, 1 - c) = \int_{0+}^{\infty} \frac{x^{c-1}}{1 + x}\,dx \qquad 0 < c < 1$$

$$= \int_{0+}^{1} \frac{x^{c-1}}{1 + x}\,dx + \int_{0+}^{1} \frac{t^{-c}}{1 + t}\,dt \qquad x = t^{-1}$$

Making use of the identity

$$\frac{1}{1 + x} = 1 - \frac{x}{1 + x}$$

we get

$$\Gamma(c)\Gamma(1 - c) = \int_{0+}^{1} x^{c-1}\,dx + \int_{0+}^{1} \frac{x^{-c} - x^c}{1 + x}\,dx$$

But by expanding the integrand of the second integral in power series, we have

$$\Gamma(c)\Gamma(1 - c) = \frac{1}{c} - \left(\frac{1}{c + 1} - \frac{1}{1 - c}\right) + \left(\frac{1}{c + 2} - \frac{1}{2 - c}\right) - \cdots$$

whence

(3) $$\Gamma(c)\Gamma(1 - c) = \frac{\pi}{\sin \pi c} \qquad 0 < c < 1$$

To justify the term-by-term integration, we may show that the remainder of the integrated series approaches zero, or that

$$\lim_{k \to \infty} \int_{0}^{1} \frac{x^{k+1}}{1 + x}(x^{-c} - x^c)\,dx = 0$$

Set

$$\max_{0 \le x \le 1} \frac{x^{1-c} - x^{1+c}}{1 + x} = M$$

Then

$$\int_0^1 \frac{x^{k+1}}{1+x}(x^{-c} - x^c)\, dx < M \int_0^1 x^k\, dx = \frac{M}{k+1}$$

whence the desired result becomes evident.

EXAMPLE D.   Again making use of the Fourier series for cos cx, we have

(4) $$\cot \pi t = \frac{2t}{\pi}\left(\frac{1}{2t^2} + \frac{1}{t^2 - 1^2} + \frac{1}{t^2 - 2^2} + \dots\right) \qquad t \ne 0, \pm 1, \pm 2, \dots$$

$$\pi \cot \pi t - \frac{1}{t} = \sum_{k=1}^{\infty} \frac{2t}{t^2 - k^2}$$

Integrating term by term from 0 to x, $-1 < x < 1$, we see that

$$\log\left(\frac{\sin \pi x}{\pi x}\right) = \sum_{k=1}^{\infty} \log\left(1 - \frac{x^2}{k^2}\right)$$

The term-by-term integration may be justified by uniform convergence. The latter equation clearly gives the following infinite product expansion of $\sin \pi x$:

(5) $$\sin \pi x = \pi x(1 - x^2)\left(1 - \frac{x^2}{2^2}\right)\left(1 - \frac{x^2}{3^2}\right) \dots \quad -1 < x < 1$$

The expansion is actually valid for all x.   In particular, when $x = \frac{1}{2}$, we have

$$\frac{\pi}{2} = \frac{2 \cdot 2}{1 \cdot 3}\frac{4 \cdot 4}{3 \cdot 5}\frac{6 \cdot 6}{5 \cdot 7} \dots$$

which is Wallis's product (§2.4, Chapter 11).

EXERCISES (4)

1. Check equation (3) for $c = \frac{1}{2}$.

2. Same problem for $c = \frac{1}{6}$.   Use tables.

3. Prove Theorem 8 for the integral involving cos xt.

4. Check equation (4) for $t = \frac{1}{2}$ by showing directly that

$$\sum_{k=1}^{\infty} \frac{1}{4k^2 - 1} = \frac{1}{2}\sum_{k=1}^{\infty}\left(\frac{1}{2k-1} - \frac{1}{2k+1}\right)$$

5. Prove by equation (5) and check roughly by a slide rule that

$$\frac{\pi}{3} = \left(\frac{6}{5}\frac{6}{7}\right)\left(\frac{12}{11}\frac{12}{13}\right) \dots \left(\frac{6n}{6n-1}\frac{6n}{6n+1}\right) \dots$$

6. Prove (3) by Theorem 18, Chapter 11.

7. Set $f(x) = 0$, $-\pi \leq x \leq 0$; $f(x) = \pi$, $0 < x \leq \pi$. What will be the sum of the Fourier series of this function at $x = -\pi$, $x = 0$, $x = +\pi$? Obtain your result both by use of Theorem 7 and by the actual Fourier series.

8. Solve the same problem for $f(x) = -\pi$ in $-\pi \leq x \leq 0$; $f(x) = x$ in $0 < x \leq \pi$.

9. Use the Fourier series for the function $f(x) = e^x$, $0 \leq x \leq 2\pi$ to find the sum of the series

$$\sum_{k=1}^{\infty} \frac{1}{k^2 + 1}$$

10. For what values of x does

$$x^2 = \frac{\pi^2}{3} + 4 \sum_{k=1}^{\infty} (-1)^k \frac{\cos kx}{k^2}?$$

11. Find the sum of the series

$$\sum_{k=1}^{\infty} \frac{(-1)^k}{k^2} \qquad \text{(2 ways)}$$

12. Show that the maximum M in Example C is not greater than 2.

13. Verify the validity of the term-by-term integration of the series in Example D.

14. Prove by use of the equation

$$\Gamma(x + 1) = x\Gamma(x)$$

that equation (3) holds for all nonintegral numbers c.

15. By means of Exercise 9, §3, show that

$$\sum_{k=1}^{n} \frac{(-1)^k}{k} - \sum_{k=1}^{n} \frac{\cos kx}{k} = \frac{1}{2} \int_{\pi}^{x} \cot \frac{t}{2} \, dt - \frac{1}{2} \int_{\pi}^{x} \frac{\cos (n + \frac{1}{2})t}{\sin t/2} \, dt$$

Hence, show that

$$\log \left(2 \sin \frac{x}{2} \right) = - \sum_{k=1}^{\infty} \frac{\cos kx}{k} \qquad 0 < x < 2\pi$$

16. Show that

$$\int_{0}^{\pi/2} \cos 2kx \log (2 \sin x) \, dx = - \frac{\pi}{4k} \qquad k = 1, 2, \ldots$$

and then show that the series of Exercise 15 is a Fourier series. *Hint:* Integrate by parts; express $\sin 2kx \cos x$ as the sum of two sines; use Theorem 4, replacing x by $2x$.

17. By use of Exercises 15 and 16, show that the sufficient conditions of Theorem 7 are not necessary.

§5. Vibrating String

In this section, we shall discuss one of the classical physical applications of Fourier series. The problem of the vibrating string may be taken as typical of the physical situation which can be analyzed by the series; in fact, it is so typical that the term *harmonic analysis* has come to be applied to the general study of Fourier series. In many physical problems it will be convenient to study functions which have periods different from 2π. Accordingly, we shall begin by considering a suitable generalization of Fourier series so that they may apply to an arbitrary interval rather than to $(-\pi, \pi)$.

5.1 FOURIER SERIES FOR AN ARBITRARY INTERVAL

Since the interval $(-l, l)$ can be reduced to the interval $(-\pi, \pi)$ by a simple change of variable, it is easy to see that the functions $\cos(k\pi x/l)$, $\sin(k\pi x/l)$, $k = 0, 1, 2, \ldots$, form an orthogonal set on the interval $(-l, l)$. Let us place the two series corresponding to the intervals $(-\pi, \pi)$ and $(-l, l)$ in juxtaposition:

$$\frac{a_0}{2} + \sum_{k=1}^{\infty} a_k \cos kx + b_k \sin kx \qquad \frac{a_0}{2} + \sum_{k=1}^{\infty} a_k \cos \frac{k\pi x}{l} + b_k \sin \frac{k\pi x}{l}$$

(1) $\qquad a_k = \frac{1}{\pi} \int_{-\pi}^{\pi} f(x) \cos kx \, dx \qquad a_k = \frac{1}{l} \int_{-l}^{l} f(x) \cos \frac{k\pi x}{l} \, dx$

$$b_k = \frac{1}{\pi} \int_{-\pi}^{\pi} f(x) \sin kx \, dx \qquad b_k = \frac{1}{l} \int_{-l}^{l} f(x) \sin \frac{k\pi x}{l} \, dx$$

EXAMPLE A. $\quad f(x) = 2hx/l \qquad\qquad\qquad\qquad\quad 0 \leq x \leq l/2$

$\qquad\qquad\quad f(x) = f(l - x) \qquad\qquad\qquad\qquad l/2 \leq x \leq l$

$\qquad\qquad\quad f(x) = -f(-x) \qquad\qquad\qquad\quad -\infty < x < \infty$

$\qquad\qquad\quad f(x + 2l) = f(x) \qquad\qquad\qquad\quad -\infty < x < \infty$

It is a simple matter to compute the Fourier coefficients of this function by formula (1). Of course, Theorem 6 will be applicable to the present function, so that

$$f(x) = \frac{8h}{\pi^2} \sum_{k=0}^{\infty} (-1)^k \frac{\sin(2k+1)\pi x/l}{(2k+1)^2} \qquad -\infty < x < \infty$$

5.2 DIFFERENTIAL EQUATION OF VIBRATING STRING

In setting up the differential equation of a stretched elastic string we make certain simplifying assumptions. One may keep in mind the situation obtaining for a piano string or for a violin string. Here the vibrations are very small and the tension is high. The force of gravity is negligible. We shall make the following assumptions.

I. *There is no gravity, air resistance, or other damping factor.*

II. *The motion is all in a single plane.*

III. *All moving points of the string move in straight lines perpendicular to the same straight line, called the line of equilibrium.*

IV. *Compared with the length of the string, the motion of any point of the string is small.*

V. *At any point, the angle between the string and the line of equilibrium is small.*

Although these conditions can never be completely realized, they are so close to actual conditions in the examples cited above that theoretical results obtained by their use will fit the observed facts extremely closely in most respects. In certain other respects the theoretical results will be quite far from the facts. For example, as a result of the first assumption, we shall see that any vibration, once started, will continue undiminished forever! It is altogether possible to introduce a damping factor, but this would bring with it mathematical complications that might obscure the method and would not alter the principal results regarding overtones, and the like.

Take the line mentioned in III, the line of equilibrium, as the x-axis. By II and III the motion will be completely described by a function $y(x, t)$, where t is, for example, the number of seconds after some initial time $t = 0$, x and y are the coordinates of a point of the string at time t. Assumption IV means that $y(x, t)$ is small. Assumption V means that $y_1(x, t) = \dfrac{\partial}{\partial x} y(x, t)$ is small, so small that the sine of the slope angle $\tan^{-1} y_1$ can be replaced by the tangent of that angle. By assumptions IV and V, the tension T in the string may be taken constant.

We now isolate a portion of the string and apply Newton's law: *Mass times acceleration equals force.* For definiteness, let us use c.g.s. units:

x, y in centimeters

t in seconds

ρ, the density, in grams per centimeter

α, the acceleration, in centimeters per second per second

T, the tension, in dynes.

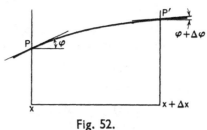

Fig. 52.

Let P and P' be two points of the curve $y = y(x, t)$ with x coordinates, x and $x + \Delta x$, and slope angles, φ and $\varphi + \Delta\varphi$, respectively. Let the center of gravity of the arc PP' have x coordinate $x + \theta \Delta x$, $0 < \theta < 1$. By V the mass of the string between P and P' is $\rho \Delta x$. There is a force at P' whose y-component, tending to increase y, is $T \sin (\varphi + \Delta\varphi)$; one at P whose y-component, tending to decrease y, is $T \sin \varphi$. The net force tending to move the segment PP' in the direction of increasing y is

$$T[\sin (\varphi + \Delta\varphi) - \sin \varphi]$$

The acceleration of the center of gravity of the segment is $y_{22}(x + \theta \Delta x, t)$. Now applying Newton's law to a particle of mass $\rho \Delta x$ at the center of gravity, we have

$$\rho \Delta x y_{22}(x + \theta \Delta x, t) = T[\sin (\varphi + \Delta\varphi) - \sin \varphi]$$

If $\sin \varphi$ is replaced by $\tan \varphi$, this equation becomes

$$\rho \Delta x y_{22}(x + \theta \Delta x, t) = T[y_1(x + \Delta x, t) - y_1(x, t)]$$
$$= T y_{11}(x + \theta' \Delta x, t) \Delta x \qquad 0 < \theta' < 1$$

Cancel Δx and let $\Delta x \to 0$:

$$(2) \qquad\qquad y_{22}(x, t) = c^2 y_{11}(x, t) \qquad\qquad c^2 = T/\rho$$

We have set $T/\rho = c^2$ because T/ρ has the dimensions of a velocity squared. Equation (2) is the partial differential equation of the vibrating string. It is linear, of the second order, and with constant coefficients. It is said to be of *hyperbolic type*.

5.3 A BOUNDARY-VALUE PROBLEM

Let us assume next that the string is fixed at the two points $(0, 0)$ and $(l, 0)$, and that it is released from rest in a distorted position, given by the curve $y = f(x)$ where $f(x)$ is small. Let us try to determine the subsequent motion. We must find a function $y(x, t)$ satisfying equation (2) and the *boundary conditions:*

1. $y(0, t) = y(l, t) = 0$ $\qquad\qquad 0 \leqq t < \infty$
2. $y(x, 0) = f(x)$ $\qquad\qquad 0 \leqq x \leqq l$
3. $y_2(x, 0) = 0$ $\qquad\qquad 0 \leqq x \leqq l$

It is clear that the given function $f(x)$ must be such that $f(0) = f(l) = 0$.

We begin by looking for functions $y(x, t)$ of a special type,

$$y(x, t) = g(x)h(t)$$

If this is to satisfy equation (2) we must have

$$\frac{g''(x)}{g(x)} = \frac{1}{c^2} \frac{h''(t)}{h(t)}$$

Since the left-hand side is a function of x and the right-hand side a function of t, this equation can hold only if both sides are constant. We may take this constant positive, zero, or negative. Setting the constant equal to α^2, 0, or $-\alpha^2$, we have ordinary equations to solve. In the three cases, we obtain

CASE I. $y(x, t) = (A \sinh \alpha x + B \cosh \alpha x)(C \sinh \alpha ct + D \cosh \alpha ct)$

CASE II. $y(x, t) = (Ax + B)(Ct + D)$

CASE III. $y(x, t) = (A \sin \alpha x + B \cos \alpha x)(C \sin \alpha ct + D \cos \alpha ct)$.

It is easy to see that, in Cases I and II, $y(x, t)$ must be identically zero if boundary condition 1 is to be satisfied. For example, in Case II,

$$y(0, t) = B(Ct + D) = 0 \qquad\qquad 0 \leq t < \infty$$

$$y(l, t) = (Al + B)(Ct + D) = 0 \qquad\qquad 0 \leq t \leq \infty$$

The first equation shows that $B = 0$, the second that $A = 0$, whence $y(x, t)$ is identically zero.

But in Case III we can find infinitely many solutions satisfying boundary conditions 1 and 3. They are

(3) $$y_k(x, t) = b_k \sin\frac{k\pi x}{l} \cos\frac{k\pi ct}{l} \qquad k = 0, \pm 1, \pm 2, \ldots$$

Here the constants b_k are arbitrary. But none of these functions will satisfy condition 2 unless $f(x)$ happens to be of the form $b_k \sin (k\pi x/l)$. But notice that the sum of any number of the functions (3) will satisfy equation (2) and conditions 1 and 3. We can hope that it may be possible to determine the constants b_k so that the sum of the series

(4) $$y(x, t) = \sum_{k=1}^{\infty} b_k \sin\frac{k\pi x}{l} \cos\frac{k\pi ct}{l}$$

will be the solution of our problem. If $t = 0$, the series is a Fourier series. Can its sum be $y(x, 0) = f(x)$? Yes, if $f(x)$ satisfies the conditions of Theorem 6 and if the b_k are determined by equations (1).

5.4 SOLUTION OF THE PROBLEM

It must not be supposed that we have proved that the function defined by equation (4) is the required solution. The sum of the infinite series (4) may conceivably fail to satisfy equation (2) even though its general term does so. In fact, we are not even certain of the convergence of the series except when $t = 0$. Let us extend the definition of $f(x)$ outside the interval

$(l, 0)$ so that $f(-x) = -f(x)$, $f(x + 2l) = f(x)$ for all x. Let $f(x) \in C$, $f(x) \in D^1$. Then by Theorem 6

$$f(x) = \sum_{k=1}^{\infty} b_k \sin \frac{k\pi x}{l} \qquad -\infty < x < \infty$$

Since

$$\sin \frac{k\pi x}{l} \cos \frac{k\pi ct}{l} = \frac{1}{2}\left[\sin \frac{k\pi}{l}(x + ct) + \sin \frac{k\pi}{l}(x - ct) \right]$$

we see that the series (4) is the sum of two convergent series and that

(5) $$y(x, t) = \frac{f(x + ct) + f(x - ct)}{2}$$

It is now evident by direct differentiation that equation (2) is satisfied at all points (x, t) such that $f''(x \pm ct)$ exists. This is all one could hope to prove. Actually, in any physical problem $f(x) \in C^2$; though in the case of the plucked string, actual conditions are very closely approximated by defining the curve $y = f(x)$ as a broken line (Example A).

Note that to plot the functions $f(x \pm ct)$ it is only necessary to translate the curve $y = f(x)$. Hence, the motion may be regarded as the sum of two others each of which is a translation of the curve $y = f(x)$ with velocity c, one to the right, the other to the left.

5.5 UNIQUENESS OF SOLUTION

In view of the rather special way in which the function (4) was found, one naturally raises the question whether there might not be other solutions. If so, we may have no reason to suppose that the solution we have obtained will be the one that fits the physical facts. Suppose there were two distinct solutions. Their difference $z(x, t)$ would be a function such that

(6) $$z_{22}(x, t) = c^2 z_{11}(x, t)$$

(7) $$z(0, t) = z(l, t) = 0 \qquad\qquad 0 \le t < \infty$$

(8) $$z(x, 0) = z_2(x, 0) = 0 \qquad\qquad 0 \le x \le l$$

Make the change of variable

$$x - ct = u, \qquad x = (u + v)/2$$
$$x + ct = v, \qquad t = (v - u)/2c$$

Equation (6) becomes

$$\frac{\partial^2 z}{\partial u\, \partial v} = 0$$

whence

$$z = \varphi(u) + \psi(v)$$

where $\varphi(u) \in C^1$, $\psi(v) \in C^1$ and are otherwise arbitrary. That is,

$$z = \varphi(x - ct) + \psi(x + ct)$$

By equations (8) $$\varphi(x) + \psi(x) = 0 \qquad\qquad 0 \leqq x \leqq l$$

$$\varphi'(x) - \psi'(x) = 0$$

from which it is clear that $\varphi(x)$ and $\psi(x)$ are constants, $\varphi = A$, $\psi = -A$. Equations (7) become

(9) $$\varphi(-ct) + \psi(ct) = 0$$

(10) $$\varphi(l - ct) + \psi(l + ct) = 0 \qquad\qquad 0 \leqq t < \infty$$

Equation (9) shows that $\varphi = A$ also in $(-l, 0)$, and then (10) shows that $\psi = -A$ also in $(l, 3l)$. In general (9) shows that if $\psi = -A$ in $(0, ml)$ then $\varphi = A$ in $(-ml, 0)$, and (10) shows that if $\varphi = A$ in $(-ml, l)$ then $\psi = -A$ in $(l, ml + 2l)$. Proceeding step by step we see that $\psi = -A$ in $(0, \infty)$ and $\varphi = A$ in $(-\infty, l)$ so that $z = A - A = 0$ for $0 \leqq t < \infty$, $0 \leqq x \leqq l$, to produce the desired contradiction. We have thus established that the function (4) is the unique solution of the differential system consisting of equation (2) and boundary conditions 1, 2, 3.

5.6 SPECIAL CASES

Certain special cases are of particular interest.

EXAMPLE B. $f(x) = h \sin (\pi x/l)$ $0 \leqq x \leqq l$
Then

$$y(x, t) = h \sin (\pi x/l) \cos (\pi ct/l)$$

Note that the curve always keeps the shape of one arch of a sine curve, suitably scaled down. The motion is clearly periodic with period $2l/c = 2l(\rho/T)^{1/2}$ and frequency $(2l)^{-1}(T/\rho)^{1/2}$. The musical note produced by such a vibrating string is called the *fundamental* of the string. Observe that the frequency (which determines the pitch of the note) of the string is inversely proportional to the length, proportional to the square root of the tension, and inversely proportional to the diameter of the string. These facts are all used in the construction of a piano, or harp, for example. Of course, h must be so small that the original assumptions are valid. This constant determines the *intensity* of the note.

EXAMPLE C. $f(x) = h \sin (k\pi x/l)$ $k = 1, 2, \ldots$
Here

$$y(x, t) = h \sin (k\pi x/l) \cos (k\pi ct/l).$$

The frequency is now found to be k times its value in Example B. The musical note produced is said to be the $(k-1)$st *overtone* of the string. If the fundamental has the pitch of C, the various overtones have the following pitch:

| k | 1 | 2 | 3 | 4 | 5 | 6 | 7 | 8 | 9 | 10 | 11 | 12 | ... |

musical note

| | C | C | G | C | E | G | B♭ | C | D | E | F♯ | G | ... |

Note that the frequencies corresponding to the notes C, E, G are in the ratio $4 : 5 : 6$, a familiar fact for the so-called *just scale*.

EXAMPLE D.   The plucked string.   Here we assume that $f(x)$ is defined as in Example A. Then

$$y(x, t) = \frac{8h}{\pi^2} \sum_{k=0}^{\infty} \frac{(-1)^k}{(2k+1)^2} \sin (2k+1) \frac{\pi x}{l} \cos (2k+1) \frac{\pi ct}{l}$$

Notice that theoretically the musical note corresponding to this motion of the string could be reproduced by combining fundamental and overtones with suitable intensities.   It is this principle which is used in the construction of certain musical instruments, such as the electric organ.   The note is said to be *analyzed* into its various overtones.   Hence, the term *harmonic analysis*.

EXERCISES (5)

Find the Fourier series for the functions of Exercises 1–4 and find the sum of the series.

1. $f(x) = x, 0 \leq x < 1$; $f(-x) = f(x), f(x+2) = f(x)$, for all x.

2. $f(x) = x, 0 \leq x < 1$; $f(x+1) = f(x), -\infty < x < \infty$

3. $f(x) = x^2, 0 < x \leq 3$; $f(x+3) = f(x), -\infty < x < \infty$.

4. $f(x) = 1, 0 \leq x \leq \pi/2$; $f(x) = 0, \pi/2 < x < \pi$; $f(x+\pi) = f(x)$, $-\infty < x < \infty$.

5. Give the details in Example A.

6. Plot the position of the plucked string after $\frac{1}{8}$ of the period has expired. Use equation (5).

7. A stretched string has its ends fastened at points with rectangular coordinates $(0, 0)$ and $(\pi, 0)$ and is held initially in a curve with equation $y = x - (x^2/\pi)$. When the string is released, what will be the ratio of the intensity of the fundamental tone to that of the first nonvanishing overtone?

8. Show that Case I, §5.3, is useless for the boundary-value problem.

9. Give the details of the change of variable outlined in §5.5.

10. Discuss the hammered string:

$$y(0, t) = y(\pi, t) = y(x, 0) = 0, \quad y_2(x, 0) = F(x)$$

$$F(x) = 0 \qquad\qquad\qquad 0 \leq x < \frac{\pi}{2} - \delta, \quad \frac{\pi}{2} + \delta < x \leq \pi$$

$$F(x) = h \qquad\qquad\qquad \frac{\pi}{2} - \delta \leq x \leq \frac{\pi}{2} + \delta$$

$$F(-x) = -F(x), \quad F(x + 2\pi) = F(x) \qquad\qquad -\infty < x < \infty$$

Give your result first in the form of an infinite series. Then reduce to the following forms:

$$y(x, t) = \frac{1}{2} \int_0^t [g(x - cu) + g(x + cu)] \, du$$

$$= \frac{1}{2c} \int_{x-ct}^{x+ct} g(u) \, du$$

11. Compare the maximum velocities of the mid-points of the strings in Examples B and D. In Example D use equation (5).

12. In Example D, when the string was plucked at its middle point, the first and all odd numbered overtones were missing; that is, the Fourier series involved had all terms missing in which k was even. At what point may the string be plucked so as to eliminate the rth overtone ($k = r + 1$, $2r + 2, \ldots$)?

13. Show that series (4) may be differentiated term by term twice with respect to x, at least if $f(x)$ satisfies the conditions of Exercise 14, §1, with $p = 4$.

14. By use of the previous exercise show that $y(x, t)$ defined by (4) satisfies equation (2).

15. If $\sin \varphi$ is not replaced by $\tan \varphi$ in §5.2, show that (2) becomes

$$y_{22}(x, t) = c^2 y_{11}(x, t)[1 + (y_1(x, t))^2]^{-3/2}$$

§6. Summability of Fourier Series

We have seen that the Fourier series of certain discontinuous functions converge. One might be tempted to suppose that the Fourier series of a continuous function surely converges. This is not the case. It was for functions of class D^1 that we proved convergence. But there are functions of class C which are not of D^1 (see Figure 51). In fact, Fejér gave in 1910 the first example of a continuous function whose Fourier series diverges. This does not mean that every function of class C, not of class D^1, has a divergent Fourier series.   This is far from being the case.   The conditions of Theorem 7 are sufficient but not necessary.   That is, the actual region of convergence, Figure 51, is much larger than the region D^1, but certainly does not include *all* the region C.   If $f(x) \in C$ and if no further property of $f(x)$ is known, then the Fourier series for $f(x)$ may diverge and we resort to summability methods.   Fejér showed in 1904 that the Fourier series of a continuous function is summable $(C, 1)$ to the function.   We now prove this result.

6.1 PRELIMINARY RESULTS

Theorem 9.   For $-\infty < x < \infty$; $n = 1, 2, \dots$

$$(1) \qquad \sum_{k=0}^{n} D_k(x) = F_n(x) = \frac{\sin^2 (n + 1)(x/2)}{2 \sin^2 (x/2)}$$

Here $D_k(x)$ is the Dirichlet kernel;   the new function $F_n(x)$ is called the *Fejér kernel*.   The limit value is of course intended when the denominator (1) vanishes.   The proof of the theorem is like that of Theorem 4 and is omitted.

Corollary 9.

$$\frac{1}{\pi} \int_{-\pi}^{\pi} F_n(t)\, dt = n + 1 \qquad\qquad n = 0, 1, 2, \dots$$

This follows from Corollary 4.2.

Theorem 10.   1. $f(x) \in P$

2. $f(x) \in C$

$$(2) \quad \Rightarrow \sigma_n(x) - f(x) = \frac{1}{\pi(n + 1)} \int_{-\pi}^{\pi} [f(x + t) - f(x)]F_n(t)\, dt$$

$$n = 0, 1, \dots$$

Here

$$(3) \qquad \sigma_n(x) = \frac{1}{n + 1} \sum_{k=0}^{n} S_k(x)$$

where $S_k(x)$ is defined by equation (3), §2.   By use of equation (3) and Theorem 5, we have

$$\sigma_n(x) - f(x) = \frac{1}{n + 1} \sum_{k=0}^{n} [S_k(x) - f(x)]$$

$$= \frac{1}{\pi(n + 1)} \int_{-\pi}^{\pi} [f(x + t) - f(x)] \sum_{k=0}^{n} D_k(t)\, dt$$

Then by Theorem 9 the proof is completed.

6.2 FEJÉR'S THEOREM

It should be noted that the kernel in the integral remainder formula (2), $[\sin^2 (n + 1)t/2]/[2 \sin^2 (t/2)]$ is never negative. It is this important fact that makes the proof of Fejér's theorem essentially simpler than that of Theorem 6. No preliminary result comparable to Corollary 3 is now necessary. It should be carefully observed where the positiveness of the kernel intervenes in the following proof.

Theorem 11.   1. $f(x) \in C$

2. $f(x) \in P$

$\Rightarrow \cdot \qquad\qquad f(x) = \sum_{k=0}^{\infty} \overset{\cdot}{C}_k(x) \qquad\qquad (C, 1)$

We have only to prove that $\sigma_n(x) \to f(x)$ as $n \to \infty$. Let x_0 be an arbitrary constant. Since $f(x) \in C$ at x_0, there corresponds to an arbitrary positive ϵ a number δ such that when $|t| \leqq \delta$ we have $|f(x_0 + t) - f(x_0)| < \epsilon$, Express the integral (2) with x replaced by x_0 as the sum of three others. I_1, I_2, I_3, corresponding to the intervals $(-\pi, -\delta)$, $(-\delta, \delta)$, (δ, π). Then

$$|I_2| < \frac{\epsilon}{\pi(n + 1)} \int_{-\delta}^{\delta} F_n(t)\, dt \qquad\qquad n = 0, 1, \ldots$$

and the right-hand side is less than ϵ by Corollary 9 (replacing δ by π only strengthens the inequality). If M is the maximum value of $|f(x)|$, then, since $\sin^2 (x/2) \in \downarrow$ in $(-\pi, -\delta)$, we have

$$|I_1| \leqq \frac{2M}{\pi(n + 1)} \int_{-\pi}^{-\delta} \frac{dt}{2 \sin^2 (t/2)} \leqq \frac{M}{(n + 1) \sin^2 (\delta/2)} \qquad n = 0, 1, \ldots$$

Also I_3 has the same upper bound. Hence,

$$|\sigma_n(x_0) - f(x_0)| < \epsilon + \frac{2M}{(n + 1) \sin^2 (\delta/2)}$$

$$\overline{\lim_{n \to \infty}} \, |\sigma_n(x_0) - f(x_0)| \leqq \epsilon$$

$$\lim_{n \to \infty} \sigma_n(x_0) = f(x_0)$$

Since x_0 was arbitrary, the proof is complete.

6.3 UNIFORMITY

Theorem 12.   1. $f(x) \in C$

2. $f(x) \in P$

$\Rightarrow \qquad \lim_{n \to \infty} \sigma_n(x) = f(x)$ *uniformly in the interval* $-\pi \leqq x \leqq \pi$.

Since $f(x)$ is uniformly continuous in the interval $-2\pi \leq x \leq 2\pi$, then corresponding to an arbitrary $\epsilon > 0$ there is a δ such that the inequalities

(4) $-2\pi \leq x' \leq 2\pi$, $-2\pi \leq x'' \leq 2\pi$, $|x' - x''| < \delta$

imply $|f(x') - f(x'')| < \epsilon$

Now choose $x' = x$, $x'' = x + t$. If $|t| < \delta$ and $-\pi \leq x \leq \pi$, then surely inequalities (4) are satisfied (assuming as we may that $\delta < \pi$). Hence, the integrals I_1, I_2, I_3, with x_0 replaced by x, satisfy the same inequalities as before. Hence, we can certainly find an integer m, *independent of x in the interval* $-\pi \leq x \leq \pi$ such that

$$|\sigma_n(x) - f(x)| < 3\epsilon \qquad n > m, \quad -\pi \leq x \leq \pi$$

This completes the proof.

EXAMPLE A. $\displaystyle \lim_{n \to \infty} \frac{1}{n} \int_{-1}^{1} (1 - \cos t) t^{-4} \sin^2 (nt/2)\, dt = \frac{\pi}{4}$

Take $f(t) = t^{-4}(1 - \cos t) \sin^2 (t/2)$ in $(-1, 1)$, defining it arbitrarily in the rest of the interval $(-\pi, \pi)$ so as to make it continuous and periodic. It is the behavior of $f(t)$ in a neighborhood of $t = 0$ only, that is important for the limit.   By Theorem 11

$$\lim_{n \to \infty} \frac{1}{2\pi n} \int_{-\pi}^{\pi} f(t) \frac{\sin^2 (nt/2)}{\sin^2 (t/2)}\, dt = f(0) = \frac{1}{8}$$

EXERCISES (6)

1. $\displaystyle \lim_{n \to \infty} \frac{1}{n} \int_{-\pi}^{\pi} e^t \frac{\sin^2 (n + 1)t/2}{\sin^2 (t/2)}\, dt = ?$

2. $\displaystyle \lim_{n \to \infty} \frac{1}{n} \int_{-\pi}^{\pi} \cos^{-1} t \frac{\sin^2 (nt/2)}{\sin^2 (t/2)}\, dt = ?$

3. $\displaystyle \lim_{n \to \infty} \frac{1}{n} \int_{-\pi/2}^{\pi/2} \log |x + t| \frac{\sin^2 nt}{\sin^2 t}\, dt = ?$ $(x \neq 0)$

4 $\displaystyle \lim_{n \to \infty} \frac{1}{n} \int_{0}^{2\pi} \frac{\sin^2 nt}{\sin^2 t}\, dt = ?$

5. $\displaystyle \lim_{n \to \infty} \log \left(1 + \frac{2}{n}\right) \int_{0}^{2\pi} \frac{\cot t}{t} \sin^2 nt\, dt = ?$

6. Prove Theorem 9.

7. Prove: $\displaystyle 0 = \frac{1}{2} + \sum_{k=1}^{\infty} \cos kx$ $(C, 1),\ x \neq 0, \pm 2\pi, \pm 4\pi, \ldots$

8. Prove: $\displaystyle \frac{1}{2} \cot \frac{x}{2} = \sum_{k=1}^{\infty} \sin kx$ $(C, 1),\ x \neq 0, \pm 2\pi, \pm 4\pi, \ldots$

9. Prove by use of the test-ratio test that the following series converges:

$$\sum_{k=1}^{\infty} \frac{k!\, \cos^{2k} \delta}{\Gamma(k + \frac{1}{2})} \qquad 0 < \delta < 1$$

10. Prove:

$$H_n = \int_{-\pi/2}^{\pi/2} \cos^{2n} x \, dx \quad \Rightarrow \quad \lim_{n \to \infty} \frac{\cos^{2n} \delta}{H_n} = 0 \qquad 0 < \delta < 1$$

Hint: Use No. 498, Peirce's Tables, and Exercise 9.

11. Prove that, if H_n is defined as in Exercise 10 and if $f(x) \in C$ in $-\infty < x < \infty$, then

$$\lim_{n \to \infty} \frac{1}{H_n} \int_{-\pi/2}^{\pi/2} f(x + u) \cos^{2n} u \, du = f(x) \qquad -\infty < x < \infty$$

12. In Exercise 11, show that the limit is uniform in $-\pi \leq x \leq \pi$.

13. Prove:

$$\frac{1}{\pi} \int_{-\pi}^{\pi} [\sigma_n(x) - f(x)]^2 \, dx = \frac{1}{\pi} \int_{-\pi}^{\pi} f^2(x) \, dx - \frac{a_0^2}{2}$$
$$- \sum_{k=1}^{n} (a_k^2 + b_k^2) + \frac{1}{(n+1)^2} \sum_{k=1}^{n} k^2 (a_k^2 + b_k^2)$$

14. Prove:

$$f(x) \in C, P \quad \Rightarrow \quad \lim_{n \to \infty} \frac{1}{(n+1)^2} \sum_{k=1}^{n} k^2 (a_k^2 + b_k^2) = 0.$$

Hint: Use Theorem 2, Theorem 12, and Exercise 13.

§7. *Applications*

We shall derive in this section several interesting consequences of Fejér's theorem. One application is Parseval's theorem, which states that the infinite series of Corollary 2 has for its sum

$$\frac{1}{\pi} \int_{-\pi}^{\pi} f^2(x) \, dx$$

To prove this, we need to investigate the relation of Fourier series to the method of least square approximation.

7.1 TRIGONOMETRIC APPROXIMATION

Theorem 13.   1. $f(x) \in C$ $\qquad\qquad\qquad\qquad\qquad -\pi \leq x \leq \pi$

  2. $f(-\pi) = f(\pi)$

⇒·   *There corresponds to every positive ϵ a trigonometric polynomial*

(1) $$T_n(x) = \frac{A_0}{2} + \sum_{k=1}^{n} A_k \cos kx + B_k \sin kx$$

such that

$$|T_n(x) - f(x)| < \epsilon \qquad\qquad\qquad -\pi \leq x \leq \pi$$

For, $f(x)$ can be defined outside the interval $(-\pi, \pi)$ so as to belong to P. Hence, the result follows by Theorem 12.   One has only to note that $\sigma_n(x)$ is a function of the form (1) since

$$\sigma_n(x) = \sum_{k=0}^{n} \left(1 - \frac{k}{n+1}\right) C_k(x)$$

Observe that we will not in general be able to take $A_k = a_k$ and $B_k = b_k$, for we have pointed out that in general $S_n(x)$ does not approach $f(x)$, much less uniformly, if $f(x)$ is merely continuous.   It could be shown that, if we added the hypothesis $f(x) \in D^1$, then we could take $A_k = a_k$, $B_k = b_k$.

7.2 WEIERSTRASS'S THEOREM ON POLYNOMIAL APPROXIMATION

The following application of Fejér's theorem, was provided by Weierstrass in 1885 by other methods.

Theorem 14.   1. $f(x) \in C$ $a \leqq x \leqq b$

$\Rightarrow$ *There corresponds to every positive ϵ a polynomial*

$$P_n(x) = \sum_{k=0}^{n} C_k x^k$$

such that

$$|f(x) - P_n(x)| < \epsilon \qquad\qquad a \leqq x \leqq b$$

Make a transformation $x = ct + d$, $c \neq 0$, which will carry the interval (a, b) into $(-\pi/2, \pi/2)$, and set

$$g(t) = f(ct + d) \qquad\qquad -\pi/2 \leqq t \leqq \pi/2$$

Complete the definition of $g(t)$ in $(-\pi, \pi)$ so that it satisfies the conditions of Theorem 13.   Then corresponding to the given ϵ of the present theorem there exists $T_m(t)$ such that

$$(2) \qquad\qquad |T_m(t) - g(t)| < \epsilon/2 \qquad\qquad -\pi \leqq t \leqq \pi$$

But $T_m(t)$ is a sum of trigonometric functions each of which has a Maclaurin expansion that converges uniformly in any finite interval.   Clearly, $T_m(x)$ has a similar expansion.   The partial sums of this expansion are polynomials which approximate uniformly to $T_m(t)$.   That is, there exists a polynomial $Q_n(t)$ such that

$$(3) \qquad\qquad |T_m(t) - Q_n(t)| < \epsilon/2 \qquad\qquad -\pi \leqq t \leqq \pi$$

Combining inequalities (2) and (3), we have

$$|g(t) - Q_n(t)| < \epsilon \qquad\qquad -\pi \leqq t \leqq \pi$$

$$|g([x - d]/c) - Q_n([x - d]/c)| < \epsilon \qquad\qquad a \leqq x \leqq b$$

$$|f(x) - P_n(x)| < \epsilon \qquad\qquad a \leqq x \leqq b$$

where $P_n(x) = Q_n([x - d]/c)$

Certainly, $P_n(x)$ is a polynomial of the same degree as $Q_n(t)$.

7.3 LEAST SQUARE APPROXIMATION

A function $g(x, A, B, C)$ is said to be a least square approximation to $f(x)$ on (a, b) if the parameters A, B, C are determined in such a way that the integral

$$\int_a^b [f(x) - g(x, A, B, C)]^2 \, dx$$

has its smallest possible value. The definition could be extended in an obvious way to include functions g of any number of parameters. The definition is clearly analogous to one given in Chapter 4, §3.3 involving approximation at a finite number of points.

EXAMPLE A.   Find the least square approximation by a function of the form $Ax + B$ to the function $\sin x$ on $(0, \pi)$. We have to minimize the function

$$F(A, B) = \int_0^\pi (\sin x - Ax - B)^2 \, dx$$

Equating the two partial derivatives to zero, we obtain

$$2 \int_0^\pi (\sin x - Ax - B)x \, dx = 0$$

$$2 \int_0^\pi (\sin x - Ax - B) \, dx = 0$$

The solution of this pair of equations is $A = 0$, $B = 2/\pi$. The graph of the required function is a straight line parallel to the x-axis and a distance $2/\pi$ above it.

Theorem 15.   1. $f(x) \in D$

2. $T_n(x) = \dfrac{A_0}{2} + \sum_{k=1}^n A_k \cos kx + B_k \sin kx$

(4) $\Rightarrow$ $\displaystyle\int_{-\pi}^\pi [f(x) - S_n(x)]^2 \, dx \leq \int_{-\pi}^\pi [f(x) - T_n(x)]^2 \, dx$

Here $S_n(x)$ is defined by equation (3), §2. The right-hand side of inequality (4) is a function of the $(2n + 1)$ parameters $A_0, A_1, B_1, \ldots, A_n, B_n$. Differentiating partially with respect to each of these and equating the result to zero, we have

$$2 \int_{-\pi}^\pi [f(x) - T_n(x)] \cos kx \, dx = 0 \qquad k = 0, 1, \ldots, n$$

$$2 \int_{-\pi}^\pi [f(x) - T_n(x)] \sin kx \, dx = 0 \qquad k = 1, 2, \ldots, n$$

By use of the orthogonality relations, these equations reduce to

$$A_k = a_k \qquad\qquad k = 0, 1, \ldots, n$$

$$B_k = b_k \qquad\qquad k = 1, 2, \ldots, n$$

This would conclude the proof if we knew that the minimum existed. Conceivably, the point we have found may be a maximum or even a saddle-point. Rather than complete the proof by use of second derivative tests, we give an algebraic proof which is of interest in itself. Clearly,

$$(5) \quad \int_{-\pi}^{\pi} [f(x) - T_n(x)]^2 \, dx = \int_{-\pi}^{\pi} [f(x) - S_n(x)]^2 \, dx$$

$$+ 2\int_{-\pi}^{\pi} [f(x) - S_n(x)][S_n(x) - T_n(x)] \, dx + \int_{-\pi}^{\pi} [S_n(x) - T_n(x)]^2 \, dx$$

The middle integral on the right is zero by virtue of the orthogonality relations and equation (4), §3. That is, the left-hand side is the sum of two nonnegative terms, and is hence not less than either. This completes the proof.

7.4 PARSEVAL'S THEOREM

Theorem 16.   1. $f(x) \in C$ $-\pi \leqq x \leqq \pi$

               2. $f(-\pi) = f(\pi)$

$$(6) \quad \Rightarrow \qquad \frac{1}{\pi} \int_{-\pi}^{\pi} f^2(x) \, dx = \frac{a_0^2}{2} + \sum_{k=1}^{\infty} (a_k^2 + b_k^2)$$

As we saw in the proof of Bessel's inequality,

$$\frac{1}{\pi} \int_{-\pi}^{\pi} [f(x) - S_n(x)]^2 \, dx = \frac{1}{\pi} \int_{-\pi}^{\pi} f^2(x) \, dx - \frac{a_0^2}{2} - \sum_{k=1}^{n} (a_k^2 + b_k^2)$$

$$n = 0, 1, 2, \ldots$$

If $\epsilon > 0$, determine $T_n(x)$ by Theorem 13. Then by Theorem 15

$$\frac{1}{\pi} \int_{-\pi}^{\pi} [f(x) - S_n(x)]^2 \, dx \leqq \frac{1}{\pi} \int_{-\pi}^{\pi} [f(x) - T_n(x)]^2 \, dx < 2\epsilon^2$$

That is, for some integer n and *a fortiori* for any larger n, $(a_k^2 + b_k^2 \geqq 0)$,

$$0 \leqq \frac{1}{\pi} \int_{-\pi}^{\pi} f^2(x) \, dx - \frac{a_0^2}{2} - \sum_{k=1}^{n} (a_k^2 + b_k^2) < 2\epsilon^2$$

Allowing n to become infinite:

$$0 \leqq \frac{1}{\pi} \int_{-\pi}^{\pi} f^2(x) \, dx - \frac{a_0^2}{2} - \sum_{k=1}^{\infty} (a_k^2 + b_k^2) < 2\epsilon^2$$

Since ϵ was arbitrary, this implies the equality (6).

Corollary 16.　1. $f(x)$, $\varphi(x) \in C$　　　　　　　　　　$-\pi \leqq x \leqq \pi$

2. $f(-\pi) = f(\pi)$, $\varphi(-\pi) = \varphi(\pi)$

3. $\alpha_k = \dfrac{1}{\pi} \displaystyle\int_{-\pi}^{\pi} \varphi(x) \cos kx\, dx$　　　　　　$k = 0, 1, 2, \ldots$

　　$\beta_k = \dfrac{1}{\pi} \displaystyle\int_{-\pi}^{\pi} \varphi(x) \sin kx\, dx$　　　　　　$k = 1, 2, 3, \ldots$

$\Rightarrow$　　$\dfrac{1}{\pi} \displaystyle\int_{-\pi}^{\pi} f(x)\varphi(x)\, dx = \dfrac{a_0 \alpha_0}{2} + \sum_{k=1}^{\infty} (a_k \alpha_k + b_k \beta_k)$

The proof is easily supplied by expanding the integral

$$\frac{1}{\pi} \int_{-\pi}^{\pi} [f(x) - \varphi(x)]^2 \, dx$$

7.5 UNIQUENESS

An important question, which we can now answer, is whether a given trigonometric series can be the Fourier series of more than one continuous function. If there were two distinct continuous functions, their difference would be a function all of whose Fourier coefficients would be zero and this would imply, by the following theorem, that their difference is identically zero. Hence, the answer is in the negative.

Theorem 17.　1. $f(x) \in C, P$　　　　　　　　　　$-\pi \leqq x \leqq \pi$

2. $\displaystyle\int_{-\pi}^{\pi} f(x) \cos kx\, dx = 0$　　　　　　$k = 0, 1, 2, \ldots$

　　$\displaystyle\int_{-\pi}^{\pi} f(x) \sin kx\, dx = 0$　　　　　　$k = 1, 2, 3, \ldots$

$\Rightarrow$　　　$f(x) = 0$　　　　　　　　　　$-\pi \leqq x \leqq \pi$

Let ϵ be an arbitrary positive number. By Theorem 13, determine $T_n(x)$ corresponding to it. Then if M is the maximum of $|f(x)|$ in the interval $-\pi \leqq x \leqq \pi$,

$$\left| \int_{-\pi}^{\pi} f(x)[f(x) - T_n(x)]\, dx \right| \leqq 2\pi M \epsilon$$

By hypothesis 2, this inequality is equivalent to

$$\int_{-\pi}^{\pi} f^2(x)\, dx \leqq 2\pi M \epsilon$$

Since the left-hand side does not depend on ϵ, it must be zero. Now suppose

$f(x_0) \neq 0$, $-\pi < x_0 < \pi$.   Since $f(x) \in C$, there is a neighborhood of x_0, say, $-\pi < x_0 - \delta \leq x \leq x_0 + \delta < \pi$ where $f^2(x) > 0$.   Hence

$$0 = \int_{-\pi}^{\pi} f^2(x)\, dx \geq \int_{x_0-\delta}^{x_0+\delta} f^2(x)\, dx > 0$$

This is a contradiction.   Hence, $f(x) = 0$ in $-\pi < x < \pi$.   By continuity, $f(\pi) = f(-\pi) = 0$ also.   This completes the proof.   By use of Exercise 15 below, the assumed periodicity of $f(x)$ could be dispensed with.

Corollary 17.   *A trigonometric series cannot be the Fourier series of more than one continuous function.*

EXERCISES (7)

1. Determine the constants c and d used in the proof of Theorem 14.

2. Show that the Maclaurin series for $\sin kx$ or $\cos kx$ converges uniformly in any finite interval.

3. Give the details of the proof that the middle integral on the right-hand side of equation (5) is zero.

4. Find the least square approximation of x^2 on $(0, 1)$ by a function of the form $A + Bx$.   It is unnecessary to prove the existence of the minimum.

5. Solve the same problem for x^3 on $(0, 1)$ by $A + Bx + Cx^2$.

6. Solve the same problem for x on $(0, 1)$ by $A + Be^x$.

7. Show that if $f(x) \in C^2$, P in $-\pi \leq x \leq \pi$, the Fourier series of $f(x)$ converges uniformly in that interval.   By use of Theorems 1 and 17, show that the sum of the series is $f(x)$.   *Hint:* Use Exercise 14 of §1.

8. Apply Parseval's theorem to Example D, §1.

9. Theorem 16 is not applicable to Example A, §1.   Show directly that the conclusion of the theorem is none the less true.   Hence, show that the hypotheses are not necessary.

10. Solve the same problem for Example C, §1.

11. Prove Corollary 16.

12. Use Corollary 16 to obtain the value of the integral

$$\frac{1}{\pi} \int_{-\pi}^{\pi} |x| x^2 \, dx$$

Hint: Use No. 889 and No. 891 of Peirce's Tables.

13. Same problem for

$$\frac{1}{\pi} \int_{-n}^{n} |x| \sin^2 x \, dx$$

Check by direct integration.

14. The integral in (6) is clearly $\geq a_0^2/2$. Prove this directly by the Schwarz inequality.

15. Show that if $F \in D$ and $\epsilon > 0$, there exists $G \in C$ such that

$$\left| \int_{-\pi}^{\pi} (F - G) \, dx \right| < \epsilon$$

Hint: Assume first that F has a single finite jump and replace F near the jump by a linear function so as to make the altered function G continuous.

16. By use of Exercise 15 prove that equation (6) holds for every $f \in D$. *Hint:* Replace f by $g \in C \cdot P$ in such a way as to alter the kth term of the series by less than $\epsilon/2^k$.

§8. *Fourier Integral*

In order for it to be possible that a function should have an expansion in a Fourier series, one essential property of the function is its periodicity. If a function fails to have this property, it is possible in many cases to give it an integral representation analogous to the Fourier series expansion. For this representation the function should be defined from $-\infty$ to $+\infty$. If it is known only in a finite interval, its definition can be completed, usually by defining it as zero in the rest of the range.

8.1 ANALOGIES WITH FOURIER SERIES

To set forth the analogies between Fourier series and Fourier integrals, we arrange them side by side below. The signs Σ and $\int$, the integer k and the variable y, the intervals $(-\pi, \pi)$ and $(-\infty, \infty)$, correspond.

$$\frac{a_0}{2} + \sum_{k=1}^{\infty} (a_k \cos kx + b_k \sin kx) \qquad \int_0^{\infty} (a(y) \cos yx + b(y) \sin yx) \, dy$$

$$a_k = \frac{1}{\pi} \int_{-\pi}^{\pi} f(t) \cos kt \, dt \qquad a(y) = \frac{1}{\pi} \int_{-\infty}^{\infty} f(t) \cos ty \, dt$$

$$b_k = \frac{1}{\pi} \int_{-\pi}^{\pi} f(t) \sin kt \, dt \qquad b(y) = \frac{1}{\pi} \int_{-\infty}^{\infty} f(t) \sin ty \, dt$$

If we insert the integral expressions for a_k, b_k, $a(y)$, $b(y)$ into the series and integral, we obtain

$$\frac{a_0}{2} + \sum_{k=1}^{\infty} \frac{1}{\pi} \int_{-\pi}^{\pi} f(t) \cos k(x - t) \, dt \qquad \frac{1}{\pi} \int_0^{\infty} dy \int_{-\infty}^{\infty} f(t) \cos y(x - t) \, dt$$

These relations make the form of the Fourier integral easy to remember. The sum of the Fourier series and the value of the Fourier integral is $f(x)$ for a very general class of functions.

8.2 DEFINITION OF A FOURIER INTEGRAL

Definition 9. *The Fourier integral of a function* $f(x)$ *is the iterated integral*

$$\frac{1}{\pi} \int_0^\infty dy \int_{-\infty}^\infty f(t) \cos y(x - t) \, dt$$

There is no question here of the convergence of the integral. Of course, we hope to be able to impose conditions on $f(x)$ which will guarantee that the integral converges to $f(x)$.

EXAMPLE A. $f(t) = 1, |t| \leq 1; f(t) = 0, |t| > 1.$ The Fourier integral of $f(x)$ is

$$\frac{1}{\pi} \int_0^\infty dy \int_{-1}^1 (\cos xy \cos ty + \sin xy \sin ty) \, dt$$

$$= \frac{2}{\pi} \int_0^\infty \cos xy \, dy \int_0^1 \cos ty \, dt = \frac{2}{\pi} \int_0^\infty \frac{\sin y \cos xy}{y} \, dy$$

By No. 500 of Peirce's Tables this integral is equal to $f(x)$ except at points of discontinuity, where its value is the average of the right-hand and left-hand limits of $f(x)$.

8.3 A PRELIMINARY RESULT

Theorem 18.   1. $f(x) \in D^1$

2. $f(x) \in C$ at $x = x_0$

$$\Rightarrow \qquad \lim_{R \to \infty} \frac{1}{\pi} \int_{-M}^M f(x_0 + t) \frac{\sin Rt}{t} \, dt = f(x_0) \qquad 0 < M < \infty$$

Since

$$\lim_{R \to \infty} \frac{1}{\pi} \int_{-M}^M \frac{\sin Rt}{t} \, dt = 1$$

we need only show that

$$\lim_{R \to \infty} \frac{1}{\pi} \int_{-M}^M [f(x_0 + t) - f(x_0)] \frac{\sin Rt}{t} \, dt = 0$$

As in §3.4, hypothesis 1 implies that $[f(x_0 + t) - f(x_0)]/t \in D$. The conclusion now follows by Theorem 8.

Theorem 19.   1. $f(x) \in D^1$

2. $\int_{-\infty}^\infty |f(x)| \, dx < \infty$

3. $f(x) \in C$ at $x = x_0$

$$(1) \quad \Rightarrow \qquad \lim_{R \to \infty} \frac{1}{\pi} \int_{-\infty}^\infty f(x_0 + t) \frac{\sin Rt}{t} \, dt = f(x_0)$$

Let ϵ be an arbitrary positive number. We can determine M so large that

(2)
$$\frac{1}{\pi} \int_M^\infty \frac{|f(x_0 + t)|}{t} dt + \frac{1}{\pi} \int_{-\infty}^{-M} \frac{|f(x_0 + t)|}{t} dt < \epsilon$$

This is possible by hypothesis 2. Set the integral on the left-hand side of equation (1) equal to $I(R)$ and write it as the sum of three integrals $I_1(R)$, $I_2(R)$, $I_3(R)$ corresponding to the three intervals $(-\infty, -M)$, $(-M, M)$, (M, ∞). Then by inequality (2)

$$|I(R) - f(x_0)| < \epsilon + |I_2(R) - f(x_0)|$$

Now let $R \to \infty$. By Theorem 18, we see that $I_2(R) \to f(x_0)$.

Hence,
$$\varlimsup_{R \to \infty} |I(R) - f(x_0)| \leq \epsilon$$

$$\lim_{R \to \infty} I(R) = f(x_0)$$

This completes the proof of the theorem.

8.4 THE CONVERGENCE THEOREM

Theorem 20.   1. $f(x) \in D^1$

2. $\displaystyle\int_{-\infty}^\infty |f(x)|\, dx < \infty$

3. $f(x) \in C$ at $x = x_0$

(3) $\Rightarrow$ $\displaystyle f(x_0) = \frac{1}{\pi} \int_0^\infty dy \int_{-\infty}^\infty f(t) \cos y(x_0 - t)\, dt$

By the Weierstrass M-test for integrals, the integral

$$\int_{-\infty}^\infty f(t) \cos y(x_0 - t)\, dt$$

converges uniformly in the interval $0 \leq y \leq R$. Hence, we may interchange the order of integration in the following integral:

(4) $\displaystyle\frac{1}{\pi} \int_0^R dy \int_{-\infty}^\infty f(t) \cos y(x_0 - t)\, dt$

$$= \frac{1}{\pi} \int_{-\infty}^\infty f(t)\, dt \int_0^R \cos y(x_0 - t)\, dy$$

$$= \frac{1}{\pi} \int_{-\infty}^\infty f(t) \frac{\sin R(x_0 - t)}{x_0 - t} dt = \frac{1}{\pi} \int_{-\infty}^\infty f(x_0 + t) \frac{\sin Rt}{t} dt$$

As R becomes infinite, the left-hand side of equation (4) approaches the Fourier integral of $f(x)$ and, by Theorem 19, the right-hand side approaches $f(x_0)$.

8.5 FOURIER TRANSFORM

In case $f(x)$ is even or odd, equation (3) takes a somewhat simpler form as follows:

$$f(x) = f(-x) \qquad f(x) = \frac{2}{\pi} \int_0^\infty \cos xy \, dy \int_0^\infty f(t) \cos yt \, dt$$

$$f(x) = -f(-x) \qquad f(x) = \frac{2}{\pi} \int_0^\infty \sin xy \, dy \int_0^\infty f(t) \sin yt \, dt$$

Definition 10. *A function $g(x)$ defined by the equation*

$$(5) \qquad g(x) = \sqrt{\frac{2}{\pi}} \int_0^\infty f(t) \cos xt \, dt$$

is the Fourier cosine transform of $f(x)$.

Definition 11. *A function $g(x)$ defined by the equation*

$$g(x) = \sqrt{\frac{2}{\pi}} \int_0^\infty f(t) \sin xt \, dt$$

is the Fourier sine transform of $f(x)$.

We can now state the following consequence of Theorem 20.

Corollary 20. *If $f(x)$ is an even function satisfying the conditions of Theorem 20, then equation (5) defines the Fourier cosine transform $g(x)$ of $f(x)$ for all x. Moreover $f(x)$ is the Fourier cosine transform of $g(x)$.*

A similar statement holds for odd functions and Fourier sine transforms.

EXAMPLE B. $f(x) = e^{-x^2}$

$$g(x) = \sqrt{\frac{2}{\pi}} \int_0^\infty e^{-t^2} \cos xt \, dt = \frac{1}{\sqrt{2}} e^{-x^2/4}$$

The Fourier cosine transform of e^{-x^2} is $2^{-1/2} e^{-x^2/4}$.   By Corollary 20, we should have

$$e^{-x^2} = \sqrt{\frac{2}{\pi}} \int_0^\infty g(t) \cos xt \, dt = \sqrt{\frac{1}{\pi}} \int_0^\infty e^{-t^2/4} \cos xt \, dt$$

and this is also verified by No. 523 of Peirce's Tables.   Using the Fourier integral, we have

$$e^{-x^2} = \frac{2}{\pi} \int_0^\infty \cos xy \, dy \int_0^\infty e^{-t^2} \cos ty \, dt \qquad -\infty < x < \infty$$

EXERCISES (8)

1. Find the Fourier cosine transform of e^{-x^2}.   (Use tables.)

2. Same problem for $1/\sqrt{x}$.

3. Show that $e^{-x^2/2}$ is its own Fourier cosine transform except for a constant factor.   (Use tables.)

4. Use Exercise 3 to check Theorem 20.

5. By use of the Fourier integral show that

$$\int_0^\infty \frac{y \sin xy}{1 + y^2}\, dy = \begin{cases} \dfrac{\pi}{2} e^{-x} & x > 0 \\[2mm] -\dfrac{\pi}{2} e^{x} & x < 0 \end{cases}$$

6. Prove:

$$\int_0^\infty \frac{\cos xy}{1 + y^2}\, dy = \frac{\pi}{2} e^{-|x|} \qquad -\infty < x < \infty$$

7. $\displaystyle\int_0^\infty dy \int_{-\infty}^\infty \frac{\sin^2 t}{t^2} \cos y(x - t)\, dt = ?$ (all x)

8. $\displaystyle\int_0^\infty \cos xy\, dy \int_0^1 t^2 \cos ty\, dt = ?$ (all x)

9. $\displaystyle\int_0^\infty \sin xy\, dy \int_0^1 t^2 \sin ty\, dt = ?$ (all x)

10. $\displaystyle\lim_{R\to\infty} \int_{-1}^1 \frac{\sin \pi t \sin Rt}{t^2}\, dt = ?$

11. $\displaystyle\lim_{R\to\infty} \int_0^1 \frac{\sin Rx}{\sqrt{x}}\, dx = ?$

12. $\displaystyle\lim_{R\to\infty} \int_{-\infty}^\infty \frac{\sin^2 t \sin Rt}{t^3}\, dt = ?$

13. Find the Fourier cosine transform of $f(x) = \cos x$, $|x| < \pi$; $f(x) = 0$, $|x| > \pi$.

14. $\displaystyle\int_0^\infty \frac{x \sin \pi x \cos xy}{1 - x^2}\, dx = ?$ (all x)

Hint: Use Exercise 13.

15. If hypothesis 3 is omitted in Theorem 20, show that

$$\frac{f(x_0+) + f(x_0-)}{2} = \frac{1}{\pi} \int_0^\infty dy \int_{-\infty}^\infty f(t) \cos y(x_0 - t)\, dt \qquad -\infty < x_2 < \infty$$

Hint: Use the function of Example A as the function $g(x)$ was used in the proof of Theorem 7.

16. $\displaystyle\frac{1}{\pi} \int_0^\infty dy \int_{-\pi}^\pi e^{-t} \cos t \cos y(x - t)\, dt = ?$ (all x)

13

The Laplace

Transform

§1. *Introduction*

In this chapter, we shall consider theoretic aspects of the Laplace transform, reserving for Chapter 14 the application of the subject to the solution of linear differential equations. This transform is defined by the equation

$$(1) \qquad\qquad f(s) = \int_{0+}^{\infty} e^{-st}\varphi(t)\,dt$$

It may be thought of as transforming one class of functions into another. Thus, the function $\varphi(t)$ is replaced by the function $f(s)$ by use of equation (1). The advantage in the operation is that under certain circumstances it replaces complicated functions by simpler ones. For example, it replaces the transcendental function $\varphi(t) = e^{-t}$ by the rational function $f(s) = (s + 1)^{-1}$. If we establish rules whereby we can pass easily from the class of functions $\varphi(t)$ to the class of functions $f(s)$ and back again, then a problem originally given to us in one of the classes may be solved in the other, sometimes much more easily. Of course, for the success of such a method it is important that the correspondence between two functions of the two classes should be unique. It is clear from equation (1) that a given function $\varphi(t)$ leads to

436

at most one function $f(s)$. Later, we shall show that for a given function $f(s)$ there is essentially only one function $\varphi(t)$. We say "essentially," for it is clear that if the definition of $\varphi(t)$ were altered at a finite number of points, $f(s)$ would not be changed at all. But we can show that there will be at most one *continuous* function $\varphi(t)$ corresponding to a given function $f(s)$. It must not be supposed that one may set down an arbitrary function in one class and expect it to have a mate in the other. For example, if $\varphi(t) = e^{t^2}$, then the integral (1) diverges for all s. Again if $f(s) = s$, there will be no function $\varphi(t)$ corresponding. We will show this later. But even now we can see that there can be no *absolutely* converging integral (1) representing s. For if there were such an integral, converging absolutely for $s = c$, we should have

$$|s| \leqq \int_0^\infty e^{-(s-c)t} e^{-ct} |\varphi(t)|\, dt$$

$$|s| \leqq \int_0^\infty e^{-ct} |\varphi(t)|\, dt \qquad\qquad c \leqq s < \infty$$

This is clearly absurd, since s becomes infinite in the range indicated.

1.1 RELATION TO POWER SERIES

At first sight, the integral (1) appears to be of a very special nature. Although the function $\varphi(t)$ may be chosen very generally—we insist only that the integral should converge for some value of s—the other factor of the integrand is indeed of a specific nature. Why should we choose the exponential function rather than any other? The answer to this is that the integral (1) may be regarded as a generalization of a power series. These series occur in Maclaurin's and in Taylor's expansions and are of fundamental importance in analysis. We shall now show how the Laplace transform may be evolved from a power series.

Consider the power series

$$(2) \qquad\qquad F(x) = \sum_{k=0}^\infty a_k x^k$$

If it converges at all for $x \neq 0$, then it converges in an interval, $|x| < h$, extending equal distances on either side of the origin, and diverges outside the interval. The points $x = h$ and $x = -h$ may or may not be included in this interval of convergence depending upon the particular sequence $\{a_k\}_{k=0}^\infty$ involved. Of course, we may have $h = \infty$, when the series (2) converges for all x. Or the series may diverge for all x except $x = 0$. One natural way of generalizing the series (2) would be to replace the sequence of integers which appear as the exponents of x by a more general sequence. Let $\{\lambda_k\}_{k=0}^\infty$ be such a sequence that

$$0 \leqq \lambda_0 < \lambda_1 < \lambda_2 < \dots, \qquad \lim_{k \to \infty} \lambda_k = \infty$$

With this sequence as exponents of x, we obtain

$$F(x) = \sum_{k=0}^{\infty} a_k x^{\lambda_k}$$

But there is now some ambiguity.   At least if λ_k is nonintegral, say $\frac{1}{2}$, it may not be clear which root of x is the natural one to take.   To avoid this difficulty, make the change of variable $x = e^{-s}$;   ($x = e^s$ would be equally good). We are thus led to the *Dirichlet series*

(3)
$$F(e^{-s}) = \sum_{k=0}^{\infty} a_k e^{-s\lambda_k}$$

And now it is quite natural to replace the sequence $\{\lambda_k\}_{k=0}^{\infty}$ by a continuous variable t which ranges from 0 to ∞.   We would then replace the summation sign by an integral sign and the sequence $\{a_k\}_{k=0}^{\infty}$ by a function $a(t)$:

$$F(e^{-s}) = \int_0^{\infty} a(t)e^{-ts}\, dt$$

If we replace $F(e^{-s})$ by $f(s)$ and $a(t)$ by $\varphi(t)$, we arrive in this way at the integral (1).   Except then for the unimportant exponential change of variable, the Laplace integral (1) may be regarded as a generalized power series, the sequence of integral exponents having been replaced by a continuous variable in the generalization.

1.2 DEFINITIONS

We now turn to the formal definition of the transform.

Definition 1.   *The function $f(s)$ is the Laplace transform of $\varphi(t)$, the relation being indicated by*

(4)
$$L\{\varphi(t)\} = f(s)$$

$\Leftrightarrow$ *equation* (1) *holds, the integral converging for some value of s.*

Equation (4) is sometimes written

(5)
$$L^{-1}\{f(s)\} = \varphi(t)$$

We have already indicated that the relationship between $f(s)$ and $\varphi(t)$ is in some sense one to one; that is, each is essentially determined by the other. In equation (4) we think of $\varphi(t)$ as given and $f(s)$ as determined from it. In equation (5) it is $f(s)$ that is given.   Accordingly, equation (4) defines the *direct* transform, equation (5), the *inverse* transform.   At present it is not clear how the inverse operation L^{-1} is to be performed, whereas the direct transform L is accomplished by evaluating the improper integral (1).

Definition 2. *The function $f(s)$ in equation* (1) *or in equation* (4) *is the generating function.*

Definition 3. *The function $\varphi(t)$ in equation* (1) *or in equation* (4) *is the determining function.*

EXAMPLE A. $L\{1\} = s^{-1}$, $L^{-1}\{s^{-1}\} = 1$

For, if $0 < s < \infty$, we have

$$\int_0^\infty e^{-st}1\,dt = \lim_{R\to\infty}\int_0^R e^{-st}\,dt = \lim_{R\to\infty} s^{-1}(1 - e^{-sR}) = s^{-1}$$

Here $1/s$ is the generating function, $\varphi(t) = 1$ is the determining function. Observe that the determining function must be defined for $0 \leq t < \infty$. So far as Definition (1) is concerned, the generating function need be defined only "for some value of s." But we shall see later that, if the integral (1) converges for some value of s, it converges for all larger values. Hence, the generating function will always be defined on some right half-line (or on the whole s-axis).

EXAMPLE B. $L\{te^{-t}\} = (s + 1)^{-2}$ $\qquad\qquad\qquad\qquad s > -1$

Here we have indicated at the right of the equation the region of convergence of the integral (1) with the present determining function for $\varphi(t)$. The integral can be evaluated by integration by parts.

EXAMPLE C. If $-\infty < c < \infty$, $a > -1$,

$$L\{t^a e^{-ct}\} = \Gamma(a + 1)/(s + c)^{a+1} \qquad\qquad s > -c$$

For, we have

$$\int_0^\infty e^{-st}t^a e^{-ct}\,dt = \int_0^\infty e^{-(s+c)t}t^a\,dt$$

and this integral was evaluated in Theorem 9, Chapter 11.

EXAMPLE D. Find $L^{-1}\{(s^2 - 1)^{-1}\}$. By use of partial fractions we have

$$\frac{1}{s^2 - 1} = \frac{1}{2}\left(\frac{1}{s - 1} - \frac{1}{s + 1}\right) = \frac{1}{2}L\{e^t\} - \frac{1}{2}L\{e^{-t}\} = L\left\{\frac{e^t - e^{-t}}{2}\right\}$$

Here the two integrals involved converge for $s > 1$ and $s > -1$, respectively. Hence,

$$L\{\sinh t\} = 1/(s^2 - 1) \qquad\qquad\qquad s > 1$$

$$L^{-1}\{(s^2 - 1)^{-1}\} = \sinh t \qquad\qquad\qquad 0 \leq t < \infty$$

EXERCISES (1)

Find the Laplace transforms of Exercises 1–13, indicating the region of convergence of the integrals involved.

1. $L\{\sqrt{t}\}.$

2. $L\{e^t/\sqrt{t}\}.$

3. $L\{\sinh ct\}.$

4. $L\{e^{at} \sinh t\}.$

5. $L\{e^{at} \cosh ct\}.$

6. $L\{e^{at} \sin ct\}.$

7. $L\{e^{at} \cos ct\}.$

8. $L^{-1}\{1/\sqrt{s}\}.$

9. $L^{-1}\{s^a\} \qquad a < 0.$

10. $L^{-1}\{s^{-1}(s + 1)^{-2}\}.$

11. $L^{-1}\{s/(s^2 + a^2)\}.$

12. $L^{-1}\{3/(s^2 + 9)\}.$

13. $L^{-1}\{s^{-1}(s^2 + 9)^{-1}\}.$

14. Prove that the integral (1) diverges for all s if $\varphi(t) = e^{t^2}$.

15. Show that no constant except zero can be a generating function, at least if the Laplace integral is to be absolutely convergent at $s = 0$. *Hint:*
$$|f(s)| \leqq \int_0^R |\varphi(t)| \, dt + e^{-sR} \int_R^\infty |\varphi(t)| \, dt \text{ for any positive numbers } s \text{ and } R.$$
Now choose s and R so as to make the right-hand side less than the positive constant $|f(s)|$.

16. Same problem for any polynomial.

17. Show that if $\varphi(t)$ is constant in each of the intervals $(\lambda_k, \lambda_{k+1})$, $k = 0, 1, 2, \ldots,$ of §1.1, then $sL\{\varphi(t)\}$ is formally a Dirichlet series. Determine the value of $\varphi(t)$ in each interval if the series is to reduce to series (3).

§2. *Region of Convergence*

Since the Laplace integral may be regarded as a generalization of a power series, we can predict in what sort of region that integral is likely to converge. Recall that in §1.1 we made the exponential change of variable $x = e^{-s}$. If x and s are real, this transformation will be useful only for half the interval of convergence of the power series (2), §1.1. Since the power series converges for $0 < x < h$, we should expect the Dirichlet series (3) and the Laplace integral (1), §1, to converge in the interval $\log(1/h) < s < \infty$. This assumes, of course, that making the sequence of exponents more dense does not affect the type of region of convergence. We shall show that this is, in fact, the case; that the Laplace integral, if it converges at all, converges on a right half-line or on a whole line (corresponding to the case $h = \infty$).

2.1 POWER SERIES

In §9.1 of Chapter 9 we showed that a power series

$$\text{(1)} \qquad \sum_{k=0}^{\infty} a_k x^k$$

converges in an interval $-R < x < R$ extending equal distances on the two sides of the origin (allowing for the special cases $R = 0$ and $R = \infty$). Our proof involved a formula for the determination of the radius of convergence R in terms of the coefficients a_k. There is another more elementary way of discussing the convergence of the series (1). We present it here as preliminary to our study of the convergence of the Laplace integral, since an analogous method will be used in the proof of Theorem 1.

Theorem A.   1. *Series* (1) *converges at* $x = x_0 \neq 0$

$\Rightarrow$ *Series* (1) *converges absolutely for* $|x| < |x_0|$.

Since the series (1) converges at $x = x_0$, the general term approaches zero,

$$\text{(2)} \qquad \lim_{k \to \infty} a_k x_0^k = 0$$

Now use a limit test for absolute convergence, Theorem 9, Chapter 9. We have for $|x| < |x_0|$

$$\lim_{k \to \infty} k^2 a_k x^k = \lim_{k \to \infty} \left(\frac{k^2 x^k}{x_0^k} \right)(a_k x_0^k) = 0$$

and the theorem is proved.

Corollary A.   1. *Series* (1) *diverges at* $x = x_0 \neq 0$

$\Rightarrow$ *Series* (1) *diverges for* $|x| > |x_0|$

For, by Theorem A, if the series (1) converged for some x_1 such that $|x_1| > |x_0|$ then it would converge at every point nearer the origin than x_1 and hence at x_0, contrary to hypothesis 1.

By use of Theorem A and Corollary A, we see that there must be some interval of convergence for the power series (1). It may be a single point, a finite interval, or the whole x-axis. The three examples $a_k = k!$, $a_k = 1$, $a_k = 1/k!$ show that all three cases actually arise.

2.2 CONVERGENCE THEOREM

A strict analogy with power series does not hold in the proof of the next theorem, and it is easy to see why. In deriving equation (2) we used the fact that the general term of a convergent series tends to zero. We know that the integrand of a convergent integral need not tend to zero as the

independent variable becomes infinite (§1.2, Chapter 10). Recall that
$\varphi(t)$ may be changed at isolated points without changing $f(s)$ and without
affecting the convergence properties of the Laplace integral. This fact
alone would prevent us from trying to establish an equation analogous to
(2). But an indefinite integral of $\varphi(t)$ is unchanged if $\varphi(t)$ is altered at
isolated points. Hence, we may hope to deal with such an indefinite integral,
which may probably be introduced by an integration by parts.

Throughout the remainder of the chapter let us assume without further
statement that $\varphi(t) \in C$ except at isolated points. In particular, $\varphi(t)$ may be
discontinuous at $t = 0$, so that the Laplace integral may be improper of
Type III as well as of Type I. Any discontinuity inside the interval $(0, \infty)$
will be assumed to be a finite jump; that is, right-hand and left-hand limits
will exist. Now consider the Laplace integral

$$(3) \qquad \int_{0+}^{\infty} e^{-st}\varphi(t)\, dt$$

Theorem 1. 1. *Integral* (3) *converges at* $s = s_0$

$\Rightarrow$ *Integral* (3) *converges for* $s > s_0$.

Set
$$\alpha(t) = \int_{0+}^{t} e^{-s_0 u}\varphi(u)\, du \qquad\qquad 0 < t < \infty$$

Clearly $\alpha(0+) = 0$ and $\alpha(\infty)$ exists by virtue of hypothesis 1. Choose ϵ
and R so that $0 < \epsilon < R$. Integration by parts gives

$$\int_{\epsilon}^{R} e^{-st}\varphi(t)\, dt = \int_{\epsilon}^{R} e^{-(s-s_0)u}\alpha'(u)\, du$$
$$= \alpha(R)e^{-(s-s_0)R} - \alpha(\epsilon)e^{-(s-s_0)\epsilon} + (s - s_0)\int_{\epsilon}^{R} e^{-(s-s_0)u}\alpha(u)\, du$$

Now let $\epsilon \to 0+$. Both terms on the right which depend on ϵ approach a
limit and

$$(4) \qquad \int_{0+}^{R} e^{-st}\varphi(t)\, dt = \alpha(R)e^{-(s-s_0)R} + (s - s_0)\int_{0}^{R} e^{-(s-s_0)u}\alpha(u)\, du$$

Notice that the integral on the right is certainly not improper since $\alpha(0+) = 0$.
Now let $R \to \infty$. If $s > s_0$, the first term on the right approaches zero
and we shall have

$$(5) \qquad \int_{0+}^{\infty} e^{-st}\varphi(t)\, dt = (s - s_0)\int_{0}^{\infty} e^{-(s-s_0)u}\alpha(u)\, du \qquad\qquad s > s_0$$

if the integral on the right converges. But it does converge absolutely as we
see by use of a limit test, Theorem 4, Chapter 10. For, we have

$$\lim_{u \to \infty} u^2 e^{-(s-s_0)u}\alpha(u) = 0 \cdot \alpha(\infty) = 0$$

This completes the proof of the theorem. It must not be supposed that the integral on the left-hand side of equation (5) converges absolutely just because the one on the right-hand side does so.

Corollary 1.1.   1. *Integral* (3) *diverges at* $s = s_0$

$\Rightarrow$ *Integral* (3) *diverges for* $s < s_0$.

Corollary 1.2. *The region of convergence of the integral* (3) *is a right half-line or a whole line.*

Corollary 1.3. $f(+\infty) = 0$.

That is, every generating function vanishes at $+\infty$. To show this, determine, corresponding to an arbitrary $\epsilon > 0$, a number δ such that $|\alpha(t)| < \epsilon$ for $0 < t \leqq \delta$. Then from equation (5) we have for $s > s_0 + 1$

$$|f(s)| \leqq \epsilon(s - s_0) \int_0^\delta e^{-(s-s_0)t}\, dt + (s - s_0)e^{-\delta(s-s_0-1)} \int_\delta^\infty e^{-t}|\alpha(t)|\, dt$$

Now let $s \to +\infty$:

$$\varlimsup_{s \to +\infty} |f(s)| \leqq \epsilon$$

Hence

$$\lim_{s \to +\infty} f(s) = 0$$

As a consequence of this result, it is clear that a polynomial which is not identically zero cannot be a generating function.

By virtue of Theorem 1 and Corollary 1.1 three cases may arise:

(a) the integral (3) converges for all s;

(b) the integral (3) diverges for all s;

(c) there exists a number s_c such that the integral (3) converges for $s > s_c$ and diverges for $s < s_c$.

The number s_c is called the *abscissa of convergence*. In case (a) we write $s_c = -\infty$ and in case (b), $s_c = +\infty$.

2.3 EXAMPLES

To show that the three cases of §2.2, which are logically possible, actually occur we exhibit three examples:

(a) $\displaystyle \int_0^\infty e^{-st} e^{-t^2}\, dt$ $s_c = -\infty$

(b) $\displaystyle \int_0^\infty e^{-st} e^{t^2}\, dt$ $s_c = +\infty$

(c) $\displaystyle \int_0^\infty e^{-st} t\, dt$ $s_c = 0$

The facts here asserted are easily established by use of limit tests.

EXAMPLE A. Find s_c if $\varphi(t) = (\sin t)/t$.

Here we certainly have convergence for $s \geqq 0$ since the integral

$$\int_0^\infty \frac{\sin t}{t}\, dt$$

converges. Hence, $s_c \leqq 0$. But $s_c = 0$. For if we have $s = -a < 0$, the integral (3) becomes

$$\sum_{k=0}^\infty \int_{k\pi}^{(k+1)\pi} g(t) \sin t\, dt, \qquad g(t) = e^{at}/t$$

The general term of this series is in absolute value greater than $2g(k\pi)$. The series cannot converge since its general term does not tend to zero.

This, with example (c) above, shows that the end point of the interval of convergence may or may not belong to the region of convergence. In example (c) the integral diverges at $s = s_c$; in Example A the integral converges at $s = s_c$.

EXERCISES (2)

Find s_c in Exercises 1–10.

1. $\varphi(t) = 1 + e^{-t}$

2. $\varphi(t) = \sin t + e^t$

3. $\varphi(t) = \cos 2t$

4. $\varphi(t) = t^{-3/2} \sin 3t$

5. $\varphi(t) = t^{-1/2} \cos t$

6. $\varphi(t) = (t + 1)^{-3/2} \cosh t$

7. $\varphi(t) = t^{-1/2} \sin t$

8. $\varphi(t) = t^{-3/2}$

9. $\varphi(t) = t^{-3/2} e^{3t} \sin 3t$

10. $f(s) = (s^2 - 3s + 2)^{-1}$

11. Prove that, if $\varphi(t)$ is bounded, then $s_c \leqq 0$.

12. Prove that, if $\varphi(t)e^{-at}$ is bounded, then $s_c \leqq a$.

13. Find s_c for Examples (a), (b), (c).

14. Prove that if the integral (3) converges at $s = s_0 > 0$, then

$$e^{-s_0 t} \int_{0+}^t \varphi(u)\, du$$

is bounded in the interval $0 < t < \infty$. *Hint:* Use equation (4) with $s = 0$ and use the fact that $\alpha(t)$ is bounded.

15. Pattern a proof of Theorem A (conditional convergence only) after that of Theorem 1. *Hint:* Set

$$s_n = \sum_{k=0}^n a_k x_0^k$$

and show that

$$\sum_{k=0}^n a_k x^k = s_n(x/x_0)^n + \sum_{k=0}^{n-1} s_k(x/x_0)^k (1 - [x/x_0])$$

16. Prove that the region of convergence of a Dirichlet series is a right half-line or a whole line.

17. What is the relation of a determining function $\varphi(t)$ to the class of functions D defined in Chapter 12?

§3. *Absolute and Uniform Convergence*

We saw in §2 that a power series converges absolutely at any point inside the interval of convergence (boundary points of the interval excluded). The analogous result for the Laplace transform is false. The integral

$$(1) \qquad \int_{0+}^{\infty} e^{-st}\varphi(t)\,dt$$

need not converge absolutely in any part of its interval of convergence. On the other hand, it may in some cases converge absolutely in part or in all of that interval. This leads us to define an abscissa of absolute convergence. By means of a discussion of the uniform convergence properties of the integral (1), we shall show that any generating function belongs to C^{∞} for $s > s_c$.

3.1 ABSOLUTE CONVERGENCE

Theorem 2.   1. *Integral* (1) *converges absolutely at* $s = s_0$

$\Rightarrow$ *Integral* (1) *converges absolutely for* $s \geqq s_0$.

The proof may be obtained from that of Theorem 1 by replacing $\varphi(t)$ by $|\varphi(t)|$. However, a much simpler proof is available in the present case. Since

$$(2) \qquad e^{-st}|\varphi(t)| \leqq e^{-s_0 t}|\varphi(t)| \qquad\qquad s_0 \leqq s < \infty$$

we have our result at once by comparison, Theorem 1, Chapter 10.

This result enables us to define an *abscissa of absolute convergence*, s_a. The integral (1) will converge absolutely for $s > s_a$, will fail to do so for $s < s_a$, may or may not do so at $s = s_a$. In particular, we may have $s_a = -\infty$ or $s_a = +\infty$. Since absolute convergence implies convergence, it is clear that $s_c \leqq s_a$. The following example will show that s_a does not always coincide with s_c.

EXAMPLE A. $\varphi(t) = e^t \sin e^t$.   Set $e^t = u$.   Then

$$f(s) = \int_1^{\infty} \frac{\sin u}{u^s}\,du$$

The integral converges absolutely for $s > 1$ by a limit test.   It converges conditionally for $0 < s \leq 1$ and diverges for $s = 0$.   Hence, $s_c = 0$, $s_a = 1$. By replacing e^t by e^{kt} in this example, we obtain $s_a = k$ and thus see that s_a and s_c may differ by any positive number.   Example (a) of §2 shows that s_a may be $-\infty$.

EXAMPLE B. $\varphi(t) = e^t e^{e^t} \sin e^{e^t}$

Here $s_c = 0$ and $s_a = \infty$.

3.2 UNIFORM CONVERGENCE

Theorem 3.   1. *Integral* (1) *converges absolutely at* $s = s_0$

$\Rightarrow$ *Integral* (1) *converges uniformly for* $s_0 \leq s \leq R$, *where R is arbitrary.*

Note first that the integral (1) is the sum of an improper integral of Type I and an integral of Type III.   Inequality (2) is sufficient for the application of Weierstrass's M-test in either case.

It can be shown that Theorem 3 remains true if the word "absolutely" is omitted in hypothesis 1 (compare Exercise 12, §6, Chapter 10).

3.3 DIFFERENTIATION OF GENERATING FUNCTIONS

We shall now show that it is always permissible to differentiate a Laplace integral under the sign of integration, thus establishing the fact that $f(s) \in C^\infty$ for $s > s_c$.

Theorem 4.   1. $f(s) = \int_{0+}^{\infty} e^{-st}\varphi(t)\, dt$ $s > s_c$

$$(3) \Rightarrow \qquad f'(s) = -\int_{0+}^{\infty} e^{-st} t \varphi(t)\, dt \qquad\qquad s > s_c$$

Let $s = s_0 > s_c$.   By equation (5), §2,

$$(4) \qquad\qquad f(s) = (s - s_0)\int_{0}^{\infty} e^{-(s-s_0)t}\alpha(t)\, dt \qquad\qquad s > s_0$$

$$\alpha(t) = \int_{0+}^{t} e^{-s_0 u}\varphi(u)\, du \qquad\qquad 0 < t < \infty$$

Hence,

$$(5) \qquad f'(s) = \int_{0}^{\infty} e^{-(s-s_0)t}\alpha(t)\, dt - (s - s_0)\int_{0}^{\infty} e^{-(s-s_0)t} t \alpha(t)\, dt$$

provided that it is permissible to differentiate the integral (4) under the integral sign. But this operation is valid by Theorem 14, Chapter 10. The integral

$$\int_0^\infty e^{-(s-s_0)t}t\alpha(t)\,dt$$

converges uniformly for $s_0 < s_0 + \epsilon \leq s \leq R$, where ϵ and R are arbitrary, by Weierstrass's M-test:

$$e^{-(s-s_0)t}t|\alpha(t)| \leq e^{-\epsilon t}tM \qquad 0 \leq t < \infty, \quad s_0 + \epsilon \leq s$$

Here M is an upper bound for $|\alpha(t)|$, which must exist since $\alpha(\infty)$ exists. Since

$$\int_0^\infty e^{-\epsilon t}tM\,dt < \infty$$

the integral (4) may be differentiated under the sign of integration in the interval $(s_0 + \epsilon, R)$. On account of the arbitrary nature of ϵ and R, the process is valid for $s > s_0$. Finally, if we integrate the integral (3) by parts, we obtain

$$-\int_{0+}^\infty e^{-st}t\varphi(t)\,dt = -\int_{0+}^\infty e^{-(s-s_0)t}t\alpha'(t)\,dt$$

$$= \int_0^\infty \alpha(t)[e^{-(s-s_0)t}t]'\,dt \qquad s > s_0$$

If we perform the indicated differentiation, we obtain the right-hand side of equation (5). Since s_0 was arbitrary, the proof is complete.

Corollary 4. $f(s) \in C^\infty$ $\qquad\qquad s > s_c$

We apply the theorem successively and prove by induction that for each positive integer k

$$f^{(k)}(s) = \int_0^\infty e^{-st}(-t)^k\varphi(t)\,dt \qquad s > s_c$$

EXAMPLE C. Find $L\{t\sin ct\}$. We differentiate with respect to s the equation

$$\frac{c}{s^2 + c^2} = \int_0^\infty e^{-st}\sin ct\,dt \qquad s > 0, \quad -\infty < c < \infty$$

and obtain

$$\frac{2cs}{(s^2 + c^2)^2} = \int_0^\infty e^{-st}t\sin ct\,dt$$

$$L\{t\sin ct\} = 2cs(s^2 + c^2)^{-2} \qquad s > 0, \quad -\infty < c < \infty$$

EXERCISES (3)

Find s_c and s_a in Exercises 1–7.

1. $\varphi(t) = \sin t$

2. $\varphi(t) = t^{-3/2} \sin t$

3. $\varphi(t) = t^{-3/2} \cos t$

4. $\varphi(t) = e^{2t} \sin e^{2t}$

5. $\varphi(t) = \sin e^{2t}$

6. $\varphi(t) = t^{-1/2} e^{\sqrt{t}} \sin e^{\sqrt{t}}$

7. $\varphi(t) = te^{t^2} \sin e^{t^2}$

8. Give details in Example B.

9. Prove: $\varphi(t) \in C, |\varphi(t)| \leqq Me^{ct}, \ 0 \leqq t < \infty \Rightarrow$ integral (1) converges uniformly in $c + \epsilon \leqq s \leqq R$, where ϵ is positive and R is arbitrary.

10. Give an example to show that ϵ cannot be zero in Exercise 9. May we infer uniform convergence in $c < s \leqq R$?

11. Use Theorem 4 to obtain $L\{t^k\}, \ k = 1, 2, 3, \ldots$. Check by use of the gamma function.

12. $L^{-1}\left\{\dfrac{d^k}{ds^k} \dfrac{s}{s^2 + a^2}\right\} = ?$ $k = 1, 2, 3, \ldots$

13. $L^{-1}\left\{\dfrac{d^k}{ds^k} \dfrac{1}{s^2 + a^2}\right\} = ?$ $k = 1, 2, 3, \ldots$

14. $L\{t \sinh t\} = ?$

15. $L\{t^2 \cos t\} = ?$

16. $L\{te^{at} \sin ct\} = ?$

17. $L\{(t - 1)^2 e^{-2t}\} = ?$

§4. *Operational Properties of the Transform*

Skill in manipulating the Laplace transform will be greatly increased by the study of the effect on a given function produced by certain elementary operations performed on its mate. We have already observed that differentiation of the generating function $f(s)$ corresponds to the multiplication of the determining function $\varphi(t)$ by $-t$. It is such operational considerations which we now discuss.

4.1 LINEAR OPERATIONS

Let $$f(s) = L\{\varphi(t)\}, \qquad g(s) = L\{\psi(t)\}$$
both integrals converging for $s > s_0$. If a and b are arbitrary constants, then

$$L\{a\varphi(t) + b\psi(t)\} = aL\{\varphi(t)\} + bL\{\psi(t)\} = af(s) + bg(s)$$

The integral on the left will certainly converge for $s > s_0$ and perhaps in a larger region. These facts are evident from the definition of the transform. Obviously,

$$L^{-1}\{af(s) + bg(s)\} = aL^{-1}\{f(s)\} + bL^{-1}\{g(s)\} = a\varphi(t) + b\psi(t)$$

4.2 LINEAR CHANGE OF VARIABLE

If $a > 0$, we have by setting $t = au$

$$(1) \qquad f(s) = \int_{0+}^{\infty} e^{-st}\varphi(t)\, dt = a \int_{0+}^{\infty} e^{-asu}\varphi(au)\, du \qquad\qquad s > s_c$$

whence

$$(2) \qquad\qquad L\{\varphi(at)\} = \frac{1}{a} f\left(\frac{s}{a}\right) \qquad\qquad s > s_c, \quad a > 0$$

If $\varphi(t) = 0$ for $-\infty < t < 0$, we have for $b > 0$

$$\int_{0}^{\infty} e^{-st}\varphi(t - b)\, dt = e^{-bs} \int_{-b}^{\infty} e^{-su}\varphi(u)\, du = e^{-bs} \int_{0+}^{\infty} e^{-st}\varphi(t)\, dt \qquad s > s_c$$

Consequently, for $b > 0$ and $\varphi(t) = 0$ in $(-\infty, 0)$

$$(3) \qquad\qquad L\{\varphi(t - b)\} = e^{-bs}L\{\varphi(t)\} = e^{-bs}f(s) \qquad\qquad s > s_c$$

Let us next make corresponding changes of variable in the generating function. We obtain easily

$$f(as) = L\left\{\frac{1}{a}\varphi\left(\frac{t}{a}\right)\right\} \qquad\qquad a > 0, \quad s > s_c$$

$$f(s - b) = L\{e^{bt}\varphi(t)\} \qquad\qquad -\infty < b < \infty$$

In the latter equation we must have $s > s_c + b$ since $f(s)$ is defined for $s > s_c$ only.

4.3 DIFFERENTIATION

We have already seen that

$$f'(s) = -L\{t\varphi(t)\} \qquad\qquad s > s_c$$

Let us investigate next the effect of differentiating the determining function.

Theorem 5.   1. $\varphi(t) \in C^1$ $\qquad\qquad\qquad\qquad\qquad\qquad 0 \leq t < \infty$

2. $f(s) = L\{\varphi(t)\}$ $\qquad\qquad\qquad\qquad\qquad s > s_c$

3. $\lim_{t \to +\infty} e^{-st}\varphi(t) = 0$ $\qquad\qquad\qquad\qquad s > s_c$

$\Rightarrow \qquad\qquad L\{\varphi'(t)\} = -\varphi(0) + sf(s) \qquad\qquad\qquad s > s_c$

The theorem implies that the integral on the left converges for $s > s_c$. The proof follows from an integration by parts.

Corollary 5.   1. $\varphi(t) \in C^n$ $\qquad\qquad\qquad\qquad\qquad\qquad 0 \leq t < \infty$

2. $f(s) = L\{\varphi(t)\}$ $\qquad\qquad\qquad\qquad\qquad s > s_c$

3. $\lim_{t \to +\infty} e^{-st}\varphi^{(k)}(t) = 0 \qquad\qquad k = 0, 1, \ldots, n - 1; \ s > s_c$

$\Rightarrow \qquad\qquad L\{\varphi^{(n)}(t)\} = -\sum_{k=1}^{n} \varphi^{(k-1)}(0)s^{n-k} + s^n L\{\varphi(t)\}$

4.4 INTEGRATION

Theorem 6.   1. $f(s) = L\{\varphi(t)\}$ $s > s_c$

              2. $\lim\limits_{t \to \infty} e^{-st} \int_{0+}^{t} \varphi(u)\, du = 0$ $s > s_c$

$\Rightarrow$ $f(s) = sL\left\{ \int_{0+}^{t} \varphi(u)\, du \right\}$ $s > s_c$

For, set

$$\alpha(t) = \int_{0+}^{t} \varphi(u)\, du \qquad\qquad 0 < t < \infty$$

Then

$$f(s) = \int_{0+}^{\infty} e^{-st}\alpha'(t)\, dt = \alpha(t)e^{-st}\Big|_{0+}^{\infty} + s \int_{0}^{\infty} e^{-st}\alpha(t)\, dt \qquad s > s_c$$

from which the theorem is evident. It can be shown by use of Exercise 14, §2, that hypothesis 2 is redundant if $s_c > 0$. It is not so if $s_c < 0$, as may be seen by taking $\varphi(t) = e^{-t}$.

Theorem 7.   1. $f(s) = L\{\varphi(t)\}$ $s > s_a$

              2. $\int_{0+}^{1} \dfrac{|\varphi(t)|}{t}\, dt < \infty$

$\Rightarrow$ $\int_{s}^{\infty} f(x)\, dx = L\{\varphi(t)/t\}$ $s > s_a$

Choose $s_0 > s_a$ and $R > s_0$. Since the integral (1) converges absolutely at $s = s_0$, then by Theorem 3 it converges uniformly in $s_0 \leq s \leq R$. Hence,

$$(4) \quad \int_{s_0}^{R} f(x)\, dx = \int_{0+}^{\infty} \varphi(t)\, dt \int_{s_0}^{R} e^{-xt}\, dx = \int_{0+}^{\infty} \varphi(t)[e^{-s_0 t} - e^{-Rt}]t^{-1}\, dt$$

$$= \int_{0+}^{\infty} \frac{\varphi(t)}{t} e^{-s_0 t}\, dt - \int_{0+}^{\infty} \frac{\varphi(t)}{t} e^{-Rt}\, dt$$

Both integrals on the right of equation (4) converge absolutely. By Corollary 1.3 the second of these tends to zero as $R \to \infty$. Since s_0 was arbitrary, the proof is complete.

4.5 ILLUSTRATIONS

EXAMPLE A.   Verify equation (2) for the special case $\varphi(t) = \sin t$, $f(s) = (s^2 + 1)^{-1}$. We have

$$L\{\varphi(at)\} = L\{\sin at\} = \frac{a}{s^2 + a^2} = \frac{1}{a} \frac{1}{1 + (s/a)^2} = \frac{1}{a} f\left(\frac{s}{a}\right) \qquad a \neq 0$$

EXAMPLE B.   Find $L\{\cos t\}$ from the equation

$$L\{\sin t\} = (s^2 + 1)^{-1}$$

Take $\varphi(t) = \sin t$ in Theorem 5.   Then $s_c = 0$ and

$$\lim_{t \to +\infty} e^{-st} \sin t = 0 \qquad\qquad s > 0$$

Hence,

$$L\{\varphi'(t)\} = L\{\cos t\} = sL\{\sin t\} = s(s^2 + 1)^{-1}$$

EXAMPLE C.   Apply Theorem 7 to the case $\varphi(t) = \sin t$.   Here $s_a = s_c = 0$, and

$$\int_0^1 \frac{|\sin t|}{t}\, dt < \infty$$

Hence,

$$\int_s^\infty \frac{dx}{1 + x^2} = \frac{\pi}{2} - \tan^{-1} s = L\!\left(\frac{\sin t}{t}\right)$$

$$\tan^{-1}\frac{1}{s} = \int_0^\infty e^{-st} \frac{\sin t}{t}\, dt \qquad\qquad s > 0$$

EXERCISES (4)

1. Show that $L\{f + g\}$ may have a smaller abscissa of convergence than $L\{f\}$ or $L\{g\}$.

2. Expand $L\{\varphi(at + b)\}, a > 0$.   What assumptions are you making about $\varphi(t)$?

3. Show that

$$L^{-1}\{f(as)\} = \frac{1}{a} L^{-1}\{f(s)\}|_{t/a} \qquad s > s_c, \quad a > 0$$

Illustrate by $f(s) = (s - 1)^{-1}$.

4. Expand $L^{-1}\{f(as + b)\}, a > 0$.   Illustrate by $f(s) = (s + 1)^{-2}$.

5. Prove Theorem 5.

6. Prove Corollary 5 by induction.

7. Prove:

1. $\varphi(t) \in C^1$ $0 < t < \infty$

2. $\displaystyle\int_{0+}^\infty e^{-at}\varphi(t)\, dt$ converges

3. $\displaystyle\int_{0+}^\infty e^{-at}\varphi'(t)\, dt$ converges

$\Rightarrow \lim_{t \to 0+} \varphi(t)$ and $\lim_{t \to +\infty} e^{-at}\varphi(t)$ exist.

8. If $s_c < 0$ in Theorem 6, show that $f(0) = 0$.   Illustrate by $\varphi(t) = (1 - t)e^{-t}$. If $\varphi(t) = e^{-t}$ then $s_c < 0$; why does not $L\{e^{-t}\}$ vanish at $s = 0$?

9. Give details in the proof that the integrals (4) converge absolutely.

10. Find $L\{\sin t\}$ by use of Corollary 5.

11. Solve the same problem for $L\{\cos at\}$.

12. Solve the same problem for $L\{\sinh at\}$ and $L\{\cosh at\}$.

13. Apply Theorem 7 to $\varphi(t) = 1 - \cos t$.

14. Solve the same problem for $\varphi(t) = 1 - e^t$.

15. $\displaystyle\int_0^\infty e^{-st}(e^{at} - e^{bt})t^{-1}\, dt = ?$

§5. *Resultant*

One important and fundamental operation on generating functions was not discussed in the preceding section. Let us inquire if the product of two generating functions will be itself a generating function and if so what the relation between the corresponding determining functions will be.

5.1 DEFINITION OF RESULTANT

Definition 4. *The resultant of two determining functions* $\varphi(t)$ *and* $\psi(t)$ *is*

$$(1) \qquad \omega(t) = \int_{0+}^{t-} \varphi(u)\psi(t - u)\, du = \varphi * \psi \qquad 0 < t < \infty$$

Other terms sometimes used for $\omega(t)$ are *convolution* and *faltung*, the latter term being taken directly from German.

EXAMPLE A.   Find $t * \sin t$.   Equation (1) becomes

$$t * \sin t = \int_0^t u \sin (t - u)\, du = t - \sin t \qquad 0 < t < \infty$$

Observe that $\varphi * \psi = \psi * \varphi$, since the change of variable $t - u = y$ gives

$$\int_{0+}^{t-} \varphi(u)\psi(t - u)\, du = \int_{0+}^{t-} \varphi(t - y)\psi(y)\, dy$$

EXAMPLE B.   Find $t^{-1/2} * t^{-1/2}$.   Here

$$\omega(t) = \int_{0+}^{t-} \frac{1}{\sqrt{u}} \frac{1}{\sqrt{t - u}}\, du = \int_{0+}^{1-} u^{-1/2}(1 - u)^{-1/2}\, du = B(\tfrac{1}{2}, \tfrac{1}{2}) = \pi$$

This example shows that the resultant of two variable functions may be a constant.

5.2 PRODUCT OF GENERATING FUNCTIONS

Theorem 8.   1. $f(s) = L\{\varphi(t)\}$ *converges absolutely at* $s = a$

2. $g(s) = L\{\psi(t)\}$ *converges absolutely at* $s = a$

$\Rightarrow$ $\qquad f(s)g(s) = L\{\varphi * \psi\}$ $\qquad\qquad\qquad s \geq a$

Let $b \geq a$. Set

$$I(R) = \int_{0+}^{R} e^{-bt}\varphi(t)\,dt \int_{0+}^{R} e^{-bu}\psi(u)\,du = \iint_{S} e^{-b(t+u)}\varphi(t)\psi(u)\,dt\,du$$

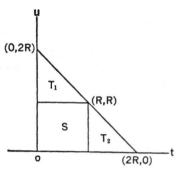

where the double integral is extended over the square S of Figure 53. Consider the double integrals of the same integrand over the triangles T_1 and T_2. Clearly,

$$\left|\iint_{T_1}\right| \leq \int_{0+}^{R} e^{-bt}|\varphi(t)|\,dt \int_{R}^{2R-t} e^{-bu}|\psi(u)|\,du$$

$$\leq \int_{0+}^{\infty} e^{-bt}|\varphi(t)|\,dt \int_{R}^{2R} e^{-bu}|\psi(u)|\,du$$

Hence, the double integral over T_1 approaches zero as $R \to +\infty$. The same is true of the integral over T_2. If $T = S + T_1 + T_2$, we have

Fig. 53.

$$\lim_{R \to +\infty} \iint_{T} = \lim_{R \to +\infty} \iint_{S} = \lim_{R \to +\infty} I(R) = f(b)g(b)$$

But $\qquad \displaystyle\iint_{T} = \int_{0+}^{2R} \varphi(t)\,dt \int_{0+}^{2R-t} e^{-(t+u)b}\psi(u)\,du$

$$= \int_{0+}^{2R} \varphi(t)\,dt \int_{t+}^{2R} e^{-by}\psi(y - t)\,dy$$

$$= \int_{0+}^{2R} e^{-by}\,dy \int_{0+}^{y-} \varphi(t)\psi(y - t)\,dt = \int_{0+}^{2R} e^{-by}\omega(y)\,dy$$

Now letting $R \to +\infty$ we have the desired result, since b was arbitrary.

EXAMPLE C.   By Theorem 8 and Example A we should have

$$L\{t\} \cdot L\{\sin t\} = L\{t - \sin t\} \qquad\qquad s > 0$$

Indeed

$$\frac{1}{s^2}\frac{1}{s^2 + 1} = \frac{1}{s^2} - \frac{1}{s^2 + 1}$$

EXAMPLE D.   By use of Example B we should have

$$L\{t^{-1/2}\} \cdot L\{t^{-1/2}\} = L\{\pi\} \qquad\qquad s > 0$$

Indeed

$$\sqrt{\pi/s}\sqrt{\pi/s} = \pi/s$$

5.3 APPLICATION

We may use Theorem 8 to prove the following modified form of Theorem 6.

Theorem 6*.   1. $f(s) = L\{\varphi(t)\}$ $\hspace{3cm}$ $s > s_a$

$$\Rightarrow \qquad f(s) = sL\left\{\int_{0+}^{t} \varphi(u)\, du\right\} \qquad\qquad s > s_a, \quad s > 0$$

Notice that hypothesis 2 of Theorem 6 is now missing. On the other hand, we are now assuming an abscissa of *absolute* convergence. The proof is immediate if we set $\psi(t) = 1$ in Theorem 8.   Then $g(s) = 1/s$ and

$$1 * \varphi(t) = \int_{0+}^{t} \varphi(u)\, du$$

EXAMPLE E.   Take $\varphi(t) = \psi(t) = \sin at$ at Theorem 8.   Then

$$\sin at * \sin at = \frac{1}{2a}(\sin at - at\cos at)$$

$$(2) \qquad \frac{a^2}{(s^2 + a^2)^2} = L\left\{\frac{\sin at}{2a} - \frac{t\cos at}{2}\right\} \qquad -\infty < a < \infty, \quad s > 0$$

EXAMPLE F.   Take $\varphi(t) = \sin at$, $\psi(t) = \cos at$.   Then

$$(3) \qquad \frac{s}{(s^2 + a^2)^2} = L\left\{\frac{t}{2a}\sin at\right\} \qquad -\infty < a < \infty, \quad s > 0$$

EXERCISES (5)

1. In the proof of Theorem 8 show that

$$\lim_{R\to\infty} \int\int_{T_2} = 0$$

2. Give the details in the computation of the resultants of Examples E and F.

3. Verify equation (3) by use of Theorem 4.

4. Take $\varphi(t) = \psi(t) = \cos at$ in Theorem 8 and thus obtain

$$L^{-1}\{s^2(s^2 + a^2)^{-2}\}$$

5. Compute the transform of Exercise 4 by means of partial fractions.

6. Show that $\varphi * (\psi * \chi) = (\varphi * \psi) * \chi$. It may be assumed that all integrals involved are proper.

7. Is it true that $\varphi * \psi \chi = \varphi \psi * \chi$?

8. Do Exercise 7 if ψ is a constant.

9. Is $\varphi * (\psi + \chi) = (\varphi * \psi) + (\varphi * \chi)$?

10. Prove that if $\varphi(t), \psi(t) \in C^1$ in $0 \leq t < \infty$, then

$$\frac{d}{dt}[\varphi(t) * \psi(t)] = \varphi(0)\psi(t) + [\varphi'(t) * \psi(t)]$$

$$L\{\omega'(t)\} = sf(s)g(s)$$

What are you assuming about the convergence of the Laplace integrals involved?

11. Set $\sigma_a(t) = 0, t < a$; $\sigma_a(t) = 1, t > a$. Prove:

$$\sigma_a(t) * \varphi(t) = 0 \qquad\qquad\qquad 0 < t < a$$

$$= \int_{0+}^{t-a} \varphi(u)\, du \qquad\qquad t > a$$

12. For $\sigma_a(t)$ as in Exercise 11, $L\{\sigma_a(t)\} = ?$ $a > 0$

13. $L^{-1}\{e^{-as}s^{-1}f(s)\} = ?$

14. Verify Theorem 8 in the special case $\varphi(t) = t^a$, $\psi(t) = t^b$; $a, b > -1$.

15. Solve the same problem for $\varphi(t) = e^{-t}$, $\psi(t) = t$.

16. $L^{-1}\{s^{-2}(s^2 + 1)^{-1}\} = ?$ (2 ways)

17. $L^{-1}\left\{\dfrac{1}{s^2} \dfrac{s}{(s^2 + 1)^2}\right\} = ?$ (2 ways)

18. Use equations (2) and (3) to find $L^{-1}\{(s - s^2)(s^2 + 1)^{-2}\}$.

§6. *Tables of Transforms*

For the practical use of Laplace transforms in the solution of differential equations it is convenient to have a table of transforms. We append a brief table at the end of this chapter. It will be found adequate for the solution of the problems of the present text. More extensive tables are available and should be used if a great number of differential equations are to be solved. In the present section we shall derive a few of the transforms, especially those involving nonelementary integrals.

6.1 SOME NEW FUNCTIONS

Many functions, such as e^{-x}/x have indefinite integrals which cannot be expressed by use of a finite number of the elementary functions. Many of these integrals occur so frequently that they have been given names and have been tabulated as new functions. We define a few of them.

Definition 5. $\text{Ei}(x) = \displaystyle\int_x^\infty \frac{e^{-t}}{t}\,dt$ $0 < x < \infty$

Definition 6. $\text{Si}(x) = \displaystyle\int_x^\infty \frac{\sin t}{t}\,dt$ $-\infty < x < \infty$

Definition 7. $\text{Ci}(x) = \displaystyle\int_x^\infty \frac{\cos t}{t}\,dt$ $0 < x < \infty$

Definition 8. $\text{Erf}(x) = \dfrac{2}{\sqrt{\pi}} \displaystyle\int_0^x e^{-t^2}\,dt$ $-\infty < x < \infty$

Definition 9. $L_n(x) = \dfrac{e^x}{n!}(x^n e^{-x})^{(n)}$ $n = 0, 1, \ldots;$ $-\infty < x < \infty$

The function $\text{Ei}(x)$ is called the *exponential integral;* $\text{Si}(x)$ is the *sine integral;* $\text{Ci}(x)$ is the *cosine integral;* $\text{Erf}(x)$ is the *error function;* $L_n(x)$ is the *Laguerre polynomial* of degree n. Here, as usual, the superscript (n) means the nth derivative. Let us develop a few of the properties of these functions, which we shall need in the computation of their Laplace transforms.

Note that $\text{Ei}(0+) = +\infty$. But for any positive number l

(1) $$\lim_{x \to 0+} x^l \int_x^\infty \frac{e^{-t}}{t}\,dt = 0$$

Also

(2) $$\lim_{x \to +\infty} xe^x \int_x^\infty \frac{e^{-t}}{t}\,dt = 1$$

In a similar way,

(3) $$\lim_{x \to 0+} x^l \text{Ci}(x) = 0$$

Integration by parts gives

(4) $$\text{Ci}(x) = -\frac{\sin x}{x} + \frac{\cos x}{x^2} - 2\int_x^\infty \frac{\cos t}{t^3}\,dt \qquad 0 < x < \infty$$

and

(5) $$\left| 2\int_x^\infty \frac{\cos t}{t^3}\,dt \right| \le 2\int_x^\infty \frac{dt}{t^3} = \frac{1}{x^2} \qquad 0 < x < \infty$$

Equation (4) and inequality (5) show the behavior of $\text{Ci}(x)$ at $x = +\infty$.

Note that $\text{Si}(0) = \pi/2$ and $\text{Erf}(+\infty) = 1$. Finally, we observe that $L_n(x)$ is, as its name implies, a polynomial of degree n. The derivative of order n of the function $x^n e^{-x}$ is clearly a polynomial of degree n multiplied by e^{-x}. The usefulness of these polynomials results chiefly from their orthogonality properties:

(6) $$\int_0^\infty e^{-x} L_n(x) L_m(x)\, dx = 0 \qquad\qquad m \neq n$$

(7) $$= 1 \qquad\qquad m = n$$

6.2 TRANSFORMS OF THE FUNCTIONS

EXAMPLE A.   Find $L\{\mathrm{Ei}\ (t)\} = f(s)$. By equations (1) and (2) the integral

$$\int_{0+}^\infty e^{-st} \mathrm{Ei}\ (t)\, dt$$

converges absolutely for $s > -1$ and diverges for $s = -1$, so that $s_c = s_a = -1$. Integration by parts gives

$$\int_{0+}^\infty e^{-st} t \mathrm{Ei}'\ (t)\, dt = -\int_{0+}^\infty e^{-st}(1 - st)\mathrm{Ei}\ (t)\, dt \qquad -1 < s < \infty$$

$$\int_{0+}^\infty e^{-st} e^{-t}\, dt = f(s) + s f'(s)$$

Accordingly, $f(s)$ satisfies the differential equation

$$[sf(s)]' = \frac{1}{s+1}$$

Hence, $$sf(s) = \log(s+1) + C$$

where C is a constant. By setting $s = 0$ we see that $C = 0$. Consequently, $f(s) = s^{-1} \log(s+1)$. This is formula 13 of the table.

EXAMPLE B.   Find $L\{\mathrm{Si}\ (t)\}$. By Example C, §4, and Theorem 6, we have

$$L\{\mathrm{Si}\ (t)\} = L\left\{\frac{\pi}{2} - \int_0^t \frac{\sin u}{u}\, du\right\}$$

$$= \frac{\pi}{2s} - L\left\{\int_0^t \frac{\sin u}{u}\, du\right\}$$

$$= \frac{\pi}{2s} - \frac{1}{s}\tan^{-1}\frac{1}{s} = \frac{1}{s}\tan^{-1} s \qquad 0 < s < \infty$$

EXAMPLE C.   Find $L\{\mathrm{Ci}\ (t)\} = f(s)$. As in Example A, we have by equations (3), (4) and inequality (5) that

(8) $$f(0) = \int_{0+}^\infty \mathrm{Ci}\ (t)\, dt$$

Then

$$f(s) + sf'(s) = \int_{0+}^{\infty} e^{-st} \cos t \, dt = \frac{s}{s^2 + 1}$$

$$sf(s) = \tfrac{1}{2} \log (s^2 + 1) + C$$

The constant of integration C is again zero and

$$L\{\text{Ci}\,(t)\} = \frac{1}{2s} \log (s^2 + 1) \qquad\qquad 0 < s < \infty$$

EXAMPLE D.   Find $L\{\text{Erf}\,\sqrt{t}\}$.   By Theorem 8 we have

$$\frac{1}{s}\frac{1}{\sqrt{s+1}} = L\left\{1 * \left(\frac{e^{-t}}{\sqrt{\pi t}}\right)\right\} \qquad\qquad 0 < s < \infty$$

$$1 * \left(\frac{e^{-t}}{\sqrt{\pi t}}\right) = \frac{1}{\sqrt{\pi}} \int_0^t \frac{e^{-u}}{\sqrt{u}} \, du = \text{Erf}\,(\sqrt{t})$$

Hence,

$$L\{\text{Erf}\,(\sqrt{t})\} = s^{-1}(s + 1)^{-1/2} \qquad\qquad s_c = s_a = 0$$

EXAMPLE E.   Find $L\{L_n(t)\}$.   We have by Theorem 5

$$\int_0^{\infty} e^{-st} e^t (t^n e^{-t})^{(n)} \, dt = (s - 1) \int_0^{\infty} e^{-(s-1)t} (t^n e^{-t})^{(n-1)} \, dt$$

$$= (s - 1)^n \int_0^{\infty} e^{-(s-1)t} t^n e^{-t} \, dt$$

$$= (s - 1)^n n! s^{-n-1} \qquad\qquad 0 < s < \infty$$

Consequently,

$$L\{L_n(t)\} = \frac{(s - 1)}{s^{n+1}} \qquad\qquad s_c = s_a = 0$$

EXERCISES (6)

1. Prove equation (1).

2. Prove equation (2).

3. Prove equation (3).

4. Prove that, if $P(x)$ is a polynomial of degree less than n,

$$\int_0^{\infty} e^{-x} P(x) L_n(x) \, dx = 0$$

Prove equation (6).

5. Prove $L_n(0) = 1$ and $\lim_{x\to\infty} x^{-n} L_n(x) = (-1)^n/n!$

6. Prove equation (7).

7. Prove that the integral $L\{\text{Ei}\,(t)\}$ converges uniformly in the interval $0 \leqq s \leqq 1$.

8. Prove that the integral (8) converges.

9. Prove formula 8 of the table. *Hint:* Differentiate formula 1 of the table result with respect to a. Then set $a = 1, c = 0$. The validity of differentiation under the integral sign need not be verified.

10. Prove that $L^{-1}\{(s^{3/2} - s^{1/2})^{-1}\} = e^t \operatorname{Erf} \sqrt{t}$.

11. Prove that $L^{-1}\{s^{1/2}(s - 1)^{-1}\} = \dfrac{1}{\sqrt{\pi t}} + e^t \operatorname{Erf}(\sqrt{t})$.

12. Prove that $L^{-1}\{(\log s)(s - a)^{-1}\} = e^{at}[\log a + \operatorname{Ei}(at)]$, $\qquad a > 0$.

13. Prove that $L\{L\{e^{-t}\}\} = e^s \operatorname{Ei}(s)$.

14. If $\operatorname{Erfc}(x) = 1 - \operatorname{Erf}(x)$, find $L\{e^t \operatorname{Erfc}(\sqrt{t})\}$.

§7.　Uniqueness

We come next to the important problem of uniqueness mentioned in §1. We shall show that a given generating function cannot have more than one continuous determining function.

7.1 A PRELIMINARY RESULT

We prove first the following result.

Theorem 9.　1. $\alpha(t) \in C$ $\qquad\qquad\qquad\qquad 0 \leqq t \leqq 1$

$\qquad\qquad$ 2. $\displaystyle\int_0^1 t^n \alpha(t)\, dt = 0$ $\qquad\qquad n = 0, 1, 2, \ldots$

$\Rightarrow \qquad\qquad \alpha(t) \equiv 0$ $\qquad\qquad\qquad\qquad 0 \leqq t \leqq 1$

For, by Theorem 14, Chapter 12, there corresponds to an arbitrary $\epsilon > 0$ a polynomial $P(t)$ such that

(1) $\qquad\qquad\qquad |\alpha(t) - P(t)| < \epsilon \qquad\qquad\qquad 0 \leqq t \leqq 1$

By hypothesis 2 we have

$$\int_0^1 \alpha^2(t)\, dt = \int_0^1 \alpha(t)[\alpha(t) - P(t)]\, dt$$

By inequality (1)

$$\int_0^1 \alpha^2(t)\, dt \leqq \epsilon \int_0^1 |\alpha(t)|\, dt$$

That is,

$$\int_0^1 \alpha^2(t)\, dt = 0$$

It follows as in the proof of Theorem 11, Chapter 12, that $\alpha(t) \equiv 0$.

7.2 THE PRINCIPAL RESULT

Theorem 10.   1. $\varphi(t) \in C$ $0 < t < \infty$

   2. $f(s) = L\{\varphi(t)\}$ $s > s_c$

   3. $f(s_0 + nl) = 0$ *for some* $l > 0$ $n = 0, 1, 2, \ldots$

$\Rightarrow$ $\varphi(t) \equiv 0$ $0 < t < \infty$

Set

(2) $\displaystyle \alpha(t) = \int_{0+}^{t} e^{-s_0 u} \varphi(u)\, du$

Then by hypothesis 3 for $n = 0$, we have $\alpha(\infty) = 0$. By integration by parts we obtain

$$f(s_0 + nl) = nl \int_0^\infty e^{-nlt} \alpha(t)\, dt = 0 \qquad n = 1, 2, \ldots$$

Now if we set $e^{-lt} = u$, this becomes

$$\int_0^1 u^{n-1} \alpha\left(\frac{1}{l} \log \frac{1}{u}\right) du = 0 \qquad n = 1, 2, \ldots$$

If we define the function α to be zero at $u = 0$ and $u = 1$, it becomes continuous in the closed interval $0 \le u \le 1$. By Theorem 9 it is identically zero. Hence,

$$\alpha'(t) = e^{-s_0 t} \varphi(t) \equiv 0 \qquad 0 < t < \infty$$

and the theorem is proved.

Corollary 10.1. *If a generating function vanishes at an infinite set of points in arithmetic progression, it is identically zero.*

Corollary 10.2. *A generating function cannot have more than one continuous determining function.*

For, if φ, ψ, $\in C$ and

$$f(s) = L\{\varphi\} = L\{\psi\}$$

then $L\{\varphi - \psi\} \equiv 0$

By Theorem 10, $\varphi \equiv \psi$.

Notice that hypothesis 1 can be relaxed. If $\varphi(t)$ is any determining function of the type admitted in §2.2 and if hypotheses 2 and 3 are satisfied, then $f(s)$ is still identically zero. For, $\alpha(t)$ will still be a continuous function and hence identically zero. Hence,

$$f(s) = (s - s_0) \int_0^\infty e^{-(s-s_0)t} \alpha(t)\, dt \equiv 0 \qquad s > s_0$$

It is only by virtue of this uniqueness theorem that the use of tables of transforms is justified. For example, let it be required to find the function $L^{-1}\{(s+1)^{-1}\}$. We know that $L\{e^{-t}\} = (s+1)^{-1}$. By Corollary 10.2 there is only one continuous determining function corresponding to $(s+1)^{-1}$. Hence, $L^{-1}\{(s+1)^{-1}\}$ is e^{-t}.

EXERCISES (7)

1. Show that Theorem 9 holds if the interval $(0, 1)$ is replaced by $(0, a)$.

2. Prove that
$$(x^2 + 2x + 2)(x^2 - 2x + 2) - x^4 = 4$$
Hence, show that every fourth coefficient in the Maclaurin expansion of $4(x^2 + 2x + 2)^{-1}$ is zero.

3. Show that
$$\frac{d^{4n+3}}{dx^{4n+3}} \frac{1}{x^2+1}\bigg|_{x=1} = 0 \qquad n = 0, 1, 2, \ldots$$
Hint: The Taylor expansion of $(x^2 + 1)^{-1}$ about the point $x = 1$ can be had from the Maclaurin expansion of Exercise 2 by a change of variable.

4. Prove that
$$\int_0^\infty e^{-t} t^{(4n+3)} \sin t \, dt = 0 \qquad n = 0, 1, 2, \ldots$$
Hint: Express $(s^2 + 1)^{-1}$ as a Laplace integral and compute thereby its successive derivatives at $s = 1$.

5. Show that Theorem 9 is no longer valid if the interval $(0, 1)$ is replaced by $(0, \infty)$. *Hint:* Make the change of variable $t^4 = u$ in Exercise 4.

6. Use the table of transforms to find $L^{-1}\{(s^3 + s)^{-1}\}$.

7. $L^{-1}\{s^{-1}(s^2 - 1)^{-2}\} = ?$

8. $L^{-1}\{(s^2 - 1)(s^2 + 1)^{-2}\} = ?$

9. $L^{-1}\{(s^2 + 2s + 2)^{-2}\} = ?$

10. $L^{-1}\{s(s^2 + 2s + 2)^{-2}\} = ?$

§8. *Inversion*

Thus far we have been obliged to evaluate the inverse Laplace transform $L^{-1}\{f(s)\}$ by reducing $f(s)$ to some combination of function each of which can be recognized, by tables or otherwise, as the direct transform of some known determining function. There are, however, several direct formulas for computing $\varphi(t)$ from $f(s)$. The one of these discovered first involves a knowledge of $f(s)$ for values of s which are not real. We shall give here another formula which depends only on $f(s)$ for real s.

8.1 PRELIMINARY RESULTS

We begin with the definition of an operator which we shall denote by $L^{-1}[f(s)]$. The notation differs from that of the inverse transform defined in §1 only in the use of a square bracket rather than the brace. We shall see that it is an explicit inversion of the Laplace transform.

(1) **Definition 10.** $L^{-1}[f(s)] = \lim\limits_{k \to \infty} \dfrac{(-1)^k}{k!} \left[f^{(k)}\left(\dfrac{k}{t}\right) \right] \left(\dfrac{k}{t}\right)^{k+1}$

Observe that the operator is applicable to such functions $f(s)$ that belong to C^∞ for all s greater than some constant (which may be arbitrarily large since k/t becomes infinite with k). The function $f(s)$ must also be of such a nature that the limit (1) exists.

EXAMPLE A. $L^{-1}[s] = 0$. For all the derivatives of s beyond the first are zero. This example shows that $L^{-1}[f]$ may exist when $L^{-1}\{f\}$ does not.

EXAMPLE B. $L^{-1}[s^{-2}] = \lim\limits_{k \to \infty} \left(1 + \dfrac{1}{k}\right) t = t.$ $0 < t < \infty$

Note that $L\{t\} = s^{-2}$ as predicted. Observe that the simplest procedure for the computation of $L^{-1}[f(s)]$ is to begin by calculating $(-1)^k f^{(k)}(s)s^{k+1}$ and then to set $s = k/t$.

Lemma 11.1. $\dfrac{k^{k+1}}{k!} \displaystyle\int_0^\infty (e^{-u}u)^k \, du = 1$ $k = 1, 2, \ldots$

This is immediate by use of $\Gamma(x)$.

Lemma 11.2. $\lim\limits_{k \to \infty} \dfrac{k^{k+1}}{k!} (e^{-a}a)^k = 0$ $a > 0, \quad a \neq 1$

Set the function of k whose limit we wish to evaluate equal to u_k. Then

$$\frac{u_{k+1}}{u_k} = e^{-a}a(1 + k^{-1})^{k+1}$$

(2) $$\lim_{k \to \infty} \frac{u_{k+1}}{u_k} = e^{1-a}a$$

We see geometrically or by use of the law of the mean that

$$e^x > 1 + x \qquad\qquad\qquad x \neq 0$$

so that $\qquad\qquad\qquad e^{a-1} > a \qquad\qquad\qquad\qquad a \neq 1$

Hence, the limit (2) is less than 1. Therefore, by the ratio test u_k is the general term of a convergent series and consequently tends to zero with $1/k$.

8.2 THE INVERSION FORMULA

Theorem 11.   1. $\varphi(t) \in C$ $0 < t < \infty$

2. $f(s) = L\{\varphi(t)\}$ $s > s_a$

$\Rightarrow$ $L^{-1}[f(s)] = \varphi(t)$ $0 < t < \infty$

By Theorem 4

$$L^{-1}[f(s)] = \lim_{k \to \infty} \frac{1}{k!}\left(\frac{k}{t}\right)^{k+1} \int_{0+}^{\infty} e^{-ku/t} u^k \varphi(u)\, du \qquad t > 0,\ \ \frac{k}{t} > s_a$$

(3) $$= \lim_{k \to \infty} \frac{k^{k+1}}{k!} \int_{0+}^{\infty} (e^{-u}u)^k \varphi(ut)\, du$$

It will be sufficient to prove that this limit is $\varphi(1)$ when $t = 1$. For, if $t_0 \ne 1$ we can replace $\varphi(ut)$ by $\psi(u) = \varphi(ut_0)$. Then, applying the result assumed proved, we get for the limit (3) $\psi(1) = \varphi(t_0)$. Hence, we set $t = 1$ in the integral (3). By Lemma 11.1 it is clear that we need only prove that

(4) $$\lim_{k \to \infty} \frac{k^{k+1}}{k!} \int_{0+}^{\infty} (e^{-u}u)^k [\varphi(u) - \varphi(1)]\, du = 0$$

Now let ϵ be an arbitrary positive number. Choose numbers a and b so near to 1 ($0 < a < 1 < b < \infty$) that

(5) $$|\varphi(u) - \varphi(1)| < \epsilon \qquad\qquad a \le u \le b$$

This is possible since $\varphi \in C$ at $u = 1$. Now write the integral (4) as the sum of three others, I_1, I_2, I_3 corresponding, respectively, to the intervals $(0, a)$, (a, b), (b, ∞).

Now by Lemma 11.1 and inequality (5) we have

$$|I_2| \le \frac{k^{k+1}}{k!} \int_a^b (e^{-u}u)^k \epsilon\, du < \epsilon \qquad\qquad k = 1, 2, \ldots$$

Since $e^{-u}u \in \uparrow$ in $(0, a)$ it follows that

$$|I_1| \le (e^{-a}a)^k \frac{k^{k+1}}{k!} \int_{0+}^a |\varphi(u) - \varphi(1)|\, du$$

By Lemma 11.2 this tends to zero when $k \to \infty$. Finally, since $e^{-u}u \in \downarrow$ in (b, ∞) we see that

(6) $$|I_3| \le \frac{k^{k+1}}{k!} (e^{-b}b)^{k-k_0} \int_b^{\infty} e^{-k_0 u} u^{k_0} |\varphi(u) - \varphi(0)|\, du \qquad k > k_0$$

where k_0 is a positive integer greater than s_a. It is clear that the integral on the right of inequality (6) converges. By Lemma 11.2 we see that I_3 also tends to zero with $1/k$. Hence,

$$\varlimsup_{k \to \infty} \frac{k^{k-1}}{k!} \left| \int_{0+}^{\infty} (e^{-u}u)^k [\varphi(u) - \varphi(1)] \, du \right| \leqq \varlimsup_{k \to \infty} (|I_1| + |I_2| + |I_3|) \leqq \epsilon$$

so that the theorem is proved.

EXAMPLE C. Show that $L^{-1}[(s + 1)^{-1}] = e^{-t}$. Simple computations give

$$L^{-1}[(s + 1)^{-1}] = \lim_{k \to \infty} \left(1 + \frac{t}{k} \right)^{-k-1} = e^{-t} \qquad 0 < t < \infty$$

EXERCISES (8)

1. $L^{-1}[e^{-s}] = ?$ $\hfill 0 < t, \quad t \neq 1$

2. What can be said of the limit (1) for Exercise 1 when $t = 1$?

3. Prove Lemma 11.2 by use of Stirling's formula.

4. $L^{-1}[(s + a)^{-3}] = ?$

5. $L^{-1}[s^{-n}] = ?$ $\hfill n = 1, 2, \ldots$

6. Show $(-1)^k \left(\dfrac{1}{\sqrt{s}} \right)^{(k)} = \dfrac{\Gamma(k + \frac{1}{2})}{\Gamma(\frac{1}{2})} s^{-k-1/2}$ $\hfill k = 0, 1, \ldots$

Then compute $L^{-1}[s^{-1/2}]$. Stirling's formula,

$$\Gamma(x + 1) \sim (x/e)^x \sqrt{2\pi x} \qquad x \to +\infty$$

may be assumed to hold for nonintegral x.

7. Prove that, if $L\{\varphi(t)\}$ converges absolutely for $s > s_a$, the same is true for $L\{t^k \varphi(t)\}$, $k = 1, 2, \ldots$. Hence, show that the integral on the right of equation (6) converges absolutely.

8. L^{-1} [polynomial] $= ?$

9. Is it true that $L\{L^{-1}[f(s)]\} = f(s)$ for every $f(s)$ for which the operators are defined?

10. Is it true that, if $L^{-1}[f(s)]$ and $L^{-1}[g(s)]$ both exist and are continuous and equal for $0 < t < \infty$, then $f(s)$ is equal to $g(s)$?

11. Under the conditions of Theorem 11, prove that

$$\lim_{k \to \infty} \left. \frac{(-1)^k}{k!} f^{(k)}(s) s^{k+1} \right|_{s = \frac{k}{t} + c} = \varphi(t) \qquad -\infty < c < \infty, \quad 0 < t < \infty$$

Hint: $L^{-1}[f(s + c)] = e^{-ct}\varphi(t)$.

12. Prove that

$$\lim_{k\to\infty} \frac{k^{k+1}}{k!} \int_0^a (e^{-u}u)^k\, du = \begin{cases} 0 & 0 < a < 1 \\ 1 & 1 < a < \infty \end{cases}$$

13. Prove that $L^{-1}[af(s) + bg(s)] = aL^{-1}[f(s)] + bL^{-1}[g(s)]$. What are you assuming about $f(s)$ and $g(s)$?

14. Prove that $L^{-1}[f(as)] = \dfrac{1}{a} L^{-1}[f(s)]\big|_{t/a}$, $a > 0$. What are you assuming about $f(s)$?

§9. Representation

We have seen that not all functions are generating functions. Certainly, a generating function must belong to C^∞ and vanish at $s = +\infty$. But not all functions with these properties are generating functions. For example, $s^{-1}\sin s$ is not the Laplace transform of any determining function since it vanishes at infinitely many points in arithmetic progression. The problem of characterizing completely the class of all generating functions is a difficult one. Here we shall develop only a few elementary but useful sufficient conditions.

9.1 RATIONAL FUNCTIONS

Theorem 12. 1. $R(s)$ is rational

2. $R(\infty) = 0$

$\Rightarrow$ $R(s)$ is a generating function.

The function $R(s)$ is the ratio of two polynomials. The degree of the denominator is greater than that of the numerator by hypothesis 2. By the theory of partial fractions, $R(s)$ is a finite sum of functions of the form

$$(1) \qquad \frac{A}{(s - a)^n}, \qquad \frac{B(s - b) + C}{[(s - b)^2 + c^2]^m}$$

where a, b, c, A, B, C are real constants and m, n are positive integers.

But both the functions (1) are generating functions. The first appears as formula 1 in the table. If $m = 1$, the proof is concluded by use of formulas 2 and 3 of the table. If $m > 1$, we have only to observe that the product of a finite number of generating functions ($s_a < \infty$) is itself a generating function in order to conclude the proof.

9.2 POWER SERIES IN 1/s

Another very important class of generating functions consists of functions which can be expanded in a convergent series of powers of $1/s$, the constant term being zero.

Theorem 13.   1. $f(s) = \sum\limits_{k=0}^{\infty} \dfrac{A_k}{s^{k+1}}$ $s > r$

$\Rightarrow$ $f(s) = L\{\varphi(t)\}$ $s_a \leq r$

(2) $\varphi(t) = \sum\limits_{k=0}^{\infty} A_k \dfrac{t^k}{k!}$ $0 \leq t < \infty$

Let s_0 be an arbitrary number $> r$.   Then by Theorem A, §2, we have

(3) $\sum\limits_{k=0}^{\infty} \dfrac{|A_k|}{s_0^{k+1}} < \infty$

Since the general term of series (3) approaches zero as $k \to \infty$, there must exist a constant M such that

$$|A_k| < M s_0^{k+1} \qquad\qquad k = 0, 1, 2, \ldots$$

Hence, the series (2) converges uniformly in $0 \leq t \leq R$ for any $R > 0$. Consequently,

(4) $\int_0^R e^{-st} \varphi(t)\, dt = \sum\limits_{k=0}^{\infty} \dfrac{A_k}{k!} \int_0^R e^{-st} t^k\, dt$

for any real s.   In like manner,

(5) $\int_0^R e^{-st} |\varphi(t)|\, dt \leq \sum\limits_{k=0}^{\infty} \dfrac{|A_k|}{k!} \int_0^R e^{-st} t^k\, dt \leq \sum\limits_{k=0}^{\infty} M \left(\dfrac{s_0}{s}\right)^{k+1}$ $s > s_0$

so that the integral $L\{\varphi(t)\}$ converges absolutely for $s > s_0$.   Hence, $s_a \leq r$. Let n be a positive integer.   Then from equation (4)

$$\left| \int_0^R e^{-st} \varphi(t)\, dt - \sum\limits_{k=0}^{n} \dfrac{A_k}{k!} \int_0^R e^{-st} t^k\, dt \right| \leq \sum\limits_{k=n+1}^{\infty} M \left(\dfrac{s_0}{s}\right)^{k+1} \qquad s > s_0$$

The left-hand side of this inequality tends to a limit as $R \to \infty$, so that

$$\left| \int_0^\infty e^{-st} \varphi(t)\, dt - \sum\limits_{k=0}^{n} \dfrac{A_k}{s^{k+1}} \right| \leq \sum\limits_{k=n+1}^{\infty} M \left(\dfrac{s_0}{s}\right)^{k+1} \qquad s > s_0$$

Since the right-hand side of this inequality tends to zero as $n \to \infty$, the theorem is proved.

9.3 ILLUSTRATIONS

EXAMPLE A.   Find $L^{-1}\{(s-1)^{-1}\}$ by Theorem 13.   We have

$$\frac{1}{s-1} = \frac{1}{s} + \frac{1}{s^2} + \frac{1}{s^3} + \cdots \qquad\qquad s > 1$$

Hence, series (2) becomes

$$L^{-1}\{(s-1)^{-1}\} = \varphi(t) = \sum_{k=0}^{\infty} \frac{t^k}{k!} = e^t$$

Here $s_a == r = 1$.

EXAMPLE B. Find $L^{-1}\{(s^2+1)^{-1}\}$ by Theorem 13.   In this case,

$$\frac{1}{s^2+1} = \frac{1}{s^2} - \frac{1}{s^4} + \frac{1}{s^6} - \cdots \qquad\qquad s > 1$$

so that

$$\varphi(t) = t - \frac{t^3}{3!} + \frac{t^5}{5!} - \cdots = \sin t$$

In the present example $s_a = 0 < r$.

EXAMPLE C. Find $L^{-1}\{s^{-1}e^{-1/s}\}$.   The series expansion is

$$\frac{1}{s}e^{-1/s} = \sum_{k=0}^{\infty} \frac{(-1)^k}{k!}\frac{1}{s^{k+1}}$$

Hence

$$\varphi(t) = \sum_{k=0}^{\infty} \frac{(-1)^k}{k!}\frac{t^k}{k!}$$

This function can be expressed in terms of a Bessel's function, which we now define.   The *Bessel's function of order n* is

$$J_n(t) = \sum_{k=0}^{\infty} \frac{(-1)^k}{k!(k+n)!}\left(\frac{t}{2}\right)^{2k+n} \qquad n = 0, 1, 2, \ldots$$

Clearly,

$$\varphi(t) = J_0(2\sqrt{t})$$

EXERCISES (9)

1. $L^{-1}\{s(s^2+1)^{-3}\} = ?$

2. $L^{-1}\{(s^2+1)^{-3}\} = ?$

3. Prove that $L^{-1}\{(s-1)(2s^2+1)s^{-3}(s^2+1)^{-2}\}$ is equal to the function

$$-\frac{t^2}{2} + t + \frac{t}{2}(\sin t - \cos t) - \frac{1}{2}\sin t.$$

4. Give details of the proof that series (2) converges uniformly in $0 \leq t \leq R$.

5. Give details of the proof of inequalities (5).

6. Under the hypotheses of Theorem 13, show that $f(s)/\sqrt{s}$ is a generating function.

7. Solve the same problem for $\sqrt{s}f(s)$.

8. Solve the same problem for $s^p f(s)$, $-\infty < p < 1$.

9. Find $L^{-1}\left\{\log\left(1 + \dfrac{1}{s}\right)\right\}$ by Theorem 13.

10. Solve the same problem for $\tan^{-1}(1/s)$.

11. Prove $L\{J_0(t)\} = (s^2 + 1)^{-1/2}$.

12. Prove $L\{J_1(t)\} = (\sqrt{s^2+1} - s)/\sqrt{s^2+1}$.

13. Prove:

$$e^t \operatorname{Erf}(\sqrt{t}) = \frac{1!}{\sqrt{\pi}}\int_0^t \frac{e^{t-u}}{\sqrt{u}}\,du = \sum_{k=0}^{\infty}\frac{1}{k!}\int_0^t \frac{(t-u)^k}{\sqrt{u}}\,du$$

$$= \sum_{k=0}^{\infty}\frac{t^{(2k+1)/2}}{\Gamma([2k+3]/2)}$$

Hint: Use the beta function to evaluate the general term of the series.

14. Using Exercises 6 and 13 prove that

$$L\{e^t \operatorname{Erf}(\sqrt{t})\} = (s-1)^{-1}s^{-1/2}$$

The result may be checked by Example D of §6.

§10. *Related Transforms*

We conclude this chapter by brief mention of several other transforms which are closely related to the Laplace transform. We shall make no attempt to develop the general theory of these transforms.

10.1 THE BILATERAL LAPLACE TRANSFORM

The integral

$$(1) \qquad f(s) = \int_{-\infty}^{\infty} e^{-st}\varphi(t)\,dt$$

is called the *bilateral Laplace transform*. It is easy to see what the region of convergence of such an integral is. For, we have

$$(2) \qquad f(s) = \int_0^{\infty} e^{-st}\varphi(t)\,dt + \int_{-\infty}^0 e^{-st}\varphi(t)\,dt$$

The first of these integrals converges on a right half-line. The second integral becomes by the change of variable $t = -u$

$$\int_{-\infty}^0 e^{-st}\varphi(t)\,dt = \int_0^{\infty} e^{su}\varphi(-u)\,du$$

and the latter integral is an ordinary Laplace integral, sometimes called *unilateral*, in which s has been replaced by $-s$. Consequently, its region of convergence is a left half-line. The common part of the two half-lines of convergence of the integrals (2) will be the region of convergence of the integral (1). Accordingly, it will often be a finite interval but may be a right half-line, a left half-line, the whole s-axis, a single point, or fail to exist.

EXAMPLE A. Express $(s^2 + s)^{-1}$ as a bilateral Laplace integral. We have

$$\frac{1}{s(s+1)} = \frac{1}{s} - \frac{1}{s+1}$$

$$\frac{1}{s+1} = \int_0^\infty e^{-st}e^{-t}\, dt \qquad\qquad s > -1$$

$$\frac{1}{s} = \int_0^\infty e^{-st}\, dt = \int_{-\infty}^0 e^{st}\, dt \qquad\qquad s > 0$$

$$= \int_{-\infty}^0 e^{-st}(-1)\, dt \qquad\qquad s < 0$$

Hence, $\varphi(t) = -e^{-t}$, when $0 < t < \infty$; $\varphi(t) = -1$ when $-\infty < t < 0$; the interval of convergence is seen to be $-1 < s < 0$.

Note that this function can also be expressed as a unilateral Laplace integral of either of the types (2).

$$\frac{1}{s(s+1)} = \int_0^\infty e^{-st}(1 - e^{-t})\, dt \qquad\qquad s > 0$$

$$= \int_{-\infty}^0 e^{-st}(e^{-t} - 1)\, dt \qquad\qquad s < -1$$

Let us determine formally what the form of the bilateral resultant should be. Let $f(s)$ be defined by equation (1) and let

$$(3) \qquad\qquad g(s) = \int_{-\infty}^\infty e^{-st}\psi(t)\, dt$$

Then, if the change of order of integration is valid, we have

$$f(s)g(s) = \int_{-\infty}^\infty dt \int_{-\infty}^\infty e^{-s(t+u)}\varphi(t)\psi(u)\, du$$

$$= \int_{-\infty}^\infty dt \int_{-\infty}^\infty e^{-sy}\varphi(t)\psi(y - t)\, dy$$

$$= \int_{-\infty}^\infty e^{-sy}\omega(y)\, dy$$

$$(4) \quad \text{where} \qquad \omega(y) = \int_{-\infty}^\infty \varphi(t)\psi(y - t)\, dt$$

It can be shown, somewhat as was done in §5, that the above formal procedure is valid whenever the two integrals (1) and (3) have a common region of absolute convergence. The function $\omega(y)$ defined by the integral (4) is called the *bilateral resultant* or *bilateral convolution* of $\varphi(t)$ and $\psi(t)$.

EXAMPLE B.   Do Example A by use of the bilateral convolution.   Take

$$\varphi(t) = e^{-t} \text{ in } (0, \infty), \qquad \psi(t) = 0 \text{ in } (0, \infty)$$
$$= 0 \text{ in } (-\infty, 0), \qquad\quad = -1 \text{ in } (-\infty, 0)$$

Then $f(s) = (s + 1)^{-1}$ and $g(s) = s^{-1}$.   By equation (4) we compute

$$\omega(t) = \int_{-\infty}^{\infty} \varphi(t - u)\psi(u)\, du = -\int_{-\infty}^{0} \varphi(t - u)\, du$$

$$= -\int_{t}^{\infty} \varphi(u)\, du = -e^{-t} \qquad\qquad 0 < t$$

$$= -\int_{0}^{\infty} \varphi(u)\, du = -1 \qquad\qquad t < 0$$

10.2 LAPLACE-STIELTJES TRANSFORM

The Stieltjes integral

$$(5) \qquad\qquad f(s) = \int_{0}^{\infty} e^{-st}\, d\alpha(t)$$

is known as the *Laplace-Stieltjes transform*.   In particular, if

$$\alpha(t) = \int_{0}^{t} \varphi(u)\, du \qquad\qquad 0 < t < \infty$$

then equation (5) becomes

$$f(s) = L\{\varphi(t)\} = \int_{0}^{\infty} e^{-st}\varphi(t)\, dt$$

On the other hand, if the sequence $\{\lambda_k\}_{k=0}^{\infty}$ is defined as in §1.1, and if

$$\alpha(t) = 0 \qquad\qquad\qquad -\infty < t \leq \lambda_0$$
$$= s_n \qquad\quad \lambda_n < t \leq \lambda_{n+1}, \quad n = 0, 1, 2, \ldots$$
$$s_n = \sum_{k=0}^{n} a_k \qquad\qquad\qquad n = 0, 1, 2, \ldots$$

then equation (5) becomes

$$(6) \qquad\qquad f(s) = \sum_{k=0}^{\infty} a_k e^{-\lambda_k s}$$

Thus, equation (5) includes as special cases not only the classical Laplace transform defined in §1 but also an arbitrary Dirichlet series.

The properties of a function $f(s)$ defined by the integral (5) are somewhat different from those of the generating functions defined in §1. For example, $f(s)$ defined by equation (5) need not vanish at infinity. Indeed a constant may be a Laplace-Stieltjes integral, as we see by taking $\lambda_0 = 0$, $a_0 \neq 0$, $a_k = 0$ ($k = 1, 2, \ldots$) in equation (6).

EXAMPLE C. Find $f(s)$ if $\alpha(t) = t$, $0 \leq t \leq 1$, $\alpha(t) = 0$, $1 < t < \infty$. We have

$$f(s) = s \int_0^1 e^{-st} t \, dt = \frac{1}{s}(1 - e^{-s}) - e^{-s}$$

The region of the convergence is the entire s-axis.

10.3 THE STIELTJES TRANSFORM

Let us iterate the Laplace transform:

$$f_2(s) = L\{L\{\varphi(t)\}\} = \int_{0+}^\infty e^{-su} \, du \int_{0+}^\infty e^{-tu} \varphi(t) \, dt$$

$$= \int_{0+}^\infty \varphi(t) \, dt \int_{0+}^\infty e^{-(s+t)u} \, du$$

(7)
$$f_2(s) = \int_{0+}^\infty \frac{\varphi(t)}{s + t} \, dt$$

Regardless of the validity of the above change in the order of integration, equation (7) is said to define the *Stieltjes transform*. It can be shown that the region of convergence of the integral (7) is a half-line which includes the positive s-axis. In particular, it may include some of the negative s-axis, as is the case if $\varphi(t) = t^{-1}$ in the interval $(1, \infty)$ and is zero elsewhere.

EXAMPLE D. Find the Stieltjes transform of $1/\sqrt{t}$. In this case, the integral (7) clearly converges for $s > 0$ and diverges for $s \leq 0$. It is easily seen by use of Peirce's Tables or by iteration of the Laplace transform of $1/\sqrt{t}$ that $f_2(s) = \pi/\sqrt{s}$.

EXERCISES (10)

1. Show that $\Gamma(s)$ is a bilateral Laplace transform. What is the region of convergence?

2. Solve the same problem for $B(s, 1 - s)$.

3. As in Example A, obtain three integral representations of the function $(s - a)^{-1}(s - b)^{-1}$, determining the region of convergence in each case.

4. Solve the same problem for $s^{-3}(s - 1)^{-2}$.

5. Find four integral representations for the function $2/(s^3 - s)$ corresponding to the intervals of convergence $(-\infty, -1)$, $(-1, 0)$, $(0, 1)$, $(1, \infty)$.

6. Find the inverse bilateral Laplace transform of $(s - a)^{-1}(s - b)^{-1}$ by use of the resultant.

7. Solve the same problem for $s^{-3}(s - 1)^{-2}$.

8. Solve the same problem for $2/(s^3 - s)$ in the interval $0 < s < 1$.

9. What is the region of convergence of the integral (1) if $\varphi(t)$ is the function e^{-t^2}?

10. Solve the same problem if $\varphi(t) = (\sin t)/t$.

11. Find the nth iterate of the Laplace transform of $1/\sqrt{t}$.

12. Find the Stieltjes transform of $(t + 1)^{-1}$.

13. Find the Stieltjes transform of $t^{-1/3}$.

14. Evaluate the integral (5) if $\alpha(t) = 0$ for $n \leq t < n + 1$ when n is even, if $\alpha(t) = 1$ in that interval when n is odd.

15. Solve the same problem if the interval is changed to $n < t \leq n + 1$.

16. Sum the series in Exercises 14 and 15 and find the region of convergence.

	GENERATING FUNCTIONS	DETERMINING FUNCTIONS	CONDITIONS
1.	$\dfrac{\Gamma(a)}{(s-c)^a}$	$t^{a-1}e^{ct}$	$a > 0,\ s_c = s_a = c$
2.	$\dfrac{a}{(s-c)^2 + a^2}$	$e^{ct}\sin at$	$s_c = s_a = c$
3.	$\dfrac{s-c}{(s-c)^2 + a^2}$	$e^{ct}\cos at$	$s_c = s_a = c$
4.	$\dfrac{a}{(s-c)^2 - a^2}$	$e^{ct}\sinh at$	$s_c = s_a = c + |a|$
5.	$\dfrac{s-c}{(s-c)^2 - a^2}$	$e^{ct}\cosh at$	$s_c = s_a = c + |a|$
6.	$\dfrac{2a^3}{(s^2 + a^2)^2}$	$\sin at - at\cos at$	$s_c = s_a = 0$
7.	$\dfrac{2as}{(s^2 + a^2)^2}$	$t\sin at$	$s_c = s_a = 0$
8.	$\dfrac{\log s}{s}$	$\Gamma'(1) - \log t$	$s_c = s_a = 0$
9.	$\log\left|\dfrac{s-a}{s-b}\right|$	$\dfrac{e^{bt} - e^{at}}{t}$	$s_c = s_a = \max(a, b)$
10.	$\tan^{-1}\dfrac{1}{s}$	$\dfrac{\sin t}{t}$	$s_c = s_a = 0$
11.	$\dfrac{(s-1)^n}{s^{n+1}}$	$L_n(t)$	$s_c = s_a = 0$
12.	$\dfrac{1}{s\sqrt{s+1}}$	$\operatorname{Erf}\sqrt{t}$	$s_c = s_a = 0$
13.	$\dfrac{\log(s+1)}{s}$	$\operatorname{Ei}(t)$	$s_c = s_a = -1$
14.	$\dfrac{1}{s}\tan^{-1}s$	$\operatorname{Si}(t)$	$s_c = s_a = 0$
15.	$\dfrac{1}{2s}\log(s^2 + 1)$	$\operatorname{Ci}(t)$	$s_c = s_a = 0$
16.	$\dfrac{1}{\sqrt{s^2 + 1}}$	$J_0(t)$	$s_c = s_a = 0$
17.	$\dfrac{e^{-1/s}}{s}$	$J_0(2\sqrt{t})$	$s_c = s_a = 0$

473

14

Applications
of the Laplace
Transform

§1. *Introduction*

In this chapter we shall give a few of the more important applications of the Laplace transform. Those chosen are: the evaluation of definite integrals; the solution of linear differential equations, ordinary and partial; and the solution of linear difference equations. We have already observed in Chapter 11, §3.2, how the transform may be useful in the evaluation of definite integrals provided that a factor of the integrand is a generating function. In some cases, the transform is also useful if one factor is a determining function. We shall illustrate both methods.

1.1 INTEGRANDS THAT ARE GENERATING FUNCTIONS

Let us suppose we have an integral of the form

(1)
$$\int_0^\infty f(s)\psi(s)\,ds$$

to evaluate, where $f(s)$ is a generating function and $\psi(s)$ is a determining function,

$$f(s) = L\{\varphi(t)\}, \qquad g(s) = L\{\psi(t)\}$$

Then, if interchange in the order of integration is permissible, we have

$$\int_0^\infty f(s)\psi(s)\, ds = \int_0^\infty \psi(s)\, ds \int_0^\infty e^{-st}\varphi(t)\, dt$$

$$= \int_0^\infty \varphi(t)\, dt \int_0^\infty e^{-st}\psi(s)\, ds$$

(2) $$\int_0^\infty \psi(s)\, L\{\varphi(t)\}\, ds = \int_0^\infty \varphi(s)\, L\{\psi(t)\}\, ds$$

In many cases, the integral on the right-hand side of equation (2) is more easily evaluated than the integral (1).

EXAMPLE A.   Evaluate the integral

(3) $$\int_{0+}^\infty \mathrm{Ei}\,(s)\, ds$$

In equation (2) take $\psi(s) = e^{-s}$, $\varphi(t) = (1 + t)^{-1}$. Then, if $u = s(t + 1)$,

$$f(s) = \int_0^\infty \frac{e^{-st}}{1 + t}\, dt = e^s \int_s^\infty \frac{e^{-u}}{u}\, du = e^s\, \mathrm{Ei}\,(s)$$

$$g(s) = L\{\psi(t)\} = (s + 1)^{-1}$$

Equation (2) becomes

$$\int_{0+}^\infty \mathrm{Ei}\,(s)\, ds = \int_0^\infty \frac{1}{(s + 1)^2}\, ds = 1$$

To make the proof rigorous, we must show that

$$\int_{0+}^\infty e^{-s}\, ds \int_0^\infty \frac{e^{-st}}{1 + t}\, dt = \int_0^\infty \frac{dt}{1 + t} \int_0^\infty e^{-(t+1)s}\, ds$$

The integral

$$\int_0^\infty e^{-(t+1)s}\, ds$$

converges uniformly in the interval $0 \le t \le R$, so that

$$\int_0^R \frac{dt}{1 + t} \int_0^\infty e^{-(t+1)s}\, ds = \int_0^\infty e^{-s}\, ds \int_0^R \frac{e^{-st}}{t + 1}\, dt$$

Our result will be established if

(4) $$\lim_{R \to \infty} \int_{0+}^\infty e^{-s}\, ds \int_R^\infty \frac{e^{-st}}{t + 1}\, dt = 0$$

But

$$\int_{0+}^{\infty} e^{-s} ds \int_{R}^{\infty} \frac{e^{-st}}{t+1} dt = \int_{0+}^{\infty} \text{Ei}(sR + s) ds = \frac{1}{R+1} \int_{0+}^{\infty} \text{Ei}(s) ds$$

so that equation (4) is obviously true.

EXAMPLE B. Take $\varphi(t) = t$, $\psi(t) = t \sin t$ in equation (2). Then

$$\int_{0}^{\infty} \frac{\sin s}{s} ds = -\int_{0}^{\infty} s \frac{d}{ds}\left(\frac{1}{s^2+1}\right) ds$$

$$= -\frac{s}{s^2+1}\Big|_{0}^{\infty} + \int_{0}^{\infty} \frac{ds}{s^2+1} = \frac{\pi}{2}$$

We omit details concerning the interchange in the order of integration.

1.2 INTEGRANDS THAT ARE DETERMINING FUNCTIONS

Let it be required to evaluate an integral of the form

(5) $$\psi(t) = \int_{0}^{\infty} \varphi(xt)h(x) \, dx$$

where $h(x)$ is an arbitrary function and $f(s) = L\{\varphi(t)\}$. Assuming that we may integrate under the integral sign, we obtain

$$g(s) = L\{\psi(t)\} = \int_{0}^{\infty} h(x) \, dx \int_{0}^{\infty} e^{-st} \varphi(xt) \, dt$$

(6) $$g(s) = \int_{0}^{\infty} \frac{h(x)}{x} f\left(\frac{s}{x}\right) dx$$

The integral (6) may in some cases be more easily evaluated than the integral (5).   Once it is calculated, we have $\psi(t) = L^{-1}\{g(s)\}$.

EXAMPLE C. Evaluate the integral

$$\psi(x) = \int_{0}^{\infty} \frac{e^{-xt}}{t+1} dt$$

Take $\varphi(x) = e^{-x}$, $h(x) = (x+1)^{-1}$, $f(s) = (s+1)^{-1}$ in equations (5) and (6). Then

$$g(s) = L\{\psi(x)\} = \int_{0}^{\infty} \frac{1}{x+1} \frac{1}{x+s} dx = \frac{1}{s-1} \log s$$

Since, by formula 13 of the table following Chapter 13,

$$g(s+1) = s^{-1} \log(s+1) = L\{\text{Ei}(t)\}$$

we must have

$$\psi(t) = L^{-1}\{g(s)\} = e^t \text{Ei}(t)$$

We obtained this same result by a change of variable in Example A.

EXERCISES (1)

Evaluate the integrals of Exercises 1–14 by the methods of the present section. It is not required to justify the changes in the order of integration that are employed, but care should be taken to deal with no divergent integrals.

1. $\displaystyle\int_0^\infty se^s \, \text{Ei} \, (2s) \, ds$

2. $\displaystyle\int_{0+}^\infty e^s \cos s \, \text{Ei} \, (s) \, ds$

3. $\displaystyle\int_0^\infty e^{-s} f(s) \, ds, f(s) = \int_0^1 e^{-st} \log t \, dt$

4. $\displaystyle\int_{0+}^\infty s^{-1/2} e^s \, \text{Ei} \, (s) \, ds$

5. $\displaystyle\int_{0+}^\infty \text{Si} \, (s) f(s) \, ds, f(s) = L\{t(t+1)^{-2}\}$

6. $\displaystyle\int_{0+}^\infty \text{Ci} \, (s) \, \text{Ei} \, (s) \, ds$

7. $\displaystyle\int_{0+}^\infty f(s) \, \text{Ei} \, (s) \, ds, f(s)$ as in Exercise 5

8. $\displaystyle\int_{0+}^\infty [\text{Ei} \, (s)]^2 \, ds$

9. $\displaystyle\int_{0+}^\infty e^{-xt} t^{-1/2} \, dt.$ Take $\varphi(t) = e^{-t}$ in equation (5)

10. $\displaystyle\int_0^\infty (\cos xt)(x^2 + 1)^{-1} \, dx$

11. $\displaystyle\int_0^\infty x \, (\sin xt)(x^2 + 1)^{-1} \, dx$

12. $\displaystyle\int_0^\infty \left(\frac{\sin x}{x}\right)^2 dx$

13. $\displaystyle\int_0^\infty x^{-1/2} \sin xt \, dx$

14. $\displaystyle\int_{0+}^\infty x^{-1/2} \cos xt \, dx$

15. Show that the integral (3) converges.

§2. *Linear Differential Equations*

We consider next the solution of linear differential equations with constant coefficients. The Laplace transform is especially well adapted to this problem, particularly when all the boundary conditions are concerned with the values of the unknown function and of its derivatives at a *single* point. The method consists of taking the Laplace transform of the given differential equation. That is, each term of the equation is taken to be a determining function. The resulting equation is algebraic. The inverse Laplace transform of a solution of this algebraic equation is the required function.

2.1 FIRST ORDER EQUATIONS

Consider the differential system

(1) $$y'(t) + ay(t) = \varphi(t)$$

(2) $$y(0) = A$$

where a and A are constants, $y(t)$ is the unknown function, and $\varphi(t)$ is any determining function whose Laplace transform has an abscissa of absolute convergence. Now

(3) $$\int_0^\infty e^{-st} y'(t)\, dt = -y(0) + s \int_0^\infty e^{-st} y(t)\, dt$$

This equation assumes that for some value of s the two integrals converge and that $y(t)e^{-st}$ tends to zero as $t \to \infty$. If we set $f(s) = L\{\varphi(t)\}$ and $Y(s) = L\{y(t)\}$, we obtain from equations (1), (2), (3) that

$$\int_0^\infty e^{-st}[y'(t) + ay(t)]\, dt = -A + sY(s) + aY(s) = f(s)$$

so that $$Y(s) = \frac{A + f(s)}{a + s}$$

Suppose that

$$\frac{f(s)}{a + s} = L\{\beta(t)\}$$

Then $$Y(s) = L\{Ae^{-at} + \beta(t)\}$$

and by Corollary 10.2, Chapter 13, we obtain

(4) $$y(t) = Ae^{-at} + \beta(t)$$

It is incorrect to suppose that we have *proved* that equation (4) yields the solution of the system (1), (2). We have proved only that, if there is a solution having all the properties assumed, it must have the form (4). However, we avoid all difficulties by showing directly that the function $y(t)$ defined by equation (4) satisfies the given system.   First

$$(Ae^{-at})' + aAe^{-at} = 0$$
$$Ae^{-at}|_{t=0} = A$$

It remains only to show that $\beta(0) = 0$ and that $\beta(t)$ satisfies equation (1). By Theorem 8, Chapter 13,

$$\frac{f(s)}{s+a} = L\{\varphi(t) * e^{-at}\}$$

the integral on the right converging for s sufficiently large.   Hence, by Corollary 10.2, Chapter 13,

$$\beta(t) = \varphi(t) * e^{-at} = e^{-at}\int_0^t e^{au}\varphi(u)\,du$$

From this explicit expression the desired result is evident.

EXAMPLE A.   Solve the differential system

$$y'(t) + y(t) = 1$$
$$y(0) = 2$$

by two methods.   Since we have proved that the method of the Laplace transform is valid for any such system of the first order, we may now use it without verification.   We have

$$\int_0^\infty e^{-st}y'(t)\,dt = -2 + sY(s)$$

$$\int_0^\infty e^{-st}[y'(t) + y(t)]\,dt = -2 + sY(s) + Y(s) = s^{-1}$$

so that

$$Y(s) = \frac{1}{s} + \frac{1}{s+1} = L\{1 + e^{-t}\}$$

(5)
$$y(t) = 1 + e^{-t}$$

As a second method, we may use the specific formula obtained above in the general case.   We have

$$y(t) = Ae^{-at} + \beta(t) = 2e^{-t} + \beta(t)$$

$$\beta(t) = e^{-t}\int_0^t e^u\,du = 1 - e^{-t}$$

This again gives equation (5).

2.2 UNIQUENESS OF SOLUTION

We have shown that the function $y(t)$ defined by equation (4) is a solution of the system (1), (2). We must prove that there can be no other. Suppose there were two different ones. Their difference would satisfy the homogeneous system

$$y'(t) + ay(t) = 0$$
$$y(0) = 0$$

But such a function must be identically zero. For,

$$[e^{at}y(t)]' = e^{at}[y'(t) + ay(t)] = 0$$
$$e^{at}y(t) = C$$

where C is a constant. It must be zero since $y(0) = 0$. The uniqueness of the solution is thus established.

2.3 EQUATIONS OF HIGHER ORDER

The method works equally well for equations of higher order. We illustrate by several examples. We reserve for a later section the verification of the method. Here we may show directly that the function obtained by the method actually satisfies the system.

EXAMPLE B.   Solve the system and check:

$$y'' + y = 2e^t$$
$$y(0) = y'(0) = 2$$

Using the notation employed above, we have by Corollary 5, Chapter 13,

$$L\{y''\} = -2 - 2s + s^2 Y(s)$$

$$\frac{2}{s-1} = -2 - 2s + (s^2 + 1)Y(s)$$

$$Y(s) = \frac{2s^2}{(s-1)(s^2+1)} = \frac{1}{s-1} + \frac{s+1}{s^2+1}$$

By use of the table of transforms, we see that

$$Y(s) = L\{e^t + \cos t + \sin t\}$$
$$y(t) = e^t + \cos t + \sin t$$

That this function actually satisfies the given system we see by inspection.

EXAMPLE C.   Find the general solution of the system

$$y'' + y' = 0$$

We may require a solution of this equation for which $y(0) = A$, $y'(0) = B$, where A and B are arbitrary. Here

$$L\{y''\} = -B - As + s^2 Y(s)$$

$$L\{y'\} = -A + sY(s)$$

$$0 = -A - B - As + (s^2 + s)Y(s)$$

$$Y(s) = \frac{A + B}{s} - \frac{B}{s + 1} = L\{A + B - Be^{-t}\}$$

$$y(t) = C + De^{-t}, \qquad C = A + B, \qquad D = -B$$

This may be checked directly.

EXERCISES (2)

Solve the following differential systems. If the order is greater than one the solution should be checked. Systems of order one should be done two ways.

1. $y'(t) + 2y(t) = 0$, $y(0) = 1$

2. $y'(t) + 2y(t) = t$, $y(0) = -1$

3. $y'(t) + y(t) = \cos t$, $y(0) = 0$

4. $y'(t) - y(t) = \sin t$, $y(0) = -1$

5. $y'(t) - y(t) = 0$, general solution

6. $y'(t) - y(t) = \cos t$, general solution

7. $y'(t) + ay(t) = 1 + e^t$, general solution

8. $y''(t) - y(t) = 0$, $y(0) = 0$, $y'(0) = 2$

9. $y''(t) - y(t) = t$, $y(0) = 0$, $y'(0) = 1$

10. $y'' + 2y' + y = 1$, $y(0) = y'(0) = 0$

11. $y'' + y' + y = 1$, $y(0) = y'(0) = 0$

12. $y''' - y' = e^{-t}$, $y(0) = y'(0) = y''(0) = 0$

13. $y''' + 17y'' - 10y' + y = 0$, $y(0) = y'(0) = y''(0) = 0$

14. $y''' + y' = t - 1$, $y(0) = 2$, $y'(0) = y''(0) = 0$

15. Treat the general system

$$y'' + a_1 y' + a_0 y = 0, \quad y(0) = A_0, \quad y'(0) = A_1$$

as was done in §2.1.   Consider three cases according as $a_1^2 - 4a_0$ is positive, negative, or zero.   Show that the solution obtained actually satisfies the system.

16. Solve the same problem for the general system

$$y'' + a_1 y' + a_0 y = \varphi(t), \quad y(0) = A_0, \quad y'(0) = A_1$$

§3. *The General Homogeneous Case*

We saw in Examples B and C of §2.3 that the Laplace method may be applied to differential systems of order higher than the first. In order to avoid the necessity of checking each answer, we shall now show that the function obtained by the method is, in fact, always the desired solution. In this section we treat the general homogeneous linear equation with constant coefficients, reserving the nonhomogeneous case for the following section.

3.1 THE PROBLEM

Define a linear differential operator H as follows:

$$H\{y(t)\} = y^{(n)}(t) + a_{n-1}y^{(n-1)}(t) + \ldots + a_1 y'(t) + a_0 y(t)$$

Here $a_0, a_1, \ldots, a_{n-1}$ are given constants. For example, if $n > 2$,

$$H\{t^2\} = 2a_2 + 2a_1 t + a_0 t^2$$

We wish to solve the system

(1) $$H\{y(t)\} = 0$$

(2) $$y^{(k)}(0) = A_k \qquad k = 0, 1, \ldots, n-1$$

for the unknown function $y(t)$. The A_k $(k = 1, 2, \ldots, n-1)$ are given constants.

We begin by defining, by use of the two sequences of constants $\{a_k\}$ and $\{A_k\}$, the following polynomials:

$$q_k(s) = A_0 s^k + A_1 s^{k-1} + \ldots + A_k$$

$$M(s) = q_{n-1}(s) + a_{n-1}q_{n-2}(s) + \ldots + a_1 q_0(s)$$

$$N(s) = s^n + a_{n-1}s^{n-1} + \ldots + a_1 s + a_0$$

The formal solution of the problem is easily expressed in terms of these functions. If $Y(s)$ is the Laplace transform of the desired solution $y(t)$ we see by Corollary 5, Chapter 13, that

$$L\{y^{(k)}(t)\} = -q_{k-1}(s) + s^k Y(s) \qquad k = 1, 2, \ldots, n$$

If $q_{-1}(s)$ is defined as 0, this equation also holds for $k = 0$. Multiply it by a_k and sum from 0 to n $(a_n = 1)$. This gives

$$0 = L\{H\{y(t)\}\} = -M(s) + N(s)Y(s)$$

so that $Y(s)$ is $M(s)/N(s)$. Since the quotient $M(s)/N(s)$ is a rational function which vanishes at infinity, we know by Theorem 12, Chapter 13, that it is a generating function:

$$(3) \qquad \frac{M(s)}{N(s)} = \int_0^\infty e^{-st}\alpha(t)\,dt$$

We shall show that $\alpha(t)$ is the desired solution of the system (1), (2).

3.2 THE CLASS E

We now introduce a class of functions whose derivatives do not become infinite at $+\infty$ more rapidly than exponential functions.

Definition 1. $\alpha(t) \in E$

$\Leftrightarrow$ A. $\alpha(t) \in C^\infty$ $0 \le t < \infty$

          B. *Constants M_k and σ exist such that*

$$|\alpha^{(k)}(t)| \le M_k e^{\sigma t} \qquad k = 0, 1, 2, \ldots ; \quad 0 \le t < \infty$$

Notice that the constants M may vary with k but that σ may not. For example, if $\alpha(t) = \sin 2t$ we may take $\sigma = 0$ and $M_k = 2^k$. Clearly, any polynomial belongs to E. Any linear combination of functions in E is also in E.

Theorem 1.   1. $\alpha(t) \in E$

              2. $\beta(t) \in E$

$\Rightarrow$ $\alpha(t)\beta(t) \in E$

For, let

$$(4) \qquad |\alpha^{(k)}(t)| \le M_k e^{\sigma t}$$

$$(5) \qquad |\beta^{(k)}(t)| \le M_k e^{\sigma t} \qquad k = 0, 1, \ldots ; \quad 0 \le t < \infty$$

There is no restriction in using the same constants M and σ for the two functions. For, if two of the corresponding constants differed initially, we could replace the smaller by the larger, retaining the inequality *a fortiori*. By Leibniz's rule for the derivative of a product we have

$$[\alpha(t)\beta(t)]^{(k)} = \sum_{j=0}^{k} \binom{k}{j} \alpha^{(j)}(t)\beta^{(k-j)}(t)$$

$$|[\alpha(t)\beta(t)]^{(k)}| \le N_k e^{2\sigma t} \qquad k = 0, 1, \ldots ; \quad 0 \le t < \infty$$

$$N_k = \sum_{j=0}^{k} \binom{k}{j} M_j M_{k-j}$$

This completes the proof.

Theorem 2.   1. $\alpha(t) \in E$

2. $\beta(t) \in E$

$\Rightarrow$ $\alpha(t) * \beta(t) \in E$

From the definition of the resultant we have by differentiation that

$$[\alpha(t) * \beta(t)]' = \alpha(t)\beta(0) + \alpha(t) * \beta'(t)$$

Then by induction we have for every positive integer k

(6) $$[\alpha(t) * \beta(t)]^{(k)} = \sum_{j=0}^{k-1} \alpha^{(j)}(t)\beta^{(k-1-j)}(0) + \alpha(t) * \beta^{(k)}(t)$$

By inequalities (4), (5),

$$|[\alpha(t) * \beta(t)]^{(k)}| \leq e^{\sigma t} \sum_{j=0}^{k-1} M_j M_{k-j-1} + \int_0^t M_0 M_k e^{\sigma u} e^{\sigma(t-u)} \, du$$

$$\leq N_k e^{(\sigma+1)t} \qquad k = 1, 2, \ldots; \quad 0 \leq t < \infty$$

$$N_k = \sum_{j=0}^{k-1} M_j M_{k-j-1} + M_0 M_k$$

Here we have used the fact that $t < e^t$ for $0 \leq t \leq \infty$. The proof is complete.

3.3 RATIONAL FUNCTIONS

We shall show next that any rational generating function is the Laplace transform of a function of E.

Theorem 3.   1. $R(s)$ is rational

2. $R(s) = L\{\varphi(t)\}$

$\Rightarrow$ $\varphi(t) \in E$

As we saw in the proof of Theorem 12, Chapter 13, $\varphi(t)$ is a linear combination of functions of the forms

$$L^{-1}\left\{\frac{1}{s-a}\right\}, \quad L^{-1}\left\{\frac{c}{(s-b)^2 + c^2}\right\}, \quad L^{-1}\left\{\frac{s-b}{(s-b)^2 + c^2}\right\}$$

and others obtained from these by the process of convolution. These three functions are e^{at}, $e^{bt} \sin ct$, and $e^{bt} \cos ct$. Our result is now a consequence of Theorems 1 and 2 since we may see by inspection that e^{at}, e^{bt}, $\cos ct$, $\sin ct$ all belong to E. As a corollary to this theorem, we see that the function $\alpha(t)$ defined by equation (3) belongs to E.

3.4 SOLUTION OF THE PROBLEM

To prove that the function $\alpha(t)$ defined by equation (3) satisfies the system (1), (2), we need a preliminary result.

Lemma 4. $\dfrac{M(s)}{N(s)} = \dfrac{A_0}{s} + \dfrac{A_1}{s^2} + \ldots + \dfrac{A_{n-1}}{s^n} - \dfrac{P_{n-1}(s)}{s^n N(s)}$

where $P_{n-1}(s)$ is a polynomial of degree $n-1$, at most.

Let $P_k(s)$ be the polynomial consisting of the last $k+1$ terms of $q_{n-1}(s)$, $k = 0, 1, \ldots, n-2$:

$$P_k(s) = A_{n-k-1}s^k + \ldots + A_{n-2}s + A_{n-1}$$

Then

$$q_{n-1} - 0 = q_{n-1}$$

$$q_{n-1} - P_0 = sq_{n-2}$$

$$q_{n-1} - P_1 = s^2 q_{n-3}$$

$$\ldots\ldots\ldots\ldots\ldots$$

$$q_{n-1} - P_{n-2} = s^{n-1}q_0$$

$$q_{n-1} - q_{n-1} = s^n \cdot 0$$

Multiply the first equation by 1, the second by $s^{-1}a_{n-1}$, the last by $s^{-n}a_0$ and add:

$$\frac{q_{n-1}(s)N(s) - P_{n-1}(s)}{s^n} = M(s)$$

$$P_{n-1}(s) = a_{n-1}s^{n-1}P_0(s) + \ldots + a_1sP_{n-2}(s) + a_0q_{n-1}(s)$$

Clearly, each term of $P_{n-1}(s)$ is a polynomial whose degree is at most $n-1$. Since

$$\frac{q_{n-1}(s)}{s^n} = \frac{A_0}{s} + \frac{A_1}{s^2} + \ldots + \frac{A_{n-1}}{s^n}$$

the proof is complete.

Theorem 4.   1. $\dfrac{M(s)}{N(s)} = L\{\alpha(t)\}$

$\Rightarrow$ A. $H\{\alpha(t)\} = 0$

   B. $\alpha^{(k)}(0) = A_k$ $\qquad\qquad\qquad k = 0, 1, \ldots, n-1$

Let us prove B first. Since $\alpha(t) \in E$, we see by Theorem 5, Chapter 13, that

(7) $\qquad L\{\alpha'(t)\} = -\alpha(0) + sL\{\alpha(t)\}$ $\qquad\qquad s > \sigma$

By Corollary 1.3, Chapter 13, the left-hand side of equation (7) tends to zero as $s \to \infty$. By Lemma 4

$$\lim_{s \to \infty} sL\{\alpha(t)\} = \lim_{s \to \infty} \frac{sM(s)}{N(s)} = A_0$$

so that $\alpha(0) = A_0$. Proceed by induction. Assume B for $k = 0, 1, \ldots,$ $m - 1$, where $m < n$. Then

$$L\{\alpha^{(m+1)}(t)\} = -\alpha^{(m)}(0) - A_{m-1}s - \ldots - A_0 s^m + s^{m+1}L\{\alpha(t)\}$$

By Lemma 4 the right-hand side tends to $-\alpha^{(m)}(0) + A_m$, whereas the left-hand side tends to zero as $s \to \infty$. Hence, $\alpha^{(m)}(0) = A_m$, and the induction is complete. Accordingly,

(8) $$L\{\alpha^{(k)}(t)\} = -q_{k-1}(s) + s^k L\{\alpha(t)\} \qquad k = 1, 2, \ldots, n$$

Combining these n equations, we obtain

$$L\{H\{\alpha(t)\}\} = -M(s) + N(s)L\{\alpha(t)\}$$

By equation (3) the right-hand side is zero. Hence, by Theorem 10, Chapter 13, we see that A holds. The proof is complete.

EXERCISES (3)

1. Which of the following functions belong to E?

   $\sin^2 (3t)$, $\log (1 + t)$, $e e^t$, $(t + 1)(t - 1)^{-1}$, $(t - 1)(t + 1)^{-1}$

2. Show that $te^t \in E$. Is there a smallest value for σ? If σ is chosen as 2, what is the smallest possible value of M_k? Answer the same question if $\sigma = \frac{3}{2}$.

3. Show that $t * e^t \in E$ and that we may take $\sigma = M_k = 1$.

4. Show that $e^{-t} \sin t \in E$ and that we may take $\sigma = -1$.

5. Complete the induction in the proof of Theorem 2.

6. If $M/N = (s^2 + 1)^{-2}$, find $\alpha(t)$ explicitly and show that $\alpha(t) \in E$.

7. Solve the same problem if $M/N = s(s + 1)^{-1}(s^2 + 1)^{-2}$.

8. In Exercise 6 show that $\alpha(0) = 0$.

9. In Exercise 7 show that $\alpha(0) = \alpha'(0) = \alpha''(0) = 0$.

10. Assuming that $M(s)/N(s)$ can be expanded in a convergent series

    $$\sum_{k=0}^{\infty} \frac{B_k}{s^{k+1}}$$

    find $\alpha(t)$ explicitly by use of Theorem 13, Chapter 13. Thus, prove conclusion B of Theorem 4.

11. Find $M(s)$, $N(s)$, $\alpha(t)$ if $n = 3$, $a_2 = a_0 = 0$, $a_1 = 1$, $A_0 = A_1 = A_2 = 1$. Show that $\alpha(t) \in E$ and that $\alpha(t)$ satisfies the system (1), (2).

12. Expand the function $M(s)/N(s)$ of Exercise 11 in powers of $1/s$ through 5 terms by long division. Show that the three first coefficients are A_0, A_1, A_2, as predicted by the theory (Lemma 4 and Exercise 10).

13. Show that the system (1) (2) cannot have more than one solution of class E.

§4. *The Nonhomogeneous Case*

In this section we shall solve the nonhomogeneous linear system corresponding to the system (1), (2), of §3.1. That is, we shall replace the right-hand side of equation (1), §3.1, by an arbitrary function $\varphi(t)$ which has an absolutely convergent Laplace transform, retaining equations (2), §3.1, as they were.

4.1 THE PROBLEM

We wish to solve the system

(1) $$H\{y(t)\} = \varphi(t), \qquad L\{\varphi(t)\} = f(s) \qquad\qquad s > s_a$$

(2) $$y^{(k)}(0) = A_k \qquad\qquad k = 0, 1, \dots, n-1$$

It will be sufficient to solve this system with all the $A_k = 0$. For, if $\beta(t)$ is a solution of this modified system and if $\alpha(t)$ is the function of §3, then the sum $\alpha(t) + \beta(t)$ is a solution of the system (1), (2).

4.2 SOLUTION OF THE PROBLEM

Let $N(s)$ be defined as in §3.1. Then by Theorem 3 the function $1/N(s)$ is a generating function, the Laplace transform of a function of E:

$$\frac{1}{N(s)} = \int_0^\infty e^{-st}\delta(t)\,dt \qquad\qquad \delta(t) \in E$$

This integral converges absolutely for $s > \sigma$. By Theorem 8, Chapter 13, $f(s)/N(s)$ is also a generating function, the Laplace transform of $\varphi(t) - \delta(t)$. This resultant is the required function $\beta(t)$.

Theorem 5. 1. $f(s) = L\{\varphi(t)\}$ $\qquad\qquad\qquad\qquad\qquad s > s_a$

2. $\dfrac{1}{N(s)} = L\{\delta(t)\}$ $\qquad\qquad\qquad\qquad\qquad s > \sigma$

3. $\beta(t) = \varphi(t) * \delta(t)$

4. $\dfrac{M(s)}{N(s)} = L\{\alpha(t)\}$

$\Rightarrow$ A. $H\{\alpha(t) + \beta(t)\} = \varphi(t)$

B. $\alpha^{(k)}(0) + \beta^{(k)}(0) = A_k$ $\qquad\qquad k = 0, 1, \dots, n-1$

We show first that

(3) $$\delta^{(k)}(0) = 0 \qquad\qquad k = 0, 1, \dots, n-2$$

By Theorem 3

$$\frac{s^{n-1}}{N(s)} = \int_0^\infty e^{-st}\gamma(t)\, dt \qquad\qquad \gamma(t) \in E$$

But $1/s^{n-1} = L\{t^{n-2}/(n-2)!\}$. Hence, by Theorem 8, Chapter 13, we have

$$\delta(t) = \frac{t^{n-2}}{(n-2)!} * \gamma(t) = \int_0^t \frac{(t-u)^{n-2}}{(n-2)!}\gamma(u)\, du$$

Now equations (3) are evident by inspection.

Next we show that

(4) $$\beta^{(k)}(0) = 0 \qquad\qquad k = 0, 1, \ldots, n-1$$

By virtue of equation (6), §3, and by equations (3) we have

$$\beta^{(k)}(t) = \varphi(t) * \delta^{(k)}(t)$$

$$= \int_0^t \varphi(t-u)\delta^{(k)}(u)\, du \qquad k = 0, 1, \ldots, n-1$$

so that equations (4) become obvious.

It remains only to show that $\beta(t)$ satisfies equation (1). By virtue of equation (8), §3, we see that

$$L\{\beta^{(k)}(t)\} = s^k L\{\beta(t)\} \qquad\qquad k = 0, 1, \ldots, n$$

$$L\{H\{\beta(t)\}\} = N(s)L\{\beta(t)\}$$

$$= N(s)f(s)\frac{1}{N(s)} = L\{\varphi(t)\}$$

The proof is now completed by use of Corollary 10.2, Chapter 13.

4.3 UNIQUENESS OF SOLUTION

Theorem 6. *There is only one solution of the system* (1), (2).

As in §2.2 we see that it will be sufficient to show that the only solution of the system (1), (2), modified so that $\varphi(t) = 0$ and $A_k = 0$ ($k = 0, 1, \ldots, n-1$), is $y(t) = 0$. Since $N(s)$ can always be factored into real linear and quadratic factors (some of which may be repeated), it is clear that the differential expression $H\{y(t)\} = N(D)\{y(t)\}$ can be written symbolically as a "product" of "factors" of the form

$$(D - a), \quad (D - b)^2 + c^2$$

Here a, b, c are real constants, and D is the symbol for differentiation with respect to t. The order of the symbolic factors can be changed at will. Suppose that

(5) $$N(D)\{y(t)\} = (D - a)\{z(t)\} = 0$$

so that $z(t)$ is a linear combination of $y, y', \dots, y^{(n-1)}$. By equations (1) and (2) with $\varphi(t) = A_k = 0$ ($k = 0, 1, \dots, n-1$) it is clear that $y^{(n)}(0)$ is also zero, so that $z(0) = z'(0) = 0$. But we showed in §2.2 that these boundary conditions applied to equation (5) imply that $z(t)$ is identically zero. In this way we can eliminate step by step all the linear factors in $N(D)$. Clearly, our proof will be complete if we establish the following result.

Theorem 7.   1. $[(D - b)^2 + c^2]\{y(t)\} = 0$

2. $y(0) = y'(0) = 0$

$\Rightarrow$ $y(t) = 0$ $-\infty < t < \infty$

It is easy to see that

(6) $$y'' - 2by' + (b^2 + c^2)y = \frac{e^{bt}}{\cos ct} \frac{d}{dt} \left\{ \cos^2 ct \frac{d}{dt} \frac{y}{e^{bt} \cos ct} \right\}$$

from which we have our result by two integrations. In each case the constant of integration is zero by virtue of hypothesis 2.

EXERCISES (4)

1. If $n = 2$, $a_0 = 1$, $a_1 = 0$, $A_0 = A_1 = 1$, $\varphi(t) = t$, find $M(s)$, $N(s)$, $f(s)$, $\alpha(t)$, $\beta(t)$, $\delta(t)$. Check directly that $\alpha(t) + \beta(t)$ satisfies the given system.

2. In Exercise 1 find $\gamma(t)$. Check directly that $\delta(t) = 1 * \gamma(t)$. Find a function $\psi(t) = L^{-1}\{1/M(s)\}$. Show that $\delta(t) = \alpha(t) * \psi(t)$.

3. In Exercise 11, §3, add the condition $\varphi(t) = 1$. Solve Exercise 1 of the present section with these data.

4. Do Exercise 2 with the data of Exercise 3.

5. Let $P_n(s)$ denote a polynomial of degree n. Show that, if k is a positive integer, the function $L^{-1}\{P_n(s)/P_{n+k+2}(s)\}$ vanishes with its first k derivatives at the origin.

6. Illustrate Exercise 5 by use of the following functions

$$\frac{1}{s^2 + s + 1}, \qquad \frac{1}{s^2 + 2s + 1}, \qquad \frac{1}{s^4 + s^2}$$

7. In Exercise 1, use a series expansion of $f(s)/N(s)$ to show that $\beta(0) = \beta'(0) = 0$. Why is $\beta''(0)$ also zero in this case? Compare Exercise 12, §3.

8. The corresponding problem for Exercise 3.

9. Verify equation (6).

10. Give the details in the proof of Theorem 7.

11. Show by expanding that $(D - a)(D - b)[(D - c)^2 + d^2]y(t)$ is equal to $(D - b)[(D - c)^2 + d^2](D - a)y(t)$.

12. Let $y_1(t)$, $y_2(t)$ be two solutions of the equation

$$K\{y(t)\} = y'' + p(t)y' + q(t)y = 0$$

such that $$W(t) = y_2'y_1 - y_1'y_2 \neq 0$$

Prove that $$K\{y\} = \frac{1}{y_1}\frac{d}{dt}\left[\frac{y_1^2}{W}\frac{d}{dt}\frac{y}{y_1}\right]$$

13. Use Exercise 12 to prove a uniqueness theorem for general second order linear differential equations. State the theorem.

14. Use Exercise 12 to obtain equation (6).

15. Replace equation (6) by one involving sin ct.

§5. *Difference Equations*

A difference equation is analogous to a differential equation, the operation $\Delta y_n = y_{n+1} - y_n$ in the former corresponding to the operation of differentiation $Dy(t) = y'(t)$ in the latter. In a difference equation the unknown is a sequence $\{y_n\}$, whereas in a differential equation it is a function $(y)t$. In §1, Chapter 13, we showed how a Laplace integral may be regarded as a generalized power series, in which the sequence of integral powers has been replaced by a continuous variable. Since the Laplace transform was so useful in solving linear differential equations where the unknown is a function of the continuous variable, it is natural to conjecture that power series would play an analogous role in the solution of linear difference equations where the unknown is a sequence—that is, a function of a variable that takes on only integral values. It is this point of view which we shall adopt. It would be possible to use the Laplace transform instead of the power series transform. The present section should be regarded as a means of giving insight into the method described in the previous sections.

5.1 THE PROBLEM

The general linear difference system with constant coefficients has the form

(1) $$H\{y_k\} = \Delta^n y_k + a_{n-1}\Delta^{n-1}y_k + \ldots + a_1\Delta y_k + a_0 y_k = \varphi_k$$

(2) $$\Delta^j y_0 = A_j \qquad\qquad j = 0, 1, \ldots, n-1$$

Here $\{y_k\}_0^\infty$ is the unknown sequence, $\{\varphi_k\}_0^\infty$ is a given sequence, and $a_0, a_1, \ldots,$ $a_{n-1}, A_0, A_1, \ldots, A_{n-1}$ are given constants. Furthermore,

$$\Delta y_k = y_{k+1} - y_k$$

$$\Delta^2 y_k = \Delta(\Delta y_k) = y_{k+2} - 2y_{k+1} + y_k$$

(3) $$\Delta^n y_k = \Delta(\Delta^{n-1}y_k) = \sum_{j=0}^n (-1)^{j+1}\binom{n}{j}y_{k+j}$$

By virtue of equation (3) the system (1), (2) can be replaced by

(4) $H\{y_k\} = y_{k+n} + b_{n-1}y_{k+n-1} + \dots + b_1 y_{k+1} + b_0 y_k = \varphi_k$

(5) $\qquad\qquad\qquad\qquad y_j = B_j \qquad\qquad j = 0, 1, \dots, n - 1$

In fact, the constants b_j and B_j can be expressed explicitly in terms of the constants a_j and A_j, and conversely. Hence the systems (1), (2) and (4), (5) are interchangeable.

EXAMPLE A. Convert the system

$$y_{k+2} - 3y_{k+1} + 2y_k = -1$$
$$y_0 = 2, \qquad y_1 = 4$$

into its equivalent form involving Δ. Since

$$y_{k+1} = y_k + \Delta y_k$$
$$y_{k+2} = y_k + 2\,\Delta y_k + \Delta^2 y_k$$

we have for the equivalent form

$$\Delta^2 y_k - \Delta y_k = -1$$
$$y_0 = \Delta y_0 = 2$$

It is easy to see by direct computation that $y_k = 2^k + k + 1$ satisfies both systems.

5.2 THE POWER SERIES TRANSFORM

A power series

(6) $$f(s) = \sum_{k=0}^{\infty} \varphi_k s^k$$

may be regarded as a transform which carries the sequence of the coefficients $\{\varphi_k\}$ into the function $f(s)$ which is the sum of the series. The sequence may be called the *determining* sequence; the sum function, the *generating* function.

Definition 2. *The function $f(s)$ is the power series transform of the sequence $\{\varphi_k\}_0^{\infty}$, the relation being indicated by*

$$f(s) = l\{\varphi_k\}$$

$\Leftrightarrow$ *equation (6) holds, the series converging for some $s \neq 0$.*

EXAMPLE B. Find $l\{k\}$. We have

$$f(s) = \sum_{k=0}^{\infty} ks^k = s\frac{d}{ds}\sum_{k=0}^{\infty} s^k$$
$$l\{k\} = s(1 - s)^{-2} \qquad\qquad -1 < s < 1$$

EXAMPLE C. $l\{1/k!\} = e^s$ $-\infty < s < \infty$

EXAMPLE D. $l\{\rho^k\} = (1 - \rho s)^{-1}$ $|s\rho| < 1$

5.3 A PROPERTY OF THE TRANSFORM

We now obtain a relation analogous to that established in Corollary 5, Chapter 13.

Theorem 8. $l\{\varphi_k\} = \varphi_0 + \varphi_1 s + \ldots \varphi_{p-1} s^{p-1} + s^p l\{\varphi_{k+p}\}$

For, if $p = 0, 1, \ldots$, we have

$$s^p l\{\varphi_{k+p}\} = \sum_{k=p}^{\infty} \varphi_k s^k = l\{\varphi_k\} - \sum_{k=0}^{p-1} \varphi_k s^k$$

EXAMPLE E.   Find $l\{k + 2\}$.   Here $\varphi_k = k$ and

$$l\{\varphi_{k+2}\} = -\frac{\varphi_0}{s^2} - \frac{\varphi_1}{s} + \frac{1}{s^2} l\{\varphi_k\}$$

$$l\{k + 2\} = -\frac{1}{s} + \frac{1}{s(s-1)^2} = \frac{2-s}{(1-s)^2}$$

We can check this result directly by expanding the rational function in Maclaurin's series.

5.4 SOLUTION OF DIFFERENCE EQUATIONS

We shall make no attempt to solve the general problem, but we shall illustrate the method by solving a number of particular difference systems.

EXAMPLE F.   Solve the system

$$\Delta y_k + 2y_k = 0 \qquad\qquad y_0 = 1$$

This is equivalent to

$$y_{k+1} + y_k = 0 \qquad\qquad y_0 = 1$$

Set $f(s) = l\{y_k\}$.   By Theorem 8 with $p = 1$

$$l\{y_{k+1}\} = -\frac{y_0}{s} + \frac{1}{s} l\{y_k\}$$

$$l\{y_{k+1} + y_k\} = 0 = -\frac{1}{s} + \frac{f(s)}{s} + f(s)$$

so that

$$f(s) = \frac{1}{s+1} = \sum_{k=0}^{\infty} (-1)^k s^k \qquad\qquad -1 < s < 1$$

Since a function cannot be expanded in powers of s in more than one way, we must have

$$y_k = (-1)^k \qquad\qquad k = 0, 1, \ldots$$

We can verify directly that this sequence satisfies the given system.

EXAMPLE G.   Solve the system

$$y_{k+1} + y_k = 1 \qquad\qquad y_0 = 1$$

In this case, we obtain as before

$$-\frac{1}{s} + \left(\frac{1}{s} + 1\right) f(s) = \frac{1}{1-s}$$

$$f(s) = \frac{1}{1-s^2} = \sum_{k=0}^{\infty} s^{2k}$$

$$y_k = 0 \qquad\qquad k \text{ odd}$$

$$y_k = 1 \qquad\qquad k \text{ even}$$

Another form of the answer, which puts into evidence that the answer is the sum of a solution of the homogeneous system of Example F and a solution of the modified nonhomogeneous system ($y_0 = 0$), is

$$y_k = (-1)^k + [1 + \cos(k+1)\pi]/2 \qquad\qquad k = 0, 1, \ldots$$

EXAMPLE H.   Solve the system of Example A.

$$l\{y_{k+2}\} = -\frac{2}{s^2} - \frac{4}{s} + \frac{f(s)}{s^2}$$

$$l\{y_{k+1}\} = -\frac{2}{s} + \frac{f(s)}{s}$$

$$-\frac{1}{1-s} = \left(\frac{1}{s^2} - \frac{3}{s} + 2\right) f(s) - \frac{2}{s^2} + \frac{2}{s}$$

$$f(s) = \frac{2 - 4s + s^2}{(1-s)^2(1-2s)} = \frac{1}{(1-s)^2} + \frac{1}{1-2s}$$

$$= l\{k+1\} + l\{2^k\}$$

$$y_k = 2^k + k + 1 \qquad\qquad k = 0, 1, \ldots$$

EXERCISES (5)

Convert the systems in Exercises 1–7 to the other form.

1. $y_{k+1} - 2y_k = k$ \qquad\qquad $y_0 = 0$

2. $\Delta y_k - 2y_k = 0$ \qquad\qquad $y_0 = 1$

3. $y_{k+2} - y_k = k + 1$ $y_0 = y_1 = 0$

4. $\Delta^2 y_k + 4 \Delta y_k + 4y_k = 0$ $y_0 = 1, \quad \Delta y_1 = 0$

5. $\Delta^3 y_k + 2 \Delta^2 y_k = 1$ $y_0 = 1, \quad \Delta y_0 = \Delta^2 y_0 = 0$

6. $y_{k+3} + 3y_{k+2} - 4y_k = -k$ $y_0 = y_1 = y_2 = 0$

7. $y_{k+4} + y_{k+3} + y_{k+2} + y_{k+1} = 1$ $y_0 = \Delta y_0 = \Delta^2 y_0 = \Delta^3 y_0 = 0$

8. Solve the system of Exercise 1.

9. Solve 2.   **11.** Solve 4.   **13.** Solve 6.

10. Solve 3.   **12.** Solve 5.   **14.** Solve 7.

15. Show that $l\{\varphi_k\} \cdot l\{\psi_k\} = l\{\varphi_k * \psi_k\}$, where

$$\varphi_k * \psi_k = \sum_{j=0}^{k} \varphi_j \psi_{k-j}$$

Questions of convergence may be omitted.

16. $l\{\Delta y_k\} = ?$

17. $l\{\Delta^2 y_k\} = ?$

18. Show that the general solution of the equation

$$y_{k+2} - (\rho_1 + \rho_2)y_{k+1} + \rho_1\rho_2 y_k = 0 \qquad \rho_1 \neq \rho_2$$

is $y_k = A_1\rho_1^k + A_2\rho_2^k$.   Use the method of the present section.

19. What is the general solution in the previous exercise if $\rho_1 = \rho_2$?

20. Prove equation (3) by induction.

21. Express the B's of equation (5) in terms of the A's of equations (2) and express the A's in terms of the B's.

22. Solve the same problem for the a's and the b's.

§6. *Partial Differential Equations*

The method of the Laplace transform may be used to solve partial differential equations. A first application transforms the equation to an ordinary equation. A transformation of the latter equation converts it into an algebraic equation, which is then solved. Two inverse Laplace transforms give the desired solution. We illustrate by the problem of the vibrating string solved in Chapter 12.

6.1 THE FIRST TRANSFORMATION

Since we have already solved the problem when set with general constants, let us here specialize the constants to simplify the writing.   Let us solve the system

(1) $$\frac{\partial^2 y(x, t)}{\partial t^2} = \frac{\partial^2 y(x, t)}{\partial x^2}$$

(2) $y(0, t) = y(2, t) = 0$ $0 \leq t < \infty$

(3) $y(x, 0) = f(x)$ $-\infty < x < \infty$

(4) $\frac{\partial y}{\partial t}(x, 0) = 0$ $-\infty < x < \infty$

where $f(x) \in D^1$ (Definition 8, Chapter 12) and

$$f(x + 4) = f(x)$$ $-\infty < x < \infty$

$$f(-x) = -f(x)$$ $-\infty < x < \infty$

Note that, if $f(x)$ had the period 2π instead of the period 4, we should have precisely the conditions of Theorem 7, Chapter 12.
Set

$$Y(x, s) = \int_0^\infty e^{-st} y(x, t)\, dt$$

When we are thinking of s as a constant, we shall write

$$Y(x) = Y(x, s)$$

Transforming in the usual way, our new system is

(5) $Y''(x) - s^2 Y(x) = -sf(x)$

(6) $Y(0) = Y(2) = 0$

When s is kept constant, this is a linear system with constant coefficients. However, the boundary conditions involve more than one point. We may still apply the method, as we did in earlier examples, involving the general solution of an equation.

6.2 THE SECOND TRANSFORMATION

Set

$$u(z) = \int_0^\infty e^{-zx} Y(x)\, dx, \qquad v(z) = \int_0^\infty e^{-zx} f(x)\, dx$$

Now replace conditions (6) by

(7) $Y(0) = 0, \qquad Y'(0) = A$

where A is a constant that will later be determined so that $Y(2) = 0$. The transform of the system (5), (7) is

$$u(z) = \frac{A - sv(z)}{z^2 - s^2}$$

Hence,

$$Y(x) = \frac{A}{s} \sinh sx - f(x) * \sinh sx$$

$$= \frac{A}{s} \sinh sx - \int_0^x f(w) \sinh s(x - w) \, dw$$

After determining the constant A so that $Y(2) = 0$, this becomes

$$(8) \quad Y(x) = \frac{\sinh sx}{\sinh 2s} \int_0^2 f(w) \sinh s(2 - w) \, dw - \int_0^x f(w) \sinh s(x - w) \, dw$$

By virtue of the identity

$$(9) \quad \begin{vmatrix} \sinh s(2 - w) & \sinh s(x - w) \\ \sinh 2s & \sinh sx \end{vmatrix} = \sinh sw \sinh s(2 - x)$$

equation (8) becomes

$$(10) \quad Y(x) = \frac{\sinh sx}{\sinh 2s} \int_x^2 f(w) \sinh s(2 - w) \, dw$$

$$+ \frac{\sinh s(2 - x)}{\sinh 2s} \int_0^x f(w) \sinh sw \, dw$$

From this form of the solution it may be checked directly that equations (5) and (6) are satisfied by the function $Y(x)$ and that

$$(11) \qquad\qquad Y(2 - x) = Y(x)$$

whenever $f(x)$ has that property. We have proved the following result.

Theorem 9. *The function $Y(x)$ defined by equation (10) is the solution of the system (5), (6).*

6.3 THE PLUCKED STRING

Let us now specialize the function $f(x)$ as follows:

$$(12) \qquad\qquad f(x) = x \qquad\qquad 0 \leqq x \leqq 1$$

$$(13) \qquad\qquad = 2 - x \qquad\qquad 1 \leqq x \leqq 2$$

Then from equation (8) or equation (10) we have

$$Y(x) = \frac{x}{s} - \frac{1}{s^2} \frac{\sinh xs}{\cosh s} \qquad\qquad 0 \leqq x \leqq 1$$

To obtain $Y(x)$ in the interval $(1, 2)$, we have only to use equation (11). To find the inverse transform of $Y(x, s) = Y(x)$, considered now as a function of s, we need a preliminary result.

Lemma 10.1. 1. $\omega(t) = 0$ $\qquad\qquad\qquad -\infty < t \leqq 0$

$\qquad\qquad\quad = t \qquad\qquad\qquad\qquad\qquad 0 \leqq t \leqq 2$

$\qquad\qquad\quad = 2 \qquad\qquad\qquad\qquad\qquad 2 \leqq t < 4$

$\qquad\quad$ 2. $\omega(t + 4) = \omega(t) + 2 \qquad\qquad\quad 0 \leqq t < \infty$

$\Rightarrow \cdot \qquad\qquad L\{\omega(t)\} = \dfrac{1}{s^2(1 + e^{-2s})} \qquad\qquad 0 < s < \infty$

For,

$$\int_{4k}^{4k+4} e^{-st}\omega(t)\,dt = e^{-4ks} \int_0^4 e^{-st}\omega(t + 4k)\,dt$$

Hence,

$$L\{\omega(t)\} = \int_0^4 e^{-st}\omega(t)\,dt \sum_{k=0}^{\infty} e^{-4ks} + \int_0^4 e^{-st}\,dt \sum_{k=0}^{\infty} 2k e^{-4ks}$$

$$= \frac{1}{1 - e^{-4s}} \int_0^4 e^{-st}\omega(t)\,dt + \frac{2e^{-4s}}{(1 - e^{-4s})^2} \int_0^4 e^{-st}\,dt$$

The result is now obtained by evaluating the integrals.

Lemma 10.2. $\dfrac{1}{s^2}\dfrac{\sinh xs}{\cosh s} = L\{\omega(t - 1 + x) - \omega(t - 1 - x)\}$

$$0 < s < \infty$$

For,

$$\frac{\sinh xs}{s^2 \cosh s} = \frac{e^{xs-s}}{s^2(1 + e^{-2s})} - \frac{e^{-xs-s}}{s^2(1 + e^{-2s})}$$

We have only to use Lemma 10.1 and equation (3), §4.2, Chapter 13, to obtain the result.

Theorem 10. *The solution of the system* (1), (2), (3), (4) *with* $f(x)$ *defined by equations* (12), (13) *is, for* $0 \leqq t \leqq 1$,

$$y(x, t) = x \qquad\qquad\qquad 0 \leqq x \leqq 1 - t$$

$$\quad = 1 - t \qquad\qquad 1 - t \leqq x \leqq 1 + t$$

$$\quad = 2 - x \qquad\qquad 1 + t \leqq x \leqq 2$$

From equation (11) and the definition of $Y(x, s)$ we see that

(14) $\qquad\qquad y(2 - x, t) = y(x, t) \qquad 0 \leqq t < \infty, \ 0 \leqq x \leqq 2$

Hence, it will be sufficient to prove our result for $0 \leqq x \leqq 1$. By virtue of Lemma 10.2 we can find the inverse transform of $Y(x, s)$. It is

$$y(x, t) = x - \omega(t - 1 + x) + \omega(t - 1 - x)$$

Now, if $0 \leq t \leq 1, 0 \leq x \leq 1$, we have from the definition of $\omega(t)$ that

$$y(x, t) = x \qquad\qquad\qquad 0 \leq x \leq 1 - t$$
$$= x - (t - 1 + x) \qquad 1 - t \leq x \leq 1$$

This establishes our result.   It agrees with the result obtained in Chapter 12, where it was verified that $y(x, t)$ actually satisfied the given system.

EXERCISES (6)

1. Prove identity (9).

2. Prove equation (10).

3. Prove equation (11).

4. Give details in the proof of Theorem 9.

5. Solve the system (5), (6) by the method of variation of parameters.

6. If $\varphi(t + a) = \varphi(t), 0 \leq t < \infty, 0 < a$, show that

$$L\{\varphi(t)\} = (1 - e^{-as})^{-1} \int_0^a e^{-st}\varphi(t)\, dt \qquad 0 < s < \infty$$

7. Illustrate Exercise 6 by taking $\varphi(t) = 1, 0 < t < a/2, \varphi(t) = -1, a/2 < t < a$.

8. Do the same problem if $\varphi(t) = t, 0 \leq t \leq a/2, \varphi(a - t) = \varphi(t)$.

9. If $\varphi(t + a) = \varphi(t) + b, 0 \leq t < \infty, 0 < a, 0 < b$, find $L\{\varphi(t)\}$ in a form analogous to that obtained in Exercise 6.

10. Prove Lemma 10.1 by use of Exercise 9.

11. Give details in the proof of equation (14).

12. Show that the function $y(x, t)$ of §6.3 has the form
$$y(x, t) = \tfrac{1}{2} f(x + t) + \tfrac{1}{2} f(x - t)$$

13. Find $y(x, t)$, §6.3, for $1 < t < 2$.

14. Solve the same problem for $2 < t < 3$.

15. Show that $y(x, t)$, §6.3, has the period 4 in t.

Selected Answers

CHAPTER I

§1

1. $[y \cos xy \cos (x + y) + \sin xy \sin (x + y)] \cos^{-2} (x + y)$.
3. $f_2(1, 2) = 8 \tan 3 \sec^2 3$. **5.** 0, 2, 1. **10.** $u_y = -y/u, u \neq 0$.

§2

1. 0, 0, ∞. **3.** $0 < x < 1, 0 \leqq x < 1$. **5.** 0. **7.** 3/8. **11.** No.

§3

1. $2r, 0$. **4.** 1/2, 1/2.
5. $\theta_1 = \log (e - 1), \theta_2 = (1/2)[\log (e^2 - 1) - \log 2]$. **14.** 1/2.

§4

2. (a) $n = 1/2$; (b) $n = 0$; (c) $n = 2$; (d) No; (e) $n = 1$.
6. $e^v x \cos (xyz) - e^v x^2 yz \sin (xyz) + e^v x^2 yz \cos^2 xyz$.
12. $-2x f_{11} + (4xy - 1) f_{12} + 2y f_{22}$.

§5

1. $1/(x + u - 1)$. **3.** $-(y^2 + y)/(u + yu \log u), y/1 + y \log u)$.
5. -2π. **11.** $f_1/(1 - f_1 - y f_2), u f_2/(1 - f_1 - y f_2)$.

§6

1. $2uv + xw - xyv - x - z$. **9.** $u_y = f_2 g_2/[(1 - f_1)(1 - g_2) - f_2 g_1]$.
11. Yes. **12.** No.

§7

1. $(xz + uy)/(u^2 - z^2)$, $(y^2 - x^2)/(yu + xz)$.   5. $2x$, $2x + 2yzx^{z-1}$,
8. $-f_3/f_2$, $(f_1g_1 - f_3g_3)/f_2g_3$.

§8

1. $5, -5$.
7. $\cos \alpha + 4 \sin \alpha$, magnitude $= \sqrt{17}$, $\sin \alpha = 4/\sqrt{17}$, $\cos \alpha = 1/\sqrt{17}$.
12. $-8/\sqrt{65}$.   16. $n!\,(\cos \alpha + \sin \alpha)^n$.

§9

1. -7, $6h^2k$, $(x - 1)^3 f_{111} + 3(x - 1)^2 y f_{112} + 3(x - 1)y^2 f_{122} + y^3 f_{222}$.
4. $1/2$.   11. -2.   14. $f_1(0, 0)/g_2(0, 0)$; $f, g \in C^1$, $g_2(0, 0) \neq 0$.
15. $(f_{11} - f_2)/(f_{33} + f_2)$ at $(0, 1, 1)$; $f \in C^2$, $f_{33} \neq -f_2$.

§10

1. $2xr$.   6. $J = -2$, $j = -1/2$.

§11

1. Δx^3, $8 \Delta x \Delta y - 2 \Delta x \Delta y^2$.   6. $0, 0$.   9. Yes.

§12

1. $x, -x$; both.   4. $\log z \pm 2\sqrt{\log z}$, $\exp(\sqrt{z + 1} - 1)^2$.

CHAPTER 2

§1

1. $(-4, -2, 2)$, $(3, 0, 1)$, $(1, 2, -3)$, -1, $(1, -5, -3)$, 4.
6. $k = -(r \cdot r)/(r \cdot s)$ if $r \cdot s \neq 0$; k is arbitrary if $\vec{r} = 0$.

§2

1. $\sqrt{138}$.   5. $(k\vec{r} + l\vec{s} + m\vec{t})/(k + l + m)$.   12. $(rtu)(svw) - (rvw)(stu)$.

§3

1. $\vec{x} = \vec{a}t - \vec{b}t + \vec{b}$ passes through $\vec{a}$ and $\vec{b}$.
2. $x_1 = 2 \cos t$, $x_2 = 2 \sin t$, $x_3 = 4\sqrt{2}\, t/(8n + 1)\pi$, $n = 0, \pm 1, \pm 2, \ldots$.
5. $\sin^2 \theta$, $\sin \theta$, $\cos \theta$.   8. $(2, 4, 8)$.   15. $A \cos t + B \sin t + C$.

§4

1. $a \cos u \cos v$, $a \sin u \cos v$, $b \sin v$.   5. An ellipsoid of revolution.

§5

1. $2\vec{x}/|x|^2$, $2/|x|^2$.   2. 3, $(0, 0, 0)$.   7. $2/r$, 0.

§6

1. $\ddot{x} = \ddot{\alpha}x_1' + \ddot{\beta}x_2' + \ddot{\gamma}x_3'$.   **5.** $(0, t, t)$, $\cos^{-1}(1/3)$.

CHAPTER 3

§1

1. $(a, a\sqrt{3}\cos\theta, a\sqrt{3}\sin\theta)$; $2\pi a\sqrt{3}$.   **6.** $\cos^{-1}(\sqrt{21}/42)$.

8. $(t_0 + r, f(t_0) + rf'(t_0), g(t_0) + rg'(t_0))$; $X_1 + X_2 f'(t_0) + X_3 g'(t_0)$
$$= t_0 + f(t_0)f'(t_0) + g(t_0)g'(t_0).$$

10. $(1 + r, 1 + (r/2), 1 - (r/2))$; $2X_1 + X_2 - X_3 = 2$.

§2

1. $X_1 + X_2 = 1$.   **6.** 0, 1, 2 with planes $x_1 = 0$, $x_2 = 0$, $x_3 = 0$.   **7.** 3.

§3

1. 1/2.   **2.** 1/2.   **9.** $2(1 + 9t^2 + 9t^4)^{1/2}(1 + 4t^2 + 9t^4)^{-3/2}$, $a = b = c = 1$.

§4

1. $\alpha' = (-x_1/2, -x_2/2, 0)$; $\beta = (-x_1, -x_2, 0)$; $1/R = 1/2$.

2. $\beta' = -\alpha/R - \gamma/T$, $\gamma' = \beta/T$.   **6.** $(-x_1, -x_2, x_3)$.

§5

1. $2X_1 + X_2 - 2X_3 = 1$; $(1 + 2t, 1 + t, 1 - 2t)$.   **3.** $t = 0, 1, -4$; 0 at $t = 0$.

5. $3X_3 \pm 2\sqrt{3}\,X_2 = 3$.   **7.** $X_1 + \sqrt{3}X_3 = 2a$.   **13.** 0 at both points.

§6

3. $e = -\cos^2 v, f = 0, g = -1$.   **4.** $-1/a$ if radius is a.   **5.** $-2/a, 1/a^2$.

§7

1. $a\int (F_1^2 + F_2^2\cos^2\varphi)^{1/2}/F_2\,d\theta$, $\int (F_1^2 + a^2F_2^2)^{1/2}/F_2\,d\theta$.   **5.** $\pi a\sqrt{1 + \lambda^{-2}}$.

11. $\tan^{-1}[(2 - \cos^2 1)^{-1}3\cos 1]$, $\tan^{-1}[(4 + \cos^2 2)^{-1}3\cos 2]$.

CHAPTER 4

§1

1. Absolute max, 6; absolute min, -2.

3. Min, $e^{-1/e}$ (absolute by Theorem A); absolute max $= 1$.

7. Point of inflection.   **10.** Walk all the way; walk 60° of arc, swim the rest.

11. 12 mph; path makes an angle of $\cos^{-1}(3/5)$ with the tracks.

§2

1. 1/27.   **3.** A cube.   **9.** $8\sqrt{3}\,abc/9$.

10. Angle at ridge pole is 90°, sides are squares.   **11.** $A^{3/2}(2\pi)^{-1/2}5^{-1/4}/3$.

§3

1. Min at $(-4, 2)$.   **3.** Min at $(\sqrt{2}/8, \sqrt{2}/4)$ and $(-\sqrt{2}/8, -\sqrt{2}/4)$; sp at $(0, 0)$.
7. Max at $(1, 1)$; min at $(-1, -1)$.   **9.** $\sqrt{2}/2$.   **10.** $\sqrt{6}$.

§4

1. Max.   **4.** $220y = 45x^2 - 21x + 112$.
5. The altitudes of the rectangles and triangles mentioned are equal; the floor is square; the angle at the ridge pole is $120°$.

§5

1. $\cos^{-1}(3/\sqrt{41})$.   **5.** Major axis makes an angle of $45°$ with the positive x-axis.
14. $\sqrt{2}/2$.   **15.** $\sqrt{6}/9$.

§6

1. $y^2 = 4x$.   **5.** $|xy| = k$.
7. $x = g - h'F, y = h + g'F, F = (g'^2 + h'^2)/(g'h'' - g''h')$.
11. Two planes: $y = z, y = z + 4$.

§7

1. Characteristic lines: $x_2^2 + x_3^2 = 1, x_1 = t$; envelope: $x_2^2 + x_3^2 = 1$.
4. Characteristic lines: $x_1 = -x_3 \sin \theta, x_2 = x_3 \cos \theta$; envelope: $x_1^2 + x_2^2 = x_3^2$.
6. $x = x(s) + (u\alpha/T) + (u\gamma/R)$.   **9.** $(x_1x_2 - x_3)^2 - 4(x_1^2 - x_2)(x_2^2 - x_1x_3) = 0$.

CHAPTER 5

§1

3. $f(b)[\alpha(b) - \alpha(b-)]$.   **4.** $f(x_0)[\alpha(x_0+) - \alpha(x_0-)]$.   **6.** $2/3$.   **12.** $1/2$.   **15.** $3/4$

§2

6. -2.   **8.** 0.   **10.** 32.   **13.** $-f(1), f(1)$.

§3

1. $(e^2 + 1)/2$.   **3.** 60.   **5.** 2.   **7.** 0.
8. $g(2) - g(1) + g(-1) - g(-2), g = (1 + e^{-1/x})^{-1}$.   **10.** No.

§4

13. $g = (x + 2)^{-1} \sin x, f = (x + 2)(x + 1)^{-1}$.   **14.** $f(a)[\alpha(a+) - \alpha(a)]$.   **15.** 0.

§5

1. $8 \sin^2(s/4)$.   **2.** $s^2/2$ on $(0, \pi), (4\pi s - s^2 - 2\pi^2)/2$ on $(\pi, 2\pi)$.   **5.** $2/3$.   **7.** $17/3$.

§6

1. I_c, $c = k/8(k = 0, 1, ..., 6)$; add $k = 7$ or $c = 1$.   **3.** $x(2 - [x + 1])$, $a = 0$, $b = 1$.
4. $1 - (3/\sqrt{10})$.
12. lub $= \infty, 1, 1/e, \infty, \sqrt{2}$;   max $=$ none, 1, $1/e$, none, $\sqrt{2}$;   glb $= -\infty, 0, 0, 0, -1$;
   min $=$ none, none, 0, none, none.

§7

1. $\pi 2^{-n}[1 + \cot (\pi 2^{-n-1})]$.

CHAPTER 6

§1

1. (a) Points below the line $2y = 3x + 1$; (b) points outside the unit circle; (c) no;
   (d) points inside the unit circle, center at (1, 2).
4. $R(-5, -3, -\sqrt{25 - x^2}, \sqrt{25 - x^2}) + R(-3, 0, 4x/3, \sqrt{25 - x^2}) + R(0, 15/4, 4x/3, 5)$.
7. $R(-2, 2, g(x), \sqrt{4 - x^2}) + R(-2, 2, -\sqrt{4 - x^2}, -g(x))$; $g(x) = \sqrt{2 - x^2}$, $x^2 < 2$;
   $g(x) = 0$, $x^2 \geqq 2$.

§2

4. $1/3$.   **5.** $\pi/4$.   **7.** $f = c(1 - a^{-1}x - b^{-1}y)$, $R[0, a, 0, b(a - x)/a]$.

§3

1. $abc/6$.   **4.** $kab(3a^2 + b^2)/12$.

§4

1. $2\pi ka^3/3$.   **2.** $ka^3[\sqrt{2} + \log (1 + \sqrt{2})]/6$.   **5.** $a^2/2$.   **13.** 4/9.

§5

1. $R[0, 1, 0, \sqrt{y}]$.
4. $R[1/9, 4/9, 1/3, \sqrt{y}] + R[4/9, 1/\sqrt{3}, 1/3, 2/3] + R[1/\sqrt{3}, \sqrt{2/3}, y^2, 2/3]$.
5. $R[0, \pi/4, a, a \sec x]$.
11. $R[\csc^{-1} 4, \pi/4, \csc \theta, 4] + R[\pi/4, \sec^{-1} 4, \sec \theta, 4]$
                                           $- [0, \pi/4, 1, \sec \theta] - [\pi/4, \pi/2, 1, \csc \theta]$.

§6

2. $(0, 0)$.   **3.** $(\pi/2) \log (1 + a^4)$.   **11.** $M(3a^2 + 4b^2)/4$.
13. Points within distance $a\sqrt{5}/3$ from center.   **17.** $12\pi a^3/25$.

§7

1. $2a^2(\pi - 2)$.   **2.** $\pi[a\sqrt{a^2 + 1} + \log (a + \sqrt{a^2 + 1})]$.
4. $4\pi a^2 - 16a^2 \sin^{-1} a(2a^2 - 2b^2)^{-1/2} + 16ab \tan^{-1} b(a^2 - 2b^2)^{-1/2}$.
6. $[(1 + a^2)^{3/2} - 1]2\pi/3$.

§8

1. $M(b^2 + c^2)/10$.   **2.** $(a/4, b/4, c/4)$.   **5.** $V_{xx} = V[R_z, z, x]$, $R_z = R[0, a, z, a]$.
8. Cone of altitude 1, vertex at $(0, 1, 0)$; base is $x^2 + z^2 = 1$, $y = 0$.

§9

1. $2Ma^2/5$.   **2.** $3Ma^2/10$.   **3.** $2\pi k\, ph[1 - h(a^2 + h^2)^{-1/2}]$.
15. $V_{\theta r} = V[R_\theta, 0, 1 - r\cos\theta - r\sin\theta]$, $R_\theta = R[0, \pi/2, 0, (\cos\theta + \sin\theta)^{-1}]$.

§10

1. $f(x, y) = x$ when $|x| < 1$ and $|y| \leqq 1$; $f(x, y) = 0$ when $|x| = 1$.
2. $f(x, y) = 1/x$ when $0 < x < 1$ and $0 < y < 1$.   **4.** The example of Exercise 2.

CHAPTER 7
§1

1. $(1/6) + (\pi/4)$, $1/2$.   **2.** 0.   **3.** $-2/3$.   **4.** -1.
8. 0; the component of the force in the direction of motion is zero.
14. $\displaystyle\int_\Gamma fg_1\, dx + fg_2\, dy, \int_a^b f(\varphi, \psi)g_1(\varphi, \psi)\, d\varphi + \int_a^b f(\varphi, \psi)g_2(\varphi, \psi)\, d\psi$.

§2

1. -1.   **3.** 0.   **7.** πab.   **8.** $2a^2 - (\pi a^2/2)$.   **9.** $(3/2)a^2$.

§3

1. $-e - 1$.   **3.** 0.   **4.** $xy^{-1} + yx^{-1} - ab^{-1} - ba^{-1}$; path must not cross an axis.
5. 0; $F = -\tan^{-1}(x^2 + y^2 - 1)(2y)^{-1}$.   **6.** 2π.   **9.** $\displaystyle\int_{a,b}^{x,y} -u_2\, dx + u_1\, dy$.
10. $3x^2y - y^3$, $-e^y\sin x$, $x(x^2 + y^2)^{-1}$.

§4

1. $e - (5/2)$.   **2.** 0.   **4.** $(29\pi/4)\sqrt{2}$.   **6.** $(2/3)Mr^2$.

§5

1. $\pi/4$.   **3.** $R_{uv} = R_v = R[1, 2, \cos^{-1}(v/2), 2\pi - \cos^{-1}(v/2)]$.
8. $R_{uv} = R_v = R[0, 2^{-1/2}, (1 - v^2)^{1/2}, (1 + v^2)^{1/2}] + R[2^{-1/2}, 1, v, (2 - v^2)^{1/2}]$.
9. $3/2$.
19. $2\pi l$; the surface is generated by moving a unit circle normal to, and with center on, the curve $x = x(s)$, $y = y(s)$, $z = 0$.

§6

1. 0.   **2.** -4π.   **3.** 6.   **4.** 1.   **12.** 0.   **13.** $xy = cz$.   **14.** $xy + \log z = c$.

CHAPTER 8

§1

1. 1. **4.** 1. **11.** $-\infty, +\infty$. **12.** $f^{(4)}(x)$.

§2

1. 0. **3.** 1/2. **4.** $-\infty$. **6.** $+\infty$. **7.** $+\infty, +\infty$. **13.** 0. **14.** Does not exist.

§3

1. $-2/\pi$. **3.** $1/e$. **4.** 1. **5.** $e^{1/3}$. **7.** 1. **10.** 0. **18.** $\partial^2 f/\partial \xi_\alpha^2$ at $(0, 0)$.

§4

1. 1/4. **3.** $-1/9$. **7.** 1/10. **8.** 2/9. **10.** 0. **11.** $+\infty$. **12.** $e/2$.
14. The second is strongest; the third is weakest.

§5

1. $\overline{\lim} = 1$, $\underline{\lim} = -1$; no others. **4.** $\overline{\lim} = +\infty$, $\underline{\lim} = 0$; no others.
10. $\overline{\lim} = +\infty$, $\underline{\lim} = -\infty$; 0 is a limit point. **23.** e^2. **25.** $1/e$. **26.** $4/e$.

CHAPTER 9

§1

1. c. **2.** d. **3.** d. **5.** c. **10.** c.

§2

1. d. **4.** c. $(p > 1)$; d. $(p \leqq 1)$. **7.** d. **9.** c. $(p > 0)$; d. $(p \leqq 0)$.

§3

1. a.c. $(p < -1)$; c.c. $(-1 \leqq p < 0)$; d. $(0 \leqq p)$. **4.** c.c. **5.** c.c.
12. Same as Exercise 1.

§4

1. c., d., d. **4.** c. $(\alpha > 0)$; d. $(\alpha \leqq 0)$. **8.** c. **9.** False. **10.** True.

§5

1. Yes. **3.** Yes. **5.** Yes. **9.** No.

§6

2. 0. **3.** 1/2. **4.** Yes. **5.** No.

§7

1. 0. **3.** Not summable $(C, 1)$. **4.** 1/2. **5.** 3/4.

§8

1. $8^2 < 9^2$; $\sqrt{34} < 6$.   **4.** $49a^2$.   **11.** $a_k = 1, b_k = (-1)^k$.

§9

1. $R = 1/3$.   **3.** $R = 0$.   **5.** $-1/(e - 1) \leq \lambda \leq 1/(e + 1)$.   **7.** 1.   **8.** $1/\sqrt{2}$.
13. $(3 - 2x)(1 - x)^{-2}$.   **18.** $\pi/4$.

CHAPTER 10

§1

1. d.   **4.** c.   **6.** c.   **7.** Yes.   **8.** No.   **9.** Yes.   **10.** ∞, π.
12. $2\zeta(2) - 2$.   **13.** Same as Ex. 12.   **16.** Yes; (1) with $f = x^{-2}, x = 1/t$.

§2

1. d., d., c.   **4.** c. (all x).   **7.** c. ($\beta > 1$), d. ($\beta \leq 1$).   **11.** True.

§3

1. c.c.   **2.** c.a.   **4.** c.c.   **5.** d.   **13.** 2π.

§4

1. c.   **2.** d.   **4.** d.   **5** p. ($\alpha \leq 0$), c. ($\alpha > 0$).   **7.** c.

§5

1. d. (all α).   **3.** c.c.   **7.** d.   **12.** No.   **14.** 0.

§6

8. $\displaystyle\int_{0+}^{1} xt^{-1/2}(x^2 + t)^{-1}\, dt$ on $0 \leq x \leq 1$.   **9.** No.   **10.** Yes.

§7

1. $2(1 - x^2)/(x^4 + x^2 + 1)$.   **4.** $-8t^3 y^2\, dt - 4y(t^4 + y^4 + 2y^2)\, dy$.

6. $3x^{-1} \sin x^3 \left(\displaystyle\int_0^{x^2} t^{-1} \sin xt\, dt \right)^{-1}$.

8. $(1/4)x^{-1/2} \exp(4xy^2 - x^2) + (1/2) \displaystyle\int_{\sqrt{x}}^{2y} t^2 \exp(4xy^2 - xt^2)\, dt$.

11. $f(0)/3$.   **12.** 0; no.   **14.** 3.   **15.** 0.

§8

1. $F = (1/2)\sqrt{\pi/x}$.   (Proved in Chapter 11.   This explicit result checks the theory.)

2. $F' = -(1/4)\sqrt{\pi/x^3}$.   **3.** $\displaystyle\int_0^{\infty} (e^{-t^2} - e^{-2t^2})t^{-2}\, dt$.   **6.** $F(0) = 0, F(0+) = \pi/2$.

§9

1. $1/a$ $(a \neq 0)$; 0 $(a = 0)$.   **3.** $\cos a$.   **4.** 0.   **7.** No.

§10

1. $1 < (\pi/2)(\pi/2)$.   **2.** $\sqrt{\pi} < \sqrt{\pi/2} + \sqrt{\pi/2}$.
13. $[\int f^2\, dt \int g^2\, dt - (\int fg\, dt)^2]^{1/2}/[\int g^2\, dt]^{1/2}$, if the denominator $\neq 0$.

CHAPTER 11

§1

1. $-2\sqrt{\pi} = -3.545$, $(15/8)\sqrt{\pi} = 3.32$, 1.065, 0.259.   **3.** $0.903s^{-5/3}$.   **8.** 2.
9. 0.   **16.** $\sqrt{2\pi}$.

§2

1. $1/140$.   **4.** $\Gamma(1/4)\Gamma(3/4)/2$.   **6.** π.   **9.** $(\sqrt{\pi}/2)\Gamma(t)/\Gamma(t + 2^{-1})$.   **17.** $3\pi/8$.

§4

1. 2.61, 2.53, $\sqrt{2\pi} = 2.51$.   **5.** e.   **6.** -1.   **8.** $4^n/\sqrt{\pi n}$.   **18.** 1.

CHAPTER 12

§1

1. $a_0 = 2\pi^2/3$, $b_k = 0$, $a_k = (-1)^k 4k^{-2}$ for $k = 1, 2, \ldots$.   **4.** $(1 - \cos 2x)/2$.
10. Yes, since $2k^2 - k + 7$ is an integer and since the series converges uniformly.
12. $a_k' = kb_k$, $b_k' = -ka_k$, $a_k'' = -k^2 a_k$, $b_k'' = -k^2 b_k$.

§2

1. All; all but P; all but P.   **2.** C, D.   **3.** All but P.   **6.** All.   **8.** D, D^1.
10. None.   **12.** $g(x) \equiv 0$ (trivially); $f = \tan^{-1}(1/x)$, $g = x$.   **17.** $\tan^{-1}(1/x)$; $x \sin(1/x)$.

§3

1. $1 - 2^{-2} + \ldots + k^{-2} \leq \pi^2/6$; verify by §3.3, Chapter 11.   **3.** 0.
5. $\Gamma^2(3/4)\sqrt{2/\pi}$.   **7.** $\pi/2$.   **14.** All but 0.   **15.** All but 0.

§4

7. $\pi/2, \pi/2, \pi/2$.   **9.** $\pi(e^{2\pi} - 1)^{-1} + (\pi - 1)/2$.   **10.** $|x| \leq \pi$.   **11.** $-\pi^2/12$.

§5

1. $(1/2) - (4/\pi^2) \sum\limits_{k=0}^{\infty} (2k + 1)^{-2} \cos(2k + 1)\pi x$; $f(x)$ for all x.

2. $(1/2) - (1/\pi) \sum\limits_{k=1}^{\infty} k^{-1} \sin 2k\pi x$; $f(x)$ except at the integers, where the sum is $1/2$.

7. $b_1/b_3 = 27$.   **11.** $\pi hc/l$; $2ch/l$.

§6

1. 2π.   **3.** $\pi \log |x|$.   **4.** 2π.   **5.** π.

§7

1. $(b - a)/\pi$, $(a + b)/2$.   **4.** $6y = 6x - 1$.   **8.** $\sum_1^\infty (2k - 1)^{-4} = \pi^4/96$.

§8

1. $(1/\sqrt{2})e^{-x^2}/4$.   **7.** $\pi(\sin x)^2 x^{-2}$.   **11.** 0.   **12.** π.
16. 0 for $x < -\pi$, $-e^\pi/2$ for $x = -\pi$, $e^{-x} \cos x$ for $x > -\pi$.

CHAPTER 13

§1

1. $(\sqrt{\pi}/2)s^{-3/2}$, $s > 0$.   **3.** $c/(s^2 - c^2)$, $s > |c|$.
5. $(s - a)/[(s - a)^2 + c^2]$, $s > a + |c|$.   **10.** $1 - e^{-t} - te^{-t}$, $s > 0$.
12. $\sin 3t$, $s > 0$.

§2

1. 0.   **3.** 0.   **9.** 3.   **10.** 2.

§3

1. $0, 0$.   **3.** ∞, ∞.   **5.** $-2, 0$.   **12.** $(-t)^k \cos at$.   **17.** $(s^2 + 2s + 2)(s + 2)^{-3}$.

§4

2. $e^{bs/a}f(s/a)/a$; $\varphi(t) = 0$, when $t < b$.   **4.** $(1/a)e^{-bt/a}[L^{-1}\{f(s)\}]_{t/a}$.
13. $L\{(1 - \cos t)/t\} = (1/2) \log (1 + s^{-2})$.   **15.** $\log [(s - b)/(s - a)]$; $s > a, s > b$.

§5

4. $(\sin at + at \cos at)/2a$.   **7.** No; $\varphi = 1$, $\psi = \chi = t$.   **8.** Yes.   **16.** $t - \sin t$.

§6

14. $1/(s + \sqrt{s})$.

§7

6. $1 - \cos t$.   **9.** $e^{-t}(\sin t - t \cos t)/2$.

§8

1. 0, since $\log t > (t - 1)/t$.   **2.** ∞.   **8.** 0.   **9.** No.   **10.** No.

§9

1. $\sum_{k=0}^{\infty} \left(\dfrac{k+2}{2}\right) \dfrac{(-1)^k t^{2k+4}}{(2k+4)!} = t\,(\sin t - t \cos t)/8.$

9. $(1 - e^{-t})/t.$ 10. $(\sin t)/t.$

§10

1. $\varphi = \exp(-e^{-t}),\, s > 0.$ 2. $\varphi = 1/(1 + e^{-t}),\, 0 < s < 1.$ 9. $-\infty < s < \infty.$
10. $s = 0.$ 12. $(s - 1)^{-1} \log s.$

CHAPTER 14

§1

1. $1 - \log 2.$ 3. $-\pi^2/12.$ 7. $1.$ 9. $(\pi/x)^{1/2}.$ 12. $\pi/2.$ 13. $\sqrt{\pi/(2t)}.$

§2

1. $e^{-2t}.$ 3. $(\cos t + \sin t - e^{-t})/2.$ 5. $Ae^t.$ 9. $-t + 2 \sinh t.$
11. $1 - e^{-t/2} \cos(\sqrt{3}t/2) - (1/\sqrt{3})e^{-t/2} \sin(\sqrt{3}t/2).$

§3

1. $1, 2, 5.$ 2. No; $M_k = e^{(k-1)}$; $M_k = 2e^{(k-2)/2}.$ 6. $(\sin t - t \cos t)/2,\, \sigma > 0.$

§4

1. $M = 1 + s,\, N = 1 + s^2,\, f = 1/s^2,\, \alpha = \sin t + \cos t,\, \beta(t) = t - \sin t,\, \delta = \sin t.$
2. $\gamma = \cos t,\, \psi = e^{-t}.$

§5

1. $\Delta y_k - y_k = k.$ 3. $\Delta^2 y_k + 2\,\Delta y_k = k + 1,\, y_0 = \Delta y_0 = 0.$ 8. $2^k - k - 1.$
10. $(2k^2 - 1 + (-1)^k)/8.$ 16. $-y_0 s^{-1} + (1 - s)s^{-1} l\{y_k\}.$

§6

7. $s^{-1} \tanh(sa/4).$ 9. $b(se^a - s)^{-1} + (1 - e^{-sa})^{-1} \displaystyle\int_0^a e^{-st}\varphi(t)\,dt.$

10. $\displaystyle\int_0^4 e^{-st}\omega(t)\,dt = s^{-2}(1 - e^{-2s}) - 2s^{-1}e^{-4s}.$

13. $\cdots x$ on $0 \leq x \leq t - 1,\, 1 - t$ on $t - 1 \leq x \leq 3 - t,\, x - 2$ on $3 - t \leq x \leq 2.$

Index of Symbols

INDEX

A